"Torres uncovers and analyzes, in novel ways, remarkable developments in philosophy and science, new thinking that undermines everything preceding. We have hardly begun to grasp the meaning of this upheaval in how people perceive humanity's future. Torres's engagingly written book, admirably thorough, is a good start for facing our awesome responsibilities."

Spencer Weart, former director of the Center for History of Physics of the American Institute of Physics and author of *Nuclear Fear: A History of Images* and *The Discovery of Global Warming*

Human Extinction

This volume traces the origins and evolution of the idea of human extinction, from the ancient Presocratics through contemporary work on "existential risks."

Many leading intellectuals agree that the risk of human extinction this century may be higher than at any point in our 300,000-year history as a species. This book provides insight on the key questions that inform this discussion, including when humans began to worry about their own extinction and how the debate has changed over time. It establishes a new theoretical foundation for thinking about the ethics of our extinction, arguing that extinction would be very bad under most circumstances, although the outcome might be, on balance, good. Throughout the book, graphs, tables, and images further illustrate how human choices and attitudes about extinction have evolved in Western history. In its thorough examination of humanity's past, this book also provides a starting point for understanding our future.

Although accessible enough to be read by undergraduates, *Human Extinction* contains new and thought-provoking research that will benefit even established academic philosophers and historians.

Émile P. Torres is a philosopher whose research focuses on existential threats to civilization and humanity. They have published widely in the popular press and scholarly journals, with articles appearing in the *Washington Post, Aeon, Bulletin of the Atomic Scientists, Metaphilosophy, Inquiry, Erkenntnis*, and *Futures*.

Routledge Studies in the History of Science, Technology and Medicine

43 Lifestyle and Medicine in the Enlightenment
The Six Non-Naturals in the Long Eighteenth Century
Edited by James Kennaway and H.G. Knoeff

44 Germs in the English Workplace, c.1880–1945
Laura Newman

45 Pathogens Crossing Borders
Global Animal Diseases and International Responses, 1860–1947
Cornelia Knab

46 Evidence in Action between Science and Society
Constructing, Validating, and Contesting Knowledge
Edited by Sarah Ehlers and Stefan Esselborn

47 George Wilson's Vision of Early Victorian Science and Technology
Unity in Variety
David F. Channell

48 Transforming American Science
Universities, the Government, and the Cold War
Jonathan Engel

49 Human Extinction
A History of the Science and Ethics of Annihilation
Émile P. Torres

For more information about this series, please visit: www.routledge.com/Routledge-Studies-in-the-History-of-Science-Technology-and-Medicine/book-series/HISTSCI

Human Extinction

A History of the Science and Ethics of Annihilation

Émile P. Torres

Routledge
Taylor & Francis Group
NEW YORK AND LONDON

First published 2024
by Routledge
605 Third Avenue, New York, NY 10158

and by Routledge
4 Park Square, Milton Park, Abingdon, Oxon, OX14 4RN

Routledge is an imprint of the Taylor & Francis Group, an informa business

Library of Congress Cataloging-in-Publication Data
Names: Torres, Émile P., author.
Title: Human extinction : a history of the science and ethics of annihilation / Émile P. Torres.
Description: New York, NY : Routledge, 2023. | Series: Routledge studies in the history of science, technology, and medicine | Includes bibliographical references and index. | Contents: An apocalypse without kingdom—Beginnings of "the end"—'Till entropy death do us part—The invention of omnicide—Mother nature wants to kill us—The perfection of evil—What is human extinction?—Early ruminations—Ethical innovations of the postwar era—Astronomical value and the harm of existence—Recent developments—Looking forward to the future.
Identifiers: LCCN 2022061817 (print) | LCCN 2022061818 (ebook) | ISBN 9781032159065 (hardback) | ISBN 9781032159089 (paperback) | ISBN 9781003246251 (ebook)
Subjects: LCSH: Human beings—Extinction—Research—History. | Human ecology—Research—History.
Classification: LCC QH78 .T677 2023 (print) | LCC QH78 (ebook) | DDC 576.8/4—dc23/eng/20230405
LC record available at https://lccn.loc.gov/2022061817
LC ebook record available at https://lccn.loc.gov/2022061818

ISBN: 978-1-032-15906-5 (hbk)
ISBN: 978-1-032-15908-9 (pbk)
ISBN: 978-1-003-24625-1 (ebk)

DOI: 10.4324/9781003246251

Typeset in Bembo
by Apex CoVantage, LLC

Dedicated to the curious little scientists Lucy and Zach, my niece and nephew, whom I love more than anything.

Also dedicated to Brian Deebel and three other acquaintances who have tragically passed on. So long as I am living, you will not die a "second death."

Contents

Preface and Acknowledgments

This is a long book that hardly scratches the surface of its subject: human extinction. I examine the origins and evolution of this idea from a primarily Western perspective and hence neglect entire universes of thought from other regions of the world, and from other (for example, Indigenous) points of view. Yet even from this Western perspective, my limitations as a historian and philosopher will no doubt frustrate those with expertise on particular historical and philosophical ideas. Despite these shortcomings, I hope this book outlines a useful and perhaps compelling theoretical framework for thinking about how our understanding of humanity's existential predicament has changed over time, and how Western intellectuals have thought about the normative issues surrounding the possibility of our species' disappearance. Given the breadth of this work, there should be something of interest to people in many different fields: philosophers may learn something about history and science; scientists may learn something about philosophy and history; and historians may learn something about science and philosophy. At the very least, I hope this book encourages people to take seriously the large-scale threats to our continued existence and collective wellbeing that now confront us. Tentatively, my plan is to follow this up with one or more subsequent publications exploring the same topic from non-Western perspectives. The present work may thus be seen as the first volume of a larger project.

The main themes of this book will make the most sense—unsurprisingly—if one reads it entirely and in order. However, Part I and Part II are modular to an extent, meaning that they could be understood somewhat independently of each other. To those primarily interested in how thinking about the ethical and evaluative implications of our extinction developed over time, I recommend reading Chapter 1 for an overview, which may be sufficient to make sense of Part II. On the flip side, what one makes of the particular "existential mood" that defines our current moment will depend in part on one's views about the ethical aspects of our disappearance, and hence one's reading of Part I may be enriched by Part II. The two halves can be decoupled, in other words, but are best thought of as a marriage of overlapping and interacting narratives.

I am deeply grateful for feedback from and conversations with many extraordinary people over the past four years. The list includes Fred Adams, Peter Bowler, Gerry Canavan, Lewis Coyne, Oswaldo Chinchilla, Zoe Cremer, Roger Crisp, James Dator, Jason Dawsey, Paul Ehrlich, Kyle Evanoff, Debbie Felton, Elizabeth Finneron-Burns, Bennett Gilbert, Walter Glannon, Martin Glazier, Pavel Gregoric, Thomas Hornigold, Tom Hurka, Erika Juhlin, Aatu Koskensilta, Bas Leijssenaar, James Lenman, Adrienne Mayor, Theresa Morris, Thomas Moynihan, Ingo Müller, Ian Myles, Jan Narveson, Morton Paley, Michael Rampino, Toni Rønnow-Rasmussen, Chase Roycroft, Bart Schultz, Will Steffen, Stephen Self, Susan Schneider (who secured an office for me at the United States Library of Congress on two occasions, during which I wrote drafts of this manuscript), Christian Tornau, and Robert Wicks, as well as the Centre for the Study of Existential Risk (CSER), which hosted me for several months in 2019, when the idea for this project was born. Special thanks to Frances Flannery for helping me understand early Christian beliefs, Spencer Weart for insightful feedback on Part I, Luke Kemp for comments on an early draft that significantly influenced how I rewrote the book, Simon Knutsson and S. J. Beard for many incisive comments on Part II, and Mathias Frisch at Leibniz Universität Hannover and Ralph Stoecker at Universität Bielefeld for supervising my Ph.D. dissertation, which is largely coextensive with this book. I am most indebted to Daniel Deudney and Dan Zimmer for extensive, detailed comments on and criticisms of early and later drafts. Dan Zimmer, in particular, helped with edits right up to when a final draft was submitted. I cannot express in words my appreciation to everyone above: thank you.

Finally, thanks to my father John Paul, sister Sylvia, and brother-in-law Chris for supporting me when I was jobless and giving me a home when I had nowhere to stay. Compassion is the glue that holds the world together—without it, there would be only isolation and despair. Through laughter and tragedy, the ups and the downs, good people are there for each other.

I take full responsibility for all errata, which I will catalogue here: www.xriskology.com/book-errata.

1 An Apocalypse Without Kingdom

Too Much Algae

"Oh yes, that could happen," Malcolm declared. "A meteor could strike Earth. Climate change could destroy all plant life, causing animals and humans to starve to death. And if we catch too many fish, there will be too much algae filling up the ocean, and without water, we die."[1]

If some of this sounds implausible, there is a good reason: Malcolm is a seven-year-old Swedish boy responding to a question posed by a colleague of mine, at my behest: "Could humanity go extinct like the dinosaurs or dodo?" I had initially posted this on social media, asking friends if they would be willing to query their children about it and relay the answers given. Several replied, and in every case the answer was "Yes, our extinction is possible," often followed by some imaginative account of how this might happen. The most common means of annihilation involved asteroid strikes, although other children mentioned climate change and evil robots. One clever child even huffed back at her parent that my question was confusing since only the *non-avian* dinosaurs perished 66 million years ago. The *avian* dinosaurs, which descended from the Theropoda clade that boasts of charismatic reptiles like *T. rex* and the velociraptor, survive among us as modern-day birds. In fact, she was right to object, and so I must apologize for not being clearer!

The point of this evidence gathering via anecdotal survey was to get a sense of how commonplace the concept or idea of *human extinction* is today.[2] Although no large-scale surveys have yet been conducted on the topic, I suspect that most people in the West nowadays would acknowledge that our extinction is at least *possible*.[3] If this is correct, it points to an extraordinary fact, since for much of Western history the concept of human extinction would have struck nearly everyone as (P1) unintelligible, incoherent, or self-contradictory, not unlike the concepts of *married bachelor* and *circles with corners*, and (P2) denoting an outcome that could *not possibly obtain*, just as there are no married bachelors or circles with corners. These phenomena are related but distinct: a concept might be unintelligible to some people but still denote a real possibility in the world, and there are many impossible outcomes that are nonetheless intelligible, such as pigs flying, which we can easily imagine despite pigs lacking the ability to take flight. The idea of our

DOI: 10.4324/9781003246251-1

extinction, though, wouldn't have made sense if people in the past had considered it, and just about everyone would have claimed that it could never happen. As Malcolm shows, this is no longer the case. What changed?

To better answer this question, it will help to first outline a preliminary account of the idea of *human extinction* so that we know what we are talking about. For now, let's define human extinction as having occurred if there are no more tokens of the type "humanity" in the world. A token is the instantiation of a type, and hence this definition states that if there are no living members of *Homo sapiens* at some point in the future, then *Homo sapiens* will have gone extinct. This is intended to be a *naturalistic* definition, one that precludes humanity from "living on" in an afterlife of some sort; it is the kind of extinction that the dinosaurs and dodos underwent—they existed and now they don't. Notice right away that this contrasts with religious conceptions of humanity's future. On these accounts, the end of the world is not the end of our story but, in a profound sense, the *beginning*. Religious views anticipate a future *transformation*, whereas naturalistic extinction entails our complete *termination*. As the German philosopher Günther Anders wrote in 1959, extinction would be "a naked apocalypse," an "apocalypse without Kingdom."[4]

One of the main contentions of this book is that for approximately 1,500 years, between the fourth and fifth centuries of the Common Era (CE) and the nineteenth century, the idea of *human extinction* was almost entirely "blocked," as I will say, from the minds of most people in the West. This differs in important ways from other conclusions defended in the nascent literature on human extinction, which as of this writing consists of only a handful of books and articles.[5] For example, the Oxford historian Thomas Moynihan argues in an article for *Aeon* that "as ideas go, human extinction is a comparatively new one," having "emerged first during the 18th and 19th centuries."[6] I disagree: the idea that humanity could disappear from the universe entirely—in other words, go extinct as defined above—turns out to have been entertained by ancient Greek philosophers and gestured at by mythological systems of even earlier provenance. There are, indeed, ample references to human extinction in the ancient world, as when the Akkadian epic poem of *Atrahasis*, which dates back to the eighteenth century BCE, describes one god attempting to completely annihilate humanity, or when the Presocratic philosopher Xenophanes posited one stage of cosmic evolution as entailing the elimination of all human life on Earth. To oversimplify somewhat, I will argue that the idea of human extinction first made an appearance before the Common Era, was subsequently blocked during the roughly 1,500-year period specified above, reemerged in the nineteenth century (especially the second half), and has since steadily grown in prominence up to the present.

By "prominence," I mean the extent to which the idea is salient on the cultural landscape. A proxy measure of prominence can be obtained using the Google Ngram Viewer, which its creators describe as enabling "scholars to make powerful

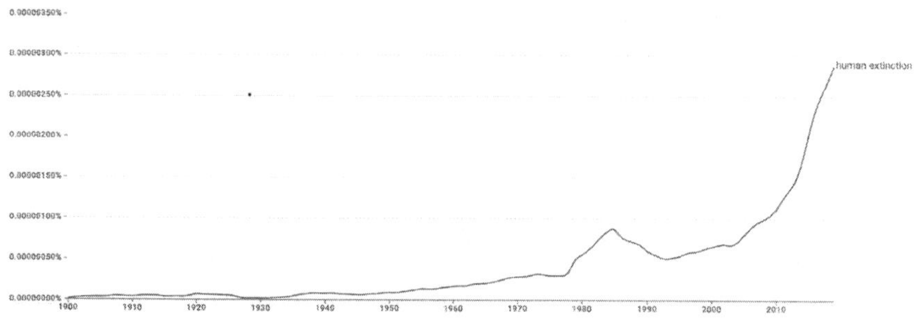

Figure 1.1 Google Ngram Viewer results for the keyword "human extinction."

inferences about trends in human thought" by combing through Google's text corpora of roughly 8 million digitized books (see Figure 1.1).[7] Although 8 million books amounts to only about 6 percent of all the books published between 1500 and 2019, this is currently the best tool available for understanding the salience of ideas along a diachronic dimension, and I will rely on it now and then to buttress certain conclusions of mine. In sum, human extinction is an old idea, but it disappeared from sight for much of Western history, only to reappear more recently—though not so much in the eighteenth century as in the one that followed.

The Great Chain, Personal Death, and the End of the World

This leads to the question of *why* the idea of human extinction was blocked for so long. The answer concerns two clusters of beliefs that became central to Christianity around the fourth or fifth centuries CE, each of which was sufficient to render our extinction unthinkable.[8] I have separated these into clusters according to which concept in *human extinction* they target. That is to say, *human extinction* consists of two concepts—*human* and *extinction*—and hence my claim is that one cluster of beliefs specifically concerns the first concept, while the other concerns the second. Let's take a closer look at this:

To understand why extinction seemed unthinkable, we must begin with the Great Chain of Being, a model of reality whereby all things, living and nonliving, are ordered in a linear and immutable hierarchy. First articulated by the Neoplatonists in the third century CE, it became enormously influential within the West after Christian writers like Saint Augustine (354–430 CE) incorporated it into the Christian tradition. The Great Chain asserts that there are no gaps in nature: everything that *can* exist *does* exist, now and forever. This is just the fundamental structure of reality, however odd it may strike contemporary readers. It follows that since no links in the chain can ever go missing, extinction of any sort is impossible, which means that our own extinction is impossible as well. In other words, by precluding the disappearance of *any kind* of thing in the universe, the Great Chain implies that we, too, can never disappear. As we will see, this model of reality collapsed in the early nineteenth century, which thus removed one major barrier to imagining our collective demise.

The second cluster of beliefs concerns the essential *nature* of humanity and our unique *role* in the unfolding of God's grand plan for the world. Most Christians since the middle of the first millennium have accepted a dualistic anthropology according to which human beings are composed of both material and immaterial parts: a body and a soul. The soul is immortal, although at the end of time it will be reunited with a physical resurrection body, which will also be immortal. This has been the standard Christian view of what is called "personal eschatology," which concerns our fate as individuals rather than the cosmos as a whole. Importantly, it yields a second reason that human extinction cannot occur: since each individual human is immortal, and since humanity is just the sum total of all humans, it follows that humanity itself is immortal. Let's call this the *ontological thesis*, since it concerns the ontological status of human beings as body-soul composites that, once created by God, will never cease to exist. It also explains why the concept of *human extinction* would have struck many as incoherent or self-contradictory: contained within the idea of *human* is the idea of *immortality*, since to be human is to be immortal. Consequently, asserting that "humanity can go extinct" would be like saying "an immortal kind of thing can undergo a process that only mortal kinds of things can undergo," which is an obvious logical contradiction.[9] This is why I likened *human extinction* to concepts like *married bachelor* and *circles with corners*, as these are also self-contradictory.

In contrast, "cosmic eschatology" is about "the ultimate resolution of the entire creation," and thus concerns end-times events like the Second Coming of Christ (*Parousia*), Battle of Armageddon, Final Judgment of humanity, and creation of a new heavens and Earth.[10] An important idea here is that cosmic eschatology is ultimately about balancing the scales of justice by punishing the wicked and rewarding the righteous. In other words, it is about *theodicy*, a term coined by Gottfried Wilhelm Leibniz in the early eighteenth century to denote the problem of vindicating God given the presence of evil in the world. As the New Testament scholar Craig Hill writes about this,

> at heart, all eschatologies are responses if not quite answers to the problem of evil. . . . Eschatologies differ in how they conceptualize God's triumph, but they are essentially alike in asserting God's victory as the supreme reality against which all seemingly contrary realities are to be judged.[11]

This yields a third reason that human extinction cannot happen: since there can be no balancing of the scales of justice without humanity surviving beyond

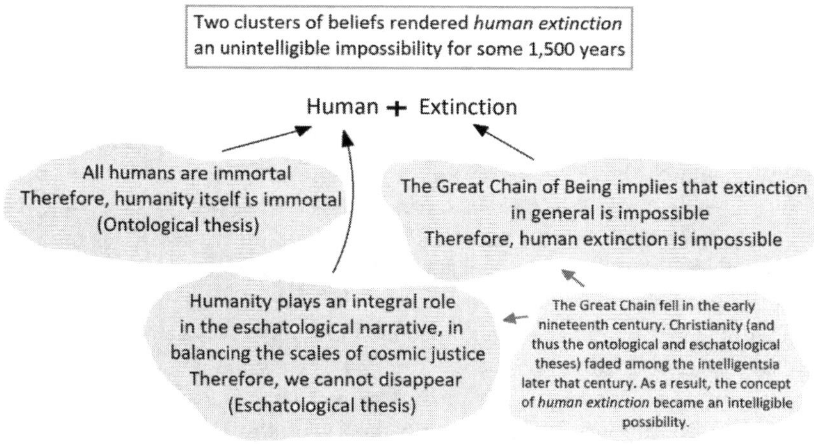

Figure 1.2 The three beliefs that "blocked" thinking about human extinction.

the end of the world, it is inconceivable that we might cease to exist. How could God's grand plan unfold without us? How would good prevail over evil? Human extinction simply isn't on the cards for us; it just isn't the way our story ends. Let's call this the *eschatological thesis*, noting that it, like the ontological thesis, specifically concerns the idea of *human* rather than *extinction*, whereas the Great Chain makes a general claim about *extinction*.

We thus have three reasons that our extinction is fundamentally impossible: first, because it is metaphysically impossible, given the Great Chain model of reality. Second, because it is ontologically impossible, given that humanity is immortal. And third, because it is eschatologically impossible, as we are the main characters in the cosmic drama of good and evil: without us, the show cannot go on, and since the show must go on, we cannot die out.[12] Because these beliefs became central to the Christian worldview beginning in the early first millennium, the rise and fall of Christianity will be an integral part of this book's account of the origins and evolution of *human extinction* in a naturalistic sense. However, we will also see that even before Christianity took root in the West, there was nonetheless a widespread assumption that our species is indestructible, though the reasons tended to be peculiar to the various philosophical, religious, and mythological systems that people accepted at the time.

Kill Mechanisms

Yet the intelligibility of the concept of *human extinction* and the possibility of the outcome it denotes—that is, propositions (P1) and (P2) above—form only half of the story. The other half concerns the distinct question of whether our extinction could *actually happen* in our particular world, assuming that it is possible *in principle*. That is to say, it could be that, as a matter of fact, our world contains no "kill mechanisms," which I will define as a means of elimination capable of precipitating our complete non-existence. In fact, for much of the twentieth century, the scientific community almost unanimously agreed that the universe doesn't contain any *natural* kill mechanisms that pose risks to our survival (aside from the Second Law of thermodynamics, discussed in Chapter 3). In other words, nearly everyone believed that we live on a very safe planet in a very safe universe—not on an individual level, of course, since any one of us could be eaten by lions, crushed by a falling boulder, or catapulted into the grave by a deadly virus, but on the species level: the natural world does not pose any real threats to *humanity*. This view dominated the Earth sciences from at least the 1850s until it was overturned in the 1980s and 1990s, after it became clear that the non-avian dinosaurs died out because a large asteroid struck the Yucatán Peninsula in southeastern Mexico (as the young respondent mentioned earlier is no doubt aware). The implications of this were ominous: if the dinosaurs could be annihilated by naturally occurring hazardous phenomena, then so could humanity. The point is that even if extinction is possible in principle, if there is no reason to believe it could actually happen, then there may be no particular reason to take the idea seriously.

Glancing across the horizons of history, one is struck by a dazzling array of proposed kill mechanisms. Some were associated with supernatural deities or events; others were built into the natural cycles of cosmic evolution; and still others involved idiosyncratic speculations about phenomena like comets and floods. For the purposes of our study, what matters are kill mechanisms that we could describe as "scientifically credible," that is, means of elimination that were widely accepted by the community of scientists (or natural philosophers) based on compelling empirical evidence. The first kill mechanism of this sort was the aforementioned Second Law of thermodynamics, which physicists in the mid-nineteenth century immediately recognized as posing a long-term threat to humanity: Earth will become increasingly inhospitable as the sun cools down, until the flickering flames of all forms of life are snuffed out. Since then, a whole constellation of credible kill mechanisms has been discovered and created—in the case of anthropogenic threats—and there is no good reason to believe that more won't be discovered or created in the future, as science pushes back the envelope of human ignorance and technology enhances the violence capacities of state and nonstate actors.

The Mood of the Times

The identification of kill mechanisms thus constitutes the second half of the story that spans Part I of this book. Since the collapse of the Great Chain and retreat of religion enabled the idea of human extinction to become an intelligible possibility, let's refer to them as *enabling conditions*. And since, as we will see, the discovery, creation, and even mere anticipation of new kill mechanisms often triggered qualitatively novel understandings of our existential vulnerability in the universe, let's refer to them as *triggering factors*. These two phenomena bring us to one of the most important ideas of Part I, namely, that of an *existential mood*, which provides the organizing principle behind the periodization outlined in the first half of this book.

One way to approach the idea of an existential mood is as follows: by combining the phenomena of enabling conditions and triggering factors, one can construct an explanatory-predictive hypothesis that, I argue, accounts for the historical record of thinking about human extinction and provides insights—predictions—about how this thinking could change in the future. (The predictive aspect of this hypothesis will occupy us in Chapter 12.) What, then, does the historical record show? It shows a number of major shifts in how people thought about our existential predicament. More specifically, these shifts corresponded to different sets of answers to crucial questions like: Is our extinction possible? If so, how could it happen? How many kill mechanisms are there? Are they natural or anthropogenic? Do they pose risks in the near term or distant future? How probable is our extinction? Is this probability rising or falling? Is our extinction inevitable? And so on. Let's define an existential mood as proceeding from a *particular set of answers* to these questions, where "mood" is understood in a collective rather than individual sense, as in a "public mood" or the "mood of the times." As Erik Ringmar explains this idea, beginning with the more familiar moods had by individual people:

> To be in a certain mood . . . is to attune oneself to the situation in which one finds oneself. A mood answers a question of how we feel and thereby reports on the state of our attunement. A public mood would thereby be a question of how a public attunes itself to the situation in which it finds itself.

Public moods are something akin to an "atmosphere" that imbues society (or some segment of society), thereby "coloring everything we see around us in a certain hue." The 1950s, for example, "was allegedly characterized by a mood of optimism but also of anxiety; the 1960s by a mood of liberation, rebellion, and experimentation; the 1970s by disillusionment and lost hopes, and so on."[13]

An *existential* mood thus arises from the situation in which people find themselves given some set of epistemologically robust answers to the questions above about the possibility, probability, etc. of our extinction. The result is a general *outlook* on our collective future—on whether we will have a future—that colors everything we see right now and up ahead in a certain hue. An existential mood is an atmosphere that permeates the thoughts and expectations of large numbers of people in the same general way, leading them to similar beliefs about where humanity is and might be going. Also relevant to my conception of an existential mood is the etymology of its second term, which derives from the Old English word *mod* meaning "heart, frame of mind; courage, arrogance, pride; power, violence," all of which capture some aspect of being "in" one existential mood or another. Of note is that there may also be an etymological connection between "mood" and "moral," as the latter comes from the Latin *mos*, meaning, in plural, "mores, customs, manners, morals."[14] For reasons hinted at below and elaborated in subsequent chapters, this too will be pertinent. As for "existential," I take its definiens to be "relating to existence" rather than "concerned with existentialism," the mid-twentieth-century cultural and philosophical movement associated with thinkers like Simone de Beauvoir and John-Paul Sartre. This sense of the word has become common today, due in part to the attention that the notion of *existential risks* has received among academics and within the popular media.

The Five Moods

Turning back to the historical record, I will argue that it reveals *five distinct* existential moods, thus yielding a five-part periodization of Western thinking about our extinction. There are several points to make about this: first, each existential mood was highly *stable* during its corresponding period of time. Second, the shifts between one to another existential mood have all been quite *abrupt*, unfolding over a matter of years or, at most, just over a decade. These transitions were in every case accompanied by figurative, and sometimes literal, gasps, as they marked fundamentally new understandings of our existential predicament in the universe. Third, once human extinction became an intelligible possibility during the nineteenth century, each shift was triggered by the discovery of one or more novel kill mechanisms, with one notable exception: the most recent shift. Fourth, given that it usually took some time for these shifts to unfold, I will distinguish between when an existential mood first *emerged* and when it fully *solidified*, at which point the mood, or outlook, became stable. And fifth, the emergence and solidification of new existential moods was in all cases, except for one, *cumulative*, that is, each built upon rather than replaced the previous existential moods; here, the metaphor of a palimpsest may be useful. The only exception was the initial shift in the 1850s, when the new existential mood *superseded* the earlier one rather than building on it, although we will see in Chapter 12 that the first existential mood could very well reappear in the future. Each of the five existential moods will receive a chapter of its own. In brief, these moods are:

(1) *Indestructibility* (ancient times to the 1850s). The notion that human beings are in some sense a permanent fixture of reality—that we are fundamentally indestructible—is found in many cosmological theories and mythological

systems dating back at least to the Presocratics of Ancient Greece. This does not mean that some ancient peoples never imagined the universe without us. Those who did, though, almost always believed that this would only be a temporary state of affairs. In other words, they accepted the possibility of our extinction in a minimal sense, but rejected the idea that we could disappear forever, which is why I said above that the idea dates back to the ancient world. Nonetheless, the belief in our indestructibility took on a more radical form once Christianity came to dominate the Western worldview. During this ~1,500-year period, naturalistic extinction *of any sort* would have been seen by virtually everyone as impossible in the three senses specified earlier—metaphysical, ontological, and eschatological. This offered a reassuring sense—a feeling of "Comfort" and "perfect security," to quote two notable figures writing at the end of this period—that no matter what might happen in the future, no matter what catastrophes might befall humanity, we will ultimately endure forever. (Chapter 2.)

(2) *Existential Vulnerability and Cosmic Doom* (1850s to the mid-twentieth century). This was initiated by the discovery of the first scientifically credible kill mechanism, that is, the Second Law of thermodynamics, which entails that our planetary and/or cosmic abode will become increasingly inhospitable to life, until no life at all is possible. The Second Law thus stamped an expiration date on humanity's forehead, though physicists did not expect this to happen for many millions of years. Nonetheless, not only did it become clear to many that our extinction is, in fact, *possible*, but the fundamental laws of nature implied that this outcome is ultimately *inevitable*—a double trauma that led many to despair about the purpose or meaning of life (see Chapter 8). The background condition for this shift in existential moods was, of course, the loosening of religion's stranglehold on conceptions of human nature and the future of humanity (by the time this mood descended, the Great Chain had already been mortally wounded, but the ontological and eschatological theses remained largely intact). As we will see, the decline of Christianity and the discovery of the first credible kill mechanism swung open the floodgates for all sorts of fascinating and creative speculations about how humanity might die out, although only the Second Law was widely accepted by scientists (or natural philosophers) as posing an *actual* threat to our survival. (Chapter 3.)

(3) *Impending Self-Annihilation* (1945/mid-1950s to the 1980s/early 1990s). The emergence of this shift coincided with the onset of the Atomic Age in 1945, although it did not solidify until the second half of the 1950s, after it became clear to many leading scientists that even a relatively small-scale thermonuclear conflict could blanket the entire planet with lethal quantities of radioactive particles. The following decades witnessed a proverbial explosion of credible new anthropogenic catastrophe scenarios, some relating to nuclear weapons (for example, the nuclear winter hypothesis), others associated with

environmental contamination and degradation caused by pollution and overpopulation, along with the possibility of runaway climate change, and still others linked to more hypothetical threats from biological weapons, self-improving artificial intelligence, and atomically precise nanotechnology. Suddenly, human extinction was not merely inevitable in the very long run but terrifyingly probable in the *near term*, due not to one but a *multiplicity* of distinct threats. (Chapter 4.)

(4) *Nature Could Kill Us* (1980/early 1990s to the late 1990s/early 2000s). This was initiated by the realization that natural phenomena like asteroids, comets, and volcanic supereruptions can affect the entire planet and precipitate mass extinctions, during which large numbers of species perish on geologically brief timescales. Prior to this, since at least the 1850s and stretching through most of the Cold War period, it was widely believed that naturally occurring catastrophes were always localized affairs, limited to circumscribed regions of our planet. The shift to this mood, which took longer than any other, coincided with the dramatic implosion of an Earth-sciences paradigm known as *uniformitarianism* during the 1980s, due in large part to novel research showing that the non-avian dinosaurs died out 66 million years ago after a large asteroid struck Earth. Suddenly, with uniformitarianism replaced by an unsettling new paradigm called *neo-catastrophism*, it became clear that we do *not* in fact live on a safe planet in a safe universe but are no less vulnerable to sudden annihilation from natural hazards than the dinosaurs were. Sooner or later, Nature will try to commit filicide. (Chapter 5.)

(5) *The Worst Is Yet to Come* (late 1990s/early 2000s to the present). Unlike the previous three shifts in mood, this wasn't driven by the discovery of any new kill mechanisms. Instead, it was catalyzed by two developments: first, a radical new philosophical perspective on the moral importance of avoiding our extinction, which directly inspired efforts to outline a maximally comprehensive picture of our existential predicament, or what I will call the "threat environment." This involved, in part, a "futurological pivot" toward the various emerging and anticipated future risks arising from advancements in biotechnology, synthetic biology, molecular nanotechnology, and artificial intelligence (including "artificial superintelligence"). The second triggering factor concerned new research in the environmental sciences showing that human-caused climate change, global biodiversity loss, and the sixth major mass extinction event pose dangers that are far more urgent and catastrophic than was previously known. This coincided with the idea that humanity (specifically the Global North) has initiated a new geological epoch called the "Anthropocene," in which our actions have permanently altered the geological record. At the heart of this mood was the frightening suspicion that however perilous the twentieth century may have been, the

twenty-first century will be *even more so*. In other words, the worst is yet to come. (Chapter 6.)

As stated, the two main components of the explanatory-predictive hypothesis that underlies the above periodization are enabling conditions and triggering factors. A key idea that connects these components—expounded in much more detail below—is what I will call an "existential hermeneutics." This denotes the interpretive lens through which an empirical model of the physical universe and everything it contains (including us) yields a picture of the *threat environment*. For example, a certain kind of *religious* existential hermeneutics might lead one to minimize or entirely dismiss the risk of an asteroid collision: if God is in control, if humanity is immortal, and if the future must unfold according to his prewritten plan, then we have no reason to worry. (There are, in fact, many historical examples of precisely this sort of reasoning, as we will see.) However, a *secular* existential hermeneutics according to which there is no omnibenevolent God watching out for us, humanity is no less vulnerable to annihilation than any other species, and there is no grand narrative of cosmic history in which humanity plays a central role might lead one to a rather different conclusion: we should be extremely worried if NASA announces that a large asteroid will intersect with Earth's orbit. Hence, even if people accept the same empirical data about potentially hazardous phenomena in our vicinity of the cosmos, different existential hermeneutics could lead to wildly different mappings of the threat environment. Even more, this has obvious *practical* implications: some who accept a religious hermeneutics might see an incoming asteroid as an occasion for elation, since, for Christians, the other side of the apocalypse is paradise. This could thus promote passivity in the face of danger. In contrast, those who accept a secular hermeneutics might see the asteroid as requiring immediate action to divert the incoming mass away from our planet. In a morally indifferent universe, it is our—and entirely our—responsibility to ensure the continued survival of our species.

Right and Wrong, Good and Bad

So far, our focus has been entirely on the possibility, probability, etiology, and so on of our extinction, which I bundled together under the concept of an *existential mood*. But there is a cluster of additional questions about our extinction that concern a related but distinct matter, namely, the *ethical and evaluative* implications of disappearing—questions that we can place within a philosophical subfield that I will call "Existential Ethics."[15] Examples include: Would causing or allowing our extinction be right or wrong? For what reasons? Under which conditions? Would our extinction, however it comes about, be good or bad, better or worse, or perhaps just neutral? For what reasons? Under which conditions? Is everything meaningless if extinction is inevitable? Would extinction undermine the significance of past progress? Would knowledge of our imminent extinction deprive our lives of important sources of value? Do we have an obligation to create future people? Do the unborn have a right to exist? Are there moral obligations to past individuals that would make dying out wrong? And so on.

As we will see, philosophers have proposed a fascinating array of answers to these questions. According to some, causing or allowing our extinction would be very bad, and therefore wrong, because of the associated *opportunity costs* of no longer existing, such as the loss of all future generations, or the loss of further scientific and moral development. I will classify these as "further-loss views," which can take many different forms and be interpreted in many different ways. Others believe that the badness or wrongness of our extinction boils down entirely to the *manner* in which it is brought about: if there is nothing bad or wrong with *how* we go extinct, then there is nothing bad or wrong with extinction, period. This has the intriguing implication that human extinction does not pose any *unique moral problem*. It is different in degree rather than kind from other effects our actions might have or other catastrophes that might befall us. I will call the class of positions that accept this idea "equivalence views" and defend a version of it in Chapter 11. Still others maintain that the non-existence of our species would be *less bad* than continuing to exist—or perhaps even positively *good*—since it would mean the absence of all human suffering. This has led some to argue that we should actively strive to produce this very outcome, preferably through voluntary, peaceful means such as by refusing to have children. Let's label these "pro-extinctionist views," which, like the other two main categories above, can be fleshed out in a variety of ways.

Because the questions of Existential Ethics are distinct from those linked to existential moods, how philosophers have answered them over time constitutes its own unique history. For ease of exposition, I will refer to the history of existential moods as "History #1" and the history of Existential Ethics as "History #2."[16] These are of course *causally* related: it wasn't until our extinction was seen as possible (during the nineteenth century, but especially after the second existential mood solidified) that philosophers began to seriously contemplate the topic, and not until the latter twentieth century (coinciding with the third and fourth existential moods, then continuing into the fifth) that it became the focus of sustained and systematic investigation. This makes sense, of course, since what reason is there for pondering the ethics of something that one believes is impossible? We don't, for obvious reasons, write books about the ethicality of eating unicorns raised in deplorable conditions on factory farms. Hence, only once it became clear that human extinction is really possible and, later, that we could actually bring this about did philosophers begin to seriously consider the normative aspects of extinction. History #1 thus provides, as it were, the background context of History #2: in certain important respects, the contours of History #1 shaped those of History #2, although we will see that the causal relationship between these histories *reversed* around the turn of the twenty-first century, when

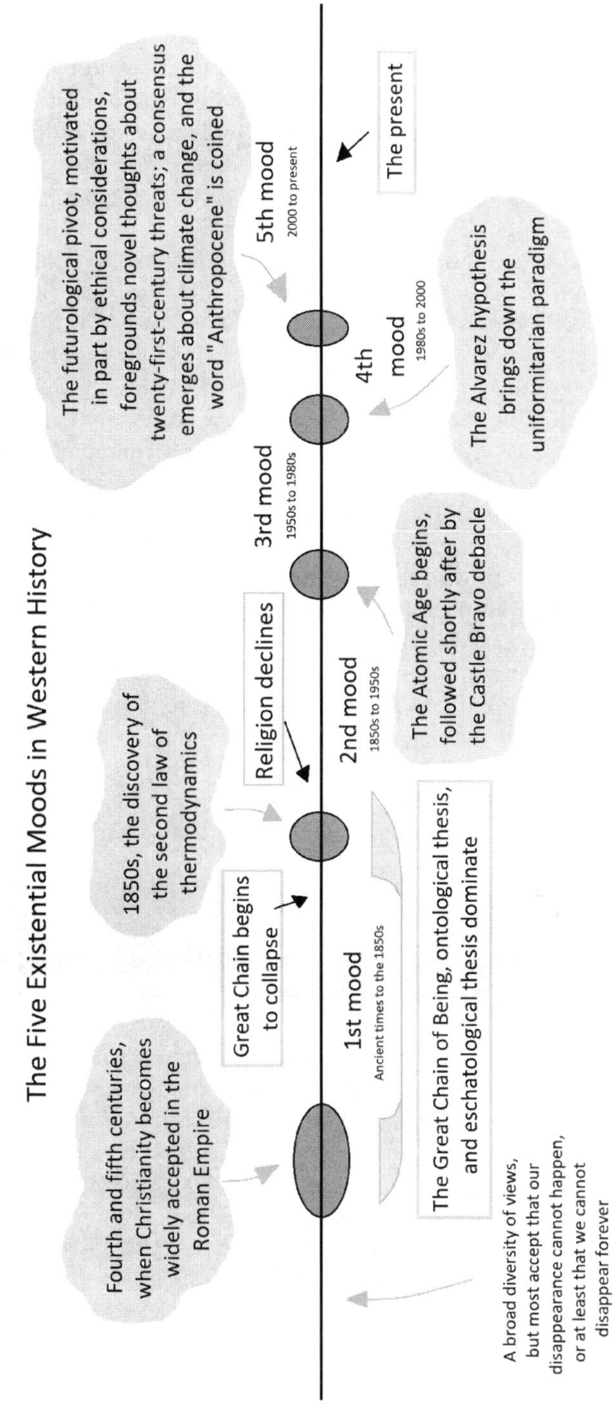

Figure 1.3 Rough outline of the five existential moods in Western history.

developments in Existential Ethics played a crucial role in triggering the shift to the fifth existential mood, our current mood.

As with the first history, this second history can also be partitioned into segments, which I will refer to as "waves." There are three points to make about this. First, I use "waves" instead of "periods" because, in contrast to the well-defined periods of History #1, with each existential mood emerging and solidifying in response to specific, datable events, the boundaries between these waves are often vague and overlapping. Furthermore, the transitions from one to another exhibit a kind of ebb and flow of ideas, unlike the rapid shifts between existential moods. Second, whereas I have defined existential moods as involving a certain *uniformity* of agreement about our existential predicament (that is, a *stable set* of answers to questions about the possibility, probability, and so on, of our extinction), every wave discussed below encompasses a *diversity* of viewpoints on the core questions of Existential Ethics. This does not mean that the periodization of History #2 is arbitrary: one can still identify particular waves in the ocean even if their boundaries are inherently ill-defined and the molecules they contain are jostling about in opposite directions. Third, I will count a total of four waves in History #2, where these waves do not straightforwardly correspond to the five periods of History #1. In brief, the initial wave spans the first and second existential moods; the second wave extends across the third and fourth existential moods (as the discovery that natural hazards could annihilate us did not have much effect on the way philosophers thought about our extinction); and while the third wave *coincides* with the onset of the fifth existential mood (due in part to the reversal mentioned above, whereby History #2 changed the course of History #1), the fourth wave arose only in the *past five years* or so, as of this writing. This may appear somewhat complicated at first glance, but the mismatch between periods and waves makes good sense, as Part II will attempt to show.

The Four Waves

Following the precedent set above, let's take a quick bird's-eye view of the four waves, each of which will receive its own chapter:

(1) *Calamity, Pessimism, and the Greatest Conceivable Crime* (early modern period to the 1950s). This covers ruminations about our extinction within the Western tradition until the third existential mood commenced in the mid-twentieth century. Most of the earliest examples were articulated by atheistic philosophers in the nineteenth century, especially its second half, although one can trace the roots of Existential Ethics back to at least the early 1700s. During this wave, some philosophical thinkers suggested that our collective non-existence would be bad because it would entail the loss of things we care about, such as knowledge and laughter, or because it would prevent the realization of all future happiness. Others argued that, since our lives are full of suffering, we

should actively work to bring about our extinction, if not (somehow) eliminate the very possibility of life existing anywhere in the universe. Still others, in a rather different mode of reflection, wondered whether human existence can be meaningful given that, according to the Second Law, the universe will eventually sink into a frozen state of eternal lifelessness. Does anything really matter if, in the end, all will be lost? (Chapter 8.)

(2) *An Explosion of Insights* (1950s to the ~2000s). The development of nuclear weapons provided the first credible means of self-annihilation, and hence the Atomic Age inspired a number of philosophers to consider the core questions of Existential Ethics more vigorously than intellectuals had in previous eras. A common theme among early pioneers of the field was that our newly acquired *powers of action*, our ability to affect every person on the planet as a result of nuclear weapons or environmental degradation, has rendered traditional ethical systems outdated or obsolete. We thus need a new *kind* of ethics, one specifically designed for the possibility of "omnicide," or "the murder of everyone." Some attempted to construct such an ethical framework, while many others proposed novel arguments, either within or outside a new ethics, for why our extinction might be right or wrong, good or bad, better or worse. For example, the fact that humanity could exist for millions or billions of years into the future, and that continued progress could make future lives much better than current lives, may provide reasons for safeguarding our collective survival. Or perhaps the badness of extinction concerns the fact that it would remove the only rational and moral beings in the known universe. Others argued that, on a "person-affecting" theory of ethics, while dying out in a global catastrophe would *itself* be very bad, the subsequent *outcome* would not be, since there would be no one around to suffer the loss of humanity. Some even argued from a radical environmentalist perspective that humanity should stop existing because of our deleterious impact on the biosphere, with a few fringe actors endorsing involuntary omnicide. More than any other wave, this one saw the articulation of an extraordinary range of innovative new ideas, some of which influenced views proposed during the next two waves. (Chapter 9.)

(3) *Astronomical Value, Longtermism, and a Dying-Extinction* (~2000s to the present). This wave is notable for two developments in particular. First, it witnessed the formation of the first *cohesive research program* focused on the ethics of human extinction and related scenarios, which were called "existential risks." Central to this research program were ideas drawn from the futurological vision of transhumanism, the ethics of utilitarianism, and a branch of cosmology known as physical eschatology. The result was a radical further-loss view according to which our extinction would constitute a moral tragedy of quite literally *cosmic proportions*. Hence, reducing the risk of extinction (and existential risks more generally) should be the top "global priority" for our species. Over the past decade, this has evolved into an ethic called "longtermism," which has

become influential far beyond the perimeter of academia, something that few philosophical ideas can boast of doing. Second, at precisely the time that the idea of existential risks was being developed, a diametrically opposed view took shape, namely, "antinatalism," which, on this version, claims that birth is a net harm, life is much worse than we typically believe, and we should all stop having children. The leading advocate of this view further argued that the outcome of extinction would be positively good, since it would mean no more births and no more suffering. We should thus—echoing earlier theorists—take steps to precipitate our extinction, and we should do this sooner rather than later. These were the two defining features of the third wave. (Chapter 10.)

(4) *Alternative Approaches* (~2017 to the present). The most recent wave emerged within the Analytic tradition of Western philosophy over the past five years or so, partly in response to the first development of the previous wave. A unifying feature of this wave has been an approach to Existential Ethics that is non-utilitarian or, more generally, non-consequentialist in orientation. Some have argued that, according to a *contractualist* theory, anthropogenic human extinction would be wrong *only insofar* as it causes harm to those living at the time, while others proposed that we should avoid extinction because of the particular value that humanity might possess ("final value"), or because extinction would undermine many of the activities that make our lives "value-laden."[17] However, still other philosophers made the case that our disappearance from the universe might on balance be good, and hence that if one were to see an asteroid heading toward Earth, one shouldn't try to redirect it away from us. After a critical survey of these positions, I will then outline my own views on the matter, which combine elements of the equivalence and pro-extinctionist views that together yield the rather surprising conclusion that, in practice, we should vigorously work to *avoid* our extinction. (Chapter 11.)

Readers may have noticed that this summary skips over Chapter 7, which opens Part II. This is because Chapter 7 does not discuss any historical wave within Existential Ethics, but rather provides a theoretical foundation for the second half of the book, which will enable us to understand more clearly the various positions outlined in the Existential Ethics literature. In doing so, I distinguish between multiple senses of "humanity" and six types of "extinction" that are directly relevant to ethical and evaluative assessments of our disappearance (all of which build on the definition proposed above). I will also argue that, in reflecting on what might be right or wrong, good or bad, better or worse about human extinction, it is imperative to differentiate between (a) the process or event of Going Extinct and (b) the state or condition of Being Extinct. As noted earlier, some ethical theories entail the equivalence thesis, which asserts that the wrongness/badness of extinction is wholly reducible to *how it comes about*, while other theories identify some additional loss associated with the *outcome* as contributing to the badness of our disappearance. Hence, the first concerns the details of Going Extinct, whereas the second also focuses on the opportunity costs of Being Extinct. There are, furthermore, many pro-extinctionist views that admit that Going Extinct may be very bad, as it could cause harms and cut lives short, yet maintain that the resulting state or condition of Being Extinct would be better than Being Extant (as we can say), if not in some sense good. As we will see, the equivalence view is theoretically compatible with certain pro-extinctionist positions, although one who accepts the former need not accept the latter.

Those interested in Existential Ethics are especially encouraged to look over this chapter, since the philosophical literature on the topic is replete with confusions and imprecise statements arising from a failure to recognize that "human extinction" is highly polysemous. For example, some arguments for the conclusion that, as one might see it expressed in the literature, "human extinction would be wrong" only concern *particular types* of human extinction, while other arguments target *every type* of extinction. Being clear about *what it is* we are talking about is critical for the field to progress.

In sum, Part I is largely an intellectual history, while Part II is a history of ethics. And while the periodization of History #2 is imprecise around the edges, I take the periodization of History #1 to be a more or less objective fact, something discovered rather than invented.

Explanation and Prediction

Why care about the origins and evolution of the idea of human extinction in the first place? What is the value of Part I's intellectual history and Part II's reconstruction of the development of Existential Ethics? Why, in short, does this study matter? One answer comes from Aristotle's *Politics*, written circa 350 BCE, in which he declares that "he who thus considers things in their first growth and origin, whether a state or anything else, will obtain the clearest view of them." As the opening paragraphs of this chapter indicated, *human extinction* is so commonplace today that few have any inkling that the idea is a quite recent addition to our shared library of concepts. Understanding this fact alone can give one a deeper appreciation of the idea and its significance in contemporary discourse.

Even more importantly, anticipating how the idea of *human extinction* might evolve in the future—because there is no reason to believe that its story has ended—requires some grasp of the causal factors that have shaped its journey so far. A causal story that deals with general categories of phenomena rather than particulars can provide not just an explanation of what happened in the past but predictions of what might happen in the future. To illustrate, consider that historians can give a detailed account of how World War I began: a Yugoslav nationalist named Gavrilo Princip shot and killed Archduke Franz Ferdinand; Austria-Hungary then declared war against Serbia; this led the countries allied with Austria-Hungary (Germany and Italy) to declare war on the countries allied with Serbia (United Kingdom, France, and the Russian Empire); and so on. Here we have a historically unique concatenation of causes and effects, with each effect except for the last—the explanandum of interest—functioning as the next cause in the sequence, that explains the origins of the Great War, yet does not enable one to predict the onset of future wars, at least not in any reliable way.

In contrast, the triggering factors and enabling conditions that explain the course of History #1 are sufficiently general to make rough predictions about how this history might unfold in the future with some confidence. Rather than "A caused B caused C . . .," we have "if something of type A is the case, then something of type B will be the case . . ." More specifically, if the decline of religion and the discovery of kill mechanisms account for the shifts in existential mood, then the further decline of religion and the discovery of additional kill mechanisms could induce yet another shift in the coming decades. Alternatively, if recent trends of secularization were to reverse and religion were to become more widely adopted, then we should expect the idea of human extinction to fade into the background, perhaps becoming incoherent once again, like the concept of *married bachelor*. Hence, understanding the history of *human extinction* could enable one to anticipate its evolution later this century, which may prove important. Why? Because if humanity *really is* at risk of going extinct, but if most people, or those holding the reins of power, do not believe that this outcome is even possible, then the probability of catastrophe could rise, perhaps significantly. By analogy, if one were convinced for some reason that they could never get in a bicycle accident, they may stop wearing their helmet, which would thus increase the chance of serious injury. There are, in other words, major practical implications to understanding this history.

Let's now turn to History #1.

Notes

1. Thanks to Olle Häggström for this anecdote (personal communication).
2. Note that I will use italics both for emphasis and to indicate that I am referring to a concept or idea rather than a lexical item (a word or set of words) or phenomenon in the world. Hence, *human extinction* would refer to the concept, whereas "human extinction" would refer to the term. I will also take "concept" and "idea" to be interchangeable.
3. The one survey that I am familiar with was conducted by Bruce Tonn and published in 2009, which consisted of only 600 respondents. See footnote 10 for details. Note also that by "the West," I mean the group of peoples in Western Eurasia who identified themselves as the inheritors of the legacy of classical Greece and Rome, and those regions of the world whose Indigenous cultures they supplanted through the oftentimes genocidal process of colonization.
4. Quoted in Moltmann and Kohl 2004. Similarly, Alison McQueen writes in *Political Realism in Apocalyptic Times* (2017):

> [I]n contrast to the Judeo-Christian apocalypse, there is no system of belief that renders nuclear annihilation meaningful, no theodicy that endows it with ultimate justification, and no promissory narrative that consoles the terrified and trembling. It is instead an apocalypse without redemption—an end that can only be confronted as a naked absurdity (quoted in Zimmer 2022).

5. This literature includes, most notably, David Sepkoski's *Catastrophic Thinking: Extinction and the Value of Diversity from Darwin to the Anthropocene* (2020), Thomas Moynihan's *X-Risk: How Humanity Discovered Its Own Extinction* (2020), and Dan Zimmer's exceptional 2022 dissertation *The Immanent Apocalypse: Humanity and the Ends of the World.*
6. Moynihan 2019. Unfortunately, I have found an enormous number of errors in Moynihan's 2020 book *X-Risk*. For a list of such errors in just the first few pages of his manuscript, see Torres 2021.
7. Michel et al. 2011.
8. Consistent with this, the survey conducted by Bruce Tonn found that "Christians and Jews overwhelmingly do not believe that humans will become extinct but secular and non-religious people strongly believe otherwise" (2009).
9. Put differently, "immortal human" would have been a pleonasm, whereas "human extinction" would have been a contradiction.
10. Walls 2008.
11. Hill 2002. Indeed, Bart Ehrman notes that

> The notion of individual resurrection, developed at the tail end of the Hebrew Bible period, arose principally in response to questions of theodicy. How is it fair—or, rather, how can God be just—if the wicked prosper and then die and get away with it? Or if the righteous suffer for doing God's will and then perish in misery? Surely there must be some kind of recompense when we pass from this world of mortality. As evidenced in the non-canonical book of 1 Enoch and then the canonical Daniel, Jewish thinkers developed views of the afterlife that explained it all (2021).

12. Although interpretations of Christian eschatology have changed over time, and while the study of eschatology has waxed and waned—for example, during the nineteenth and *early* twentieth centuries, little attention was given to the topic, leading Ernst Troeltsch to famously state in 1925 that "nowadays the eschatological office is closed most of the time" (quoted in Walls 2008, 7)—the centrality of cosmic eschatology within Christianity is indisputable. Historically speaking, traces of both amillennialism and premillennialism can be found in the very early period of Christianity, although amillennial eschatology became the dominant interpretation during (a) the Middle Ages, due in part to Saint Augustine's influence, and (b) among the Reformers. According to amillennialism, "the thousand years of [Revelation] 20 represent the entire Christian era, beginning with the cross, resurrection, and ascension of Christ and ending with the second coming," or *Parousia* (Neall 2005). At the end of this period, Satan will deceive the nations of Earth, and a series of battles, e.g., the battles of Armageddon and of Gog and Magog against "the beloved city" (Revelation 20:9), will occur, culminating with "the second coming of Christ, the judgment of the wicked, and the rewarding of the righteous—events which mark the end of the millennial Christian era" (Neall 2005, 186). On the other hand, postmillennialism, with its reassuring hope of continued progress over time, gained adherents most notably during the Enlightenment, during which the idea of *progress* was foregrounded by philosophers like Condorcet; however, we will see in Chapter 4 that premillennialism—especially premillennial dispensationalism—became immensely popular, especially among the general public, during the latter half

of the twentieth century, which is sometimes dubbed the "Century of Eschatology" (see Walls 2008). As Hans Urs von Balthasar noted in 1957, responding to Troeltsch's earlier statement, the eschatological office had been opened for business once again and was now "working overtime" (Schwöbel 2000).

13. Ringmar 2018.
14. See OED 2022.
15. This term appears in an excellent critique of "longtermism" (see Chapter 10) written by Zoe Cremer and Luke Kemp, although their definition is different than mine. For example, they argue that it is important to "separate[e] the study of extinction ethics (ethical implications of extinction) and existential ethics (the ethical implications of different societal forms)," both of which "should be analysed separately from" the study of extinction risks. This roughly corresponds, I believe, to my distinction between the study of the ethical and evaluative implications of extinction, on the one hand, and the study of kill mechanisms (and related phenomena) on the other. See Cremer and Kemp 2021.
16. To my embarrassment, I only became aware of Dipesh Chakrabarty's 2000 "History 1" and "History 2" after completing a draft of this book. Thanks to Dan Zimmer for apprising me of Chakrabarty's work.
17. See Scheffler 2013, 2018; see also Chapter 11.

Part I

Existential Moods

2 Beginnings of "The End"

A Deluge of Flood Myths

One of the first objections I often hear when describing a central thesis of Part I—that for much of Western history, just about everyone assumed that humanity is indestructible—is that tales of global catastrophes, even the complete obliteration of humanity, are common in religious and mythological thought. In some cases, these involve one-off catastrophes in the past associated with, for example, failed attempts by the gods to create humanity. In others, they are the unavoidable result of endless cosmic cycles that pass through stages of birth, growth, decline, death, and rebirth, extending forever in both temporal directions. In still others, they are anticipated future events embedded within eschatological narratives that culminate with devastating wars and supernatural disasters.

Let's begin by examining a number of examples, the overwhelming majority of which do not, in fact, suggest that humanity was or could be at risk of destruction. We will then turn to the ~1,500-year period during which *human extinction* was "blocked" by the three beliefs specified in Chapter 1 and show how one of these, the Great Chain of Being, was dealt a mortal blow at the turn of the nineteenth century. By virtue of the many different topics covered by this chapter, it will be the most discursive in the book. It will also occasionally deviate from the Western tradition, as doing so will help underline just how ubiquitous the assumption of our indestructibility has been across cultural space and time.

The most obvious example of the first is the flood myth, found in the written and oral traditions of peoples around the world. It is, indeed, something close to a universal motif, with the same basic structure and message.[1] One of the earliest instances comes from *Atrahasis*, an epic poem within the Akkadian literature, which was first recorded during the Late Old Babylonian Period, in the eighteenth century BCE.[2] In it the Sumerian god of wind, storms, Earth, and air named Enlil tries to destroy humanity on three occasions, each separated by 1,200 years, because our noise-making disturbed his sleep. The first attempt involved a plague, the second a drought, and the third a famine, though all failed because of the god Ea.[3] Enraged, Enlil then sends a flood

DOI: 10.4324/9781003246251-3

to exterminate humanity, although Ea once again thwarts his plan, warning a man named Atrahasis who builds an ark and survives.

While this story is little-known today, it may have been the basis of the flood myth in the *Epic of Gilgamesh*, which many consider to be one of the greatest masterpieces of world literature.[4] Its account of Enlil's wrath over humanity's noisiness similarly involves a great deluge, though in this case the (Babylonian) main character is named Utnapishtim, who builds a boat in which "all the living beings that I had," including "all my kith and kin . . . all the beasts and animals of the field" take refuge.[5] After seven nights and six days of rain, Utnapishtim and the other survivors then emerge to repopulate the planet.

The most famous Western flood myth, though, is the Noachian story of the Book of Genesis, which very likely drew from one or both of the Akkadian narratives above. Once again, the deluge results from divine wrath, in this case because "all the people on earth had corrupted their ways" (Genesis 6:12). Hence, God decides to "put an end to all people" and to "destroy all life under the heavens," but due to Noah's righteousness, God spares Noah, his wife, his sons Shem, Ham, and Japheth, and his sons' wives (Genesis 6:13, 17).

One also finds several flood myths within Greek mythology, the most well-known being the flood of Deucalion, the son of Prometheus. In this story, Zeus becomes livid after a young boy is sacrificed to him and consequently unleashes a catastrophic deluge. However, Prometheus tells his son Deucalion about this beforehand, so Deucalion builds an ark that enables him and his wife Pyrrha to survive the downpour, the only two on the planet who don't perish. According to Plato's dialogue *Timaeus*, this occurred in the tenth millennium BCE and is actually one of many disasters that have wiped out *nearly everyone*. The most devastating of these involved fire or water; in the former case those in the river valleys survive, while in the latter only shepherds in the mountains do. The deluges cause the loss of all culture, which must then be rebuilt from scratch. According to Critias, the Athenian statesman Solon—one of the Seven Sages and a cousin of Plato's mother—traveled to Egypt and was told by priests that Athens has been obliterated many times in the past, with each instance wiping out all the historical knowledge of the city that had accumulated up to that point. As one priest says, "none of you but the unlettered and uncultured" remained after the catastrophe, "so that you become young as ever, with no knowledge of all that happened in old times in this land or in your own." These stories are reiterated in (a) the subsequent unfinished dialogue *Critias*, in which Plato insists that many large floods have previously occurred, and (b) Plato's *Laws*, where he again references the "traditions about the many destructions of mankind which have been precipitated by deluges and pestilences, and in many other ways." In fact, the main topic of the *Critias* was to be the story of Atlantis, a "great and wonderful empire" that was defeated by Athens, after which both were destroyed by a great earthquake, tsunami, and flood, thus causing Atlantis to disappear under the sea.

Not only have such catastrophes occurred in the past, according to Plato, but they will also happen in the future. As the Egyptian priest tells Solon in the *Timaeus*, "there have been and there will be many and diverse destructions of mankind." This idea was subsequently elaborated by Plato's student Aristotle, who suggested the possibility of "cyclic floods" that destroy whole populations and which may be associated with the so-called *Great Year*, which Plato referred to as the "perfect year." The idea is that just as a month occurs when the moon completes its orbit around Earth (the word "month" being etymologically related to "moon"), and just as a year occurs when the sun completes its orbit around our planet (on a geocentric model of the solar system), so too is there a "great" or "perfect" year that occurs whenever the sun, moon, and six other stars or planets end up in perfect alignment relative to Earth, which was thought to happen every 36,000 years.[6] The completion of each cosmic cycle then triggers cataclysmic floods that wipe out previous civilizations, leaving only a few survivors. Aristotle thus believed that proverbs and aphorisms popular during his time were bits of wisdom from people who lived prior to the last catastrophe. Writing in the *Metaphysics*, he asserts that certain "inspired saying[s] . . . have been preserved as a relic of former knowledge," most of which was lost and therefore must be rediscovered.

Cycles and Arrows

Many other mythological systems reference worldwide disasters as part of their creation narratives. As the Egyptologist Geraldine Pinch writes, a common theme across mythological systems is God or the gods destroying "the unsatisfactory part of humanity"—i.e., cleansing the world of evil races—resulting in "several attempts at creating people before they are satisfied."[7] An example comes from the ancient Egyptian *Book of the Heavenly Cow*, which explains how Ra, the god who created the world, found humanity plotting against him. After consulting other gods about what to do, he decides to punish humanity, delegating this task to Hathor, the goddess of the sun. On the first day, Hathor slaughters many people in the desert, stating that she has "overpowered humanity and it was sweet to my heart."[8] She plans to continue the attack the following day, but for reasons that are unclear, Ra changes his mind and tricks Hathor into forgetting about the mission by getting her drunk. Consequently, humanity survives.

Another example outside the Western tradition that illustrates Pinch's point comes from the elaborate creation narrative of the Mesoamerican Aztecs (1300–1521). This involves four worlds, or "Suns," being created and destroyed prior to the emergence of current humanity. In each case, the humans who existed were annihilated: first by jaguars (First Sun), then by fierce winds of a hurricane (Second Sun), a fiery rain that transformed everyone into turkeys, butterflies, and dogs (Third Sun), and a great deluge (Fourth Sun), which two people survive (Tata and Nene) but are turned into dogs by the gods who cut off their heads and, rather humorously, attach them to their buttocks. The gods then recreate the original humans in the Fifth Sun, which is maintained through ritualistic human sacrifice. The mythologies of the Hopi and Navajo peoples, who lived around the Four Corners region of the United States, specify similar episodes of annihilation and creation, and the same basic pattern of creation and recreation is found in the "Five Ages of Man" adumbrated by the Greek poet Hesiod (750–650 BCE) in his *Works and Days* (c. 700 BCE).

Within Eastern belief systems like Hinduism and Buddhism, one finds cyclical eschatologies in which the cosmos oscillates between periods of renovation/growth and deterioration/decline. Buddhism, for example, posits cosmic cycles called "great eons," each of which consists of four phases or "incalculable eons." The first leads to the destruction of the entire universe and is associated with immorality, sickness, epidemics, famines, and wars. The universe then stagnates in a state of nonmanifestation until it "gradually comes back into being" during the third, which ushers in the final phase of cosmic existence.[9] However, the cosmos is never completely destroyed: some living beings survive and are reborn into the world once it becomes manifest again.

Cyclical models of cosmic evolution were also proposed by the ancient Greek philosophers. For example, the Presocratic philosopher Xenophanes of Colophon (c. 570–478 BCE) argued that the world alternates between two extremes: wetness/water and dryness/Earth. When the first occurs, the oceans submerge all the land, turning it into mud and causing everyone alive at the time to perish. But when dryness dominates, humanity and other living creatures reappear. Our disappearance is thus complete but temporary. A similar account was put forward by the poet-sage and vegetarian Empedocles (c. 494–434 BCE). In his poem *On Nature*, of which only fragments remain, he identifies the two extremes as corresponding to Love and Strife, the cosmic personifications of attraction/combination and repulsion/separation. "By turns," he declared, "they dominate while the time revolves." In the case of Love, everything is fused into an undifferentiated mass and no life is possible; in the case of Strife, everything is pulled apart and, once again, life becomes impossible. It is during the transition between Love and Strife, Strife and Love—on one interpretation—that living creatures emerge (through a form of natural selection).[10] As Empedocles wrote:

> A twofold tale I shall tell: at one time it grew to be one only from many, and at another again it divided to be many from one. There is a double birth of what is mortal, and a double passing away; for the uniting of all things brings one generation into being and destroys it, and the other is reared and scattered as they are again being divided. And these things never cease their continuous exchange of position, at one time all coming together into one through Love, at another again being borne away from each other by Strife's repulsion.[11]

Yet another example comes from the Stoics of the Hellenistic period (323–31 BCE), who proposed a cosmological theory involving *ekpyrosis*, whereby the cosmos is periodically consumed and purified by a great conflagration—an idea that the earlier (non-Stoic) philosopher Heraclitus may have also accepted.[12] However, once the purification-by-fire event has occurred, the cosmos starts over again, with all things unfolding exactly as they did before. Consequently, as Chrysippus of Soli, an ancient Greek Stoic from the third century BCE, wrote in his treatise *On Providence*, "it is evidently not impossible that we too, after our death, will return to the shape we are now, when certain periods of time have elapsed."[13] The German philologist Friedrich Nietzsche would later popularize this model as the "eternal return" or "eternal recurrence."[14]

The notion that the fundamental structure of time takes the shape of a circle rather than a line has, in fact, been the default assumption throughout much of history. This is captured most memorably by the Egyptian image of the Ouroboros, a snake eating its own tail, which "signified the capacity of the universe to perpetually renew itself, so that every end could also be a beginning."[15] There were, to be sure, linear narratives as well, but these were typically enfolded *within* larger cycles that stretch infinitely, or at least interminably, into the past and future. Consider the ancient Egyptian *Coffin Texts* (2100 BCE) and *Book of the Dead* (1550 BCE), which indicate that, as the creator-god Atum declares in the latter, "in the end I will destroy everything that I have created; the Earth will become again part of the Primeval Ocean, like the abyss of waters in their original state." This looks like a linear narrative: the world had a beginning and will someday come to an end. Yet the Primeval Ocean, during which only Atum and Osiris exist, will be followed by a period of renewal. The world starts over. Another example is the eschatology of Norse mythology, which originated from the North Germanic peoples (Scandinavians) in the ninth century CE. Our world is prophesied to end through a series of bloody battles, a worldwide flood (the land sinking into the sea), and a massive conflagration that engulfs the planet. This is called *Ragnarök*, meaning "final Fate of the Gods." Several gods will perish in this disaster, including the thunder-and-lightening deity who wields a hammer and protects Earth, Thor. The whole human population will also be annihilated except for two lone survivors: a man named Lif (meaning "life") and a woman named Lifthrasir (meaning "lover of life"), who then repopulate the planet. Hence, *Ragnarök* initially looks like a linear tale of future destruction, at which point the world ends, but in fact this is part of a larger undulation of destruction and rebirth, whereby the earth emerges from the waters that have covered it, humanity begins anew from a single couple, and another generation of gods arises.[16]

A genuinely linear view of time—as consisting of a definite beginning and end of the world—was thus a novel innovation that, as such, deviated from the older view of time as a loop by replacing it with an arrow. According to the historian Norman Cohn, this radical new idea originated with the ancient Persians, exemplified by the cosmogony and eschatology of Zoroastrianism, a monotheistic religion founded by the prophet Zoroaster (or Zarathustra), who may have lived during the tenth century BCE.[17] On this account, cosmic history will culminate with the appearance of a virgin-born messiah (the *Saoshyant*), an Armageddon-like battle, a bodily resurrection of the dead, and a final judgment of humanity by God (Ahura Mazda). The striking parallels between Zoroastrian eschatology and later Christian and Islamic narratives of the world's end might not be coincidental:[18] the storyline motifs of the former may have been picked up by the Jewish people during the Second Temple period, when for roughly two centuries Israel was subject to the Persian Empire, and subsequently transferred to the other Abrahamic faiths.[19] Hence, Christians anticipate the Second Coming of Christ (born of a virgin), Battle of Armageddon, one or more resurrections of the dead, and a final judgment, and both the Sunni and Shi'ite branches of Islam accept a similar sequence of end-times happenings. Within the West, this conception of time as an arrow has become the standard view of cosmic history, even among secularized eschatologies like those of Marxism and Ray Kurzweil's "singularitarianism," discussed later. It also fits with our contemporary understanding of cosmic evolution, whereby the universe began with a "big bang" and will eventually come to an "end" with the heat death of the universe, one of the main topics of the next chapter.

The Indestructibility of Humanity

The common theme shared by all but a few of the examples discussed above is that humanity never completely disappears. In the flood myths, two or more people always survive to repopulate the planet. The catastrophes discussed by Plato and Aristotle are localized events that, as such, enable some portion of the human population to persist. The same goes for the cyclical eschatologies of Hinduism and Buddhism, while the creation stories of the ancient Egyptians and Aztecs conclude with humanity enduring. In the linear narratives of Christianity and Islam, the culmination of world history coincides with humanity entering into the eternal afterlife with God in heaven or Satan in hell. In other words, we continue to exist, albeit with supernaturally enhanced bodies. The most interesting cases come from Xenophanes, Empedocles, and the Stoics, who posited stages of their cosmologies in which humanity does disappear entirely, but this is only ever a temporary state of affairs. Once the stage of dryness returns, on the account proposed by Xenophanes, our species always reemerges. Hence, even though humanity undergoes a certain type of extinction—which I will later call "demographic extinction"—we are nonetheless indestructible.

This notion of indestructibility is what underlies the overwhelming majority of ancient tales of past and future catastrophes, eschatologies involving global disasters at the end of time, and cosmic cycles that reduce humanity to only a few individuals or none at all. It is the common denominator of all the accounts above, even those that imagined our complete disappearance. Indeed, it is not even clear that humanity *could have* gone extinct in stories like the Noachian flood myth, given our special place in creation as the only beings made in the image or likeness of

God. What would become of creation if humanity were to cease existing? What would be the point of God's six days of work if the beings to which he gave dominion over all other creatures were to no longer be? As numerous scholars have noted, underlying many of our conceptions of the universe throughout history has been a certain persistent anthropocentrism, according to which humanity lies at the center of everything and hence without us the world would have no purpose or meaning. It would be incomplete, a "mere wasteland" as one philosopher would later put it. Since it is inconceivable that the world would have no purpose or be rendered incomplete, the assumption of our indestructibility follows.

Exceptions to the Rule?

However, there are some possible exceptions from this period, before the rise of Christianity in the opening centuries of the first millennium. Consider a line from the *Epic of Gilgamesh* in which Enlil, after discovering Utnapishtim's boat after the flood, becomes "furious" and "filled with rage," declaring: "Where did a living being escape? No man was to survive the annihilation!" This suggests that Enlil's aim was the total eradication of humanity, which points to a stronger conception of human extinction than Xenophanes, Empedocles, and the Stoics had in mind, whereby our destruction is both complete and permanent.[20]

A more intriguing case comes from the ancient Greek atomists, who held that everything in the universe is composed of "atoms" (literally, "uncuttable" or "indivisible"). Too small to see with the naked eye, they collide with each other while moving about the void, sometimes sticking together due to hooks and barbs on their surface. All macroscopic objects are the result of different configurations of atoms: one configuration yields a tree, another the moon, and yet another the human organism.[21] Cosmologically, the atomists believed space and time to be infinite, like the number of atoms, and consequently an endless procession of worlds, or *kosmoi*, are formed through the random interaction of these particles.[22] As Hippolytus of Rome (c. 170–235) describes the idea in his *Refutation of All Heresies*, Democritus (c. 460–370), who founded the atomist school along with his teacher Leucippus, maintains that

> in some [*kosmoi*] there is neither sun nor moon, while in others that they are larger than with us, and with others more numerous. And that [some] attain their full size, while others dwindle away and that in one quarter they are coming into existence, whilst in another they are failing; and that they are destroyed by clashing one with another.[23]

On this view, the ultimate destiny of all *kosmoi* is complete dissolution. Hence, it is only a matter of time before our own world disappears and, along with it, the human race. Although I am not aware of any ancient atomists elaborating on this idea, total human extinction is a straightforward implication of their cosmological theory. However, this theory also implies that another world exactly like ours will eventually arise somewhere within the infinite corridors of space and time. What makes this view unique is that the growth and collapse of our world is not part of a serial sequence of events—that is to say, it is not *our* particular *kosmos* that perishes and reappears but rather different *kosmoi* arising and collapsing throughout the void. Consequently, there is a sense in which the atomists' view implies that our species is *not* indestructible, since our lineage here on Earth will eventually cease to exist forever, although from a cosmic perspective, "humanity" may indeed be ineradicable, as creatures just like us will always exist.

This is the most notable example within the Western tradition, so far as I know, of a view that comes close to suggesting our destructibility. But there are also occasional references in the ancient literature to the idea of there *never having been* any humans at all, which is similar to *human extinction* in that it entails there being no tokens of the type "humanity," but different in that it is backward- rather than forward-looking. For example, consider the following passage from the second tractate in Seder Moed (of the Babylonian Talmud), titled Eruvin, in which two schools of early first-millennium-CE Jewish thought, namely, the schools of Beit Shammai and Beit Hillel, disagreed about which is better: the existence or non-existence of humanity. To quote the passage in full:

> The Sages taught the following baraita: For two and a half years, Beit Shammai and Beit Hillel disagreed. These say: It would have been preferable had man not been created than to have been created. And those said: It is preferable for man to have been created than had he not been created. Ultimately, they were counted and concluded: It would have been preferable had man not been created than to have been created. However, now that he has been created, he should examine his actions that he has performed and seek to correct them. And some say: He should scrutinize his planned actions and evaluate whether or not and in what manner those actions should be performed, so that he will not sin.[24]

Similarly, the Book of Genesis includes a curious passage in which God, after surveying the wickedness of humanity before deciding to send a flood, says he "regretted that he had made human beings on the earth," which suggests that God wished we had never existed. We will revisit this distinction between dying out and never having been in Chapter 10.[25] For now, it is worth registering that, even if our complete disappearance might have been difficult to imagine, some did entertain the idea of a universe that never contained any humans at all.

Fishes That Have Wings

Turning from this colorful mosaic of beliefs to the roughly 1,500-year period during which naturalistic extinction in even its most minimal sense became almost universally unthinkable in the West, let's begin with what the Great Chain of

Being is, when it was first articulated, and how it collapsed during this first existential mood. We will then examine the historical origins of the ontological and eschatological theses, which took root with the rise of Christianity and persisted into the nineteenth century, at which point their decline enabled the second existential mood to emerge. Recall that these are the three beliefs that "blocked" *human extinction* beginning around the fourth and fifth centuries CE—that is, they established a particular set of reasons, unique to Christianity, for *why* humanity is indestructible—and hence understanding the nature, origins, and eventual decline of these ideas is an integral part of the narrative of Part I.

According to Arthur Lovejoy's magisterial book *The Great Chain of Being*, published in 1936, the Great Chain can be decomposed into three main ingredients: (1) the principle of plenitude, which originated with Plato and asserts that there are no unrealized possibilities in the universe. If something *can* exist, it *will* exist, meaning that the world is exhaustively "full" of every *kind of thing*. (2) The principle of continuity, which arose from Aristotle's philosophy and claims that "the qualitative differences of things must . . . constitute linear or continuous series." This gave rise to the idea that all animals can be arranged into a linear hierarchy, or a *scala naturae* ("ladder of being"), based on excellence or perfection.[26] And (3) the principle of unilinear gradation, also from Aristotle, which corresponds to the idea that "living beings are linked to one another by regularly graduated affinities" that are infinitesimally small.[27] Putting these ingredients together: if something can occupy the space between two other kinds of things, then it *will* occupy that space; there are no gaps, only a continuous sequence of minute gradations from top to bottom.[28]

However peculiar this idea may seem to us today, it was profoundly influential for more than a millennium of Western history, having been first articulated by Neoplatonists like Plotinus in the third century CE. As Lovejoy details, one finds expressions of the Great Chain in the writings of an enormous number of influential philosophers across time, as when John Locke wrote in his 1689 *Essay Concerning Human Understanding* that

> in all the visible corporeal world we see no chasms or gaps, and a continued series that in each remove differ very little one from the other. There are fishes that have wings and are not strangers to the airy region, and there are some birds that are inhabitants of the water, whose blood is as cold as fishes.

"Amphibious animals," he continued, "link the terrestrial and aquatic together . . . not to mention what is confidently reported of mermaids or sea-men."[29] Indeed, one could more or less *deduce* the existence of mermaids and sea-men from the underlying principles of the Great Chain model: if there appears to be a gap, something must fill it, so why doubt the veracity of such reports? Others identified bats as "intermediate between animals that live on the ground and animals that fly," and believed that "zoophytes" like the "Vegetable Lamb of Tartary," a plant with sheep as fruit, connect the animal and plant realms.[30]

Figure 2.1 The Vegetable Lamb of Tartary, a zoophyte that some believed grew in Central Asia.

The key idea for our purposes is that because the Great Chain was taken to be an immutable and complete hierarchical ordering of everything that exists, it precluded the possibility of anything ceasing to exist *entirely*, even temporarily.[31] As the English poet Alexander Pope wrote in his enormously influential 1733/34 poem *An Essay on Man*:[32]

Vast chain of being! which from God began,
Natures aethereal, human, angel, man,
Beast, bird, fish, insect, what no eye can see,
No glass can reach; from Infinite to thee,

From thee to nothing.—On superior pow'rs
Were we to press, inferior might on ours;
Or in the full creation leave a void,
Where, one step broken, the great scale's destroy'd;
From Nature's chain whatever link you strike,
Tenth or ten thousandth, breaks the chain alike.

Notice the eighth line: "Where, one step broken, the great scale's destroy'd," which means that if even a single link were to go missing, the entire Great Chain would collapse. We can reconstruct the reasoning behind this as follows:

(p1) The completeness and immutability of the Great Chain testifies to God's absolute perfection.
(p2) The loss of any link in the chain would destroy the system, since it would no longer be complete and immutable.
(p3) If the system were destroyed, then God wouldn't be perfect.[33]
(p4) But God *is* perfect.
(c) Therefore, links cannot be lost.

In other words, broken chains are *metaphysically* impossible, and hence so is extinction.[34]

A Mammoth Discovery

The Great Chain dominated Western thought about the fundamental structure of reality from the Middle Ages to the early nineteenth century, although we will see in a later chapter that it evolved over time. Yet the implication that extinction is impossible was increasingly at odds with a growing collection of fossilized bones that did not match any known living creatures. People had been aware of fossils for millennia, of course: they jut out of cliffs, wash up on beaches, and are sometimes exposed by heavy rains. As the folklorist Adrienne Mayor observes, "the ancients collected, measured, displayed, and pondered the bones of extinct beasts, and they recorded their discoveries and imaginative interpretations of the fossil remains in numerous writings that survive today."[35] In fact, Xenophanes's cyclical cosmology of wetness/dryness was partly inspired by fossilized marine organisms found on Malta (a Mediterranean island south of Italy), which led him to infer that it was once submerged under water.[36] Some anthropologists have even argued that fossils may have been "prized possessions" among Neanderthals, as they have been uncovered alongside artifacts created by these Pleistocene-epoch humans.

But a number of fossil discoveries in North America, Siberia, and elsewhere during the eighteenth century added urgency to the question of extinction.[37] To preserve the integrity of the Great Chain, Christian naturalists proposed a range of imaginative hypotheses to explain away these peculiar formations. Some argued they were remnants of the Noachian flood, while others suggested that they mysteriously fell from the sky. The Welsh polymath Edward Lhuyd, in 1713, proposed that they had "originated from seeds that somehow grew within the rocks and thus mimicked living structures," while John Ray conjectured that they belong to species still alive in unknown regions of the world.[38] The latter view was accepted by the scientifically minded Thomas Jefferson, who believed that mastodon bones unearthed in North America belong to creatures still roaming parts of the continent, references to which are found in "the traditionary testimony of the Indians."[39] To prove this true, he instructed Meriwether Lewis and William Clark to collect evidence of the mastodon during their expedition to the Pacific Coast, writing to them in 1803: "Other objects worthy of notice will be . . . the remains & accounts of any which may be deemed rare or extinct."[40]

A few people prior to the nineteenth century did accept the reality of extinction. For example, according to Mayor, the Roman poet Lucretius (c. 99–55 BCE) "provided the clearest expression of extinction and 'survival of the fittest' in ancient literature," as when he wrote in *On the Nature of Things* that certain species incapable of protecting themselves have died out.[41] "It was open season on those brutes," he wrote, "Until Nature finally drove their species to extinction."[42] Seventeen centuries later, Robert Hooke—dubbed the "English Leonardo da Vinci"—claimed in a posthumously published treatise that "diverse Species of things" in the past have been "wholly destroyed and annihilated," citing the Platonic legend of Atlantis as evidence of violent earthquakes causing entire islands to disappear.[43] Others at the time, while rejecting the idea that extinctions can occur naturally, began warming to the idea that this might be possible if *caused by humans*, as this was seen as compatible with God having created a flawless natural order.[44] The problem was not God but human action enabled by free will, which has corrupted this order by knocking out steps in an otherwise infrangible hierarchy. Still others, such as the French Enlightenment *philosophe* Denis Diderot—once a theist, deist, pantheist, and eventually an atheist—argued in 1769 that species, including humanity, can indeed go extinct.[45] But, channeling the principle of plenitude, he immediately added that "the whole cycle of life [would begin] anew," including "man, but not as he is. First, a certain something; then another certain something; and then, after several hundreds of millions of years and so many more certain somethings, the bipedal animal who has the name of man."[46] On the whole, though, the notion that species could go extinct was widely dismissed or rejected, with rare exceptions like these.

This situation changed dramatically following the groundbreaking late-century work of Georges Cuvier, a French zoologist who was considered to be among the greatest minds of his generation (now known as the "Founding Father of Paleontology"). In a 1796 lecture, Cuvier presented research demonstrating beyond a reasonable doubt that elephantine bones found in North America and Siberia belong to species no longer present: the mammoth and mastodon, the latter of which he named in 1817. By 1800, he had identified twenty-three species that apparently went extinct, from which he extrapolated that many more must be buried in the stratigraphic graveyard of Earth, yet to be discovered.[47] Twelve years later, he proposed a novel classificatory system that organized the animal kingdom into four fundamentally different types, or *embranchements*, thus further breaking the linear Chain of Being, as Cuvier's system organized species based on "similarities in their internal structure, not on an ordered ranking of their external characters."[48] Although remnants of the Great Chain persisted for several decades, Cuvier more or less single-handedly established that extinction is a genuine feature of biological history, something that has occurred naturally, perhaps many times, and in doing so he dealt an immense blow to the Great Chain and its underlying principle of plenitude.[49] By the 1830s, then, the scientific community had largely come to agree with Cuvier that the Great Chain no longer represents the fundamental structure of reality, a momentous shift that cleared away one hurdle to imagining our own extinction.[50]

Origins of the Other Cluster of Beliefs

However, the dissolution of the Great Chain did not render *human extinction* conceivable, as the ontological and eschatological theses still implied that, even if other species can go extinct, our species cannot, since we are immortal beings who play a central role in God's grand plan for the cosmos. In other words, our place is unique within God's creation: we are different from all other creatures in kind rather than degree, and this fact ensures that our own extinction remains impossible.

But where did the ontological and eschatological theses come from? How did they develop? We have already touched upon the origins of the eschatological thesis: the narrative of Christian eschatology may have been inherited from the ancient Persians—in particular, from Zoroastrianism, which the Jewish people would have encountered during the Babylonian Exile. This narrative was then transferred to Christianity (and Islam), although theologians have given it many different interpretations over the centuries; for example, Augustine espoused an "amillennialist" eschatology, while many contemporary Christians embrace a "premillennialist" view, discussed more in Chapter 4. As for the ontological thesis, the notion of soul immortality was not originally part of the emerging Christian faith at the start of the first millennium. The early authors of the New Testament held a view closer to physicalism (or materialism) than dualism, following the earlier Hebraic tradition, which said little about the nature of persons or the question of personal eschatology. The conception of human beings as spiritually immortal came from Platonic philosophy, and was incorporated into Christianity as the faith spread across the Mediterranean region. By the fourth century, this Platonic idea had become fairly well established among Christians, as exemplified by the theological writings of Saint Augustine, who was expressly influenced by the "books of the Platonists," a reference to the Neoplatonists.[51]

One of the defining features of Neoplatonism was a belief in the immortality of the soul.

As for the rise of Christianity itself, Keith Hopkins writes that "the greatest surge in Christian numbers (in absolute terms) occurred in two stages, in the third century and fourth centuries" of the Common Era,[52] although it was during the fifth (or at least sixth) century that Christianity became widely accepted by citizens of the Roman Empire, following two significant developments: (a) the Edict of Milan in 313, issued by Constantine and Licinius, which decriminalized Christianity and many other religions, and (b) the Edict of Thessalonica in 380, issued by Theodosius I, Gratian, and Valentinian II, which made Nicene (Catholic) Christianity the "state religion" of the Roman Empire (though the term "state religion" is misleading, as there was no separation of church and state). Once established, Christianity dominated the Western tradition until the nineteenth century, when three major developments in particular caused atheism to spread among the educated classes, or intelligentsia. Religious belief then declined precipitously among the general public starting in the 1960s, which marked the beginning of the "Age of Atheism," and this Western trend of secularization continues up to now, with one study projecting that religion is racing toward "extinction" (their word) in nine Western countries.[53] We will return to these issues in the next two chapters.

Let's pause for a moment on the implications of this picture. If correct, it means that two historical figures may have been disproportionately responsible for the idea of *human extinction* being occluded from view for a large chunk of Western history. On the one hand, Plato introduced the principle of plenitude and established the doctrine of soul immortality through his influential writings, which thus gave rise to two reasons that naturalistic human extinction is impossible: (a) because extinction of *any sort* is impossible (principle of plenitude) and (b) because *our* extinction could never happen given the immortality of our souls. On the other hand, if the eschatological narrative of Christianity can be traced back to Zoroastrianism, then one could argue that Zoroaster, its founder, was responsible for outlining a vision of the future that would eventually give rise to the eschatological thesis. Hence, one can make a case that the philosophical and religious legacies of these two individuals shaped the history of thinking about our extinction in crucial and profound ways, as they introduced ideas that later became central to the Christian worldview, thus blocking the idea for some 1,500 years. This is a remarkable fact, although it is of course possible that such ideas would have arisen anyway, being invented by others.

Whereas the Great Chain collapsed in the early nineteenth century, it wasn't until later that century that the ontological and eschatological theses significantly declined in influence among the intelligentsia, as alluded to above. Since this occurred mostly during the second existential mood, we will save it for the next chapter. However, we should note that due to the rise of deism during the

Enlightenment, the eschatological thesis lost some of its force even before the nineteenth century arrived. Deists generally reject special revelation, and since biblical prophecies are based on special revelation, many deists rejected aspects of Christian eschatology. This nudged aside one barrier to taking seriously the idea of naturalistic extinction, although many in the eighteenth century—"the century of philosophy *par excellence*," as D'Alembert famously declared—still held that we are spiritually immortal, and were also influenced by the Great Chain model of reality and its principle of plenitude, an example being Diderot's speculation above. The more definitive blow to the eschatological thesis happened when Christianity as a whole became seen by many, in the latter nineteenth century, as untenable.

Catastrophism, Population Decline, and the Life Cycle of Species

Before concluding this chapter, it is worth registering a number of important developments toward the end of the first existential mood, some of which anticipated ideas and discoveries that became central to subsequent moods. First, the eighteenth and early nineteenth centuries witnessed many new thoughts about how natural phenomena (a) could devastate large portions of the globe and (b) potentially reduce the human population to zero, although none of these proposed "kill mechanisms" became widely accepted by contemporaries at the time. Second, often linked to this, a small handful of writers—typically deists, atheists, or those with nontraditional religious beliefs—began to take seriously the *possibility* of our permanent extinction, to imagine the world bereft of human beings forever. After surveying some examples of each, we will then explore how the dominant religious worldview led the overwhelming majority of people at the time to the conclusion that, even if natural catastrophes could devastate the planet, we can rest assured that humanity is safe—because we are, at bottom, indestructible.

Let's begin with the theory of *catastrophism*, which explained geological features of Earth as the result of large-scale, sudden catastrophes that had periodically taken place in the past. The theory's most influential exponent was, in fact, Cuvier, who referred to such catastrophes as *révolutions* and identified them with huge floods that caused mass extinctions. Of note is that Cuvier did not associate any of these floods with the biblical deluge: *révolutions* were wholly natural, a point we will return to in Chapter 5.[54] Even earlier, in the eighteenth century, the French Enlightenment philosopher Montesquieu pointed to various secular phenomena that, he argued, are responsible for the human population dwindling over time, until there are no more people left. This was presented in his 1721 epistolary novel *Persian Letters*, which consists of fictional missives written primarily by two Persian noblemen, Usbek and Rica, while visiting France. (By examining France through the eyes of a foreigner, Montesquieu hoped to enable readers to see their culture from a novel perspective.) In a letter to Usbek from his friend Rhedi, the latter reports that

> there are upon the earth hardly one tenth part of the people which there were in ancient times. And the astonishing thing is, that the depopulation goes on daily: if it continues, in ten centuries the earth will be a desert. Here, my dear Usbek, you have the most terrible calamity that can ever happen in the world. But we have scarcely perceived it, because it has stolen upon us gradually in the course of a great many centuries, which denotes an inward defect, a secret and hidden poison, a malady of declining, afflicting human nature.

Usbek, whose views seem most closely aligned with Montesquieu's own, wrote back to Rhedi that there have been "catastrophes, so common in history, which have destroyed whole cities and kingdoms: there are general ones which many a time have brought the human race next door to destruction." This includes, he says, floods like the one that reduced humanity to a single family—Noah's family—as well as "universal plagues which have one after the other desolated the earth." In some cases, "one degree more of corruption would have destroyed, perhaps in a single day, the whole human race." He further identified famine—"the earth tired out with providing subsistence for men"—as another possible threat, noting that not "all destructions have . . . been violent." However, Usbek speculates that the *actual* cause of the decline noted by Rhedi is the result of cultural practices common among Christians and Muslims that lead to people having fewer children. This is why "the earth is less populous than it was formerly."[55] Here we have, essentially, a causal explanation of how humanity could disappear that is cultural in nature. As David Young notes, "Montesquieu's investigation was meant as a serious population study," and this is how contemporaries read it: the total number of people on Earth is decreasing, and if this trend continues, the total number could eventually equal zero.[56]

A few decades later, David Hume penned a refutation of Montesquieu's depopulation thesis in which he contended that Montesquieu had uncritically accepted an exaggerated estimate of how large the human population was in the ancient world. There was not, in fact, a much larger population in the past than at present. Notably, though, Hume also conjectured that the human species may have a "life cycle" just like those of its members. "There is very little ground, either from reason or observation, to conclude the world eternal or incorruptible," he wrote. To the contrary, the evidence strongly suggests "the mortality of this fabric of the world, and its passage, by corruption or dissolution, from one state or order to another." Consequently, he concluded that the universe "must therefore, as well as each individual form which it contains, have its infancy, youth, manhood, and old age; and it is probable, that, in all these variations, man, equally with every animal and vegetable, will partake."[57] In other words, both the universe and our species may be subject to the same trajectories of growth and decline that characterize our individual lives. This idea—that species have life cycles just like individual organisms—has been made on numerous occasions throughout history. Indeed, one year after Hume's book, Diderot echoed it in conjecturing that "in the animal and vegetable kingdoms, an individual begins, so to speak, increases, lasts, withers and passes away; would it not be the same for whole species?"[58] Hence, the causal mechanism arises from the fact that species are *naturally mortal* no less than individuals, a theory of so-called "intrinsic" extinction that some naturalists embraced the following century.[59]

The First Last Men

Another population-decline scenario was depicted in Jean-Baptiste Cousin de Grainville's 1805 novel *Le Dernier homme*, or *The Last Man*. Published posthumously after de Grainville, who suffered from a desperate case of loneliness, committed suicide by leaping into the Canal de la Somme at two in the morning, the novel follows the peripatetic journey of Omegarus, who boards an airship to Brazil in search of the only remaining fertile woman on Earth, Syderia. Of note is that this infertility is natural in origin and hence some scholars have argued that *Le Dernier homme* "extrapolates perhaps the earliest images of a secular apocalypse," although de Grainville ultimately grounded his story "in the Biblical narratives of Genesis and the Apocalypse told in St. John's Revelations."[60] In fact, prior to this decline in fertility, the story describes another secular scenario in which human overexploitation of natural resources destroys the environment, an idea that may have been inspired by the population theory of Thomas Malthus, which has left an indelible mark on the Western imagination since its introduction in 1798. According to Malthus, the supply of food grows at an arithmetic rate, whereas the human population increases at a geometric rate, which thus leads to an inevitable collapse of the population. However, Malthus himself did not identify the outcome as extinction but instead predicted a "perpetual oscillation between happiness and misery"—yet another example of the Ouroboros motif.

This theory sparked significant debate at the time, as it was widely assumed up to and throughout the Enlightenment that, as Montesquieu had believed, the ancient world was far more populous than the contemporary world.[61] In a polemical response to Malthus published in 1820, for example, William Godwin contended that "war, pestilence, and famine" threaten catastrophic decreases in the human population, and hence that

> for any thing [*sic*] that appears from the enumerations and documents hitherto collected, it may be one of the first duties incumbent on the true statesman and friend of human kind, to prevent that diminution in the numbers of his fellow-men, which has been thought, by some of the profoundest enquirers [a reference to Montesquieu], ultimately to threaten the extinction of our species.[62]

It could, of course, be the case that war, pestilence, and famine are the *result* of a Malthusian catastrophe, and that this could lead not just to misery but our extinction—an idea vigorously promoted by some leading environmentalists in the twentieth century. But this was not Godwin's view, nor was it explored by de Grainville in his tragic novel about the last couple on Earth, Omegarus and Syderia.

As it happens, de Grainville's novel inaugurated a literary genre known as the "Last Man," which became immensely popular in the early nineteenth century, inspiring a large number of poems, paintings, short stories, and novels. The most well-known example, at least since the 1960s, is Mary Shelley's 1826 *The Last Man*, which was probably influenced by an anonymous English translation of de Grainville's story. In Shelley's telling, though, the human population collapses over a few years due to a worldwide plague transmitted (non-contagiously) through miasma.[63] While the autodiegetic narration of Lionel Verney, the dystopian story's main character, does not explicitly end with the complete disappearance of humanity—after all, extinction precludes the possibility of any narrator noting this fact, a point made by some early critics of the genre—Shelley strongly implied that when Verney dies, so too does the human species. Unlike de Grainville, Shelley did not embed her tale within a larger religio-apocalyptic narrative in which humanity ultimately "survives" the global pandemic. That is to say, not only did she specify a naturalistic *etiology* of extinction, but she gestured at a naturalistic *outcome*, whereby the human story simply comes to an end, without this initiating a new beginning.[64]

Conversations About Comets

A more commonly discussed natural catastrophe scenario in the eighteenth and nineteenth centuries involved comets. In contrast to asteroids—derogatorily dubbed the "vermin of the sky"—humans have known about these shimmering objects for millennia, as indicated by observations recorded on ancient Chinese oracle bones used by diviners who practiced pyromancy. Often considered to be "omens and bearers of bad news," comets have also triggered occasional fears that they could collide with Earth; as Duncan Steel writes, "the possibility of catastrophic impact by comets resurfaced from time to time before the modern era."[65] For example, in *Conversations of Lord Byron* (1824), the English poet Thomas Medwin recounts a conversation with Byron in which he pondered: "Do you imagine that, in former stages of this planet, wiser creatures than ourselves did not exist?," which he found plausible given that "we are at present in the infancy of science." These past creatures, Byron suggested, had been wiped out by comets colliding with Earth, leading him to worry that humanity could suffer the same violent fate. In perhaps the very first reference to a "planetary defense system" to protect Earth from these heavenly assassins, Byron declared:

> Who knows whether, when a comet shall approach this globe to destroy it, as it often has been and will be destroyed, men will not tear rocks from their

> foundations by means of steam, and hurl mountains, as the giants are said to have done, against the flaming mass.[66]

In other words, an annihilatory collision is all but guaranteed unless humanity uses the "power of steam" to catapult mountain-sized heaps of rocks at the incoming comet. Of note is that Byron was also among the very first of this late period to depict, in his 1816 poem *Darkness*, a world completely bereft of humanity. Unlike Shelley, a close friend of his, Byron did not specify a cause of our extinction, but he did paint an eerie, ominous picture of Earth having become "seasonless, herbless, treeless, manless, lifeless/A lump of death—a chaos of hard clay." However, the impetus of this poem wasn't so much to foreground the *possibility* of extinction but to make a rather unrelated point about the *nature* of human beings, using the extreme scenario of our collective non-existence as a way of undermining Enlightenment views of humanity (a point elaborated in Chapter 8). Still, Byron's speculations and poetic imaginings—followed by Shelley's 1826 novel—mark an important moment in intellectual history: after roughly 1,500 years of the idea being completely blocked from view, some in the Western tradition, toward the end of this first existential mood, were beginning to see *human extinction* as an intelligible possibility.

Other natural catastrophe scenarios proposed at the time involved astrophysical phenomena of a different sort. For example, Comte de Buffon hypothesized that Earth was created when a comet slammed into the sun and threw off a giant fragment.[67] Initially in a molten state, Earth has cooled over the course of some 75,000 years.[68] Buffon thus likened our planet to a "dying ember of the sun" that will become increasingly inhospitable due to "its gradual refrigeration, a reign of perpetual winter."[69] Even earlier, Bernard Le Bovier de Fontenelle discussed the possibility of sun spots forming a crust over the sun in his *Conversations on the Plurality of Worlds*, published in 1686.[70] He noted that "the ancients saw fixed stars in the sky which we no longer see," which he took as evidence that our sun could also stop shining.[71] This idea was popularized the following century by an anonymous Italian astronomer who predicted that on July 18, 1816, our sun would extinguish itself, which seemed to be corroborated by sunspots that became increasingly visible, not just telescopically but to the naked eye, and the peculiar weather of 1816 that led it to be dubbed the "Year Without a Summer" (the result of, unbeknownst to those at the time, the "super-colossal" eruption of Mount Tambora in Indonesia in 1815). This was called the Bologna prophecy.

A fascinating inventory of such doomsday speculations comes from a short article published on October 1, 1816, in *The New Monthly Magazine*. Written by an anonymous author going by "H" and titled "Of the End of the World," its explicit aim is "to stop the mouths of all who may be disposed to make light of so serious a subject," namely, the destruction of humanity. As the author opens the article, referring to the failed Bologna prophecy, "because the world was not destroyed on the 18th of July, we imagine that it will never be at an end, and laugh as if we had never been afraid." The author proceeds to outline Buffon's scenario, whereby "the globe is growing colder every day." He further notes that sunspots are, according to some, "scoria adhering to the surface of the luminary," and that a comet collision should be expected "in three or four thousand years at latest." Additional scenarios include the moon someday crashing to Earth and the earth falling into the sun. Before this happens, though, H claims that our planet is drying up and the oceans sinking. As it becomes increasingly desiccated, it will eventually catch fire, and "the generation now living, we shall all be burned, and our funeral pile will be kindled when there is no more water upon the earth—a consideration which ought to make us tremble now that water is become [*sic*] so scarce." "Here, then, is a very rational *end of the world!*," they write. Yet, in a passage that parallels Diderot's plenitude-based prediction above, the author argues that Earth will once again contain water, and out of these organic molecules (or "zoophytes," H states) will evolve into lobsters, lobsters into "tatoos" (i.e., armadillos), tatoos into apes, and eventually apes into men "who, after some more billions of centuries, will build cities, compose operas, and invent cosmogonies."[72] In other words, our disappearance would not be permanent.

Yankee Professors

As this brief survey shows, people throughout the eighteenth and early nineteenth centuries proposed an impressive range of creative ideas about how the world could end and/or humanity be destroyed. But recall from Chapter 1 that "kill mechanism" refers to a way that the collective whole of humanity could disappear based on established scientific principles (theories, laws, mechanisms) rather than idiosyncratic speculations or religio-philosophical belief systems. Prior to the mid-nineteenth century, none of the doomsday scenarios conjured up by people involved causal phenomena that were widely seen as posing genuine risks to humanity. In many cases, the destruction of the world was specifically tied to biblical prophecy and hence did not entail humanity dying out in the naturalistic sense; for this reason, they did not count as kill mechanisms. In other cases, there was just no particularly compelling reason to believe that the associated kill mechanisms could *actually* kill us off. An example of this comes from Edgar Allan Poe's 1839 short story "The Conversation of Eiros and Charmion," in which humanity perishes due to an "irresistible, all-devouring, omni-prevalent, immediate" conflagration caused by a comet passing by Earth. The comet extracts nitrogen from our atmosphere, leaving behind high concentrations of the combustible element oxygen, which subsequently ignites as "the nucleus of the destroyer" causes "a wild lurid light alone, visiting and penetrating all things." While this story earns points for creativity, no one was worried that this might actually happen. Furthermore, Poe linked the catastrophe to "the fiery and horror-inspiring denunciations of the prophecies of the Holy Book," and indeed the story involves a conversation between two people, Eiros and Charmion, in the afterlife.[73]

Other kill mechanisms may have appeared less outlandish by virtue of being based on rational extrapolation from known historical disasters, as with Shelley's global pandemic. If outbreaks of disease can spread across a whole continent—for example, Europe during the fourteenth-century Black Death—why not across the entire planet? Yet, even if Shelley herself was open to the possibility of humanity dying out, the majority of her readers would have understood the pandemic scenario quite differently, given the continuing influence of the ontological and eschatological theses. Still other scenarios were undermined by additional, equally valid considerations. For example, the French astronomer Jérôme Lalande calculated in 1773 that the probability of a comet striking Earth is just 1 in 76,000, while in the early nineteenth century, the mathematician François Arago put the probability even lower at roughly 1 in 281 million. This suggests that we ought not worry about such a collision, given the low probability of occurrence, and even if a comet were to strike Earth, there is no reason to believe that it would be large enough to obliterate everyone on the planet.[74] As Jefferson reportedly declared in 1807 after being told about the Weston meteorite that exploded into fragments over Connecticut: "I would more easily believe that two Yankee professors would lie than that stones would fall from heaven."[75]

Existential Hermeneutics

Here it would be useful to introduce the notion of an *existential hermeneutics*. Using the term "hermeneutics" in a more colloquial rather than technical sense—although the connection to biblical studies is not entirely accidental—this denotes an interpretive framework through which one can assess the potential consequences and probability of catastrophic phenomena. An existential hermeneutics could be religious or secular to varying degrees, and the extent to which a particular hermeneutics is one or the other will crucially shape one's beliefs about the riskiness of our world. To elaborate on an example mentioned in Chapter 1, imagine that astronomers identify an asteroid barreling toward Earth, large enough to cause global-scale damage to the planet. Our empirical model of the world thus changes in response to this discovery.

But there is a second, hermeneutical question: *what should we make* of this threat? Viewed through a religious hermeneutics, one might initially react with fear given the suffering that the collision itself would inflict—for example, to oneself and one's family and friends. But beyond this, the news of impending catastrophe may be an occasion for eschatological elation, since what lies on the other side of the apocalypse is eternal life with God in paradise. Although the Bible does not explicitly mention asteroidal impacts, eschatological narratives can be very *elastic*, able to fill, like a fluid, the various containers in which they are placed. This is partly because of the ambiguity of scriptural passages; one could easily find verses that appear to reference a collision, such as 2 Peter 3:10, which states that in the end "the heavens will disappear with a roar; the elements will be destroyed by fire, and the earth and everything done in it will be laid bare." The Bible, religious believers might say, thus predicted this event *all along*, although we were unable to see this until now. In contrast, when viewed through a secular hermeneutics, one might react quite differently to the astronomical discovery: not with excitement but terror and despair, since for atheists there is no redemption, no hope of an afterlife, when the end of the world comes. There is, in a phrase, no silver lining to the news of imminent extinction; the apocalypse is suffering and death, and what lies beyond it is nothing but the oblivion of eternal non-being.

The Impulse of Celestial Agents

The important point is that models of potentially dangerous phenomena must be interpreted, and it is this interpretation—via some existential hermeneutics—that gives rise to what I will call the *threat environment*. Thus, the threat environment is what one gets when a particular world-model is filtered through an interpretive framework. When filtered through certain religious frameworks, phenomena that might otherwise, from a secular perspective, be cause for alarm may instead appear benign, thus resulting in radically different mappings of the threat environment in which we are embedded. Indeed, one finds many examples of this during the first existential mood. Isaac Newton, for instance, rejected the idea that comets dashing about our solar system pose any real threat to humanity. Why? Because the laws of nature were crafted by the hands of an all-good, loving God. Consequently, he maintained that "comets obeying them [are] far more likely to have beneficial effects, such as replenishing the Earth's water supply from the tail during a close passage, than to bring disaster to the planet."[76] Newton also thought that gravity could cause the universe to collapse in on itself, and accumulated perturbations of the orbits of planets could throw the solar system into disarray. Yet he maintained that God occasionally intervenes to ensure that no such catastrophes occur; as he told the Scottish mathematician David Gregory in 1694, "a continual miracle is needed to keep the Sun and the fixed stars from rushing together through gravity," and this is precisely what God provides.[77] Hence, by looking at the world through the lens of a religious hermeneutics, Newton came to the conclusion that the threat environment does not, in fact, include any topographical features that correspond to these phenomena. We—God's beloved children—are safe.

Or consider the following comments from Benjamin Franklin and the British minister Thomas Dick, made roughly 80 years apart—one before the Great Chain collapsed and the other after.[78] In his 1757 *Poor Richard's Almanac*, Franklin said the following about comets:

> Should a Comet in its Course strike the Earth it might instantly beat it to Pieces, or carry it off out of the Planetary System. The great Conflagration may also, by Means of a Comet, be easily brought about. All the Disputes

> between the Powers of Europe would be settled in a Moment; the World, to such a Fire, being no more than a Wasp's Nest thrown into an Oven. But our Comfort is, the same great Power that made the Universe, governs it by his Providence. And such terrible Catastrophes will not happen till 'tis best they should.[79]

Similarly, Dick argued in *The Sidereal Heavens and Other Subjects Connected with Astronomy*, published in 1840, that

> when we consider that a Wise and Almighty Ruler super-intends and directs the movements of all the great bodies in the universe, and the erratic motions of comets among the rest; and that no event can befall our world without his sovereign permission and appointment, we may repose ourselves in perfect security that no catastrophe from the impulse of celestial agents shall ever take place but in unison with his will, and for the accomplishment of the plans of his universal providence.[80]

These passages illustrate the power of a religious hermeneutics: even if potentially catastrophic phenomena do threaten humanity, we can nonetheless take "Comfort" in a sense of "perfect security," as such catastrophes will not happen "till 'tis best they should," that is, in accordance with "the plans of his universal providence." The words "Comfort" and "perfect security" also nicely capture the essence of this first existential mood following the rise of Christianity, which brought together the three beliefs—Great Chain, ontological thesis, and eschatological thesis—that, together, provided a particular explanation of why humanity is indestructible. Nonetheless, we saw that some in the eighteenth and nineteenth centuries did begin to imagine the world without us, although these individuals were very much the exception and, in every notable case, such individuals espoused atheistic or non-traditional religious beliefs. For example, Hume could be described as an atheist; Godwin became an atheist; Byron held various deistic, agnostic, and skeptical views throughout hist life, although he was also sympathetic with Islam and Catholicism; and Shelley seriously considered the possibility that there is no divine plan and God is actively malevolent when writing *The Last Man*.[81] This is consistent with my claim that secularization played a crucial role in enabling *human extinction* to become intelligible—a claim that leads us to the next chapter.

Notes

1. Witzel 2012, 177–179. Trever Palmer notes that "the story of Noah is just one of more than 500 flood myths from around the world" (Palmer 2012).
2. Tigay 2002, 19.
3. Coleman 2007, 335; George 1999, xliii–xliv.
4. See Tigay 2002, 25; Kovacs 1989, xxvi.

5. As Alan Segal writes, the reason for Enlil's wrath "seems to be overpopulation, as the gods grow discontented over the noise that humanity is making" (Segal 2004).
6. Interestingly, a number that also appears in Hindu eschatology. As David Knipe writes:

   > The Puranas present the most fantastic calculations of cosmic circularity. The notion of a god who embodies all of space and time is not exceptional in the history of religions. The Puranas, however, stretch imagination to the limit when describing the god Brahma as a living cosmos who lasts, in just one day and night, for 1,000 cycles of four deteriorating ages, each cycle being 4,320,000 human years. At the end of 36,000 full years of these day-nights, Brahma rests or ceases, only to experience rebirth for another vast lifetime (2008).

7. Pinch 2002.
8. Quoted in Pinch 2002.
9. Nattier 2008.
10. Kingsley and Parry 2020, section 2.1–2.3.
11. Quoted in Wright 1981, 166.
12. See Christidis 2009; Rosenmeyer and Seneca 1989. For further details on this idea, see Wheelwright 1968.
13. Quoted in Long 1984.
14. Sambursky 1976. Note that while the Stoics believed that each of us will be qualitatively identical from one cycle to the next, we will not be numerically identical (Čapek 1976, XXIX). Note also that the Stoics believed in an afterlife, although this afterlife would last only as long as it takes for the next cycle to begin.
15. Pinch 2002.
16. Lindow 2002.
17. Cohn 1957; O'Brien 2010.
18. Although we should note that, as Marcia Hermansen observes, "the Islamic concept of time is frequently less linear than that of the Christian and Jewish traditions" (2008).
19. Ehrman 2021. Note that, while this was once taught to students at universities, it has become controversial. For example, Jan Bremmer writes in his book *The Rise and Fall of the Afterlife* that "there . . . is little reason to derive Jewish ideas about resurrection from Persian sources. Their origin(s) may well lie in intra-Jewish developments" (2002). The biblical scholar Bart Ehrman echoed this in a blog post, writing that "the Jews who first pronounced the idea, during the Maccabean period [which came more than 1.5 centuries after intermingling with Persian traditions], may have come up with it themselves. This appears to be the newer consensus on the matter, as seen in a more recent work on the afterlife by a New Testament scholar Outi Lehtipuu who in her book, *The Afterlife Imagery in Luke's Story of the Rich Man and Lazarus . . .*, makes the same basic point" (2017; see Lehtipuu 2007, 124). In contrast, Richard Taylor writes in his *Death and the Afterlife: A Cultural Encyclopedia* that, for example, "the Zoroastrian teachings on the 'last things' and the ultimate renovation of the world had a massive impact on the development of eschatology in early Judaism, Christianity, and Islam," adding in a subsequent section that "after prolonged exposure to ancient Near Eastern ideas, however, and particularly after the exile of the Israelites to Babylon and exposure to Zoroastrian ideas (c. 600 B.C.E.), Hebrew texts began to show a greater and greater acceptance that at least some of the dead might be resurrected and judged." Elsewhere he reports that

   > for centuries in the Middle East, Zoroastrianism had taught that during life, the forces of evil held sway; final compensation for the just and unjust generally occurred not in life but after death. In a final cosmic moment, "time" would stop.

All would be resurrected (ristakhez) from the dead, the wicked would be renovated, and all human souls would join God. A messiah known as the Saoshyant would bring on this final reckoning between the true god (Ahura Mazda) and the god of the lie (Angra Mainyu). All of this was new to the Hebrews, but after several generations in exile such ideas came to affect their theology profoundly and to explain why God, apparently without cause, had taken away the Holy Land and allowed evildoers to prosper (Taylor 2000).

20. Note that in the Noachian account, God at one point states that his aim is to "destroy all life under the heavens," although this verse appears in the very same paragraph of Genesis in which God tells Noah how to survive the disaster.
21. Barnes 1982, 268–269.
22. Dick 1982, 2.
23. Hippolytus 2012.
24. Safari 2017.
25. Specifically, we revise this in the section of Chapter 10 titled "A Radical Answer."
26. Lovejoy 1936.
27. Quoted in Lovejoy 1936.
28. If this sounds implausible, it is because it *is* implausible. The fact that there could obviously be some kind of creature between humans and apes was, in fact, later used by Voltaire as an argument against the Great Chain. In his words: "Is there not visibly a gap between the ape and man? Is it not easy to imagine a featherless biped possessing intelligence but having neither speech nor the human shape, . . . ? And between this new species and that of man can we not imagine others?" (quoted in Lovejoy 1936).
29. Locke 1877, quoted in Lovejoy 1936.
30. Lovejoy 1936.
31. In other words, demographic extinction was rendered impossible (again, see Chapter 7).
32. The poem was immensely influential at the time but today is widely regarded as an inferior piece of literature. See chapter 1 of Solomon 1993.
33. Or, as the English naturalist and ordained minister John Ray wrote in 1703, "the Destruction of any one Species" would be "a dismembering of the Universe, and rendering it imperfect" (quoted in Lovejoy 1936).
34. Intriguingly, we will see in Chapter 8 how some individuals came to imagine the complete destruction of our world as being in some sense *enabled* by one pillar of the Great Chain, namely, the principle of plenitude.
35. Mayor 2011.
36. Incidentally, the archaeologist Karl Taube notes that the Aztecs might have believed that the fossilized remains of mammoths belong to the ancient race of giants who lived during the First Sun (Taube 1993).
37. See Kolbert 2014, ch. 2.
38. Bowler 2003. Although Ray later came to accept Lhuyd's view; see Bowler 2003, 37.
39. Jefferson 1785.
40. Jefferson 1803.
41. Mayor 2011.
42. Carus et al. 2007.
43. Hooke 1705.
44. Indeed, it was reported in 1768 that the sea-cow, an aquatic mammal, had died out due to overhunting.
45. Hyman 2010, 7.
46. Quoted in Kors 2015; Crocker 2015.
47. Kolbert 2014, 34.
48. Bowler 2003.

49. Indeed, the Great Chain's influence persists into modern times, to the frustration of some evolutionary biologists (see Dawkins 1992, 261; Ruse 1996).
50. Yeo 1986, 266.
51. *Confessions* 7.13; Tornau 2019; see Segal 2004.
52. Hopkins 1998.
53. PEW 2015; Abrams et al. 2011.
54. Other naturalists, notably the French writer Benoît de Maillet, in 1748, proposed accounts of Earth's history that, rather scandalously, made no mention of the great deluge.
55. Montesquieu 2008.
56. Young 1975.
57. Hume 2021.
58. Diderot et al. 1979.
59. Sepkoski 2020, 30.
60. Ransom 2014; Alkon 1987. Although, of course, we just saw that Montesquieu proposed a kind of secular apocalyptic scenario before de Grainville, albeit caused by religious practices rather than widespread infertility.
61. Russell 1958.
62. Godwin 1820; see Spengler 1971.
63. McWhir 2002. Note that Shelley's novel received quite negative reviews at the time, in part because the Last Man theme was seen as having become hackneyed and boring by the 1820s. One reviewer even described the novel as "a sickening repetition of horrors," while another wrote that it was "the offspring of a diseased imagination, and of a most polluted taste" (quoted in Paley 1989).
64. Paley 1993; Tonn and Tonn 2009, 761.
65. Schwarz 1997; Steel 1997. Where "modern era" refers to the period from 1946 to the present.
66. Medwin 1824.
67. Mayr 1982.
68. Burchfield 1975, 5.
69. Shields 1889; Buffon et al. 1749.
70. Fontanelle 1990.
71. My translation.
72. NMM 1816.
73. Poe et al. 1978.
74. Moynihan 2020; Palmer 2012, 96.
75. Quoted in de Villiers 2010. Note that Jefferson later changed his mind. See Impey 2010, 142.
76. Palmer 2012.
77. Quoted in Kragh 2016; Tamny 1979.
78. Franklin's religious views were idiosyncratic, but largely deistic. As he wrote six weeks before he died:

    > Here is my Creed. I believe in one God, Creator of the Universe. That He governs it by His Providence. That he ought to be worshipped. That the most acceptable Service we render to him, is doing Good to his other Children. That the Soul of Man is immortal, and will be treated with Justice in another Life respecting its Conduct in this. . . . As for Jesus of Nazareth . . . I think the system of Morals and Religion as he left them to us, the best the World ever saw . . . but I have . . . some Doubts to his Divinity; though' it is a Question I do not

dogmatism upon, having never studied it, and think it is needless to busy myself with it now, where I expect soon an Opportunity of knowing the Truth with less Trouble (Fea 2011).

79. Franklin 1757/1849.
80. Dick 1840.
81. Russell and Kraal 2017, section 10; Godwin and Philip 2013, xxxi; Jones 2019; Airey 2019, ch. 4.

3 'Til Entropy Death Do Us Part

Carnot, Clausius, and Kelvin

The first shift in existential mood unfolded during the 1850s. It was triggered by the discovery of the first widely accepted, scientifically credible kill mechanism and enabled by the waning influence of Christianity throughout the West, especially among the intelligentsia. This shift was particularly traumatic to those at the time for a combination of three reasons: first, with the reality of species extinctions having been established by Cuvier and others in the early nineteenth century, secularization fostered novel conceptions of *humanity* that rendered the concept of *human extinction* intelligible to a large number of people in the Western tradition. That is to say, the decline of religion made it possible to believe that our extinction is a real possibility, at least in principle, as the three beliefs that had previously blocked the idea had lost much of their force by the end of the nineteenth century. Second, the identification of a scientifically credible means of elimination meant that our extinction wasn't just possible in principle, but could *actually happen* in our world. It could have been that, even if human extinction were conceivable, we just so happen to occupy a universe in which no kill mechanisms exist. The discovery of a kill mechanism accepted by the majority of scientists at the time confirmed that we do not live in such a world. This leads to the third reason: the specific mechanism that scientists discovered implies not only that humanity *might* go extinct in the future but that this outcome is *inevitable*. Hence, the transition to this new existential mood wasn't a small step from believing that our extinction is fundamentally impossible to believing that it could potentially occur, but a giant leap to the much more startling conclusion that our collective demise is *guaranteed* by the laws of nature. Let's begin with a look at the triggering factor behind this shift, then turn to the broader cultural changes that constituted its enabling condition, and finally examine some additional developments that anticipated the second shift in existential mood in the early years of the Atomic Age.

We have seen that leading up to the middle of the nineteenth century, people proposed a wildly diverse range of catastrophe scenarios, although none were

DOI: 10.4324/9781003246251-4

widely accepted as posing a genuine threat to our existence. This changed with the founding of thermodynamics and, in particular, the formulation of the Second Law in the very early 1850s. The origins of this law can be traced back to the groundbreaking work of Sadi Carnot, a mechanical engineer who died in 1832 from a cholera epidemic at the age of 36.[1] In his 1824 *Reflections on the Motive Power of Fire*, Carnot offered an analysis of *heat engines*, which produce work through the transfer of heat, as in the case of steam engines that convert heat generated by burning fuels into mechanical work that can be used to, for instance, drive a rail vehicle forward.[2] Whereas the standard model of the relationship between science and technology is that the former yields the latter—often referred to as the "fruits of science"—in this case it was technologies like the steam engine that preceded and inspired the relevant science. Carnot's aim was to understand the workings and efficiency limits of such engines, which he described as driving the machinery of an engine through the transfer of heat—considered to be a fluid called "caloric," according to the *caloric theory of heat*—from higher- to lower-temperature components, not unlike how water turns a waterwheel by spilling onto it from a higher to a lower position. Intriguingly, Carnot's research was almost completely ignored for more than a decade after its publication—in fact, much of his work has been completely lost because it was buried with him, due to cholera being contagious. It was his former classmate, Emile Clapeyron, who "rescued" Carnot's work, publishing an article in 1834 that "expressed Carnot's often verbal and sometimes obscure arguments in the acceptable language of mathematical analysis."[3]

Subsequent work by James Prescott Joule concluded that work and heat are interconvertible, which presented a problem: Carnot assumed that caloric (heat) is indestructible and hence conserved while performing work—again, just as the amount of water remains constant while turning a waterwheel—while Joule's notion of work-heat equivalence contradicted this idea. If heat is caloric, it cannot be *converted* into work; rather, it *performs* work by being transferred from hot to cold parts of a heat engine. However, the German physicist Rudolf Clausius realized that one could combine the framework of Carnot's theory with Joule's insights about energy conservation, which led him to propose two fundamental principles in an 1850 paper titled "On the Moving Force of Heat": first, that energy, whether manifested as heat or work, remains constant (energy conservation), and second, that heat can never be transferred from a lower- to a higher-temperature body within the system of a self-acting cyclic machine.[4] These are, in rough outline, the First and Second laws of thermodynamics. Around the same time, William Thomson—whom I will anachronistically call "Lord Kelvin," as he joined the House of Lords much later in 1892—made a similar realization, publishing an 1851 paper "On the Dynamical Theory of Heat" in which he declared that "it is impossible for a self-acting machine, unaided by any external agency, to convey heat from one body to another at a higher temperature."[5] The following year, he described this as a fundamental tendency in nature: "There is at present in the material world a universal tendency to the dissipation of mechanical energy."[6] Clausius meanwhile continued his research, and by 1865 had mathematically formulated the idea of *entropy* to describe the "transformation content" between two states of a system. He then concluded his paper as follows:

> If for the entire universe we conceive the same magnitude to be determined, consistently and with due regard to all circumstances, which for a single body I have called *entropy*, and if at the same time we introduce the other and simpler conception of energy, we may express in the following manner the fundamental laws of the universe which correspond to the two fundamental theorems of the mechanical theory of heat. 1. *The energy of the universe is constant.* 2. *The entropy of the universe tends to a maximum.*[7]

The Frozen Pond of a Heat Death

The eschatological implications of this were immediately recognized our world will become increasingly inhospitable over time until it reaches a state of thermodynamic equilibrium, at which point all life will become impossible. This notion of *entropic death* (or *an entropy death*) took two distinct forms depending on the scope of application, which we can call the *solar death* and the *heat death*, where the latter entails the former but not vice versa. The most prominent early description of the solar death comes from Kelvin, who wrote in a draft of his 1851 paper that "the tendency in the material world is for motion to become diffused, and that as a whole the reverse of concentration is gradually going on," adding that "I believe that no physical action can ever restore the heat emitted from the sun, and that this source is not inexhaustible." Consequently, as he later elaborated, "the end of this world as a habitation for man, or for any living creature or plant existing in it, is *mechanically inevitable*,"[8] where the term "world" refers specifically to our planetary system rather than the whole cosmos. But Kelvin believed that the physical universe must be infinite, which implies that there will never be a final condition—a dysteleological state, as it were—in which all matter and energy are uniformly distributed throughout. He articulated this idea in an 1852 paper titled "On the Age of the Sun's Heat" as follows:

> The second great law of thermodynamics involves a certain principle of *irreversible action in Nature*. It is thus shown that, although mechanical energy is *indestructible*, there is a universal tendency to its dissipation, which produces gradual augmentation and diffusion of heat, cessation of motion, and exhaustion of potential energy through the material universe. The result would inevitably be a state of universal rest and death, if the universe were finite and left to obey existing laws. But it is impossible to conceive a limit to the extent of matter in the universe; and therefore science points rather to an endless progress, through an endless space, of action involving the transformation of

> potential energy into palpable motion and thence into heat, than to a single finite mechanism, running down like a clock, and stopping for ever.[9]

In contrast, other physicists held that the universe as a whole would indeed eventually reach a state of thermodynamic equilibrium—the "universal rest and death" mentioned by Kelvin. The initial statement of this idea came from Hermann von Helmholtz in an 1854 lecture delivered in Königsberg, Prussia. "If the universe be delivered over to the undisturbed action of its physical processes," he told his audience,

> all force will finally pass into the form of heat, and all heat will come into a state of equilibrium. Then all possibility of a further change would be at an end, and the complete cessation of all natural processes must set in. . . . In short, the universe from that time onward would be condemned to a state of eternal rest.[10]

Clausius himself reiterated this conclusion the following decade, in 1867, arguing that science had finally settled the perennial debate over whether cosmic time is fundamentally linear or cyclical. In his words:

> It is often said that the world goes in a circle . . . such that the same states are always reproduced. Therefore the world could exist forever. The second law contradicts this idea most resolutely. . . . The entropy tends to a maximum. The more closely that maximum is approached, the less cause for change exists. And when the maximum is reached, no further changes can occur; the world is then in a dead stagnant state.[11]

With the central insights of thermodynamics being more or less fully articulated by 1860, the dismal conclusion that Earth and the universe will become increasingly inhospitable to human life, until no life is possible, quickly captured the attention of scientists in other fields of study, along with philosophers, artists, and the general public. The Victorian poet Algernon Charles Swinburne, for example, concluded his 1866 poem "The Garden of Proserpine" with a depiction of entropic decay leading to a condition of permanent quiescence:

> Then star nor sun shall waken,
> Nor any change of light:
> Nor sound of waters shaken,
> Nor any sound or sight:
> Nor wintry leaves nor vernal,
> Nor days nor things diurnal;
> Only the sleep eternal
> In an eternal night.

Or consider the following passage from Charles Woodruff Shield's book *Philosophia Ultima*, which depicts

> the awful catastrophe which must ensue when the last man shall gaze upon the frozen earth, when the planets, one after another, shall tumble, as charred ruins, into the sun, when the suns themselves shall be piled together into a cold and lifeless mass, as exhausted warriors upon a battle-field, and stagnation and death settle upon the spent powers of nature.[12]

In 1876, the Austrian scientist Josef Loschmidt lamented what he evocatively called "the terroristic nimbus of the second law," which acts as "a destructive principle of all life in the universe."[13] The following decade, psychologist Henry Maudsley lamented that the sun's radiative output will gradually diminish until it has been completely "extinguished." Consequently, he declared,

> species after species of animals and plants will first degenerate and then become extinct, as the worsening conditions of life render it impossible for them to continue the struggle for existence; a few scattered families of degraded human beings living perhaps in snowhuts near the equator, very much as Esquimaux live now near the pole, will represent the last wave of the receding tide of human existence before its final extinction; until at last a frozen earth incapable of cultivation is left without energy to produce a living particle of any sort and so death itself is dead.[14]

Perhaps no one summed up the gloominess of these ideas better than Bertrand Russell in his widely circulated 1903 essay "The Free Man's Worship," initially published in *The Independent Review* and later changed to "A Free Man's Worship."[15] Because of the Second Law, he wrote,

> all the labors of the ages, all the devotion, all the inspiration, all the noonday brightness of human genius, are destined to extinction in the vast death of the solar system, and that the whole temple of Man's achievement must inevitably be buried beneath the debris of a universe in ruins.[16]

Facts in Fiction

While scientists—a term introduced to the lexicon by William Whewell in 1833—were grappling with the eschatological implications of thermodynamics, the notion of entropic death was also making its way to the public arena via science fiction novels within the "Dying Earth" genre, which was anticipated by the work of de Grainville, Lord Byron, and Mary Shelley. Of note are two novels in particular, one by Camille Flammarion titled *La fin du monde*, or *Omega: The Last Days of the World* (1894). Interestingly, the first half of this book explores scientific and religious responses to news that an incoming comet could threaten the survival of humanity—essentially, it offers an early examination of the varying gestalts corresponding to different existential hermeneutics—and then provides a scientifically informed survey of several possible kill mechanisms associated with long-term astronomical and geophysical phenomena (discussed more below). However, the novel culminates with a poignant Last Man-esque depiction of the final two humans—Omegar and Eva—amidst a frozen wasteland, as "the oblique rays of the sun [prove] insufficient to warm the soil which was frozen to a great depth, like a veritable block of ice." As Flammarion describes this atrophy into oblivion:

> The world's population had gradually diminished from ten milliards [i.e., billion] to nine, to eight, and then to seven, one-half the surface of the globe being then habitable. As the habitable zone became more and more restricted to the equator, the population had still further diminished, as had also the mean length of human life, and the day came when only a few hundred millions remained, scattered in groups along the equator, and maintaining life only by the artifices of a laborious and scientific industry.

While Flammarion believed, like Kelvin, that the universe is infinite and hence will never reach a state of thermodynamic equilibrium—rather, "the future of the universe is its past"—his account of humanity's entropic demise, its "pitiless destiny" via the solar death, helped to popularize the general idea.[17] So did H. G. Wells' celebrated novel *The Time Machine*, published in 1895, which follows the adventures of an anonymous traveler who builds a "time machine" (Wells' coinage). In the book's penultimate chapter, the protagonist ventures into the future to find an "abominable desolation that hung over the world," dimly illuminated by a dying vermillion sun. "All the sounds of man," he says, "the bleating of sheep, the cries of birds, the hum of insects, the stir that makes the background of our lives—all that was over." The chapter ends with a lugubrious image of "moaning wind," "rayless obscurity," and "awful twilight" that renders the traveler on the verge of syncope. "A horror of this great darkness came on me," the traveler recounts:

> The cold, that smote to my marrow, and the pain I felt in breathing, overcame me. I shivered, and a deadly nausea seized me. Then like a red-hot bow in the sky appeared the edge of the sun. . . . Then I felt I was fainting. But a terrible dread of lying helpless in that remote and awful twilight sustained me while I clambered upon the saddle.[18]

A New Mood Descends

As these excerpts show, the development of thermodynamics, in particular the Second Law, during the 1850s radically transformed our understanding of the existential predicament of humanity. No longer was it a matter of mere speculation whether global catastrophes could kill us; our eventual demise is, to the contrary, guaranteed by the fundamental laws of physics. There had been, as noted in the previous chapter, earlier hypotheses about the gradual refrigeration of the planet, such as Comte de Buffon's theory of Earth's formation and his prognostication that Earth will someday become intolerably frigid. Indeed, Helge Kragh notes that in the waning decades of the eighteenth century, the notion that our sun is "a huge chemical machine" from which Earth and the other planets were birthed, which then cooled to their present temperatures from an initial molten state, became common.[19] But these claims were not derived from a *law of nature*. They were instead based on empirical observations of Earth's past; as Comte de Buffon wrote, to understand the history and future of our planet, we must "dig through the archives of the world,"[20] which was, incidentally, enabled in part by steam-powered excavations of strata in Earth's crust that provided evidence of Earth being hotter in the past. It was then a simple extrapolation to the proposition that the average temperature of Earth has fallen over time and will continue to drop in the future. In Buffon's particular formulation, Earth is a "dying ember of the sun" that, as such, is cooling down over time (quoted in the previous chapter).

In contrast, the prophecies from thermodynamicists were founded on a nomological generalization understood as invariant and exceptionless across cosmic space and time no less than Newton's law of universal gravitation. It as a fundamental feature of the universe, not a contingent fact about our planet or planetary system, and it was supported by an enormous body of empirical evidence that had been collected over many decades leading up to the 1850s. As Arthur Eddington later put it, "the law that entropy always increases, holds, I think, the supreme position among the laws of Nature."[21] Albert Einstein similarly declared in his autobiographical notes that thermodynamics "is the only physical theory of a universal content which I am convinced . . . will never be overthrown."[22] It was these features of the Second Law—its fundamentality and robustness, along with its annihilatory implications—that triggered the first shift in existential mood in the second half of the nineteenth century. Whether or not the universe is infinite, Earth itself will eventually become unfit for life, which seems to imply that our extinction is not only possible but inevitable. There is no escaping the brutal dictatorship of entropy; we are imprisoned by the inviolable laws of physics, on death row awaiting our execution in the distant future.

How distant? Fortunately, everyone agreed that Earth will remain habitable for at least "many million years longer," to quote Kelvin.[23] Wells, for example, describes the "stillness" and "bitter cold" of a nearly lifeless planet over 30 million years from now. In *Omega*, Flammarion states that we should expect "a future for the Sun of at least twenty million years." By the twentieth century, some calculations pushed our ineluctable demise ever farther toward the temporal horizon. Sir James Jeans provides an example: he argued in 1929 that "if the solar system is left to the natural course of evolution, the earth is likely to remain a possible abode of life for something of the order of a million million years to come," which means that "our race may look forward to occupying the earth for a time incomparably longer than any we can imagine. . . . [A]s inhabitants of the earth, we are living at the very beginning of time."[24] This idea was humorously captured by a supposed exchange between an "old lady" and a professor following a lecture on the future. "Excuse me, Professor," she said, "but when did you say that the universe would come to an end?" "In about four billion years," he replied. "Thank God," she remarked with a sigh, "I thought that you said four million."[25]

Although such vast timespans mean that nobody within the foreseeable future will encounter our entropic extinction, the eschatology of thermodynamics nonetheless elicited powerful feelings of what Russell described in 1903 as "unyielding despair" when one takes a cosmic perspective of the human existential predicament. Our existence is now framed by a sense of *cosmic doom* rather than indestructibility, and indeed this new recognition that extinction is a matter of "when" rather than "if" was integral to the existential mood that emerged at the time, which has changed from then to the present only by what has been added to it.

Competing Perspectives on the Prospect of Doom

Yet not everyone at the time accepted the gloomy implications of the Second Law *for humanity*, even if they fully embraced the new science of thermodynamics. For some, the threat environment hadn't changed, even though our empirical model of the universe had. The issue hinges upon the extent to which the Second Law was viewed through a religious or secular hermeneutics. For example, Kelvin was a devout Christian who, in some of the same papers quoted above, admitted to being agnostic about what the solar death means for our future in the cosmos. "A state of universal rest and death" is inevitable, he wrote, "if the universe were finite and left to obey existing laws." Yet he added that it is "impossible to conceive either the beginning or the continuance of life, without an overruling creative power," which he took to imply that "no conclusions of dynamical science regarding the future condition of the earth can be held to give dispiriting views as to the destiny of the race of intelligent beings by which it at present inhabited."[26] The same year, he co-authored an article in the Presbyterian magazine *Good Works* for "the non-scientific reader," which seems to gesture at the heat death, rather than merely the solar death. It reports that, based on current scientific understanding, "all energy tends ultimately to become heat," and that "when all the chemical and gravitation energies of the universe have taken their final kinetic form, the result will be an arrangement of matter possessing no realizable potential energy, but uniformly hot . . . chaos and darkness as '*in the beginning*,'" an obvious reference to the opening verses of the Book of Genesis. The article concludes with two additional biblical references: Hebrews 1:11 and 2 Peter 3:12–13, the latter of which asserts that "the elements will melt into heat," which could be interpreted as supporting the dysteleological prediction of eventual thermodynamic equilibrium. "We have the sober scientific certainty," they write,

> that heavens and earth shall "wax old as doth a garment"; and that this slow progress must gradually, by natural agencies which we see going on under fixed laws, bring about circumstances in which "the elements shall melt with fervent heat." With such views forced upon us by the contemplation of dynamical energy and its laws of transformation in dead matter, dark indeed would be the prospects of the human race if unilluminated by that light which reveals "new heavens and a new earth."[27]

Hence, Kelvin and his co-author were still operating within the parameters of the previous existential mood and its associated hermeneutics, according to which our fate is inextricably tangled up with the eschatological narratives of holy scripture. In the end, we can expect a transformation of the universe rather than its termination, resulting in a new heaven and Earth "where righteousness dwells" (2 Peter 3:13).[28]

In contrast, those who drew less sanguine conclusions about our fate on a dying planet were mostly agnostics or atheists. For example, Swinburne was an atheist, as was Russell in his popular writings, although he stated in 1947 that he prefers the term "agnostic" when speaking to other philosophers.[29] Maudsley, Helmholtz, Jeans, and Einstein were all agnostics, while Flammarion strove to merge some form of religion with the established facts and methodology of science. Wells' religious views changed throughout his life, and although he did not endorse atheism until his geriatric years, he seems to have held an idiosyncratically theistic, albeit explicitly non-Christian, view when he composed *The Time Machine*.[30] Almost nothing is known about Clausius' personal views on religion. Nonetheless, we can see clear enough how different existential hermeneutics led to significantly different conclusions about the Second Law's implications for the future of humanity. This is one reason the widespread decline of religion during the nineteenth century, which made Kelvin something of an outlier, played an integral role in enabling the new existential mood to arise.

Drivers of Decline

But what drove this decline? Earlier, we noted that the first extinction-blocking idea to take a serious hit was the eschatological thesis during the eighteenth century, due to the rise of deism among Enlightenment philosophers. Deists tended to reject special revelation as a source of knowledge and hence also tended to dismissed eschatological propositions that were privately revealed to prophets like John of Patmos, who penned the Book of Revelation. While most deists during the Enlightenment accepted the immortality of the soul, the extent of their interest in eschatology was mostly limited to the *personal* rather than the *cosmic*—that is, to our fate as individuals in the afterlife rather than the culmination of world history with the Battle of Armageddon, and so on. However, the most significant religious transformation occurred during the nineteenth century, especially the second half. As Gavin Hyman writes, it was this period during which "atheism—and religious doubt more generally—became a central and inescapable feature of the cultural landscape."[31] Not only did this enable the Second Law to be interpreted as posing an annihilatory threat to humanity, but the new secular existential hermeneutics swung opened the door for novel speculations about other potential kill mechanisms, both natural and anthropogenic. In other words, if our survival is not assured, if we do not exist in "perfect security," then what else might trip us into the eternal grave of extinction before the Second Law renders Earth uninhabitable?

We can identify three main causes of secularization during the nineteenth century, although the origins of this transformation (arguably) date back to the beginning of theological and philosophical modernity, which one could argue were inaugurated by the work of Rene Descartes.[32] The first is the most obvious: revolutionary scientific breakthroughs, most notably Charles Darwin's theory of evolution by natural selection, which he delineated in an 1859 book *On the Origin of Species*. This synthesized a mountain of evidence for the proposition that species are not immutable or fixed types (the typological view of species, a central assumption of the Great Chain) but populations of individuals whose statistical features can change over time in response to alterations in the environment. Darwin further argued that these changes are non-teleological (that is, not directed toward a goal, or *telos*), and that biological evolution does not follow a predetermined plan—an idea known as *orthogenesis,* as exemplified by the evolutionary theory of Jean-Baptiste Lamarck. Rather, species evolve through natural selection, whereby the individuals most adapted to their selective environment will tend to have more offspring, thus increasing the prevalence of their phenotypic traits within the population. This brought together numerous ideas prominent at the time: William Paley's emphasis on "design" in nature, for which natural selection provides a naturalistic explanation; Thomas Malthus' notion of a "struggle for existence," which accounts for why better-adapted individuals reproduce more; Adam Smith's idea of the "invisible hand," which tends toward better overall outcomes through individual competition; and Charles Lyell's uniformitarianism, which posits vast stretches of geological time, thus enabling small phenotypic changes to gradualistically snowball into large differences in average phenotype.[33]

However, Darwin lacked an adequate theory of inheritance, which is one reason his theory of natural selection was largely rejected until the Modern Synthesis of the early twentieth century, which integrated the mechanism of selection with the particulate theory of heredity published by Gregor Mendel in 1866. The more *immediate* triumph of Darwin's work was that it convinced many people that evolution is a fact about the biological history of Earth. As Peter Bowler writes, "by 1875 the majority of educated people in Europe and America had accepted evolution. Even religious thinkers were now trying to come to terms with the prospect of a natural origin for humanity."[34] The implications of this for the ontological thesis were profound: if *Homo sapiens* evolved through natural processes from earlier mammals, then we are different in degree rather than kind from them; if they lacked immortal souls, then surely we do as well. In other words, there is no unbridgeable *ontological gap* between humanity and the rest of the animal kingdom, and hence the ontological thesis is untenable.

Darwin himself initially sidestepped the issue, writing to Alfred Russell Wallace—the co-discoverer of natural selection—in 1857 that "I think I shall avoid the whole subject, as so surrounded with prejudices." Hence, he reassured readers at the end of the *Origin* that there are "no good reasons why the views given in this volume should shock the religious feelings of anyone."[35] However, slightly more than a decade later, he confronted human evolution head-on in *The Descent of Man* (1871), which boldly affirmed that *Homo sapiens* is "like every other species."[36] He elaborated, emphasizing the cognitive-evolutionary unity of humanity and our primate cousins:

> It is notorious that man is constructed on the same general type or model with other mammals. All the bones in his skeleton can be compared with corresponding bones in a monkey, bat, or seal. So it is with his muscles, nerves, blood-vessels and internal viscera. The brain, the most important of all the organs, follows the same law. . . . There is no fundamental difference between man and the higher mammals in their mental faculties.

The result was a major step toward a new conception of *humanity* that is compatible with the possibility of extinction. Perhaps the single greatest blow to Christianity at the time was that, by explaining the apparent "design" of nature via material forces, Darwinism rendered God explanatorily superfluous. If one no longer needed to posit a watchmaker to explain the existence of watches (to use Paley's famous analogy), then what reason is there for believing that the watchmaker exists? As one evolutionary biologist has observed, "Darwin made it possible to be an intellectually fulfilled atheist."[37]

The second major cause behind the nineteenth-century decline in religion was biblical criticism, which had its origins in (what is now) Germany. This aims to, in the words of the Victorian theologian Benjamin Jowett, "*interpret the Scripture like any other book*."[38] A deep exploration of this field and its impact on the debate between science and religion at the time goes beyond the scope of this book. Suffice it to say that when Jowett's "precept" was followed, a flurry of problems arose that called into question the historicity and internal consistency of biblical narratives.[39] Especially noteworthy was an 1860 edited collection given the nondescript title of *Essays and Reviews*, from which the Jowett quote above is extracted. This brought together scholarship from previous decades, some from the German universities, and ignited an uproar that "in some quarters overshadowed that concerning the *Origin of Species*."[40] Or, as the theologian Bernard Reardon writes, its publication was "the most sensational theological event in England in the mid-nineteenth century—apart from the appearance a year earlier of Darwin's epoch-making treatise in the realm of biology."[41] These developments cast a shadow of doubt on the infallibility and reliability of the Bible.

The final driver of secularization arose from moral considerations, such as "How is eternal damnation an appropriate punishment for disbelief?" and "How can an omniscient, omnibenevolent, and omnipotent deity allow *any*, much less *every*, evil in the world?"[42] Human existence is drenched in suffering, much of it gratuitous, often the result of natural rather than moral evils (i.e., those outside the control of moral agents, so-called "acts of God"). Among the more salient examples was the 1755 Lisbon Earthquake that struck the Portuguese city on All Saints' Day, while churches were full, virtually destroying the city. This was a major event that many saw as posing a serious challenge to the supposed goodness of God. How could the world—especially "the best of all possible worlds," as Gottfried Wilhelm Leibniz triumphantly proclaimed in 1710—have been crafted by the hands of a Being who loves us yet, at the same time, permits such tragedies? What theodicy (Leibniz's coinage) could vindicate God from these crimes of pointless harm, the infliction of which he could have but chose not to prevent? The lack of a clear or compelling answer to these questions further chipped away at the foundation of Christianity by undermining its moral legitimacy.

The Spectacles of Secularism

Let's take stock: by the end of the nineteenth century, Christianity was quickly losing its stranglehold on our conception of human nature and understanding of humanity's place in the cosmos. God's goodness was under harsh scrutiny, the Bible was widely recognized as historically unreliable, and humanity was no longer seen as having been created in the likeness of God. Instead, according to the Darwinian worldview established the same decade that thermodynamics emerged, we are nothing more than the contingently pieced-together assemblages of bone and flesh from soulless ancestral forms in a universe devoid of inherent meaning. Or as Bertrand Russell wrote in an essay quoted above, "Man is the product of causes which had no prevision of the end they were achieving; that his origin, his growth, his hopes and fears, his loves and his beliefs, are but the outcome of accidental collocations of atoms."[43] Reflecting this sea change in religious orientation, T. H. Huxley—known as "Darwin's Bulldog" for his spirited public defenses of Darwinism—coined the term "agnostic" in 1869, which he characterized as "of the essence of science," since it "means that a man shall not say he knows or believes that which he has no scientific grounds for professing to know or believe."[44] Meanwhile, in the mid-nineteenth century, Karl Marx famously described religion—a superstructural component of the capitalist mode of production—as "the sigh of the oppressed creature . . . the *opium* of the people," and Friedrich Nietzsche no less famously declared that "God is dead! God remains dead! And we have killed him!"[45]

Concomitant with these cultural transformations was the emergence of a new secular existential hermeneutics, which by the late nineteenth century had become firmly established among many intellectuals in Europe and America. It was this interpretive framework that led so many at the time to despair (Russell's word) about our ultimate fate in a universe inexorably sinking into thermodynamic equilibrium. But, importantly, the new hermeneutics also spurred a dramatic *reassessment* of the potential hazards associated with natural and anthropogenic phenomena, resulting in revised maps of the threat environment in which we find ourselves embedded. If there is no God watching out for us, if we are no more invulnerable to extinction than any other species, then what other kill mechanisms might be hiding in the cosmic shadows? What other scientific discoveries might reveal our more immediate precarity? Which technologies might be double-edged swords that end up slicing humanity into pieces? What if civilizational "progress" is actually leading us toward the precipice of self-destruction? Without the reassurance of immortality and eschatological purpose, the *possibility of risk* suddenly began to appear everywhere: in nature, science, technology, and politics. By scanning our surroundings through the spectacles of secularism, features of the world once thought benign were all-of-a-sudden seen as candidate threats to our survival on Earth. The result was an explosion of novel fears, worries, anxieties, and speculations about how humanity could be destroyed or destroy itself before the Second Law renders Earth unsuitable for life. Although none of the proposed kill mechanisms became as widely accepted by the scientific community as the Second Law, they indicate how momentous the decline of Christianity was with respect to thinking about the possibility of our species disappearing. We can organize these speculations roughly into three categories: (i) *evolutionary*, (ii) *naturogenic*, and (iii) *technoscientific*. Let's consider them in turn.

Crouching for Its Spring

Perhaps the best starting place is a short 1893 essay by Wells titled "The Extinction of Man."[46] This is particularly notable for three reasons: first, it explicitly acknowledged that if one accepts the Darwinian worldview, then humanity is no more protected from extinction than any other biological species. If we are no different in life, then we are no different in death, either. Hence, referring to this possibility, Wells wrote that "surely it is not so unreasonable to ask why man should be an exception to the rule. From the scientific standpoint at least any reason for such exception is hard to find." He followed this with an exhortation for readers to wake up from their dogmatic slumber, to shake off their complacent assumption that "because things have been easy for mankind as a whole for a generation or so, we are going on to perfect comfort and security"—a fascinating and powerful, albeit unintentional, reference to the two words identified in the previous chapter as partly defining the previous existential mood, namely, "Comfort" and "perfect security." Wells continued:

> Even now, for all we can tell, the coming terror may be crouching for its spring and the fall of humanity be at hand. In the case of every other predominant animal the world has ever seen, I repeat, the hour of its complete ascendency has been the eve of its entire overthrow.[47]

But how might this happen? This leads to the second reason the article is notable: Wells took seriously the prospect of natural phenomena wiping out humanity in a way that people in the earlier existential mood did not, a point that we will return to just below. For example, he suggested that a devastating famine could potentially bring about our extinction, or "a plague that will not take ten or twenty or thirty per cent., as plagues have done in the past, but the entire hundred." As with Shelley's 1826 novel *The Last Man*, this was based on rational extrapolation from past incidents, although by the time Wells was writing, far more people would have found it plausible than in the early nineteenth century. But—and this is the third reason—Wells also seriously considers the possibility that we go extinct as a result of evolutionary pressures rather than a single catastrophe (for example, a pandemic). Competition with rival species for limited resources could itself constitute a kill mechanism that precipitates our demise, not in a sudden disaster event but from losing the Malthusian struggle for existence over millions of years.[48] Perhaps a ferocious new species—a "terrible monster," in his words—could evolve in the ocean or on land that out-competes humanity for food and/or space. This is a real "possibility," Wells writes, "if perhaps a remote one."[49]

Evolving Evolutionary Possibilities

It is perplexing that Darwin himself never entertained this possibility, at least not in writing, to my knowledge. He never once considered the termination of our evolutionary lineage as a result of too many individuals fighting over too few resources, or from outright violence between antagonistic species. This is especially perplexing because (a) extinction was an integral part of his theory of evolutionary change, (b) he rejected our ontological uniqueness within the animal kingdom, and (c) he emphasized that, in his own words, "of the species now living, *very few* will transmit progeny of any kind to a far distant futurity."[50] All the ingredients were there, yet the cake was never baked. The closest Darwin came to imagining our extinction was in a brief passage of his autobiography, published posthumously in 1887. This passage is interesting because, first, Darwin acknowledged the inevitable annihilation of humanity due to the Second Law, which is odd given that Kelvin was among the fiercest critics of uniformitarianism, as he believed that it is inconsistent with the laws of thermodynamics (which it was). Hence, Darwin was a uniformitarian who also accepted the rather non-uniformitarian claim that time has a direction and Earth will someday become uninhabitable. Second, the passage states one of the main theses of Part I of this book, namely, that belief in the soul's immortality confers to the believer a degree of "Comfort" (quoting Franklin) when confronted with the possibility of world-destroying catastrophes, since the destruction of the world will not entail the destruction of humanity. To quote Darwin's thoughts on these matters:

> With respect to immortality, nothing shows me how strong and almost instinctive a belief it is, as the consideration of the view now held by most physicists, namely, that the sun with all the planets will in time grow too cold for life, unless indeed some great body dashes into the sun and thus gives it fresh life. Believing as I do that man in the distant future will be a far more perfect creature than he now is, it is an intolerable thought that he and all other sentient beings are doomed to complete annihilation after such long-continued slow progress. To those who fully admit the immortality of the human soul, the destruction of our world will not appear so dreadful.[51]

Notice here Darwin's statement that "man in the distant future will be a far more perfect creature," which is ambiguous between two readings: it could mean that, relative to some *species-specific* model of excellence, *Homo sapiens* will increasingly move toward this normative ideal. Or it could mean that, given enough time, we will evolve into a *new species* that is in some respect superior to current *Homo sapiens*. If the latter is the case—and indeed Darwin elsewhere wrote in the *Origin* that "judging from the past, we may safely infer that not one living species will transmit its unaltered likeness to a distant futurity," meaning that our descendants will be very different from us—then he did anticipate our eventual extinction, but only in the sense of "phyletic extinction." As Chapter 7 discusses in detail, phyletic extinction would occur if *Homo sapiens* disappears by evolving into one or more new species without a break in our evolutionary lineage. Since Darwin expected this to yield a superior version of us, let's label this sort of extinction *progress*.[52]

It contrasts with the opposite scenario of evolutionary *degeneration*, which people began to discuss—and fret over—in the decades after the publication of Darwin's *Origin*. In an 1880 book *Degeneration: A Chapter in Darwinism*, Sir Edwin Ray Lankester defined "degeneration" as "a gradual change of the structure in which the organism becomes adapted to *less* varied and *less* complex conditions of life."[53] At the extreme, some worried that this might produce one or more inferior new species, as occurs in Wells' novel *The Time Machine*, which was almost certainly inspired by Lankester's book given that Wells was Lankester's student and friend.[54] Before venturing millions of years into the future, the anonymous time traveler arrives at 802,701 AD to discover the human lineage having bifurcated into two distinct creatures: the *Eloi*, a beautiful but intellectually stunted species that lives above ground, and the *Morlocks*, a brutish subterranean species that provides goods to the Eloi, whom they devour for sustenance on moonless nights. (Consequently, the Eloi are terrified of the new moon.) The traveler conjectures that this evolutionary split may have resulted from class divisions in society: the Eloi were the "Capitalists" and the Morlocks were the "Labourers"—which somewhat poetically complements suspicions that Darwin's theory was influenced by the economic conditions of his time, that is, free-enterprise capitalism.[55] As Wells writes: "So, in the end, above ground you must have the Haves, pursuing pleasure and comfort and beauty, and below ground the Have-nots, the Workers getting continually adapted to the conditions of their labour."[56] Yet over time the power dynamics were inverted such that the Eloi became the livestock of the more intellectually vigorous Morlocks. Hence, over hundreds of thousands of years, evolutionary forces had swapped humanity with two "lesser" species, one childlike and the other subhuman.

Yet there is another possibility that arose from the Darwinian notion that *Homo sapiens* is phylogenetically plastic. Whereas the scenarios above all involve mechanistic processes of evolutionary change that are *natural*, what if humanity were to usurp the role of natural selection by intentionally altering the statistical frequency of characteristics within its population through selective breeding? If, for example, "intelligence" is heritable, then the average "intelligence" of humanity could go up if "intelligent" people were to reproduce more than "unintelligent" people. The first to suggest this was Darwin's half-cousin Sir Francis Galton in his 1869 book *Hereditary Genius*, which introduced our modern vocabulary of "nature versus nurture," perhaps taking this distinction from the Shakespearean play *The Tempest* in which Prospero complains about his adopted son Caliban ("A devil, a born devil, on whose nature/Nurture can never stick"). Eugenics, a term that Galton coined in 1883, became one of the most horrifying ideas of the twentieth century, but before its association with the moral atrocities of the Third Reich it was enthusiastically championed by influential scientists like J. B. S. Haldane, J. D. Bernal, and Sir Julian Huxley. Huxley, for instance, was president of the British Eugenics Society and later popularized a normative ideology called *transhumanism*, which takes the aims of eugenics a step further; transhumanism is basically eugenics on steroids. Rather than trying to create the best version of humanity possible through selective breeding and other methods, thereby improving the "human stock," transhumanists like Huxley imagined using these methods to enable humanity to completely "transcend" itself by taking control of its own evolutionary trajectory.[57] As he wrote in a 1927 book titled *Religion Without Revelation*,

> civilised man is beginning to realise that he can, if he so wishes, in large measure model the world in accordance with his desires. . . . [But] there is [an] extension of the same outlook to his own nature. . . . [T]he study of heredity and population-growth, and the knowledge of eugenics and of birth-control are pointing the way to wholly new aims—to a conscious control by man of his own nature and racial destiny.[58]

This idea was echoed two years later by Bernal's *The World, the Flesh, and the Devil*, which argued that if humanity wishes to keep up evolutionarily with the transformations we have brought about in our surroundings, we will need to modify our phenotypes in equally radical ways, by either altering our genes or using technology to extend the phenotypic features of our bodies. In his words:

> Man himself must actively interfere in his own making and interfere in a highly unnatural manner. The eugenists and apostles of healthy life, may, in a very considerable course of time, realize the full potentialities of the species: we may count on beautiful, healthy, and long-lived men and women, but they do not touch the alteration of the species. To do this we must alter either the germ plasm or the living structure of the body, or both together.[59]

The point for our purposes is that the Darwinian notions that species are not fixed types (the typological view) and humanity is not the pinnacle of creation (because evolution is non-teleological) opened the door to new thoughts about how we might shape our own evolution in the future, through a kind of "intelligent redesign," which could result in one or more species of "posthuman" beings, to borrow the terminology of contemporary transhumanists.

Hence, within the first category of *evolutionary* speculations about our possible disappearance, enabled by the materialist, Darwinian worldview that gradually replaced Christianity across the nineteenth century, we find four distinct possibilities, which are differentiated by their outcomes and etiologies. These are: (1) *elimination* via competition with other species (Wells), (2) *progress* via natural processes (Darwin),[60] (3) *degeneration* via natural processes (Lankester, Wells), and (4) *transcendence* via artificial processes (Haldane, Huxley, Bernal). Each of these could be sufficient to eliminate us—that is, we could either die out like the dodo or evolve into something different via natural or artificial selection, where the novel species that replaces us could be either better (as in the cases of progress and transcendence) or worse (as in the case of degeneration). Here, then, was an additional set of novel kill mechanisms, although (a) this term fits awkwardly with the idea of becoming something better, and (b), most importantly, none of these possibilities was widely accepted at the time, especially when compared to the Second Law. To underline the latter point, these avenues to our naturalistic disappearance were very little discussed, as the overwhelming focus among evolutionists at the time was the *past* rather than the *future*—on how species came to be what they are today rather than on what they might become tomorrow.

Threats from Above and Below

With respect to the second category of *naturogenic* speculations, we have already seen how Wells warned about pandemics and agricultural failures, in addition to his evolutionary concerns. One of the most comprehensive and informed snapshots of such thinking is found in Flammarion's *Omega*, which focused primarily on exploring the potential astronomical and geophysical threats to our survival. To the contemporary reader, it may be surprising that almost none of these involved *sudden* catastrophes, but around the time that the Second Law was discovered, the Earth sciences became dominated by the aforementioned uniformitarian theory of Charles Lyell, which posited that all change in the world is very gradual, and that if catastrophes do occur, they are always localized affairs—that is, they never affect the entire planet.[61] As several characters in *Omega* put the point, "worlds do not die by accident, but of old age."

In the midst of a discussion about whether a comet heading toward Earth might destroy humanity (it doesn't, as this would involve our world dying "by accident"), a motley group of scientific experts outline what one describes as "an admirable resumé of the curious theories which modern science is in a position to offer us, upon the various ways in which our world may come to an end." In addition to the entropic death of the solar system, other potential kill mechanisms include the following:

- "The gradual leveling of the continents and their slow submergence beneath the invading waters," which is estimated to occur in about 4 million years.
- "The amount of water on the surface of our planet is decreasing from century to century," and will eventually disappear entirely, thereby killing us.
- Earth will eventually lose its atmosphere, which provides a blanket of warmth in the frigidness of space. Consequently, "the very blood would freeze in our veins and arteries, and every human heart would soon cease to beat."
- A star could "emerge from space" that becomes intertwined with our sun, causing Earth to stop spinning or orbiting the solar system. Consequently, Earth's "mechanical energy would be changed into molecular motion, and its temperature would be suddenly raise to such a degree as to reduce it entirely to a gaseous state."
- Our solar system could enter "into some kind of nebula" in space, causing the sun to explode, as has been observed in space with other stars.[62]

Of note is that these are presented as scientifically plausible doomsday scenarios, supported by empirical evidence and arithmetical calculations confidently delineated by the various scientists who defend each idea. Yet one scientist after another, for this or that reason, reject some or all of the scenarios outlined. In other words, *Omega* provides a fascinating example of how educated people in the late nineteenth century were beginning to take seriously the *scientific study of kill mechanisms*, but also how there was no consensus about whether astronomical or geophysical phenomena could *actually* bring about our collective demise. We appear to be safe from annihilation, at least in the near term, so far as the best science of the day was concerned.

Exploding Boilers, Atomic Energy

This brings us to the third category of speculations, which proved to be the most menacing, especially after World War I (1914–1918): the possibility of *technoscientific* kill mechanisms that enable humanity to destroy itself. As numerous scholars have pointed out, the nineteenth century witnessed a growing sense of foreboding about our expanding capacities to inflict mass harm and effectuate large-scale destruction. Warren Wagar, for example, argues that beginning in the Romantic era with Shelley and others there was a gradual shift of focus from natural to anthropogenic scenarios, which culminated with WWI.[63] Meanwhile, Spencer Weart traces the motif of the "mad scientist," whose actions driven by malign intention or incurable curiosity spell disaster, back to Shelley's *Frankenstein* (1818), which emerged from the same sojourn near Lake Geneva, during the Year Without a Summer, that spawned Byron's *Darkness*.[64] This evolved, Weart argues, from earlier tales of sorcerers and witches that were adapted to and shaped by the nineteenth century milieu of accelerating scientific and technological "progress," in which a rapidly growing concentration of power—for better or worse—was being placed in the hands of scientific experts and specialists.

According to Weart, no one did more to establish this new stereotype than Jules Verne, often lionized as the "Father of Science Fiction" along with his contemporary Wells. For example, Verne offered what may have been the first description of technology accidentally obliterating the planet in his 1863 novel *Five Weeks in a Balloon*.[65] As the Scotsman Dick Kennedy, a character in the book, says: "By dint of inventing machinery, men will end in being eaten up by it! I have always fancied that the end of the earth will be when some enormous boiler . . . shall explode and blow up our Globe!"[66] The same year, Samuel Butler published "Darwin among the Machines," which offered a new technological interpretation of the theme of the first "evolutionary" category above. According to Butler, our machinic creations are evolving (with humans instantiating the role of natural selection in the biological world) and, as a result, they may eventually take humanity's place atop the dominance hierarchy on Earth. "It appears to us," he writes, "that we are ourselves creating our own successors. . . . In the course of ages we shall find ourselves the inferior race. . . . [T]he time will come when the machines will hold the real supremacy over the world and its inhabitants." Hence, Butler concluded with the hortatory exclamation that "every machine of every sort should be destroyed by the well-wisher of his species. Let there be no exceptions made, no quarter shown."[67] A similar idea was later explored in Karel Čapek's 1920 science fiction play *R.U.R.*, in which a population of "robots"—Čapek's coinage—rise up to overtake humanity, founding a new world order in our place from their own Adam and Eve, named Primus and Helena.

Such fears weren't limited to science fiction. There were also rumors that actual scientists, engineers, and inventors would soon have the technological tools necessary to unilaterally destroy the world. As Weart writes, "typical of public thinking [about the dangers of science] was an 1892 rumor that Thomas Edison was building an electrical device that could annihilate a city from a distance, followed by a newspaper satire about the great inventor destroying England with a pushbutton 'doomsday machine.'"[68] Around the same time, scientists were making new discoveries that would enhance the plausibility of such apocalyptic scenarios. A case in point is Henri Becquerel's discovery of radioactivity (or radioactive decay) in 1896, followed in 1902 by the realization that such decay occurs when one type of atom *transmutes* into another, as when uranium-238 decays into lead-206 with a half-life of 4.5 billion years, or thorium-228 decays into radium-224 with a half-life of just two years. This breakthrough was the work of Frederick Soddy and Ernest Rutherford, who shortly afterwards hypothesized that a "planetary chain reaction" of radioactive decay could decimate the planet by converting all of Earth's elements into new elements like helium, which thus "provided the first superficially rational description of how a person might in fact destroy the world." Indeed, the French polymath Gustave Le Bon reported in 1903, with some hyperbole, that Rutherford himself had "playfully suggested to the writer the disquieting idea that, could a proper detonator be discovered, an explosive wave of atomic disintegration might be started through all matter which would transmute the whole mass of the globe into helium or similar gases."[69]

Other scientists quickly picked up on this idea. For example, a 1923 textbook titled *The Atom and the Bohr Theory of Its Structure*, which includes a foreword by Rutherford, explores "what would happen if it were possible to bring about artificially a transformation of elements propagating itself from atom to atom with the liberation of energy." How much energy could this liberate? The answer comes from Einstein's theory of special relativity, which introduced the notion that mass is concentrated energy—that is, these are not two distinct categories of physical phenomena but *equivalent*. However, the amount of energy per unit of mass is *huge*, expressed by perhaps the most famous equation in history: $E = mc^2$. This says that the energy (E) equals the mass (m) multiplied by the *square of the speed of light*, which is 299,792,458 meters per second. Thus, as the book puts it, "the quantities of energy which would be liberated in this way would be many, many times greater than those which we now know of in connection with chemical processes." It continues:

> There is then offered the possibility of explosions more extensive and more violent than any which the mind can now conceive. The idea has been suggested that the . . . catastrophes represented in the heavens by the sudden appearance of very bright stars [i.e., novae] may be the result of such a release of subatomic energy, brought about perhaps by the "super-wisdom" of the unlucky inhabitants themselves. But this is, of course, mere fanciful conjecture.[70]

Indeed, there was no known way of artificially inducing stable elements into becoming radioactive atoms at the time—hence the fancifulness of the conjecture.

This changed in 1934, when the wife-and-husband team of Irene and Frédéric Joliot-Curie devised a way to do precisely this, converting stable atoms of aluminum into radioactive atoms of the same element by exposing them to alpha particles produced by a (separate) radioactive source. In other words, radioactive atoms (the source) could make stable aluminum atoms radioactive. As Weart notes, this "looked like a step toward contagious radioactivity, the fateful chain reaction that Soddy and Rutherford had wondered about decades earlier."[71] It also awarded the Joliot-Curies a Nobel Prize the following year, and in his acceptance speech Frédéric explicitly warned about the potential for artificial transmutation to precipitate a worldwide catastrophe:

> If such transmutations do succeed in spreading in matter, the enormous liberation of usable energy can be imagined. But, unfortunately, if the contagion spreads to all the elements of our planet, the consequences of unloosing such a cataclysm can only be viewed with apprehension. Astronomers sometimes observe that a star of medium magnitude increases suddenly in size; a star invisible to the naked eye may become very brilliant and visible without any telescope—the appearance of a Nova. This sudden flaring up of the star is perhaps due to transmutations of an explosive character like those which our wandering imagination is perceiving now—a process that the investigators will no doubt attempt to realize while taking, we hope, the necessary precautions.[72]

The Race Between Wisdom and Power

By the 1930s, the idea of a runaway energy experiment had become so widespread that even many children were aware of it,[73] although further examination of this issue—the harnessing of "atomic energy"—will have to wait until the next chapter. For now, it is enough to note that some of these speculations were articulated after WWI, which greatly amplified worries that scientific and technological development are leading us in the wrong direction. With the mechanization of mass violence and the creation of nightmarish new weaponry—machine guns, flamethrowers, poisonous gases, tanks, and submarines—earlier questions about the overall desirability of "progress" were suddenly front and center. As a *Minnesota Alumni Weekly* article published in 1919 warned, "we cannot go farther on the road we have been taking; we have learned that. It would lead to ultimate human extinction. Because progress has furnished the key to destruction."[74] Similarly, Sigmund Freud closed the final chapter of his 1930 book *Civilization and Its Discontents* with a discussion of "the derangements of communal life caused by the human instinct of aggression and self-destruction," adding ominously that "men have brought their powers of subduing the forces of nature to such a pitch that by using them they could now very easily exterminate one another to the last man."[75] This should be seen as hyperbolic, though, perhaps the result of a Eurocentric view of the world that tended to conflate the destruction of European civilization with the extinction of humanity. Indeed, throughout the history of thinking about our extinction, the term "human extinction" has often been used loosely, even sloppily, as a sensationalized synonym of "civilizational collapse"—that civilization being Western civilization—despite these being radically different outcomes with potentially unique moral implications (see Part II).

Still, Freud's point stands: it was not difficult after WWI to imagine the enterprise of technoscience eventually carrying us over the cliffs of self-annihilation. This is the direction we seemed to be heading. As Wagar observes, two-thirds of the apocalyptic scenarios presented in works of fiction were, prior to 1914, the result of natural causes whereas after WWI, two-thirds were depicted as resulting from human action, with a whopping three-quarters involving "world wars with scientific weapons."[76] Many people understood this situation as a race between the power of our technologies and the moral character or wisdom of our species. Winston Churchill, for example, wrote in a 1924 article titled "Shall We All Commit Suicide?" that "mankind has never been in this position before. Without having improved appreciably in virtue or enjoying wiser guidance," he continued,

> it has got into its hands for the first time the tools by which it can unfailingly accomplish its own extermination. . . . Death stands at attention, obedient, expectant, ready to serve, ready to share away the people *en masse*; ready, if called on, to pulverize, without hope of repair, what is left of civilization.[77]

The same year, Haldane wrote in *Daedalus; or, Science and the Future* that "Man armed with science is like a baby armed with a box of matches," to which he added that "the future will be no primrose path. It will have its own problems. Some will be the secular problems of the past, giant flowers of evil blossoming at last to their own destruction. Others will be wholly new."[78] The general sentiment of anticipatory anxiety about what the future might hold given the trends of the past was perhaps best summarized (once again) by Russell, who argued, in a response to Haldane's essay, that the problem isn't just technology but the "political and economic institutions" that wield this newly acquired power. "The changes that have been brought about have been partly good, partly bad," Russell wrote. "Whether, in the end, science will prove to have been a blessing or a curse to mankind, is to my mind, still a doubtful question."[79]

Cosmic Doom and Existential Vulnerability

To summarize this chapter, the first shift in existential mood unfolded in the 1850s, triggered by the discovery of the very first scientifically credible, widely accepted kill mechanism: the Second Law of thermodynamics. However, without a secular existential hermeneutics, there is no reason to believe that this would have been the case—that our understanding of humanity's existential predicament in the universe would have changed. Rather, people would have merely interpreted the Second Law the way Kelvin did, as subject to the will of God, by whom the laws of nature were created. It was therefore the secularization of Western intellectual culture during the nineteenth century that enabled a new hermeneutics, a novel Gestalt, according to which the entropy death of our solar system and/or the entire universe posed an *actual* threat to our long-term survival. Yet the same secularization trends that revealed the Second Law as a genuine kill mechanism also stimulated a series of radical reassessments of the threat environment surrounding us. If our extinction is both possible in principle and could—and will—actually happen, then what other dangers might emerge to our horror through the mist of human ignorance as the march of scientific "progress" and the development of powerful new technologies continues apace? The result of these transformations was a new existential mood marked by a sense of cosmic doom and existential vulnerability. There is no guarantee that we won't encounter a near-term threat to our existence, but there *is* a guarantee that in the distant future everything that humanity has built and worked for will ultimately come crashing down in an increasingly dilapidated cosmos.

Notes

1. See Kuhn 1955.
2. Van Wylen and Sonntag 1985.
3. Cropper 1987.
4. Kragh 2016; Lavenda and Lavenda 2010.
5. Thomson 1853.
6. Thomson 1852/1857; Kragh 2016.
7. Clausius 1865.
8. Quoted in Kragh 2016.
9. Thomson 1862b.
10. Quoted in Kragh 2016.
11. Quoted in Müller and Weiss 2012.
12. Shields 1889. Note that this passage is summarizing a view expressed by Alexander Winchell in his 1870 book *Sketches of Creation*.
13. Quoted in Müller 2007, 72.
14. Maudsley 1884.
15. "The Free Man's Worship" was the original 1903 publication title. The essay was republished several times, including as a short book titled *A Free Man's Worship* in 1923. The latter is the more common title, which I will reference henceforth.
16. Russell 1903.
17. Flammarion 1894, see 279–285.
18. Wells 1895.
19. Kragh 2016.
20. Quoted in Kragh 2016; see Brake 2016.
21. Eddington 1927.
22. Quoted in Kragh 2016.
23. Kelvin 1862.
24. Jeans 1929.
25. Paraphrased from Benatar 2006, 194.

26. Thomson 1862a.
27. Quoted in Kragh 2016. I am indebted to Helge Kragh's 2016 book on the topic for some of these references.
28. Or as the pragmatist philosopher William James wrote, referencing the possibility of our world eventually freezing over, "where [God] is, tragedy is only provisional and partial, and shipwreck and dissolution not the absolutely final things" (James 1904).
29. The reason, he clarified, is that proving through "logical demonstration" that God does not exist is difficult or impossible (Russell 1947).
30. See Wells 1917.
31. Hyman 2010. One detail that I will not say much—or, to be honest, enough—about concerns the rise of deism during the Enlightenment, especially in France. As suggested in Chapter 1, deists tended to reject all forms of revelation in favor of a "natural religion" based on the use of reason, or innate knowledge accessible to everyone. Hence, many deists did not accept the incarnation of Christ or his resurrection (through which humanity was atoned for its sins), nor did they subscribe to the eschatological narratives outlined in the prophetic verses of holy scripture. In other words, while atheism rejects both the ontological and eschatological theses of Chapter 1, deism—which became influential a century before atheism spread through the intelligentsia—rejects the eschatological but not (necessarily) the ontological thesis. On this account, then, the first component of the three beliefs to undergo appreciable decline was the eschatological thesis. This was followed by the collapse of the Great Chain between roughly 1800 and 1830, after which the rise of atheism throughout that century severely wounded both the ontological thesis and (what was left of the) eschatological thesis.
32. Hyman 2010, 19–20.
33. For details on how Robert Chambers's *Vestiges of the Natural History of Creation* (1844), which I will not discuss in this book due to space limitations, introduced the idea of evolution to a large audience fifteen years prior to Darwin's publication, see Bowler 2003, 134–140.
34. Bowler 2003, 141, 274.
35. Darwin 1859.
36. Note that Wallace disagreed with this claim. Thanks to Adrian Currie for pointing this out to me.
37. Dawkins 1986.
38. Quoted in Moberly 2010, italics in original.
39. Hyman 2010.
40. Bowler 2003.
41. Reardon 1966.
42. I am here putting the problem in somewhat contemporary terms; see Mackie 1955.
43. Russell 1903.
44. Quoted in Curtis 1887.
45. Marx 1843/1977; Nietzsche 1882/2006.
46. The publication date for this comes from Burd 2001.
47. Wells 1983.
48. Note that Darwin himself emphasized competition between *individuals* within a population, not between different *species*.
49. Wells 1983.
50. Darwin 1859. Or, elsewhere: "for the manner in which all organic beings are grouped, shows that the greater number of species of each genus, and all the species of many genera, have left no descendants, but have become utterly extinct" (Darwin 1859).
51. Darwin 1887/2002.

52. See Ruse 1996 for a useful discussion of progressionism in evolutionary biology. Note that in the final—sixth—edition of *Origin*, Darwin attempted to make the non-teleological aspect of his theory more explicit, writing that "natural selection, or the survival of the fittest, does not necessarily include progressive development—it only takes advantage of such variations as arise and are beneficial to each creature under its complex relations of life" (1872).
53. Lankester 1880.
54. Barnett 2006.
55. Bowler 2003, 17.
56. Wells 1895.
57. See Huxley 1951, 1957; Harrison and Wolyniak 2015; Levin 2020.
58. Huxley 1927. In his paper "A History of Transhumanist Thought," Nick Bostrom identifies Huxley as having coined this term in 1927, but this is inaccurate. For details, see Harrison and Wolyniak 2015.
59. Bernal 1929.
60. See also Wells' "The Man of the Year Million" *The War of the Worlds*, and *The First Men in the Moon* (Eisenstein 1976).
61. See Gould 1965; Chapter 5 for discussion.
62. Flammarion 1894.
63. Wagar 1982.
64. Weart 1988.
65. That is, according to Spencer Weart.
66. Verne 1863/2008.
67. Butler 1863.
68. Weart 1988.
69. Weart 1988.
70. Kramers and Holst 1923.
71. Weart 1988.
72. Quoted in WS 1999.
73. Weart 1988, 23.
74. MAW 1919, 5.
75. Freud 1930/2004.
76. Wagar 1982; Weart 1988.
77. Churchill 1924/25.
78. Haldane 1924.
79. Russell 1924/2015.

4 The Invention of Omnicide

Traumatic Transformations

We saw in the previous chapter how the new existential hermeneutics that emerged with the secularization of Western culture throughout the nineteenth and early twentieth centuries provoked a reassessment of the potential dangers associated with natural phenomena and human activities, especially after World War I. The result was a widespread foreboding that, as the existentialist philosopher Martin Buber described it in 1949, "we were living in the initial phases of the greatest crisis humanity has ever known," one in which "what is in question . . . is nothing less than man's whole existence in the world." He continued:

> During the ages of his earthly journey man has multiplied what he likes to call his "power over Nature" in increasingly rapid tempo, and he has borne what he likes to call the "creations of his spirit" from triumph to triumph. But at the same time he has felt more and more profoundly, as one crisis succeeded another, how fragile all his glories are; and in moments of clairvoyance he has come to realize that in spite of everything he likes to call "progress" he is not travelling along the high-road at all, but is picking his precarious way along a narrow ledge between two abysses.[1]

Despite the unease that gripped many at the time—a creeping suspicion that the growing power of science and technology were nudging humanity toward one of the abysses referenced by Buber—the only widely accepted kill mechanism throughout the period was the Second Law, which threatens to drown humanity in a frozen pond of thermodynamic equilibrium many millions of years from the present. In other words, while certain technoscientific *trendlines* appeared ominous, the existential *headlines* remained heartening, at least in one important respect: there was no scientifically credible reason to believe that near-term human extinction was actually possible. We may be vulnerable—a central insight of the new hermeneutics—but we are not in any immediate danger of dying out.

The situation changed dramatically in the mid-twentieth century, between the end of World War II and the late 1950s, resulting in a qualitatively new existential

DOI: 10.4324/9781003246251-5

mood that descended upon the Western world—if not the world more generally—with a crushing thump. This momentous shift was triggered by the development of nuclear weapons, which fundamentally transformed our understanding of two important properties of the threat environment, namely, the *etiology* and *temporality* of risks to our collective survival. The first pertains to the fact that thermonuclear weapons introduced the first scientifically credible anthropogenic kill mechanism in human history. While the bombings of Hiroshima and Nagasaki *initiated* the shift in existential moods, it was the 1954 Castle Bravo debacle (paired with novel insights about the deleterious health effects of radioactivity) that *solidified* this mood by convincing many leading experts that even a relatively small-scale thermonuclear exchange could blanket Earth's surface with lethal quantities of ionizing radiation, thus bringing about a sudden end to the human story.[2] The second property concerns the realization, directly connected to the phenomenon above, that our collective demise could now occur on timescales relevant and meaningful to those living at the time. Whereas the entropic death of humanity is a distant inevitability, something that has no chance of harming one's children or grandchildren, a thermonuclear conflict could precipitate our extinction in the near term, perhaps even *tomorrow*. Together, these point to the defining feature of this mood: a widespread sense of *impending self-annihilation*, where the first term ("impending") corresponds to temporality and the second ("self-annihilation") to etiology.

But this period also differed from the previous one in another significant way: whereas the second mood of cosmic doom and vulnerability was catalyzed by the discovery of a single kill mechanism, this one witnessed a veritable explosion of scientifically plausible catastrophe scenarios that were also (a) anthropogenic and (b) threatening in the near term. The most prominent were associated with environmental degradation, which scientists beginning in the early 1960s linked to phenomena like synthetic chemicals, overpopulation, the burning of fossil fuels, and ozone depletion. These served to strongly *reinforce* the newly established mood initiated by thermonuclear weapons. Some reputable scientists at the time also began sounding the alarm about additional potential threats, such as biological warfare and future developments in artificial intelligence (AI) and atomically precise (molecular) nanotechnology, although they were not taken seriously by a significant number of prominent scholars until the 2000s. The result of these developments was that over just a few short decades, from the 1950s to the 1980s, the doomsday menu of credible anthropogenic threats expanded from *zero* to *three or four*, depending on one's counting criteria, with a small but menacing swarm of anticipated future hazards buzzing on the temporal horizon. Put differently, the threat environment underwent an additional transformation with respect to the property of *multiplicity*: suddenly, there was not just one means of self-extermination but many.

Once again, this shift in existential mood was crucially enabled by the background condition of secularization. By the end of the nineteenth century, the notion that human extinction is a real possibility was accepted by many leading intellectuals, but the contagion of disbelief had not significantly infected the general public. It was during the 1960s that Western culture as a whole underwent a rapid decline in religiosity, inaugurating what some have called the "Age of Atheism."[3] Why this occurred when it did is one of the main explananda of "secularization theory," a topic that goes beyond the scope of this book.[4] Whatever *caused* this cultural metamorphosis, the *effect* was to make possible the new epiphanies about etiology, temporality, and multiplicity that induced a radical step-change in our understanding of the existential predicament of humanity. Indeed, we will see how the lingering influence of a religious existential hermeneutics among a segment of the population may have nontrivially increased the probability of catastrophe, not just during the Cold War but continuing up to the present, given the persistent effects of past environmental policies.

Finally, before turning to the substance of this chapter, it might be worth making explicit that, as noted in Chapter 1, this new existential mood didn't supplant the previous one but expanded it. The scientific conviction that cosmic doom awaits humanity, or whatever we evolve into, in the far future remained as solid as ever. However, it was greatly eclipsed on the landscape of our collective *attention* by the flurry of near-term risks that emerged from the 1950s onward.

An Explosive Discovery

Let's begin with a brief account of how nuclear weapons were developed. Recall from the previous chapter that Soddy and Rutherford discovered that radioactive decay involves the transmutation of one type of atom into another, which led to worries about a "planetary chain reaction" of infectious decay that converts the chemical mosaic of Earth's elements into helium. In the process, a huge amount of energy would be liberated, according to the equation $E = mc^2$, which led some to speculate about the causes of nova observed in the firmament. Later, in 1934, the Joliot-Curies figured out how to convert certain stable atoms into radioactive atoms, by which time the notion of "atomic energy" had inspired a profusion of utopian and dystopian proclamations about its potential to usher in a post-scarcity world or tear the planet asunder. A notable example that combined both themes was Wells' 1914 novel *The World Set Free*, which was written the previous year and dedicated to Soddy's work on radium. This book describes a catastrophic world war (initiated by Germany in the 1950s, as it happens) that ultimately leads to the creation of a harmonious world state. What is most relevant for our purposes is that the war involves what Wells called, coining the term, "atomic bombs" that pilots fling from their cockpits on urban centers below, destroying entire cities. However, these are not like the "atomic bombs" dropped on Hiroshima and Nagasaki in 1945; rather than producing a sudden massive explosion, they utilize a fictional radioactive element called "Carolinum" to generate "a blazing continual explosion" that "is never entirely exhausted," and which would create "puffs of luminous, radio-active vapour drifting sometimes scores of miles from the bomb centre and killing and scorching all they overtook."[5]

Although this was science fiction, the idea greatly influenced one of the pioneers of nuclear weapons: a young Hungarian physicist named Leó Szilárd, who read *The World Set Free* in 1932 and included Wells within his circle of acquaintances.[6] As the now-famous story goes, Szilárd read an article in *The Times* the following year that quoted Rutherford as saying that "anyone who looked for a source of power in the transformation of the atoms was talking moonshine." The reason is that, as another newspaper article on Rutherford's talk explained, "walls of electric energy surround the nucleus. To break down wall after wall and eventually reach the holy of holies [i.e., the nucleus] in which almost incredible energy is concentrated, the physicist must lay siege to the atom. So he tries to batter it and blast it apart" by shooting alpha particles at the nuclei. The problem is that only "one particle in 10,000,000 strikes the nucleus," meaning that the process is extremely inefficient (quoting here a *New York Times* article published the same day on Rutherford's comments).[7]

Finding himself "irritated" by Rutherford's confidence—one is here reminded of Clarke's First Law[8]—Szilárd went for a walk and, standing at a street corner in London, devised a method for unlocking the vast stores of energy trapped in atomic nuclei: a *nuclear chain reaction*. Whereas earlier experiments had involved alpha particles, which consist of two protons and neutrons (the latter of which were first discovered in 1932), Szilárd instead imagined bombarding atoms with free neutrons, which unlike alpha particles have a neutral rather than positive charge. This would enable them to easily trespass the aforementioned "walls of electric energy," thus striking a greater number of nuclei. Furthermore, Szilárd reasoned that if an atom struck by a free neutron were to subsequently release two additional neutrons, the process—the chain reaction—could become exponential and *self-sustaining*. Over just a few millionths of a second, billions of atoms could be struck by and release neutrons, thereby liberating enormous quantities of energy *at once* rather than (as with natural radioactive decay) over protracted stretches of time. Szilárd quickly realized that, as he later wrote, "in certain circumstances it might become possible to set up a nuclear chain reaction, liberate energy on an industrial scale, and construct atomic bombs."[9]

This was the abstract idea, but could it work? Are there elements whose atoms release two neutrons when struck by one? If so, which elements? An important step toward answering these questions came in 1938 with the discovery of *nuclear fission* in uranium atoms by a team of scientists in Berlin, the capital of Nazi Germany, which found that irradiating uranium with neutrons causes the atoms to split into fragments. Upon hearing about this the following year, Szilárd, in his words, "saw immediately that these fragments . . . must emit neutrons, and if enough neutrons are emitted in this fission process, then it should be, of course, possible to sustain a chain reaction. All the things which H. G. Wells predicted," he continued, "appeared suddenly real to me."[10] Now the crucial question became, "Is this actually the case? Does uranium fission produce neutrons and, if so, how many?" To answer the first question—to confirm his suspicions—Szilárd conducted an experiment with his colleague Walter Zinn in March of 1939. It involved using a cathode-ray oscillograph to track the movements and kinetic energy of neutrons that might be released by uranium atoms when split by slow neutrons striking them. Flashes appearing on the oscillograph's display screen would indicate that uranium *does* indeed produce neutrons, which "in turn would mean that the large-scale liberation of atomic energy was just around the corner."[11] After initiating the experiment, Szilárd and Zinn were relieved that *no flashes* appeared, although they soon realized that the screen had been unplugged.[12] Once the screen was powered on, the two scientists "turned the switch and saw the flashes," Szilárd later recalled. "We watched them for a little while and then we switched everything off and went home," he says. "That night, there was very little doubt in my mind that the world was headed for grief."[13]

Having spent much of the 1930s anxious that atomic energy—more accurately called *nuclear energy*—could be weaponized to produce "atomic bombs," Szilárd scheduled a meeting with Einstein in a Peconic, Long Island, cottage where Einstein was staying. Szilárd explained how nuclear energy could be unlocked and turned into a bomb, to which Einstein reportedly said, "I haven't thought of that at all."[14] Worried that the world was on the brink of another war and that Nazi Germany might develop an atomic bomb, Szilárd penned a letter—now called the "Einstein-Szilard letter"—intended for US President Franklin Roosevelt to alert him of the danger. He noted that "some of the American work on uranium is now being repeated" at a Berlin-based university with connections to the German Under-Secretary of State, and that Nazi "Germany has actually stopped the sale of uranium from . . . the German Under-Secretary of State," and that "Germany has actually stopped the sale of uranium from the Czechoslovakian mines which she has taken over."[15] This letter, whose only signatory was Einstein, spurred the creation of the Manhattan Project, described by some as the first "Big Science" project in history, which aimed to design, build, and test the first atomic bombs.[16] It cost $2 billion USD ($23 billion in 2018 dollars) and involved more than 130,000 scientists, although only a handful were aware of the project's details and ultimate goals. The research arm of the endeavor, based in the top-secret Los Alamos Laboratory near Santa Fe, New Mexico, was run by the physicist, child prodigy, and chainsmoker (an incredible four to five packs per day) Robert Oppenheimer, known today as the "Father of the Atomic Bomb."

The first atomic bomb, nicknamed the "Gadget," was detonated at 5:29 in the morning on July 16, 1945, in the desert of Jornada del Muerto, sometimes translated as "Journey of the Dead Man," in New Mexico. This was the Trinity test, which created a burst of smoke and fire that rapidly rose 40,000 feet into the early morning sky. Less than a month later, on the 6th and 9th of August, the United States dropped two atomic bombs—Little Boy, a uranium bomb, and Fat Man, a plutonium bomb—on the Japanese archipelago, killing more than 100,000 people and helping, some argue, to bring World War II to an end.

A New Mood Emerges (1945–1954)

News reports of the catastrophic effects of Little Boy and Fat Man presented the public with horrifying scenes of mass death and destruction, sometimes using explicitly apocalyptic language to convey the unprecedented magnitude of the bomb's explosive power. For example, a newsreel shown in movie theaters throughout the United States described Hiroshima as having been "pulverized" and nearly "wiped off the earth" by bombs that unleashed "hellfire . . . violence described by eyewitnesses as Doomsday itself!"[17] H. V. Kaltenborn declared on NBC, in one of the first public statements about the Hiroshima bombing, that "Anglo-Saxon science has developed a new explosive 2,000 times as destructive as any known before. . . . For all we know, we have created a Frankenstein!"[18] The same day—August 6—President Harry Truman said in a televised address that the atomic bomb "is a harnessing of the basic power of the universe," and that "with this bomb we have now added a new and revolutionary increase in destruction."[19] Two days later, *Delphos Daily Herald* relayed reports from Tokyo that "practically all living things, human and animal," had been "seared to death," adding that "only a few skeletons of concrete buildings still remained [while] both the dead and wounded had been burned beyond recognition."[20] The *Freeport Journal-Standard* described Nagasaki in an August 10 article as having been "smashed" in an "inferno of smoke and flame that swirled more than 10 miles into the stratosphere and could be seen for 250 miles."[21] Shortly afterwards, major outlets like the *New York Times* and BBC began publishing images of the aftermath: whole city blocks razed to the ground, the twisted steel frames of former buildings mutilated and mangled amidst "flattened acres of debris," as one caption put it.[22] On August 20, *Life* magazine printed the first images, taking up entire pages, of ginormous mushroom clouds rising over both cities, describing Hiroshima has having been "blown . . . of the face of the earth" and Nagasaki as being "disemboweled."[23] The first Western journalist to enter (surreptitiously) Hiroshima, Wilfred Burchett, reported on September 5 that "Hiroshima does not look like a bombed city. It looks as if a monster steamroller had passed over it and squashed it out of existence." At the time, little was publicly known about the radiological aspects of atomic explosions, information about which US officials would work vigorously over the next few years to suppress through campaigns of censorship and disinformation. Hence, believing that the ground, soaked with radioactivity, was releasing a poison gas of some sort, Burchett described the weary survivors as suffering from what he called "atomic plague."[24]

The month before, many witnesses of the Trinity test found themselves staggered by the destructive forces their scientific research had unleashed. Oppenheimer described the mood as "extremely solemn," adding that "a few people laughed, a few people cried. Most were silent."[25] He himself claims to have recited a haunting passage from the *Bhagavad Gita*, a sacred Hindu scripture, which reads: "Now I am become Death, the destroyer of worlds," although his brother Frank, who also worked on the Manhattan Project, reports that what he and Robert likely said was simply, and eerily: "It worked."[26] One finds a similar sense of trepidation among some of the military officers who watched the explosion. For example, an August 7, 1945, article in the *New York Times* quotes Brigadier General Thomas Farrell as describing "a searing light with the intensity many times that of the midday sun. It was golden, purple, violet, gray, and blue. It lighted every peak, crevasse, and ridge of the near-by mountain range with a clarity and beauty that cannot be describe but must be seen to be imagined." He continued:

> Thirty seconds after the explosion came first the air blast pressing hard against the people and things, to be followed almost immediately by the strong, sustained, awesome roar which warned of doomsday and made us feel that we puny things were blasphemous to tamper with the forces heretofore reserved to the Almighty.[27]

Although the story may be apocryphal, Einstein is said to have muttered "I could burn my fingers that I wrote that letter to Roosevelt" after hearing of the casualties in Japan.[28] What we do know is that both he and Szilárd tried frantically to convince the US government to halt or at least slow down the Manhattan Project after Germany surrendered in early May, 1945; as mentioned, their explicit aim in convincing Roosevelt to fund research on atomic bombs was to beat the Germans to the finish line. With the Nazis defeated, they worried that continued work on the project would lead the United States to use the bomb anyway, which is of course precisely what happened. When Einstein was asked by a newspaper reporter on August 6 about the day's momentous news, he is quoted as saying, "Ach! The world is not ready for it."[29] The following month, a group of scientists founded the Atomic Scientists of Chicago, which began publishing the *Bulletin of the Atomic Scientists* in December of that year, a periodical aimed (in part) at educating "the public to a full understanding of the scientific, technological, and social problems arising from the release of nuclear energy."[30] In 1947, the *Bulletin* created the iconic "Doomsday Clock," which metaphorizes our collective proximity to "destroying our world" and was intended, in Eugene Rabinowitch's words, "to preserve civilization by scaring men into rationality."[31] The minute hand was initially set at 7 minutes before midnight, or doom, but was moved forward to 3 minutes before midnight in 1949 after the Soviet Union conducted its first nuclear test in August of that year.

The dire implications of the atomic bomb were thus recognized by many people around the world almost immediately. As one chapter was titled in the book *The Atomic Age Opens*, published a little more than one week after the Nagasaki bombing (consisting of news articles, politicians' statements, and editorials during that period), declared, "The Whole World Gasped."[32] This was one of the first uses of "Atomic Age," an unsettling new entry in the English lexicon, although the term is often attributed to William Laurence, who was the only journalist allowed to witness the Trinity test. As Laurence wrote in a September 26, 1945, article for the *New York Times*, the Atomic Age began in the early morning of July 16, 1945, and marks a pivotal moment in human history, comparable to "the moment in the long ago when man first put fire to work for him," later describing the explosion as "terrifying," "crushing," "ominous," "devastating," and "full of . . . great forebodings."[33] Meanwhile, the *New York Herald Tribune* wrote that "one senses the foundation of one's own universe trembling. . . . It is as though we had put our hands upon the levers of a power too strange, too terrible, too unpredictable in all of its sudden consequences."[34] The *New Republic* described a "curious new sense of insecurity, rather incongruous in the face of a military victory," an idea echoed by a Rockefeller Foundation official who characterized the country's mood after the war as gloomier than before December 1941.[35] Still others were gripped by "paralyzing fear," the bomb having "cast a spell of dark foreboding over the spirit of humanity."[36] In a 1946 article titled "Consequences of Atomic Energy," Robert Redfield wrote that "everywhere you go, this greatest of all events in the history of human technology and science has become a nightmare in the minds of men."[37] Among the more poignant descriptions of the times came from Norman Cousins' article "Modern Man Is Obsolete," which opens:

> Whatever elation there is in the world today because of final victory in the war is severely tempered by fear. It is a primitive fear, the fear of the unknown, the fear of forces man can neither channel nor comprehend. This fear is not new; in its classical form it is the fear of irrational death. But overnight it has become intensified, magnified. It has burst out of the subconscious and into the conscious, filling the mind with primordial apprehensions. It is thus that man stumbles fitfully into a new age of atomic energy for which he is as ill-equipped to accept its potential blessings as he is to counteract or control its present dangers.[38]

Ants and Spears

Hence, it was in the flickering shadows of Hiroshima and Nagasaki that a new existential mood was born, one marked by subdued panic and existential disquietude centered around the radical expansion of our ability to obliterate an entire metropolis with a single explosive. "The world would not be the same," as Oppenheimer later stated. Or to quote the German philosopher and poet Günther Anders, writing in his 1962 article "Theses for the Atomic Age," "with 6 August 1945, the Day of Hiroshima, a New Age began: the age in which at any given moment we have the power to transform any given place, on our planet, and even our planet itself, into a Hiroshima."[39] Anders had earlier suggested a new calendar organized around this date, thus arguing in 1958 that "we live in the Year 13 of the Calamity. I was born in the Year 43 before. Father, who I buried in 1938, died in the Year 7 before" (see Chapter 9 for further discussion).[40]

Yet there were hardly any explicit references at the time to human extinction. The focus instead tended to center around the possibility of civilizational destruction in another global conflict. Exceptions can be found, of course, as when an article in the *St. Louis Post-Dispatch* declared that "either the world's people—our own included—will learn to use it not for war but for peace, or else science has signed the mammalian world's death warrant and deeded an earth in ruins to the ants." But most such assertions are ambiguous in their meaning. For example, three days after Nagasaki was obliterated, Edward R. Murrow told his radio audience that "seldom, if ever, has a war ended leaving the victors with such a sense of uncertainty and fear, with such a realization that the future is obscure and that survival is not assured." But *whose* survival is in question here? The United States' or humanity's? Or consider Bertrand Russell's first public comments about the atomic bomb on August 18. "The prospect for the human race is sombre beyond all precedent," he wrote. "Mankind are faced with a clear-cut alternative: either we shall all perish, or we shall have to acquire some slight degree of common sense." Yet Russell added that if the next war involves atomic bombs, we can expect that "all large cities . . . will be completely wiped out . . . Communications will be disrupted, and the world will be reduced to a number of small independent agricultural communities living on local produce, as they did in the Dark Ages."[41] In other words, despite his initial remarks, the outcome he foresees is the destruction of civilization rather than total human extinction.

Among the more memorable expressions of this civilizational rather than extinctional focus comes from an anonymous Army lieutenant in a September 25, 1946, issue of *The Zanesville Signal*, a local Ohio newspaper. A journalist named Joseph Laitin "reports that reporters at Bikini," a coral reef in the Marshall Islands where atomic bombs were being tested at the time (called Operation Crossroads), asked the lieutenant "about what weapons would be used in the next war." He replied, "I dunno . . . but in the war after the next war, sure as hell, they'll be using spears!," which of course conveys the idea that nuclear conflict would not be fatal to the species, though it would catapult us back to the sticks and stones of the Paleolithic.[42] This quote was apparently later repeated by Einstein, to whom it is now commonly (mis)attributed.[43] To mention just one more example, consider a 1980 lecture from the Nobel laureate Richard Feynman, who worked on the Manhattan Project as a young physicist. In it, he recalls that after returning to the US Northeast from Los Alamos, he would find himself wondering what the point of building anything is when the atomic bomb could so easily destroy it. In his words:

> I'd sat in a restaurant in New York, for example, and I looked out at the buildings and how far away, I would think, you know, how much the radius of the Hiroshima bomb damage was and so forth. How far down there was down to 34th Street? All those buildings, all smashed . . . And I got a very strange feeling. . . . [T]hey're *crazy*, they just don't understand, they don't understand. Why are they making new things, it's so useless?[44]

But nowhere does Feynman indicate that he or his colleagues feared that the new atom-splitting weapons had introduced a kill mechanism that, as such, could completely exterminate the human species. There was, in the years following WWII, almost no explicit talk of what Russell would later, in 1954, call "universal death," i.e., total annihilation. One does find anticipations that the *next generation* of nuclear weapons could potentially do this, but this leads us to the next crucial development in this story, whereby the existential mood initiated by the Hiroshima and Nagasaki bombings becomes solidified.

Is Mankind Exterminable?

The solidification of this mood was catalyzed by one event in particular: the March 1, 1954, Castle Bravo test on Bikini Atoll in the Marshall Islands. This involved a thermonuclear ("hydrogen") rather than atomic weapon. Thermonuclear weapons use the fission of heavy elements—such as uranium and plutonium—to cause the fusion of lighter elements—such as hydrogen isotopes (deuterium and tritium) and lithium deuteride (lithium-7 plus deuterium)—and can produce explosions 1,000 times more powerful than atomic bombs. The first thermonuclear detonation, codenamed Ivy Mike, occurred in 1952 and produced an explosive yield of 10,400 kilotons, more than 500 times the yield of the Trinity test. The Castle Bravo test was supposed to produce a yield of 6,000 kilotons, but an unexpected reaction with lithium-7 caused the explosion to be 2.5 times larger. Within a few seconds, the fireball ballooned to be over 3 miles wide, and "for a moment it seemed to cling to the earth, but then it sprung into the sky," carrying some "ten million tons of pulverized coral debris . . . coated with radioactive fission products."[45] Prior calculations suggested that the radioactive debris resulting from the explosion would be catapulted into the stratosphere, where they would be trapped by the tropopause (the boundary between the troposphere and the stratosphere). This would prevent them from immediately falling back to Earth, thereby contaminating Earth's surface with high concentrations of dangerous particles. Rather, they would be dispersed around the globe, undergoing normal radioactive decay such that by the time much of the debris had returned to the surface, the radiological hazard would be small.

But this "stratospheric trapping" phenomenon did not occur: the relatively large particles of debris quickly fell from the stratosphere, thus raining dangerous amounts of radioactivity over a much larger region than scientists thought was possible.[46] Consequently, residents of the Rongelap and Rongerik atolls had to be evacuated, and a Japanese fishing vessel named the Lucky Dragon was covered in odorless, tasteless white flakes of radioactive coral, described by one crew member as "just like sleet," which ultimately blanketed some 7,000 square miles of the ocean.[47] By the evening, those onboard the Lucky Dragon began showing signs of sickness consistent with the symptomatology of acute radiation syndrome, and upon returning to Japan, they were found to be highly radioactive.[48] Over the next few weeks, "traces of radioactive fallout were found on the Japanese mainland, in Australia, India, parts of Europe and even the United States," and later that year one of the crew members died—the first victim of the hydrogen bomb, according to the Japanese.[49]

It quickly dawned on people that the most dangerous feature of thermonuclear weapons is not their immediate effects—the blast, shock wave, heat, fires, and so on—but the subsequent fallout, which could affect areas far from the detonation site. As an "Instructor's Guide" published in 1955 by the United States Civil Defense titled *Introduction to Radioactive Fallout* states,

> before the facts of the 1954 H-bomb explosion were announced, fallout was of little concern to us. If you lived a few miles from a possible target, you could assume that you were safe from the effects of enemy bombing. . . . That is no longer true. The 1954 tests in the Pacific showed that deadly fallout could be carried nearly 200 miles by the winds.[50]

This implied that even a relatively small-scale thermonuclear conflict could potentially cover every inhabited region of the planet with dangerous levels of radioactivity, thereby threatening the very existence of humanity. Indeed, one finds a *marked shift* in how people—scientists, philosophers, political theorists, politicians, and so on—began to describe the threats posed by nuclear weapons. As we have seen, before the Castle Bravo debacle, the primary focus was the possible destruction of civilization enabled by the amplified violence capacities of states; this was essentially an extension of the technoscientific worries expressed by Churchill (1924) and Freud (1930) in the previous chapter. Atomic bombs had simply given state actors a bigger hammer with which to smash each other. Almost immediately after the Castle Bravo debacle, the rhetoric came to emphasize the *prospect of complete self-annihilation* if a thermonuclear war were to break out. For example, in his book *Human Society in Ethics and Politics* (1954), Russell warned of "universal destruction" if present policies of interstate competition continue, and in the final chapter evocatively titled "Prologue or Epilogue?" he argued that "the future of man is at stake." Drawing from this and other writings, he penned a short but powerful radio address for the BBC titled "Man's Peril," which he delivered in December of 1954. In it, he pleaded with his listeners—an audience of 6 to 7 million people—to recognize that in a thermonuclear conflict, entire cities like London, New York, and Moscow could be utterly decimated by single bombs. But, referencing Castle Bravo,

> we now know, especially since the Bikini test, that hydrogen bombs can gradually spread destruction over a much wider area than had been supposed. It is stated on very good authority that a bomb can now be manufactured which will be 25,000 times as powerful as that which destroyed Hiroshima. Such a bomb, if exploded near the ground or under water, sends radio-active particles into the upper air. They sink gradually and reach the surface of the earth in

> the form of a deadly dust or rain. It was this dust which infected the Japanese fishermen and their catch of fish although they were outside what American experts believed to be the danger zone. No one knows how widely such lethal radio-active particles might be diffused, but the best authorities are unanimous in saying that a war with H-bombs is quite likely to put an end to the human race. It is feared that if many H-bombs are used there will be universal death—sudden only for a fortunate minority, but for the majority a slow torture of disease and disintegration.[51]

He proceeds to quote several "eminent men of science," such as Sir John Slessor, who said that "a world war in this day and age would be general suicide"; Lord Edgar Adrian, who warned that such "a fight . . . might end the human race"; and Sir Philip Joubert, who declared that "with the advent of the hydrogen bomb, it would appear that the human race has arrived at a point where it must abandon war as a continuation of policy or accept the possibility of total destruction." Russell then states that while no one will claim that "the worst results are certain,"

> what they do say is that these results are possible and no one can be sure that they will not be realized. I have not found that the views of experts on this question depend in any degree upon their politics or prejudices. They depend only, so far as my researches have revealed, upon the extent of the particular expert's knowledge. I have found that the men who know most are most gloomy. . . . Here, then, is the problem which I present to you, stark and dreadful and inescapable: Shall we put an end to the human race; or shall mankind renounce war? . . . Is our race so destitute of wisdom, so incapable of impartial love, so blind even to the simplest dictates of self-preservation, that the last proof of its silly cleverness is to be the extermination of all life on our planet?—for it will be not only men who will perish, but also the animals, whom no one can accuse of Communism or anti-Communism. I cannot believe that this is to be the end.[52]

The presentation was an incredible success and the text was widely reprinted. As Russell wrote to his cousin, it "brought an avalanche of letters, mostly sympathetic," including some from top scientists like Max Born and Frédéric Joliot-Curie. Born wrote to express interest in producing a statement co-signed by Nobel laureates warning about the profound dangers posed by thermonuclear weapons, while Joliot-Curie proposed a conference of leading scientists. The first led to what is now called the "Russell-Einstein Manifesto" and the second to the Pugwash Conferences, which Russell cofounded in 1957 with Joseph Rotblat, the only scientist to leave the Manhattan Project on moral grounds.[53]

Signed by 11 of the most prominent scientists and intellectuals at the time, including Born, Joliot-Curie, Rotblat, and Einstein (just days before his death), the Russell-Einstein Manifesto gained international attention. Presented on July

9, 1955, it largely recapitulated points made in "Man's Peril," sometimes *verbatim*. It begins with an appeal for people to consider their common humanity. "We are speaking on this occasion," it states,

> not as members of this or that nation, continent, or creed, but as human beings, members of the species Man, whose continued existence is in doubt . . . we want you, if you can, to set aside such feelings and consider yourselves only as members of a biological species which has had a remarkable history, and whose disappearance none of us can desire.

It proceeds to mention the Castle Bravo debacle, stating that

> the best authorities are unanimous in saying that a war with H-bombs might possibly put an end to the human race. It is feared that if many H-bombs are used there will be universal death, sudden only for a minority, but for the majority a slow torture of disease and disintegration. . . . We have not yet found that the views of experts on this question depend in any degree upon their politics or prejudices. They depend only, so far as our researches have revealed, upon the extent of the particular expert's knowledge. We have found that the men who know most are the most gloomy.[54]

Six days later, another consensus statement was released that included signatures from 18 Nobel laureates (a total of 34 within the next year): the Mainau Declaration, which took shape in Germany and shared many similarities with the Russell-Einstein Manifesto. Also signed by Born, who was a driving force behind its composition, it states:

> With pleasure we have devoted our lives to the service of science. It is, we believe, a path to a happier life for people. We see with horror that this very science is giving mankind the means to destroy itself. By total military use of weapons feasible today, the earth can be contaminated with radioactivity to such an extent that whole peoples can be annihilated. Neutrals may die thus as well as belligerents. . . . If war broke out among the great powers, who could guarantee that it would not develop into a deadly conflict? A nation that engages in a total war thus signals its own destruction and imperils the whole world.[55]

These statements were followed by a flurry of equally dire warnings about the possibility not merely of civilizational destruction but total human annihilation. Among the most notable voices in Europe was Anders, who argued in his 1956 paper "Reflections on the H Bomb" that "all history can be divided into three chapters, with the following captions: (1) All men are mortal, (2) All men are exterminable, and (3) Mankind as a whole is exterminable." Whereas the Holocaust triggered the shift from (1) to (2), according to Anders, the advent of thermonuclear weapons has introduced the third, even more terrifying epoch of, in a different translation, the "killability" of humanity.[56] Two years later, Karl Jaspers worried about "the total doom of mankind" in in *The Future of Mankind*, arguing that "an altogether novel situation has been created by the . . . bomb. Either all mankind will physically perish or there will be a change in the moral-political condition of man."[57] And the famed journalist Arthur Koestler warned in his 1967 book *The Ghost in the Machine* that "the bomb has given us the power to commit genosuicide," adding that "it is as if a gang of delinquent children had been locked in a room filled with inflammable material, and provided with match-boxes—accompanied by the warning not to use them."[58]

Omnicide

The shift in thinking about the existential predicament of humanity between 1945 and the late 1950s could hardly have been more pronounced. It began with the startling realization that the Atomic Age marked a fundamentally new era in human history, and culminated with the 1954 Castle Bravo test, which triggered a torrent of panicked declarations about the possibility of what some came to call "omnicide." This word is almost universally attributed to the philosopher John Somerville (1905–1994), who worked tirelessly to abolish nuclear weapons—a project he described as "preventive eschatology"—and cofounded an organization in 1983 called the "International Philosophers for the Prevention of Nuclear Omnicide" (now the "International Philosophers for Peace and the Elimination of Nuclear and Other Threats to Global Existence," or IPPNO).[59] An obituary in the *Los Angeles Times*, for example, states that "Somerville started thinking of a word that transcended suicide, genocide, infanticide—the killing of all humans—and ended up with *omnicide*. Now . . . Somerville is given credit for inventing the word, which he says is the only true description of the end result of nuclear holocaust."[60]

On Somerville's definition, "omnicide" refers to "the annihilation of all human beings by some human beings," or "the final madness of some humans killing all humans including themselves," which he described as "the logical (and terminal) extension of the series of such nouns as suicide, infanticide, homicide, genocide."[61] This, he declared, is a "crime so enormous that it could be committed only once, the sin so unspeakable it never even had a name."[62] Incidentally, such remarks echo the origin story of the word "genocide," another twentieth-century neologism. In brief: during a live BBC broadcast in 1941, Winston Churchill addressed the Nazis' mass murder of Russians, reporting to his listeners that "the whole of Europe has been wrecked and trampled down by the mechanical weapons and barbaric fury of the Nazis," with "whole districts . . . being exterminated. Scores of thousands—literally scores of thousands—of executions in cold blood are being perpetrated by the German police troops upon the Russian patriots who defend their native soil." He then declared that "we are in the presence of a crime without a name."[63] Shortly afterward, having heard Churchill's evocative phrase, Raphael Lemkin coined the word "genocide" in his 1944 book *Axis Rule in Occupied Europe*, this being quickly incorporated into the lexicon of popular discourse and, later, into international criminal law with the 1948 Genocide Convention.[64] Similarly, "omnicide" was a crime without a name after it became clear, in the mid-1950s, that a thermonuclear conflict could destroy all human life on the planet. Hence, as Somerville wrote, "we have to invent new words to express [the] actual scope and content" of self-annihilation,

> for this crime encompasses the killing not only of all people but all forms of life on the planet; it not only annihilates all present human life but all future human possibilities, as well as all the records and remains of past human achievements.[65]

My own rummaging through the postwar archives, though, indicates that the word has an earlier origin: it was first used in a 1959 article by the theater critic Kenneth Tynan, published in *The New Yorker*.[66] Following a parallel line of reasoning as Somerville's, Tynan wrote that "we have always had the ability to commit suicide and the skill to commit homicide; after many a chiliad, we mastered the art of genocide; and we are now equipped for a new crime, as yet untitled, though a good name for it would be omnicide—the murder of everyone." This was the first instance, so far as I am aware, of the term being used to denote the killing of all people on the planet, though we should note that the word "omnicide" itself had, somewhat humorously, been used earlier by a company called Superior Chemical Products, Inc. as the name for one of its insecticides.[67]

Fallout, Cobalt, and Kubrick

The Castle Bravo debacle thus established the first credible anthropogenic kill mechanism: *global thermonuclear fallout*. Worse, it was widely recognized that, given the arms race of the Cold War, thermonuclear weapons pose a danger to humanity in the *present*, unlike the kill mechanism of the Second Law, which will not destroy us for many millions of years. As President John F. Kenney declared in a 1961 address to the UN General Assembly,

> today, every inhabitant of this planet must contemplate the day when this planet may no longer be habitable. Every man, woman, and child lives under a nuclear sword of Damocles, hanging by the slenderest of threads, capable of being cut at any moment by accident or miscalculation or by madness. The weapons of war must be abolished before they abolish us.[68]

Yet global thermonuclear fallout was not the only kill mechanism associated with nuclear weapons that scientists proposed at the time. Aside from fallout, the two most credible mechanisms prior to the early 1980s were proposed by Edward

Teller, a Manhattan Project physicist known as the "Father of the Hydrogen Bomb," and Szilárd. In 1942, Teller wondered whether the first atomic explosion could trigger a "self-propagating chain of nuclear reactions" in the atmosphere that would annihilate all human life on Earth, resulting in what Arthur Compton, who won a Nobel Prize in 1927 and also worked on the Manhattan Project, described in a 1959 interview as "the ultimate catastrophe."[69] Although calculations made by Hans Bethe within a few hours of Teller suggesting this possibility showed it to be improbable, Tellers' speculations nonetheless occasioned a classified report titled "LA-602: Ignition of the atmosphere with nuclear bombs" (1946), which some have described as quite possibly "the first quantitative risk assessment of human extinction."[70] Despite reassurances that the atmosphere would not ignite, the report concluded on an unsettlingly ominous note: "There remains the distinct probability that some other less simple mode of burning may maintain itself in the atmosphere [that] might become catastrophic on a world-wide scale," adding that "the complexity of the argument and the absence of satisfactory experimental foundations makes further work on the subject highly desirable." Of course, if the report's conclusions had been wrong, there would be no one around to talk about its conclusions having been wrong. Incredibly, the Manhattan Project physicist Emilio Segrè, who was later awarded a Nobel prize, wrote that, upon witnessing the Trinity test, "for a moment I thought the explosion might set fire to the atmosphere and thus finish the earth, even though I knew that this was not possible."[71]

Later, in 1950, Szilárd participated in a roundtable discussion on the radio in which he imagined a version of the hydrogen bomb that could produce extraordinary quantities of radioactivity on purpose, such that, as one of the interlocutors (a fellow scientist) summarized the proposal, "all people on Earth could be killed under the circumstances."[72] Interestingly, every historical account of the "cobalt bomb" dates the idea to comments made by Szilárd during this radio program, although Szilárd doesn't once explicitly mention or allude to cobalt. He instead adumbrates a general mechanism by which a hydrogen bomb could spread large amounts of radioactive dust around the planet over the course of months or years. Nonetheless, as an article published later that year in the *Bulletin* notes, the only two chemical elements that could instantiate this mechanism are cobalt and zinc, and since "the yield of effective gamma radiation per neutron is eight times less for zinc than for cobalt," the optimal element for the stated purpose of omnicide is cobalt.[73] So perhaps Szilárd had this in mind after all, despite his silence about the details.

Either way, while a number of scientists argued that a bomb of this sort is not practicable,[74] others backed Szilárd's speculations. Einstein, for example, is quoted in a newspaper article written by Laurence, the journalist mentioned above, as worrying that if a cobalt bomb were successfully built, then "radioactive poisoning of the atmosphere and hence annihilation of any life on Earth, will have been brought within the range of technical possibility."[75] The following decade, the Nobel laureate and anti-nuclear testing activist Linus Pauling calculated that for only "six billion dollars—one twentieth of the amount spent on armaments each year by the nations of the world—enough cobalt bombs could be built to ensure the death of every person on Earth."[76] Such claims gave rise to the notion of a "doomsday device" or, in Herman Kahn's 1960 phraseology, a "Doomsday Machine," which was catapulted into the public consciousness by Stanley Kubrick's black comedy *Dr. Strangelove or: How I Learned to Stop Worrying and Love the Bomb* (1964). The cobalt bomb also gained recognition from Nevil Shute's 1957 novel *On the Beach*, which describes a devastating all-out nuclear war waged with cobalt bombs and was later made into a movie (in both 1959 and 2000). According to public records, no state has ever built a cobalt bomb, although the Soviet did establish a system known as "Dead Hand" that would automatically launch a barrage of nuclear intercontinental ballistic missiles (ICBMs) at the United States in the event of a preemptive attack—a kind of doomsday device. There is some speculation that Russia never discontinued the program.

Ozone and Global Cooling

Many other kill mechanisms associated with nuclear weapons were also proposed, although most were scientifically unfounded. For example, the Democratic presidential candidate Estes Kefauver claimed in 1956 that hydrogen bombs could "right now blow the earth off its axis by 16 degrees," and the following year Nikita Khrushchev supposedly declared that the Soviet Union had a bomb capable of "melt[ing] the Arctic icecap and send[ing] oceans spilling all over the world."[77] A decade later, in his 1967 book, Koestler seems to have channeled earlier worries expressed in the textbook *The Atom and the Bohr Theory of Its Structure* (1923), and by Joliot-Curie in his 1935 Nobel prize speech (both discussed in the previous chapter), in asserting that the bomb has not only "given us the power to commit genosuicide" but "within a few years we should even have the power to turn our planet into a *nova*, an exploding star."[78]

More plausible concerns centered around the possibility of ozone depletion. The immense heat produced by the nuclear fireball creates nitrogen oxides (NO_x), about 10^{32} molecules per megaton of explosive yield, which can be carried into the stratosphere as the fireball rises through convection. The possibility of NO_x depleting the ozone was identified by Paul Crutzen in 1970, who found that when NO interacts with O3 (ozone), it yields NO2 (nitrogen dioxide) and O2 (the ordinary oxygen molecule that we breath); the resulting NO2 then combines with O (monoatomic oxygen created when ozone interacts with the sun's light) to yield NO and O2 such that NO ends up being recycled, with each cycle causing more ozone depletion.[79] Subsequent research raised enough alarm to galvanize the US government to fund further research on the phenomenon, which led to a 1975 book published by the National Academy of Sciences (NAS) titled *Long-Term Worldwide Effects of Multiple Nuclear-Weapons Detonations*. Startlingly, it "confirmed the potential for stunning impoverishment of ozone in the stratosphere," leading the ACDA director at the time to worry aloud that there could be any number of additional kill mechanisms that scientists have not yet identified.[80] As he made the point during a speech to the Chicago Council on Foreign Relations the same year:

> The more we know, the more we know how little we know. . . . Each of these discoveries tore a hole in the facile assumptions that screened the reality of nuclear war. Each brought a new glimpse into the cauldron of horrors. What unexpected discovery will be next?[81]

This threat was further popularized by Jonathan Schell (1943–2014) in his magisterial 1982 bestseller *The Fate of the Earth*, which offered a comprehensive, highly compelling survey of the threats posed by thermonuclear weapons (as well as the ethical and evaluative implications of our extinction, which will be examined in Part II). There are, he argued "three grave direct global effects," which would "produce innumerable secondary effects of their own throughout the ecosystem of the earth as a whole." One is of course global fallout. Another is ozone depletion, which he notes, citing the NAS study mentioned above, could have devastating global consequences. "Without the ozone shield," he writes, "sunlight, the life-giver, would become a life extinguisher." Hence,

> in judging the global effects of a holocaust, therefore, the primary question is not how many people would be irradiated, burned, or crushed to death by the immediate effects of the bombs but how well the ecosphere, regarded as a single living entity, on which all forms of life depend for their continued existence, would hold up. The issue is the habitability of the earth, and it is in this context, not in the context of the direct slaughter of hundreds of millions of people by the local effects, that the question of human survival arises.

The last direct global consequence is the possibility that ground bursts could catapult huge quantities of dust into the stratosphere, where it could block out incoming sunlight and thus cause Earth's surface to cool. This idea had been expressed as early as 1949, and the polymath John von Neumann suggested during congressional testimony in 1955 that a large number of nuclear explosions could potentially loft enough dust into the atmosphere "to bring back the conditions of the last ice age."[82] Later, Tom Stonier calculated in his 1964 book *Nuclear Disaster* how much soil a nuclear explosion could inject into the stratosphere and examined historical data of cooling periods after volcanic eruptions, as occurred following the Tambora eruption in 1815. He concluded that "although radioactive fallout could inflict a great ecological catastrophe, it could not change the climate. Other debris injected into the atmosphere from explosions, however, did have the potential to do this." Later, the above-mentioned book by the Ehrlichs and Holdren "pointed to explosive dust injections and smoke from huge fires as potential engines of regional and global climate change," while Stephen Schneider, a climatologist at the National Center for Atmospheric Research, conjectured that "ozone depletion and dust injections into the stratosphere might cause Earth's surface to cool from a fraction of a degree to a few degrees Celsius."[83]

Fire(Storms) and Ice

However, as Lawrence Badash notes, speculations about stratospheric dust were little more than "hand-waving" at the time, given the state of scientific knowledge. Although ozone depletion did appear credible, global cooling as a result of stratospheric dust injection was not accepted as especially worrisome by many scientists, which explains why Schell did not spend much time discussing it. Yet, as fate would have it, the same year that Schell's book was published, scientists proposed a revolutionary new idea that would soon be called the *nuclear winter hypothesis*. Whereas von Neumann, Stonier, Schell, and others had focused on dust, Crutzen and John Birks explored the possible climatic effects of *smoke* released into the lower atmosphere by fires "in cities, forests, agricultural fields, and oil and gas fields," ignited by nuclear explosions.[84] This smoke would produce a thick layer of particulate matter floating in the atmosphere that could reduce "the average sunlight penetration to the ground . . . by a factor between 1 and 150 at noontime in the summer," thus greatly damaging agriculture in the Northern Hemisphere. The study also found a significant increase in average ground ozone levels after the smoke had settled, which would further harm agricultural productivity by subjecting crops "to severe photochemical pollutant stress."[85] A paper published the same year by Richard Turco, Owen Toon, James Pollack, and Carl Sagan focused on a possible climatic effect that Crutzen and Birks had ignored: a reduction in average surface temperatures. In a large number of "full-scale" nuclear war scenarios examined by the authors, the outcome would be

> a combination of stresses caused by severe climate perturbations (surface coolings of 10° C or more), radiation doses in the tens of rem, and tenfold increases in uv-B solar radiation exposures, together with widespread shortages of food and potable water, epidemics, serious injuries, and lack of medical facilities and supplies, cumulatively imply the widespread death in man and possible extinction of numerous land and marine species.[86]

The following year, the above-mentioned scientists along with Thomas Ackerman published "Nuclear Winter: Global Consequences of Multiple Nuclear Explosions" in *Science*, which introduced the term "nuclear winter," coined by Turco, into the lexicon. This has come to be called the "TTAPS" study (pronounced "tee-taps") because of the order of author names on the paper (an acronym coined by Newell Mack in 1983), and it instigated a frenzied public debate thanks to Sagan's explication of the idea in two popular articles published the same year: one in *Parade* magazine and another in *Foreign Affairs*.

Although most presentations of "nuclear winter" assume this refers to a single phenomenon—that is, the reduction of global surface temperatures due to nuclear-caused urban firestorms that produce large amounts of soot (black carbon) that become lodged in the stratosphere where it blocks incoming solar radiation—this is not entirely accurate. Rather, the term denotes an *ensemble* of effects "involving darkening, cooling, enhanced radioactivity, toxic pollution, and ozone depletion."[87] As Sagan explained in his *Foreign Affairs* article, a nuclear war would loft dust into the stratosphere and ignite firestorms that, as just mentioned, would produce dark soot; this soot would disperse mostly in the troposphere and, along with the stratospheric dust, cause subfreezing temperatures for months on end and nearly pitch-black skies at noon. Urban firestorms would also release large quantities of pyrotoxins, and once the soot and dust fell out of the atmosphere, the depletion of ozone (mentioned by Crutzen and Birks) would enable dangerous levels of ultraviolet radiation to torch Earth's surface. These factors would cause catastrophic food shortages, and the combination of fallout, pyrotoxins, and ozone depletion would increase the likelihood of global pandemics, possibly involving microorganisms with enhanced pathogenicity due to mutations induced by the shower of ultraviolet radiation. Adding to the catastrophe, months of extraordinary cold, even along the equator, would greatly reduce the availability of fresh water and, as Sagan poignantly notes, it could freeze the top meter of the ground, thereby "making it unlikely that the hundreds of millions of dead bodies would be buried, even if the civil organization to do so existed." Sagan concludes that the interacting combination of "cold, dark, radioactivity, pyrotoxins, and ultraviolet light following a nuclear war . . . would imperil every survivor on the planet. There is a real danger of the extinction of humanity."[88] He made the point in his *Parade* article like this:

> There is little question that our global civilization would be destroyed. The human population would be reduced to prehistoric levels, or less. Life for any survivors would be extremely hard. And there seems to be a real possibility of the extinction of the human species.[89]

Although the possibility of firestorms caused by nuclear explosions had been known for decades—for example, a firestorm was observed in Hiroshima roughly 20 minutes after Little Boy exploded[90]—the TTAPS study outlined a new, composite kill mechanism in which the soot produced by raging fires plays a central causal role in bringing about potentially lethal outcomes for humanity. In the decades since, studies have not only affirmed the existence of this mechanism but found, to the dismay of scientists, that even fewer nuclear weapons may be necessary to precipitate a nuclear winter than had been previously thought. For example, a 2007 study co-authored by some of the TTAPS scientists (along with some additional authors) used modern climate models to simulate the effects of 100 Hiroshima-sized bombs detonated in the subtropics, which corresponds to "less than 0.03% of the explosive yield of the current global nuclear arsenal." It found that earlier studies had inadequately represented the amount of smoke that would end up in the stratosphere, where the primary removal mechanism is gravity rather than precipitation, and hence it concluded that the effects of even a *quite small* regional nuclear war (e.g., between India and Pakistan) could cause "significant cooling and reductions of precipitation." While these effects would be less dramatic than those produced in simulations of large-scale nuclear exchanges, they would nonetheless last much longer because of the large stratospheric injections of smoke.[91]

This led Alan Robock and Toon, both of whom contributed to the aforementioned study, to introduce the notion of *self-assured destruction*, or SAD, in 2012.[92] Whereas the threat of *mutually assured destruction*, or MAD, coined by von Neumann, had terrorized the United States and Soviet Union throughout the Cold War like Dionysius' sword over Damocles, the nuclear winter hypothesis implies that even if country A were to attack country B *without* B retaliating, the result would be doom for *both* B and A. As Oppenheimer told Szilárd before the end of WWII, "the atomic bomb is shit." Why? Because, he said, "this is a weapon which has no military significance. It will make a big bang—a very big bang—but it is not a weapon which is useful in war."[93] This turns out to have been more true than Oppenheimer could have known. There is no game-theoretic strategy to navigate here; a first strike would be the last strike for all parties involved.

The Age of Apostasy

We have now outlined the various triggering factors that brought about, solidified, and reinforced the shift to a new existential mood in the postwar era, especially since the mid-1950s. But of course without a secular existential hermeneutics through which to interpret these developments, neither global thermonuclear fallout nor the nuclear winter scenario would have fundamentally altered our dominant understanding of humanity's existential predicament. As previously argued, there is no necessary connection between the identification of credible ways the world might be destroyed and the belief that humanity is vulnerable to, or in danger of, going extinct. The threat environment arises from how we answer questions like: Is our extinction fundamentally possible? If so, could it actually happen? How probable is it? How many kill mechanisms are there? What are they? And so on.[94] How we answer these questions will in turn depend upon our (a) *model* of the world and (b) *interpretation* of this model. The notion of an existential hermeneutics concerns the latter, and hence can be understand as a filter through which models of the world produce particular conceptions, or mappings, of the threat environment. To illustrate with an example from the last chapter, despite a change in the world-model that Lord Kelvin accepted following his (co-)discovery of the Second Law, his religious hermeneutics did not necessitate any major revisions to the threat environment, as he understood it. In contrast, those inclined toward more secular worldviews, such as Wells and Russell, were compelled to redraw the threat environment in fundamental ways, since from a secular perspective the Second Law did indeed imply our inevitable extinction on a planet sinking slowly into the frozen abyss of thermodynamic equilibrium. The same set of world-interpretation relationships applies no less to the present period, of course. Hence, even if one were to accept that thermonuclear fallout and nuclear winter pose genuine risks, this needn't *entail* any corresponding change to our conception of the threat environment. It all depends on how one's existential hermeneutics filters these features of our world-model.

I mentioned earlier that the second half of the twentieth century witnessed a significant decline in the prevalence of religion throughout the Western world. More specifically, surveys indicate that religious belief remained strong in the United States during the 1940s and 1950s and may have actually grown. But this changed dramatically during the 1960s, a decade of radical cultural transformation that inaugurated what Gavin Hyman, borrowing a term from Gerhard Ebeling, calls the "Age of Atheism," during which Nietzsche's "God is dead!" declaration finally came to fruition.[95] Of note is that, while the intelligentsia was already quite irreligious at this point—indeed, virtually every major contributor to the story above described themselves as either agnostic or atheist[96]—the tentacles of secularization gripped the general public like never before. As Michael Buckley writes in *At the Origins of Modern Atheism*, this period saw the emergence of a "radical godlessness" that was "as much a part of the consciousness of millions of ordinary human beings as it [was] the persuasion of the intellectual."[97]

What was it about the Sixties that catalyzed this decline in religiosity? Historians and secularization theorists have singled out a plethora of causes, including better education, lower levels of insecurity and deprivation, the spread of Marxism, second-wave feminism, the hippy counterculture, multiculturalism, and the importation of Eastern religions like Buddhism and Hinduism.[98] Whatever the underlying causes were, the important consequence of this secularity growth spurt is that it greatly increased the availability of a secular hermeneutics, thus shaping the broader cultural response to key triggers like the bombings of Hiroshima and Nagasaki, Castle Bravo debacle, and discovery of the nuclear winter phenomenon. Whereas the previous existential mood of vulnerability was mostly concentrated within the intellectual class, radiating outward into the general public via popular science articles and science fiction novels like Wells' and Flammarion's works, the mood that emerged in the postwar era percolated into almost every corner of Western society. Never before had so many people thought seriously about the prospect of our permanent disappearance; never before had the general public been so open to the possibility of our extinction; never before had the fear of impending self-annihilation haunted the Western world, if not the world more generally. As the final section of this chapter explores, the result was an unprecedented surge in the conceptual prominence of the idea of *human extinction*, as indicated by Google Ngram searches for relevant keywords, which I have compiled in Appendix 1.

Atoms and the Antichrist

Yet despite the broad trend away from religion that began during the Sixties, Christianity in one form or another remained a powerful force in society. Consequently, a significant portion of the population, including some at the highest echelons of the US government, watched the events discussed above through the interpretive lens of biblical prophecy, that is, through a religious existential hermeneutics. To quote Edward Shils' 1956 book *The Torment of Secrecy*,

> the atomic bomb was a bridge over which the phantasies ordinarily confined to restricted sections of the population . . . entered the larger society which was facing an unprecedented threat to its continuance. The phantasies of apocalyptic visionaries now claimed the respectability of being a reasonable interpretation of the real situation.[99]

Indeed, for many Christians, the development of nuclear weapons did not undermine the eschatological narratives of the Bible but were instead rapidly integrated into them, being seen as the *fulfillment* of ancient prophecy and hence as further evidence of the Bible's truth. A striking example comes from Ronald Reagan during a dinner in 1971, while he was governor of California. Because of nuclear weapons, he claimed,

> for the first time ever, everything is in place for the battle of Armageddon and the second coming of Christ. . . . It can't be long now. Ezekiel [38:22] says that fire and brimstone will be rained upon the enemies of God's people. That must mean that they'll be destroyed by nuclear weapons. They exist now, and they never did in the past.[100]

He reiterated this idea in a 1980 television interview, while campaigning for president, with the televangelist Jim Bakker, averring that "we may be the generation that sees Armageddon." This view was shared by many other evangelicals at the time, leading Andrew Lang of the Christic Institute to warn in 1984 about the dangerous ascent of what he called "nuclear dispensationalism" within the Republican Party.[101] Evangelicalism is a Protestant movement most well known for the idea that one must be "born again" to enter heaven, and it gained prominence in the United States during the 1940s and 1950s, led by preachers like Billy Graham. Dispensationalism is a framework for interpreting scripture (including the eschatological parts) that was first popularized in the 1830s by John Nelson Darby of the Plymouth Brethren. Accepted by many evangelicals, its most influential innovation is the idea of the "Rapture," which denotes a future event in which Jesus swoops down from the clouds to collect every Christian, both dead (resurrected) and alive. This is followed by the emergence of the Antichrist, a seven-year Tribulation, the Battle of Armageddon, Second Coming of Christ, and a literal 1,000-year period of peace called the Millennial Kingdom. At the end of this period, God and Satan—rather than Christ and the Antichrist, as with Armageddon—fight one last cosmic battle. God of course wins, casts Satan into the Lake of Fire, remakes the heavens and the earth, and establishes paradise, that is, Heaven on Earth, in which every believer throughout history, all now with glorified bodies, reside forever with God.

Although the eschatology of dispensationalism was widely taught by the mid-twentieth century at Bible institutes, Bible colleges, and evangelical seminaries in the United States—examples being the Moody Bible Institute, Philadelphia College of the Bible, and Dallas Theological Seminary, respectively—it gained widespread popular attention following the 1970 publication of *The Late Great Planet Earth*, written by Hal Lindsey, a graduate of the aforementioned seminary school.[102] In fact, this was the best-selling "non-fiction" book in English of the entire decade, selling some 28 million copies by 1990. One reason for the book's extraordinary success was that Lindsey superimposed the narrative of dispensationalist eschatology onto contemporary geopolitical affairs; he provided a concrete account of how postwar developments tie into the prewritten narrative of prophetic scripture. Of particular relevance to the present study was his contention that thermonuclear weapons would play a central role in the Battle of Armageddon between, he claimed, us on the one side, and Russia and the Antichrist on the other. As the journalist Grace Halsell wrote in 1986, Lindsey's main thesis was that "God has foreordained that we fight a nuclear Armageddon."[103] In fact, Reagan was almost certainly channeling Lindsey's account in the above block quote, as reports suggest that Reagan had read the book, and indeed Reagan later "invited Lindsey to speak at the Pentagon on his geopolitics of the future," an experience about which Lindsey subsequently wrote: "It seems that a number of officers and non-military personnel alike has read *Late Great* and wanted to hear more."[104] By 1984, according to a Yankelovich poll, an incredible 39 percent of the American public, equaling roughly 85 million Americans, agreed that "when the Bible speaks of the earth being destroyed by fire, this means that we ourselves will destroy our earth in a nuclear Armageddon."[105]

Hence, as I argued in Chapter 1, eschatological narratives can be simultaneously rigid and elastic, eternal and unchanging yet capable of adapting to novel world developments that neither earlier apocalypticists nor the inspired authors of biblical prophecy could possibly have imagined. Yet another example of this comes from Edgar Whisenant's 1988 book *88 Reasons Why the Rapture Will Be in 1988*, which sold some 4.5 million copies. This was, of course, written after the nuclear winter hypothesis had been proposed, and Whisenant wasted no time incorporating it into his own dispensationalist account of the world's end. In the final section of the book titled "A Message to the United States," he writes that

> nuclear winter will last five years in the northern third (60 degrees) of the earth (which covers the United States) from statements made by Carl Sagan on Nuclear Winter, plus additional statements made in the Bible. We also know the whole continent will be as dark as midnight 24 hours a day for this entire five-year period, with temperatures never rising above zero fahrenheit [*sic*]. Mass starvation and unburied bodies will result. . . . [T]he destruction [will] be so complete that you can walk from Little Rock to Dallas over ashes only. All food will be gone; all water will be radioactive, except for underground water.[106]

The postwar era thus provides a number of striking examples of how antithetical existential hermeneutics can produce radically different mappings of the threat environment. Even more, one's mapping of the threat environment can yield important *practical conclusions* about which course(s) of action one should pursue in response to the perceived threats facing humanity. Once again, the decades after WWII, during the Cold War, offer some of the most compelling examples of this.

On the one side, those with more secular worldviews, who believed that our extinction could actually happen and that this would constitute a moral tragedy for one reason or another, found themselves impelled to take urgent steps to mitigate the risk. For example, a number of Manhattan Project scientists established the *Bulletin*, as mentioned. After delivering "Man's Peril" and releasing his manifesto with Einstein, Russell cofounded the Pugwash Conferences in 1957 with Joseph Rotblat, which vigorously promoted nuclear disarmament. Einstein himself, along with many scientists, philosophers, and political theorists at the time, argued vigorously for the establishment of a world government to contain the threat of nuclear proliferation, an idea that Daniel Deudney calls "nuclear one worldism."[107] Sagan made similar claims, arguing that our extinction would be tragic not just because of those who would die in the event but because it would prevent trillions of future people from coming into existence. Although Sagan was accused of nuclear alarmism by critics on the political right, when Mikhail Gorbachev met Reagan in 1988 he specifically identified Sagan as having been "a major influence on ending proliferation."[108]

On the other side, dispensationalists like Hal Lindsay proclaimed that since a nuclear holocaust is inevitable, God's ultimate will for the world, the United States *shouldn't* pursue arms-control agreements with the "Evil Empire," as Reagan described the Soviet Union. To quote the televangelist Jim Robins, whom Reagan invited to give the 1984 Republican National Convention opening prayer, "there'll be no peace until Jesus comes. Any preaching of peace prior to this return is heresy; it's against the word of God." Similarly, the dispensationalist

Jimmy Swaggart, a friend of Reagan's, declared in 1985 that "we should not make any agreements with the Soviet Union," but should instead withdraw from the United Nations and increase our nuclear stockpile. "I wish I could say we will have peace," he said, but "Armageddon is coming. . . . They can sign all the peace treaties they want. They won't do any good. There are dark days coming. . . . It's going to get worse."[109] This perspective on future history affected not just US foreign policy but environmental policy at home, as well (see below). For example, Reagan's pick for Secretary of the Interior was a dispensationalist named James Watt. When asked "about his views on preserving natural resources for future generations" during a Senate hearing, he answered that we shouldn't worry much about destroying the natural world and overexploiting Earth's resources because "I do not know how many future generations we can count on before the Lord returns."[110] As the philosopher Jerry Walls observes,

> dispensationalist eschatology inclines its adherents not only to despair of changing the world for good, but even to take a certain grim satisfaction in the face of wars and natural disasters, events which they interpret as the fulfillment of prophecy pointing to the end of the world.[111]

But this is too weak: in many cases, the inclination was not merely to relinquish hope of ameliorative change but to adopt positions that actively contribute to the overall risk of catastrophe for the sake of accelerating the onset of the apocalypse, since on the other side of the apocalypse lies paradise.[112]

The radical secularization of Western culture that commenced in the Sixties thus crucially enabled the emergence of a new existential mood. Without a secular hermeneutics, the threat environment may not have undergone any significant, qualitative revisions, despite the unprecedented events that unfolded between 1945 and 1954. Even more, the secular recognition that nuclear conflict could bring about our extinction may have nontrivially decreased the actual probability of this outcome obtaining by impelling those who value humanity's survival to advocate for anti-proliferation policies, the establishment of a world government, and the abolition of nuclear weapons altogether, which Sagan once memorably described as "elementary planetary hygiene."[113]

The Lone Wolf

So far, our discussion has covered transformations in our understanding of all three properties of the threat environment specified at the beginning of this chapter: etiology, temporality, and multiplicity. Each was altered by a single invention, namely, nuclear weapons, which for the first time in history made human self-annihilation not only feasible but unsettlingly probable in the near future. With respect to multiplicity, nuclear weapons introduced several distinct kill mechanisms, the most scientifically credible of which were global thermonuclear fallout and nuclear winter, which includes global fallout and ozone depletion within its ensemble of effects. Yet a key feature of this period was the identification of various anthropogenic phenomena that pointed toward *additional* kill mechanisms that might be, or become, no less threatening than nuclear conflict in the coming decades or centuries. The most salient were associated with forms of environmental contamination and degradation caused by radioactive fallout from thermonuclear testing, mutagenic synthetic chemicals, exponential population growth, and (later) greenhouse gases like CO2. Some reputable scholars also sounded the alarm about the potential threats posed by modified pathogens, recursively self-improving AI systems, and self-replicating nanobots. Let's take these in turn.

The modern environmental movement can be traced back to the postwar neo-Malthusian theorists Fairfield Osborn and William Vogt, both of whom published commercially successful books in 1948 about humanity's harmful impact on the natural world: *Our Plundered Planet* and *Road to Survival*, respectively. According to Charles Mann, Vogt's work spawned what some have called *apocalyptic environmentalism*, which refers to "the belief that unless humankind drastically reduces consumption and limits population, it will ravage global ecosystems."[114] However, by far the most significant contribution to modern environmentalism was the 1962 book *Silent Spring* by Rachel Carson, described by a *New York Times* article as having "influenced the environmental movement as no one had since the 19th century's most celebrated hermit, Henry David Thoreau, wrote about Walden Pond."[115] So impactful was Carson's publication that it inspired the creation of the US Environmental Protection Agency (EPA) in 1970, and "prompted the Federal Government to take action against water and air pollution—as well as against the misuse of pesticides—several years before it otherwise might have moved," to quote the EPA's official history website.[116] Other major works included Paul and Ann Ehrlich's *The Population Bomb* (1968) and *The Limits to Growth* (1972), the latter of which was commissioned by the newly formed Club of Rome, whose stated mission was to address the interconnected constellation of problems facing humanity that they dubbed the "world problematique."[117]

Environmentalism burgeoned during the 1970s. This decade saw not only the formation of the EPA but the first Earth Day, the founding of Greenpeace, the United Nations Conference on the Environment (the first of its kind), the Endangered Species Act, and the rise of "deep ecology," an ethical view that embraces what its progenitor, Arne Naess, called *biospherical egalitarianism*, which posits that all living creatures are endowed with the same amount of intrinsic (as opposed to merely instrumental) value.[118] By the 1980s, more radical forms of environmentalism associated with *ecocentrism* started gaining traction. Whereas the main focus of most environmentalists in the 1970s was the effects of ecological destruction on humanity, ecocentrists went beyond anthropocentrism (the natural world has value *only* as a means to human ends), biocentrism (that nonhuman organisms possess at least *some* intrinsic value), and biocentric egalitarianism (the contemporary term for Naess' biospherical egalitarianism) in assigning value to *non-living* entities as well, such as the land and rivers.[119] Biocentrism and ecocentrism were popularized most notably by Earth First!, founded in 1980, which drew attention to environmentalist issues through monkeywrenching antics like returning people's trash, vandalizing roads in wilderness areas, and tree spiking, as well as other forms of "ecotage," a portmanteau of "ecological sabotage."[120] As a document published by Earth First! and signed "El Lobo Solo," which translates as "The Lone Wolf," states: "Earth First! is a verb, not a noun."[121]

From its inception, leading environmentalists were explicit that the contamination and degradation of nature could plausibly bring about the collapse of civilization, if not the complete extinction of humanity.[122] For example, Osborn wrote that humanity is at war with nature, a war far more perilous than World War II, one that "contains potentialities of ultimate disaster greater even than would follow the misuse of atomic power. . . . [I]f we continue to disregard nature and its principles the days of our civilization are numbered."[123] Similarly, Vogt declared that "excessive breeding and abuse of the land" risks "a catastrophic crash of our civilization," which might precipitate "at least three-quarters of the human race [being] wiped out."[124] The main difference between Malthusian and neo-Malthusian concerns about overpopulation is the spatial scope. Whereas Thomas Malthus, discussed in Chapter 2, focused on how the divergence between the availability of sustenance and growth of populations within particular regions of the planet would establish a "perpetual oscillation between happiness and misery," Osborn, Vogt, and the Ehrlichs claimed that *global* overpopulation could precipitate *global* catastrophes that, as such, would affect everyone on the planet.

Although parts of *The Population Bomb*—an allusion, of course, to the atomic bomb—suggested that the worst-case outcome of the growing human population would be "hundreds of millions of people [starving] to death," they also repeatedly gestured at far worse outcomes.[125] For example, in the prologue of the original edition, they warn that we must curb overpopulation and "take action to reverse the deterioration of our environment before population pressure permanently ruins our planet. . . . The birth rate must be brought into balance with the death rate or mankind will breed itself into oblivion."[126] This was reiterated in the book's forward, authored by the (co)founder of Friends of the Earth (in 1969) and Earth Island Institute (in 1982), David Brower. Environmentalist organizations, Brower declared, "have been much too calm about the ultimate threat to mankind," and hence they will need to "awaken themselves and others, and awaken them with an urgency that will be necessary to fulfillment of the prediction that mankind will survive."[127] As Adam Rome observes, referencing earlier fears of annihilation engendered by nuclear weapons, "the mounting evidence of environmental degradation in the 1960s provoked similar anxieties about 'survival,' a word that appeared again and again in environmentalist discourse."[128]

The Biocide Bomb

The connection between environmentalist concerns and nuclear weapons is best exemplified in the early literature by Carson's book, and later foregrounded in the 1980s by Jonathan Schell. Carson explicitly linked the contamination of the environment with synthetic chemicals to contamination caused by thermonuclear tests during the 1950s. It is here that the Castle Bravo disaster enters the picture once again: on the one hand, it convinced many at the time that, as mentioned above, even a relatively small-scale thermonuclear conflict could potentially poison every human being on Earth. On the other hand, it alerted the public to the idea that *testing* thermonuclear weapons, whether in the Nevada desert, the Marshall Islands, or Kazakhstan (where the Soviet's primary test site was located), could spread small amounts of radioactivity around the entire planet. The question then was whether such radioactive particles—"Death Dust," as one reporter called it—could produce adverse health effects, a possibility that the US government vigorously denied (from the very beginning, when Burchett reported on "atomic plague" in Hiroshima).[129] However, credible warnings about the deleterious effects of (ionizing) radiation for our genes, in particular our germ cells or "germ plasm," had been made for several decades, most notably by Hermann Muller, a geneticist and outspoken proponent of eugenics who won the 1946 Nobel Prize for discovering that X-rays can induce genetic mutations. Since we pass our germ cells—in contrast to our somatic cells—down from one generation to the next, a mutation in these cells will affect *all future offspring* for as long as one's genealogy persists. This means that germ-cell mutations, if sufficiently widespread, could potentially threaten the entire future of humanity. As Muller wrote in 1933

> we must remember that the thread of germ plasm which now exists must suffice to furnish the seeds of the human race even for the most remote future. We are the present custodians of this all important material and it is up to us to guard it carefully and not contaminate it for the sake of any ephemeral benefits to our own generation.

This was written as a reminder to X-ray specialists at the time, since the only significant artificial source of radiation exposure to people in the 1930s was from medical procedures, e.g., X-ray images.[130] The Castle Bravo test made it impossible to deny that nuclear tests were exposing people to a new source of radiation, a fact startlingly confirmed by the famous 1961 "Baby Tooth Survey," which found that the radioactive isotope strontium-90, a byproduct of nuclear fission, was present in the bones and deciduous teeth of babies, and that the quantity of strontium-90 in children had appreciably risen throughout the 1950s.[131] Although there was reasonable scientific debate about the health consequences of fallout exposure throughout the 1950s, many leading scientists warned that even minuscule amounts of exposure could be injurious to our genes, a point reiterated by biologists at the 1955 Atoms for Peace conference in Geneva. The following year, the Genetics Committee of a National Academy of Sciences study, which included Muller, "announced that any amount of radiation, no matter how small, would cause some genetic damage," which newspapers turned into front-page news.[132]

Carson tapped into this debate by channeling fears over genetic mutation caused by thermonuclear tests toward a cluster of distinct concerns about the effects of synthetic pesticides like DDT, which became widely used in agriculture following World War II. As Carson opened the third chapter, titled "Elixirs of Death":

> For the first time in the history of the world, every human being is now subjected to contact with dangerous chemicals, from the moment of conception until death. In the less than two decades of their use, the synthetic pesticides have been so thoroughly distributed throughout the animate and inanimate world that they occur virtually everywhere.[133]

The ubiquity of chemical exposure is dangerous for many reasons, Carson argued, one of which is that pesticides might harm "the genetic material of the [human] race by causing gene mutations." In support she quotes Muller as warning that "various chemicals (including groups represented by pesticides) 'can raise the mutation frequency as much as radiation.'" Hence, she asks the question: "We are rightly appalled by the genetic effects of radiation; how then, can we be indifferent to the same effect in chemicals that we disseminate widely in our environment?" If nuclear testing is unacceptable because of its apparent "threat to our genetic heritage"—and indeed nuclear tests above ground, under water, and in space were banned by international treaty in 1963—then surely we should take similar actions to eliminate what Carson labeled "biocides."[134] Failing to do this risks making Earth "unfit for all life." Continuing on our current path will only end in "disaster." But there is still time to change course, as we must, because this may be "our last, our only chance to reach a destination that assures the preservation of our Earth." Tying together the nuclear and pesticidal threats under the powerful theme of contamination, Carson declared:

> Along with the possibility of the extinction of mankind by nuclear war, the central problem of our age has . . . become the contamination of man's total environment with such substances of incredible potential for harm—substances that accumulate in the tissues of plants and animals and even penetrate the germ cells to shatter or alter the very material of heredity upon which the shape of the future depends.[135]

Carson's book triggered a fierce backlash, with some claiming that her warnings lacked scientific credibility. A Vanderbilt University professor named William Darby, for example, wrote a scathing review in 1962 with the overtly sexist title "Silence, Miss Carson," which appeared in *Chemical & Engineering News*, a weekly trade magazine published by the American Chemical Society. He accused Carson of writing the book for "emotional" reasons and of "ignor[ing] the sound appraisals of . . . responsible, broadly knowledgeable scientists." "In view of her scientific qualifications," Darby continued, "in contrast to those of our distinguished scientific leaders and statesman, this book should be ignored," although he concludes by exhorting scientists to do the opposite, that is, to "read this book to understand the ignorance of those writing on the subject and the educational task which lies ahead."[136] Along similar lines, *Time* magazine described the book as "hysterical" and "patently unsound."[137] However, the following year a US government report ordered by President John F. Kennedy, titled "Use of Pesticides," supported Carson's concerns about the indiscriminate and excessive use of synthetic chemicals, and the Toxic Substances Control Act of 1976 banned or severely restricted every one of the six compounds that Carson singled-out—DDT, chlordane, heptachlor, dieldrin, aldrin, and endrin—which was Carson's "greatest legal vindication."[138]

The most important contribution of Carson's work, though, wasn't drawing attention to this or that particular toxin or the chemical industry's prioritization of profit over people. Rather, it was popularizing the idea that there exists a delicate balance of biotic and abiotic forces within the various complex, interlinked ecological systems upon which our survival and flourishing depends, the sum total of which comprises the "biosphere," a word coined in 1875. As Carson articulated this insight in a 1963 CBS documentary released six weeks before the Kennedy administration's report on pesticides:

> The balance of nature is filled of a series of interrelationships between living things, and between living things and their environment. You can't just step in with some brute force and change one thing without changing many others. Now this doesn't mean, of course, that we must never interfere, but we must not attempt to tilt that balance of nature in our favor. . . . [U]nless we do bring these chemicals under better control we're certainly headed for disaster.[139]

This became quite influential within the budding movement of modern environmentalism, and it continues to shape contemporary thinking about natural systems: if nature exists in a state of balance with everything connected to everything else, and if maintaining this balance is necessary for our survival, then any change that destabilizes this balance could pose a threat to our continued existence.

Hence, the "balance of nature" model implied a new *category* of kill mechanism: whereas a thermonuclear conflict would destroy large parts of the biosphere suddenly, as the result of a single event, a similar outcome could result from the incremental accumulation over time of small ecological perturbations, none of which are individually sufficient to cause large-scale harm, but which over time add up to seriously damage the integrity of the entire system. Whether or not pesticides or overpopulation actually *instantiate* this kill mechanism is one question, and as mentioned there were plenty of reputable scientists at the time who believed that they do, or at least *would* if current demographic, technological, agricultural, and other trends were to continue into the future unabated. But the more important insight was that (1) nature's balance must be maintained if humanity is to survive on Spaceship Earth (a metaphor that became popular in the 1960s), and (2) science, technology, and population growth are making it increasingly feasible for humanity to induce precisely the sort of perturbations that could upset this balance on a planetary scale.[140] As we will see in Chapter 6, the basic insight that nature exists in a "delicate balance" was updated in the 2000s as a result of research on "tipping points," "critical thresholds," and "planetary boundaries."

Glass-Bottom Boats

Yet the multiplication of risks did not end with the identification of pollution and the exponential growth of the human population as possible ways that humanity could perish. The 1960s and 1970s also witnessed the first warnings that anthropogenic CO2 emissions could alter Earth's climatic system in deleterious ways.[141] For example, a 1958 film titled *The Unchained Goddess*, which was part of a TV series described as "among the best known and remembered educational films ever made,"[142] includes a dialogue between the actor and English professor Frank Baxter and the actor Richard Carlson, who plays "Mr. Fiction Writer." It went like this:

Baxter: Even now, man may be unwittingly changing the world's climate through the waste products of his civilization. Due to our release through factories and automobiles every year of more than six billion tons of carbon dioxide, which helps air absorb heat from the sun, our atmosphere seems to be getting warmer.

Carlson: This is bad?

Baxter: Well, it's been calculated a few degrees rise in the Earth's temperature would melt the polar ice caps. And if this happens, an inland sea would fill a good portion of the Mississippi valley. Tourists in glass-bottom boats would be viewing the drowned towers of Miami through 150 feet of tropical water. For in weather, we're not only dealing with forces of a far greater variety than even the atomic physicist encounters, but with life itself.

In 1965, during Lyndon Johnson's presidency, a team of scientists conveyed the first explicit warning about climate change to the US government. They wrote: "Man is unwittingly conducting a vast geophysical experiment. Within a few generations he is burning the fossil fuels that slowly accumulated in the earth over the past 500 million years." The outcome could be, the authors concluded, "deleterious from the point of view of human beings."[143] By the late 1970s, following a debate about whether aerosols from industrial pollution that reflect incoming sunlight could counteract the greenhouse effect, scientific opinion had largely converged on the view that warming surface temperatures could pose a significant threat in the twenty-first century.[144] Indeed, a committee convened by the US National Academy of Sciences calculated in 1979 that the average surface temperature of Earth would increase by roughly 3 degrees C if the concentration of CO2 in the ambient air were to double relative to pre-industrial levels, which was expected to happen sometime next century. The most significant event for shaping public opinion was undoubtedly James Hansen's 1988 Congressional testimony, in which he argued not only that surface temperatures are indeed rising but that "global warming has reached a level such that we can ascribe with a high degree of confidence a cause and effect relationship between the greenhouse effect and the observed warming."[145] This was extensively covered by the news media, and the issue of anthropogenic climate change, which up to that point had been generally ignored by the environmental movement, quickly became its number one concern.[146]

Over this period, climatologists and other scientists voiced a number of dire prognostications about the possible effects of tinkering with Earth's natural thermostat setting, although few experts directly linked anthropogenic climate change to human extinction, or even the collapse of civilization. The "survival" language that pervaded the environmentalist literature from Vogt and Osborn onwards was thus largely absent from this discussion. There was of course talk of *global-scale effects*, such as rising sea levels, which would occur around the entire planet. As a widely read article in *Discover* magazine about Hansen's testimony reported, sea-level rise threatens "places like the Marshall Islands in the Pacific, the Maldives off the west coast of India, and some Caribbean nations" with "national extinction," but of course national extinction is a far cry from human extinction. Other potential consequences of climate change identified at the time include devastating but survivable phenomena like more extreme weather events (droughts, floods, and wildfires), increased rates of plant and animal extinctions, refugee crises as people are forced to relocate, and geopolitical tensions over dwindling resources; however, almost everyone acknowledged that the repercussions of global warming are impossible to know given the current state of science, which they saw as a reason to worry more rather than less about what might happen. To quote the *Discover* article once more, "the unprecedented rapid change" means that "we're altering the environment far faster than we can possibly predict the consequences," and the global extent of this impact entails that we "are affecting the ecological balance of not just a region but the entire world, all at once."[147]

While the likelihood of unknown effects was indeed worrying, the most dire predictions would be best categorized as "gloomsday" rather than "doomsday," borrowing a term from a 1979 article in *Science* on the topic.[148] The one exception to this arose from the possibility of a runaway greenhouse effect. Research in the 1960s on the planetary conditions of Venus, second rock from the sun, led to the conclusion that it had undergone a runaway greenhouse effect driven by water vapor and/or CO2, resulting in its surface temperature hovering around 900 degrees F.[149] Given the similarities between Earth and Venus, this suggests that "there are circumstances in which we could change the Earth's environment so that it would run away to where Venus is," as Sagan, who wrote his 1960 dissertation mostly about the Venus greenhouse effect, told the House of Representatives Subcommittee on Space Science and Applications in 1975. "It's important to understand what went wrong on Venus," he continued, "so we know what not to do."[150] A month later, the planetary scientist Bruce Murray reiterated this point to the same subcommittee, as did Thomas Mutch, the NASA associate administrator, Office of Space Science, in 1980. As Mutch made the point, understanding how exactly the runaway greenhouse effect unfolded on Venus—for example, what role did CO2 play? Might CO2 have been the main driver?—has

> contemporary importance to us because human activities are significantly adding to the amount of carbon dioxide in the Earth's atmosphere. This build-up may lead to increases in atmospheric temperatures which, in turn, may add to the evaporation of more water vapor into the atmosphere with its additional insulating effect: conceivably the Earth could suffer a runaway effect. Since even small changes in global temperatures can have marked, if not catastrophic, effects on the environment, it is clear that we must gain an in-depth understanding of this potential problem. Understanding the Venus greenhouse is an obvious first step.[151]

Although none of these statements mentioned human extinction explicitly, the implication was obvious: human life would be impossible in Venusian temperatures hot enough to melt lead and zinc. Here, then, was a genuine kill mechanism associated with climate change: a positive-feedback loop involving CO2 that would, if triggered, inexorably bring about the complete annihilation of *Homo sapiens* along with all (or most) of our fellow creatures on the planet. Yet the scientific understanding of this phenomenon was far too impoverished at the time for anyone to make strong assertions about the probability of this occurring—of CO2 emissions pushing civilization past a critical threshold, a Rubicon of runaway warming. Hence, it was mostly raised to justify funding for the US space program. Like the possibility of igniting the atmosphere when testing the first atomic bomb, the danger was plausible but more research was needed to determine its actual credibility.

Independent of whether a runaway greenhouse effect on Earth is likely to occur if humanity continues to alter the chemical composition of the atmosphere on a global scale, the discovery of anthropogenic climate change provided additional evidence for proposition (2) in the previous section, that is, that human activities resulting from technoscientific advancements, industrial development, and a growing worldwide population could have far-reaching, long-lasting environmental consequences.

Black Death 2.0

A survey of this crucial period of radical change in our understanding of humanity's existential predicament would not be complete without mentioning a few seeds planted about the possibility of future threats associated with microbes, algorithms, and nanobots. None of these gained widespread acceptance as kill mechanisms capable of destroying humanity, though each was endorsed by reputable (if not also controversial) theorists.

Taking these in turn, the first concerns the possibility of wiping out the human species through biological warfare and/or bioterrorism. The history of pathogens being weaponized for offensive purposes goes back centuries.[152] The Mongols, for example, gathered the bodies of people who died from the plague and catapulted them over the walls of Kaffa, a Crimean city on the coast of the Black Sea, which may have introduced the plague to Europe, thus leading to the Black Death that killed up to 60 percent of the European population. During World War I, Germany infected horses shipped to the Allies with anthrax and glanders (an infectious disease that mainly affects horses), and multiple countries pursued bioweapons capabilities during WWII.[153] In many cases, the target of biological agents was crops and animals, although Imperial Japan's infamous Unit 731 dropped fleas infested with the plague and flies covered in cholera on Chinese populations, which killed up to half a million people.[154] In the last few months of WWII, the director of Unit 731, Shirō Ishii, had developed a plan to kill thousands by contaminating San Diego with plague-infected fleas, an attack that never occurred because of the events in Hiroshima and Nagasaki.

As we saw in Chapter 3, some visionaries during the nineteenth century seriously entertained the possibility of a worldwide outbreak of infectious disease causing the global population to collapse. This was central to the apocalyptic narrative of Mary Shelley's *The Last Man* (1826), and H. G. Wells singled it out in his 1893 essay "The Extinction of Man," in which he worried about a future "plague that will not take ten or twenty or thirty per cent., as plagues have done in the past, but the entire hundred."[155] At the time, the imagined possibilities were limited to natural outbreaks, perhaps enabled by global trade and travel. However, the rise of modern bacteriology in the late nineteenth century, around the time of Wells' essay, "offered new prospects for those interested in biological weapons because it allowed agents to be chosen and designed on a rational basis."[156] The same can be said about the emergence of molecular genetics in the mid-twentieth century, as this made, or would soon make, it seemed to informed observers, selecting and modifying pathogenic germs even more feasible—*terrifyingly* feasible.

Consequently, a number of leading scientists and intellectuals of the era began to fret about the potential for extremely dangerous germs to become weapons of war that devastate not just a single community or nation but the entire human population. As Russell wrote to Einstein in a letter, dated February 1955, "although the H-bomb at the moment occupies the centre of attention, it does not exhaust the destructive possibilities of science, and it is probable that the dangers from bacteriological warfare may before long become just as great."[157] The following decade, Joshua Lederberg, a pioneer in microbial genetics who won the Nobel Prize at the age of 33 for discovering that bacteria can exchange genetic material with each other, presented a statement before the US Subcommittee on National Security Policy and Scientific Developments, which was later published as "Biological Warfare and the Extinction of Man" (which curiously includes the title of Wells' essay). The development of biological weapons, Lederberg declared, "puts the very future of human life on Earth in serious peril." After discussing the potential for his research to prevent certain "serious human diseases," he said:

> However, whatever pride I might wish to take in the eventual human benefits that may arise from my own research is turned into ashes by the application of this kind of scientific insight for the engineering of biological warfare agents. . . . We simply have no way of assuring ourselves that a bacterial warfare development activity will not eventually seed a catastrophic world-wide epidemic that ignores national boundaries. . . . Unless we learn to apply our common energies against the common enemies of all mankind, we are foolish and arrogant to doubt that history will record "Black Death II," and more.

Of particular note is Lederberg's emphasis on the differences between biological and nuclear weapons. Whereas "nuclear weaponry depends on the most advanced industrial technology," bioweapons could be adopted "as a technique of aggression by smaller nations and insurgent groups," a point that will become absolutely central to discussions of the threat environment in the twenty-first century. As Lederberg added, "our continued participation in [biological weapons] development is akin to our arranging to make hydrogen bombs available at the supermarket."[158] For the present purposes, it suffices to note that a handful of scientific notables in the decades after WWII warned about the potential for future advancements in molecular genetics—followed in the 1970s by the field of genetic engineering and, more recently, synthetic biology—to empower smaller states and even non-state actors to wreak catastrophic havoc and, in doing so, unilaterally jeopardize the continued existence of humanity.

Algorithms Making Algorithms

Another possibility discussed in the postwar era concerns artificially intelligent machines. Recall from Chapter 3 that among the various "technoscientific" speculations about how we might destroy ourselves that were proposed during the previous existential mood was the possibility of machines gaining "supremacy over the world and its inhabitants,"[159] an idea subsequently explored by Karel Čapek in *R.U.R.*, which added "robot" to the lexicon. In the scenario outlined by Samuel Butler, humanity plays the evolutionary role of natural selection, crafting increasingly sophisticated and powerful machines that eventually become capable of self-regulating and "self-acting," until "the mechanical kingdom," as Butler called it, comes to dominate the Animal Kingdom. However, far more plausible mechanisms of machinic domination were proposed and elaborated by computer scientists from the 1950s onward. For example, Alan Turing presented an essay titled "Intelligent Machinery, a Heretical Theory" on a BBC radio program *The '51 Society*, in which he argued that

> it seems probable that once the machine thinking method had started, it would not take long to outstrip our feeble powers. There would be no question of the machines dying, and they would be able to converse with each other to sharpen their wits. At some stage therefore we should have to expect the machines to take control, in the way that is mentioned in Samuel Butler's *Erewhon*.[160]

(Note that *Erewhon* was a novel based in part on Butler's aforementioned essay "Darwin Among the Machines.") A central idea in Turing's work is the possibility of machines learning through experience. As he wrote in a 1950 paper that introduced the famous "Turing Test" (which he called the "imitation game"), "instead of trying to produce a programme to simulate the adult mind, why not rather try to produce one which simulates the child's? If this were then subjected to an appropriate course of education one would obtain the adult brain."[161] This idea was expanded in 1959 by I. J. Good, who considered a scenario in which machines begin to take over the activity of designing better machines, leading to what he later called an "intelligence explosion." In his words:

> Let an ultraintelligent machine be defined as a machine that can far surpass all the intellectual activities of any man however clever. Since the design of machines is one of these intellectual activities, an ultraintelligent machine could design even better machines; there would then unquestionably be an "intelligence explosion," and the intelligence of man would be left far behind.[162]

In other words, AI systems could undergo recursive self-improvement, either on themselves or by building new machines, thereby activating a positive-feedback loop that, like the splitting of the uranium atom with free neutrons, proceeds exponentially. As he wrote in 1959, "at this point an 'explosion' will clearly occur; all the problems of science and technology will be handed over to machines and it will no longer be necessary for people to work," which is why he later declared that "the first ultraintelligent machine is the *last* invention that man need ever make, provided that the machine is docile enough to tell us how to keep it under control."[163] But would it be so docile? How could we control the resulting intelligence if its capacities tower above ours to the extent that our capacities tower above those of a cockroach? "Whether this will lead to a Utopia or to the extermination of the human race," Good added, "will depend on how the problem is handled by the machines," meaning that once the critical threshold is crossed, our collective fate may no longer be within our control.[164] The machines will have gained supremacy, the mechanical kingdom will dominate, or perhaps exterminate, the biological world to which we belong.

This pointed toward a novel, superficially plausible kill mechanism. Indeed, as Marvin Minsky pointed out in 1984, even a superintelligent AI system that "wants" to fulfill our wishes could lead to catastrophic outcomes. "The first risk is that it is always dangerous to try to relieve ourselves of the responsibility of understanding exactly how our wishes will be realized," he wrote, adding that

> the greater the range of possible methods we leave to those servants, the more we expose ourselves to accidents and incidents. When we delegate those responsibilities, then we may not realize, before it is too late to turn back, that our goals have been misinterpreted, perhaps even maliciously. We see this in such classic tales of fate as *Faust*, the *Sorcerer 's Apprentice*, or the *Monkey 's Paw* by W. W. Jacobs.[165]

Hans Moravec proposed a similar idea in his 1988 book *Mind Children*, arguing that we will "have produced a weapon so powerful it will vanquish the losers and the winners alike" because this "weapon" will replace the current biological regime of earthly existence with "a postbiological world dominated by self-improving, thinking machines." "What awaits," he continued, "is a world in which the human race has been swept away by the tide of cultural change, usurped by its own artificial progeny."[166]

As we will see in Chapter 6, this possibility was taken up by later futurists like Ray Kurzweil and Nick Bostrom, and although climate change became the most-discussed threat to humanity among the general public in the 2000s, artificial superintelligence is frequently cited as the single greatest known risk within certain academic circles, a worry further bolstered by recent developments in AI like ChatGPT. Once again, for our purposes here it is enough to note that this threat—recursively self-improving AI systems with, as theorists now say, "misaligned" goals—was first articulated around the same time that global thermonuclear fallout, pollution, and overpopulation were becoming major sources of existential anxiety, decades (in the case of Good) before the nuclear winter hypothesis had been identified and the discovery of global warming was complete.

A Single Speck

This leads to the final kill mechanism speculation of this period: self-replicating nanobots. The idea of nanotechnology dates back to a lecture delivered by Feynman in 1959 titled "There's Plenty of Room at the Bottom." In it, he discussed the possibility of creating tiny machines, or nanobots, that could perform various tasks like manufacturing "small elements for computers in completely automatic factories," now called nanofactories. These nanobots could also have applications in biomedicine. As Feynman told his audience,

> although it is a very wild idea, it would be interesting in surgery if you could swallow the surgeon. You put the mechanical surgeon inside the blood vessel and it goes into the heart and "looks" around. . . . It finds out which valve is the faulty one and takes a little knife and slices it out. Other small machines might be permanently incorporated in the body to assist some inadequately-functioning organ.[167]

However, this proposal was almost entirely ignored until the 1980s, when Eric Drexler published *Engines of Creation: The Coming Era of Nanotechnology*. It explored the possibility of manufacturing macroscopic products with atomic precision, by moving single atoms or molecules at a time. Two computers, for example, produced this way would not only look identical to the human eye, but if one were to zoom in to the atomic level, one would find their constituent particles identically arranged. Of relevance to our discussion is that Drexler not only prophesied radical abundance from a future nanotech revolution but warned that self-replicating nanobots could spill into the environment, convert all organic matter into wriggling clones of themselves, and consequently destroy all human life on the planet, which he dubbed the "gray goo scenario." In his words:

> To devastate Earth with bombs would require masses of exotic hardware and rare isotopes, but to destroy all life with replicators would require only a single speck made of ordinary elements. Replicators give nuclear war some company as a potential cause of extinction.[168]

This wasn't the only serious threat posed by advanced nanotechnology, but it was the most direct. (Drexler also warned about the "basic threats to people and to life on Earth" posed by "thinking machines.") Despite capturing the imagination of journalists, science fiction writers, and the general public—even inspiring a neo-Luddite terrorist organization that has targeted and killed nanotechnologists—it was not until the early 2000s that the danger gained traction among academic futurists contemplating the array of hazards that could confront humanity in the coming century.[169]

Measuring the Mood

We have now surveyed the second shift in existential mood within the Western tradition. Whereas a key feature of the previous existential mood was the realization that humanity is existentially vulnerable, an insight enabled by the nineteenth-century decline of religious belief, the defining characteristic of this mood was that we face a growing multiplicity of anthropogenic threats in the near term. This was solidified by the identification of thermonuclear global fallout as a scientifically credible means of self-extermination, followed by similarly dire warnings from reputable scientists about mutagenic pollutants, overpopulation, nuclear winter, runaway climate change, biowarfare/bioterrorism, self-improving machines, and ecophagic ("ecosystem eating") nanobots. The mood was further amplified by the culture-wide metamorphosis that commenced in the Sixties, that is, the growth spurt in secularization that gave rise to the Age of Atheism, an era of radical godlessness, which enabled a growing portion of the Western population to interpret these threats through a secular hermeneutics. As mentioned above, never before had so many people thought so seriously about the prospect of our complete and permanent disappearance; never before had it been so clear that, to quote Buber once again, "in spite of everything he [i.e., humanity] likes to call 'progress' he is not travelling along the high-road at all, but is picking his precarious way along a narrow ledge between two abysses" (1949).[170]

With human extinction no longer an incoherent impossibility, and faced with so many novel threats, the *prominence* of this idea steadily and at times rapidly increased. Recall from Chapter 1 that the prominence of a concept refers to the extent to which it is visible on the cultural landscape, and that a useful proxy measure of this is given by Google Ngram Viewer. Hence, let's pause on what the results of Google Ngram searches of relevant keywords suggest about changing patterns of thought during this period. As Appendix 1 shows, searches for "human extinction," "extinction of humanity," "the extinction of *Homo sapiens*," "human self-extinction," "human self-annihilation," and "omnicide" reveal that all underwent an upward trend in relative frequency after WWII and the Castle Bravo debacle, followed by a sudden, significant spike during the 1980s. Why the 1980s? The most obvious answer is that this is when the nuclear winter hypothesis was proposed and popularized by Sagan, though the perceived danger of a nuclear winter/nuclear conflict may have been magnified by other developments at the time, such as the end of détente in the mid-1970s, the Soviet invasion of Afghanistan in 1979, the US presidential election of Ronald Reagan in 1980, and Schell's worldwide bestseller in 1982. (The Alvarez hypothesis, discussed in the next chapter, may have contributed to this result as well.) The confluence of these factors may help explain the shape of the curve. The more important point, though, is how these Google Ngram searches confirm the thesis that the aforementioned triggering factors and enabling conditions thrust the idea of *human extinction* into our collective consciousness like never before.

Let's now turn to the fourth existential mood in our periodization of History #1, which swung attention back toward naturogenic risks, albeit of a rather different character than the Second Law.

Notes

1. Buber 1949.
2. I am indebted to Dan Zimmer for convincing me that 1954 might be a much more significant date than 1945.
3. Hyman 2010.
4. See Swatos and Christiano 1999.
5. Wells 1914/2021.
6. Rhodes 1986, 14. Soddy himself praised the book in 1926 as exemplifying Well's "customary brilliance and insight" (1926/2018).
7. Kaempffert 1933.
8. This states that "when a distinguished but elderly scientist states that something is possible, they are almost certainly right. When they state that something is impossible, they are very probably wrong" (quoted in Risse 2023).
9. Quoted in Lanouette and Silard 2013.
10. Quoted in Lanouette and Silard 2013.
11. Weart and Szilard 1978.
12. Lanouette and Silard 2013.
13. Weart and Szilárd 1978.
14. Lanouette and Silard 2013.
15. Einstein and Szilard 1939.
16. Weinberg 1961.
17. Quoted in Weart 1985.
18. Quoted in Boyer 1994.
19. Truman 1945.
20. DDH 1945.
21. Thomas 1945.
22. SLPD 1945. However, such images were provided by the US military, which made sure not to include any corpses in the pictures. Thanks to Dan Zimmer for apprising me of this fact.
23. Quoted in Boyer 1994.
24. Burchett 1945.
25. Quoted in Rhodes 1986.
26. Else 1980.
27. Wood 1945.
28. Gimbel 2015.
29. Lanouette and Silard 2013.
30. *Bulletin* 1945.
31. *Bulletin* FAQ 2021; Rabinowitch 1951. Rather humorously, there are other things that include the word "doomsday" in their name, just like the Doomsday Clock. For example, the "Doomsday List," which sounds menacing, is a catalogue that was "created by Lighthouse Digest in 1993 . . . to draw public attention to lighthouses that were endangered [of] being lost forever" (Marilyn 2015). To quote Timothy Harrison, the editor of *Lighthouse Digest*, "since that list was created, some lighthouses" that were included as endangered "were indeed destroyed and are now lost forever." In the case of the Sabine Bank Lighthouse in Louisiana, he added, at least "the lantern room was saved" (Harrison 2011). Another example is the "Doomsday rule," which is a mnemonic device "for working out the day of the week corresponding to any given date." On this account, "Doomsday for a given year is defined to be the day of the week on which the last day of February falls" (Conway 1973). Finally, the "Doomsday Book," which in Middle English was spelled "Domesday Book," refers to "a record of a survey

of English lands and landholdings made by order of William the Conqueror about 1086" (Merriam-Webster 2022).
32. To be clear, this was the chapter title. See Boyer 1994.
33. in Zoellner 2009. Although he included hopeful responses, too. The full sentence is: "It was like the grand finale of a mighty symphony of the elements, fascinating and terrifying, uplifting and crushing, ominous, devastating, full of great promise and great forebodings" (quoted in Zoellner 2009).
34. Quoted in Keyes 1945.
35. Boyer 1994.
36. Quoted in Boyer 1994.
37. Redfield 1946.
38. Cousins 1945.
39. Quoted in Muster and Sylvest 2016; Anders 1959/1962.
40. Anders 1958.
41. Russell 1945.
42. Quoted in Winchell 1946.
43. See, for example, Jaspers 1958/1961, 2.
44. Feynman 1999.
45. Kunkle and Ristvet 2013.
46. Kunkle and Ristvet 2013; Ogle 1985.
47. Matashichi and Minear 2011; Parsons and Zaballa 2017.
48. Parsons and Zaballa 2017.
49. DeGroot 2011.
50. USCD 1955.
51. Russell 1954a.
52. Russell 1954a.
53. Pugwash 2022.
54. Russell and Einstein 1955.
55. Hahn and Born 1955.
56. Anders 1956b; Dawsey 2016.
57. Jaspers 1961.
58. Koestler 1967.
59. Somerville 1980.
60. Granberry 1986.
61. Somerville 1981, 1983.
62. Somerville 1989.
63. Quoted in Waller 2016.
64. Note that I borrow "evocative phrase" from Waller 2016, 4.
65. Somerville 1979; see also Chapter 9.
66. I was subsequently informed that this date for the coinage of the word is also confirmed by the *Oxford English Dictionary* (OED 2022).
67. TM 2022.
68. Kennedy 1961. The sword of Damocles story appeared most famously in Cicero's *Tusculanae Disputationes*, c. 45 BCE, which describes Damocles temporarily assuming the role of the tyrant—Dionysius I of Syracuse—until a "bright sword" was "let down from the ceiling, suspended by a single horse-hair, so as to hang over the head of" Damocles. At this point Damocles understood that "there can be no happiness for one who is under constant apprehensions" and thus pleaded with the king to "give him leave." Thanks to Dan Zimmer for apprising me of the quote from Kennedy.
69. Konopinski et al. 1946; Buck 1959.
70. Serber 1992; Sandberg et al. 2008.

71. Segrè 1970.
72. Bethe et al. 1950.
73. Arnold 1950.
74. See Arnold 1950.
75. Quoted in Doherty 2005.
76. Quoted in Russell 1973.
77. Quoted in Weart 1988.
78. Koestler 1967.
79. Badash 2009.
80. Badash 2009.
81. Quoted in Badash 2009.
82. Von Neumann 1955.
83. Badash 2009.
84. See also Toon et al. 2008.
85. Crutzen and Birks 1982.
86. Turco et al. 1982.
87. Sagan and Turco 1990.
88. Sagan 1983a.
89. Sagan 1983b.
90. Glasstone and Dolan 1977.
91. Robock et al. 2007.
92. Robock and Toon 2012.
93. Quoted in Rhodes 1986.
94. Hence, the threat environment and existential moods are intimately related, although the latter exists at a "higher level" than the former, so to speak. With respect to the existential mood under consideration, for example, the number of plausible kill mechanisms changed over time, as did the probability of catastrophe (e.g., during the Cuban missile crisis), without any corresponding change in the overall existential mood. Existential moods track *fundamental* rather than more *superficial* changes in the threat environment.
95. Hyman 2010; Ebeling 1964.
96. This includes Szilard, Fermi, Oppenheimer, Einstein, Anders, Joliot-Curie, Feynman, Born, Rotblat, Koestler, Teller, Pauling, Vonnegut, and Sagan, as well as Russell and Wells, discussed in the previous chapter. I am not certain that Somerville was an atheist or agnostic, but he appears to have been.
97. Buckley 1990.
98. See Stolz 2020; McLeod 2005.
99. Shils 1996; Sepkoski 2020, 128.
100. Quoted in Schorr 1988.
101. Schorr 1988; Herbers 1984.
102. Stitzinger 2002, 165; Walls 2008, 10.
103. Halsell 1986.
104. Sturm 2021; Halsell 1986; Smith 2006. For example, in the quote from Reagan above, he adds that

    > Ezekiel tells us that Gog, the nation that will lead all of the other powers of darkness against Israel, will come out of the north. Biblical scholars have been saying for generations that Gog must be Russia. What other powerful nation is to the north of Israel? None. But it didn't seem to make sense before the Russian revolution, when Russia was a Christian country. Now it does, now that Russia has become communistic and atheistic, now that Russia has set itself against God. Now it fits the description of Gog perfectly (quoted in Chidester 2001).

105. Halsell 1986.

106. Whisenant 1988.
107. Einstein et al. 1948; Deudney 2019.
108. Francis 2017.
109. Quoted in Halsell 1986.
110. Prochnau 1981; Halsell 1986.
111. Walls 2007.
112. See Flannery 2016 for some more extreme examples of this sort of "active eschatology."
113. Sagan 1985.
114. Mann 2018.
115. Griswold 2012.
116. Moyers 2007.
117. Note that Murray Bookchin, under the pseudonym "Lewis Herber," also published a book in 1962 that sounded a similar alarm about humanity's impact on the natural world, titled *Our Synthetic Environment*. However, Bookchin's focus included "a broader array of environmental problems with an impact on public health," including "chemicals, erosion, atmospheric and water pollution, radiation, waste, etc." (Pérez-Cebada 2013). In a 1963 review of Bookchin's and Carson's books for *Natural History*, Vogt praised both, writing that "these books cannot be adequately discussed in such limited space. But I should like to urge every reader: if you have time for but two books next year, read these; if only one, read one of them" (1963).
118. Naess 1973.
119. See Leopold 1949; Woodhouse 2018.
120. Foreman 1985.
121. The University of Victoria library has a copy of the document here: https://vault.library.uvic.ca/concern/generic_works/b050f5c9-ecfa-4c2b-a026-235796da859d?locale=zh.
122. See Woodhouse 2018, 71.
123. Osborn 1948, vii–viii.
124. Vogt et al. 1948, 284, 17.
125. Ehrlich 1968.
126. Ehrlich 1968.
127. Brower 1968.
128. Rome 2003.
129. Weart 1988, 201.
130. Weart 1988, 200.
131. Reiss 1961.
132. Weart 1988, 201. This led some to make somewhat exaggerated claims about the dangers posed by nuclear fallout. For example, Ernest Sternglass, a physicist at the University of Pittsburg who cofounded the Radiation and Public Health Project wrote in a 1969 article for *Esquire* that the radioactive fallout resulting from anti-ballistic missile (ABM) explosions could "cause the extinction of the human race" (Sternglass 1969). Although clearly hyperbolic, Freeman Dyson argued shortly after that while "the evidence is not sufficient to prove Sternglass is right . . . the essential point is that Sternglass may be right. The margin of uncertainty in the effects of world-wide fallout is so large that we have no justification for dismissing Sternglass's numbers as fantastic" (quoted in Fox 2014).
133. Carson 1962.
134. Specifically, Carson used this term to refer to certain pesticides. It is rather funny here to note that Superior Chemical Products, Inc. called one of their insecticides "Omnicide."

135. Carson 1962.
136. Darby 1962.
137. See Souder 2012.
138. Stoll 2020.
139. In McMullen 1963.
140. The 1960s also witnessed the development of chaos theory, according to which slight differences in initial conditions can have disproportionate consequences. The climate, for example, is a chaotic system.
141. Others had earlier suggested that burning coal, for example, could alter the global climate, although in most cases the effects were anticipated to be beneficial or neutral. See, for example, Molena 1912 and Paul Edwards' discussion of Guy Stewart Callendar in Edwards 2010, 76–81.
142. Alexander 2010, 69.
143. Revelle 1965, 127. Note that Roger Revelle was among the very first to use the metaphor of "spaceship Earth" (Weart 2022).
144. Weart 2008, 208.
145. Hansen 1988.
146. Weart 2008, 151.
147. Revkin 1988.
148. Wade 1979.
149. See Sagan and Turco 1990, 455.
150. Sagan 1975.
151. Mutch 1980.
152. Frischknecht 2003.
153. Roffey et al. 2002.
154. Lockwood 2008.
155. Wells 2018.
156. Frischknecht 2003.
157. Quoted in Feinberg and Kasrils 2013.
158. Lederberg 1969.
159. Butler 1863.
160. Turing 1951.
161. Turing and Haugeland 1950.
162. Good 1965.
163. Good 1959, 1965.
164. Good 1959.
165. Minsky 1984.
166. Moravec 1988.
167. Note that Feynman attributes this idea to a colleague. Feynman 1959.
168. Drexler 1986.
169. See Torres 2018.
170. The new existential mood was so pervasive, and loomed so large within Western intellectual culture, that some social theorists began to completely reconceptualize the nature of modern societies living under the shadow of annihilation. A notable example comes from the German sociologist Ulrich Beck's 1986 book *Risk Society: Towards a New Modernity*. Writing in West Germany during the Cold War, Beck argued that "in advanced modernity the social production of *wealth* is systematically accompanied by the social production of *risks*." Of course, individuals, tribes, and nations have always encountered uncertain and dangerous phenomena—"life is risky," as the cliche goes. But Beck recognized that the sort of risks facing societies

at the time were different in essence from those that stalked us in the past. As he put the point,

> risks are not an invention of modernity. Anyone who set out to discover new countries and continents—like Columbus—certainly accepted "risks." But these were *personal* risks, not global dangers like those that arise for all of humanity from nuclear fission or the storage of radioactive waste. In that earlier period, the word "risk" had a note of bravery and adventure, not the threat of self-destruction of all life on Earth.

Indeed, Beck argued that the risks facing societies today are not just "globalized" but uniquely *political* in nature. "What *was* until now," he wrote, "*considered unpolitical becomes political—the elimination of the causes in the industrialization process itself*" (1986). What he meant is that while industry has reduced scarcity, this has come at the cost of exposing humanity—indeed the entire biosphere—to novel kinds of increasingly dire threats. Consequently, as he put it, the new "risk society is a *catastrophic* society. In it the exceptional condition threatens to become the norm." Beck's book quickly became "one of the most influential European works of social analysis in the late twentieth century," having an outsized influence on social scientific thinking about how to ensure the safety of risk societies, which Beck identifies as the fundamental issue facing the world toward the end of the twentieth century (Lash and Wynne 1992).

5 Nature Wants to Kill Us

A Barrister and His Myth

Throughout nearly the entire Cold War period, while the traumatic developments explored above were unfolding, the scientific community as a whole was quite certain that natural geophysical and astronomical phenomena do not pose any immediate threats to our collective survival on Earth. Aside from the various risks created by humanity itself, *we live on a very safe planet in a very safe universe*—one that will ultimately murder us all but has no intention of doing so for the next millions or billions of years. The lethal dangers associated with earthquakes, hurricanes, cyclones, floods, tsunamis, tornadoes, landslides, sinkholes, blizzards, and avalanches were of course universally recognized. Our world is an obstacle course of naturogenic death traps. But these were considered to be, at most, regional in scope, while kill mechanisms are by definition global. However perilous the natural world may be to us as individuals or geographically bound groups, we could at least rest assured that *Homo sapiens* itself is not at risk of suddenly perishing in a worldwide catastrophe precipitated by natural phenomena.

This view, almost unanimously accepted within the Earth sciences for nearly one and a half centuries, was founded on a paradigm called *uniformitarianism*. (Note that "Earth sciences" is an umbrella term that subsumes fields like biology, geology, paleontology, geochemistry, ecology, and climatology.) The origins of uniformitarianism are found in the late eighteenth-century work of James Hutton (1726–1797), specifically his 1788 book *Theory of the Earth*. Often called the "Father of Modern Geology," Hutton proposed a cyclical theory of geological change that strove to explain the formation of Earth's geological features entirely in terms of causes, processes, mechanisms, and operations acting today. Hence, if a cause is not currently in operation, then we cannot invoke it to explain some past geological occurrence. On this account, the world we see around us is the product of endless cycles, perfectly balanced in dynamic equilibrium, involving erosion, deposition, consolidation, and uplift. The erosion of land deposits sediment into the sea or ocean; this sediment then consolidates and is subsequently elevated above the water by volcanoes, which are animated by "an

DOI: 10.4324/9781003246251-6

internal fire or power of heat, and a force of irresistible expansion, in the body of this Earth."[1] Furthermore, since geological processes like land erosion occur very slowly, Hutton argued that Earth must be incomprehensibly ancient. This is not to say that Earth had no beginning—Hutton himself was a deist who believed that "God made all things with creative power"[2]—only that any evidence of this has long since been permanently erased by the slow-motion churning of the machinery of nature, and hence the geologist can say nothing about Earth's origins. As he famously wrote in a passage that is frequently misinterpreted as making a metaphysical rather than epistemological claim, "we find no vestige of a beginning—no prospect of an end."[3]

Hutton's uniformitarianism—a cumbersome term he never used, as it was coined in 1832 by William Whewell—was largely ignored during his lifetime, partly because *Theory of the Earth* was a sprawling 2,138 pages long and included lengthy untranslated passages in French, one of which extended across 41 pages. The book was so rambling and unpalatable that, as Stephen J. Gould notes, it made Hutton "a man renowned . . . as the all-time worst writer among great thinkers."[4] However, the uniformitarian approach was revived the following century by Charles Lyell's 1830 publication *Principles of Geology*. This outlined a similarly cyclical theory and, in the process, convinced generations of Earth scientists that the alternative theory of catastrophism—another term coined by Whewell—was deeply unscientific, infected by religious dogmas and inclined toward supernatural, catastrophic explanations for Earth's features that harken back to a prescientific age awash in superstition. As Gould argues, "much of [Lyell's] enormous success reflects his verbal skills—not mere felicity in choice of words, but an uncanny ability to formulate and develop arguments, and to find apt analogies and metaphors for their support." Referencing Lyell's first profession as a barrister, Gould characterizes *Principles of Geology* as "the most brilliant brief ever written by a scientist."[5]

For our purposes, we can analyze Lyell's theory into four core components. The first is the Huttonian methodological constraint mentioned above: the only legitimate kinds of causes, processes, and so on for explaining past events in geological history are those currently operating in the world right now. In other words, we cannot simply *invent* new kinds of causes to explain puzzling phenomena, an idea sometimes called *actualism*. The second component goes beyond this in asserting that past and present-acting causes are also fundamentally the same with respect to their *rate* and *scope*, meaning that such causes, processes, etc. are both qualitatively (kind) and quantitatively (rate, scope) alike. This is a "substantive" rather than "methodological" thesis that allows for discontinuities in Earth's history, but only on the local or perhaps regional levels.[6] A volcanic eruption, flood, earthquake, or tsunami, for example, can cause sudden changes in the physical conditions of our planet—we know this from observation and recorded history, of course—but such events never affect Earth beyond their local vicinity. Most scholars refer to this as *gradualism*, although I find the term misleading because "gradual" suggests a temporal but not spatial dimension, whereas this idea concerns both time and space. Put differently, gradualism states that while slow geophysical changes happen locally and globally, fast changes only ever happen locally. Nonetheless, with this caveat in mind, I will follow terminological convention and call this "gradualism." The third component is also Huttonian: it states that the forces of erosion and uplift are perfectly balanced, and consequently there is no cumulative change over time; geological history has no directionality. Let's refer to this as the *steady-state model*. Finally, since fast changes are highly restricted in their spatial scope, we cannot invoke them to explain large-scale geological features like Mount Everest, the Grand Canyon, or the Great Lakes of North America. Hence, if these were produced by erosion and uplift, then since erosion and uplift are very slow-moving processes, Earth must have existed for an incredibly long period of time—so long that, as Hutton declared, we can see no evidence of a beginning or auger of an end. Let's label this the *interminability thesis*.

Despite Lyell's insistence that catastrophism is unscientific, whereas uniformitarianism places geology on firm scientific ground, the truth is just the opposite. For example, many catastrophists embraced a form of actualism, arguing that scientific theorizing should begin with known causes and that catastrophes should be invoked *only* when these causes are unable to adequately explain the past. Often, the catastrophes invoked were simply more rapid (rate) and widespread (scope) versions of phenomena known from the present, such as floods and volcanic eruptions.[7] As one of the most "radical" catastrophists of the nineteenth century, Alcide d'Orbigny, made the point: "Natural causes now in action have always existed . . . To have a satisfactory explanation of all past phenomena, the study of present phenomena is indispensable," to which he added that topographical alterations caused by violent earthquakes are "for us, on a small scale, and with effect much less marked, the same phenomenon as one of the great and general perturbations to which we attribute the end of each geological epoch."[8] Even more, catastrophists like Cuvier—who, recall from Chapter 3, believed that sudden changes in sea level have occasionally punctured Earth's history, causing many species to go extinct—adhered to what Gould calls "empirical literalism."[9] That is to say, unlike Hutton and Lyell, they read the geological record literally; and this record clearly suggests that major transition events, Cuvier's *révolutions*, have indeed occurred. As Cuvier wrote in his 1813 *Essay on the Theory of the Earth*,

> the breaking to pieces, the raising up and overturning of the older strata, leave no doubt upon the mind that they have been reduced to the state in which we now see them, by the action of sudden and violent causes; and even the force of the motions excited in the mass of waters, is still attested by the heaps of debris and rounded pebbles which are in many places interposed between the solid strata.[10]

In contrast, Hutton based key aspects of his uniformitarian theory on first principles which he then sought to support with empirical evidence, which undermines the textbook myth passed down since Lyell that he built his ideas from the ground up, by engaging in fieldwork. Hutton was thus closer to Lyell's caricature of the catastrophists than actual "catastrophists" like Cuvier and d'Orbigny.[11]

Entropy, Evolution, and the Fossil Record

Uniformitarianism also proved to be more at odds with other scientific ideas at the time than catastrophism and consequently exponents of the former view found themselves having to modify some of the core components enumerated above. Although Lyell's book made an immediate splash after its publication, uniformitarianism did not become the dominant paradigm until the middle of the nineteenth century. The 1850s, of course, are when thermodynamics emerged as a foundational subfield of physics and Darwin introduced his theory of evolution. Both undermined aspects of Lyell's view. For example, the laws of thermodynamics clearly contradict the steady-state model and interminability thesis. As Peter Bowler observes, although "Hutton's theory certainly looks modern at first sight, . . . he applied his cyclic view of the earth's history so rigorously that the earth had to be seen, in effect, as a perpetual motion machine."[12] But the Second Law, in particular, banishes perpetual motion to the ashcan of impossibility, meaning that Earth could not have existed in its current state indefinitely. Entropy is increasing and will continue to do so until the solar system, if not the entire universe, reaches the irreversible end-point of thermodynamic equilibrium. This implies that time is directional (from lower to higher states of entropy), and calculations from Lord Kelvin put Earth's age at only about 100 million years, although Kelvin later shortened his estimate of when Earth's crust had solidified to "less than 40 million years ago, and probably much nearer to 20 than 40."[13]

Throughout the 1850s, the tension between uniformitarianism and thermodynamics was mostly ignored by scientists in both camps. This changed the following decade when Kelvin launched a series of attacks against the aforementioned components of Lyell's theory, quite possibly spurred on by his reading of Darwin's *Origin*. Darwin was immensely influenced by Lyell, whom he considered a mentor, and came to embrace every aspect of Lyellian uniformitarianism except for one (see below). As he wrote in the *Origin*, mocking the use of catastrophes to explain sudden shifts in the fossil record, "so profound is our ignorance, and so high our presumption, that we marvel when we hear of the extinction of an organic being; and as we do not see the cause, we invoke cataclysms to desolate the world . . .!"[14] According to Darwin's theory, natural selection brings about changes in the frequency of traits within a population through differential reproduction, and is thus a transgenerational mechanism. As such, natural selection requires long periods of time (many generations) to operate. That of course dovetails nicely with the uniformitarian theses of interminability and gradualism: Earth has existed *long enough* for natural selection to have produced the extraordinary diversity of species observed today, and the *slow rate* of environmental change enables natural selection to ensure that species are sufficiently well-adapted to their surroundings (though environmental change can also precipitate species extinctions, as a result of Malthusian competition over resources).

While Kelvin was not opposed to biological evolution in principle, he accused Darwin and Lyell of defending a theory that violates arguably the most fundamental law of physics, and by the end of the 1860s he had convinced many that "a strictly Lyellian view was . . . untenable."[15] The abandonment of the steady-state component of Lyell's view was, in fact, also helped along by Darwin's theory, since evolution involves cumulative change over time, and cumulative change implies that history has a direction—if not *toward* an end, then at least *away from* a beginning. Hence, the only aspect of uniformitarianism that Darwin initially rejected, in 1859, was its denial that history is directional. (Other than this, his theory was crucially founded on uniformitarian principles.) Yet Darwin also shied away, to an extent, from the interminability thesis in subsequent editions of the *Origin*, shortening his estimates of how long Earth has existed. For example, he originally calculated that the denudation of the Weald in South England, a large structure formed by erosion, took up to 300 million years, which is of course much longer than Kelvin's estimate of Earth's age. After receiving fierce criticism for this, he removed the figure entirely in the third edition.

To recap, the Second Law and Darwin's theory of evolution both undermined the steady-state model, while the Second Law contradicted the interminability thesis as well. In contrast, neither posed a problem for catastrophism. We should note, though, that Kelvin's estimates turned out to be wildly inaccurate. The reason is that Earth's core contains several radioactive elements that, as a result of decay, provide an extra source of energy to warm Earth's mantle, in addition to the residual primordial heat from the planet's formation in the solar nebula. The rate of Earth's cooling is therefore much slower than Kelvin could have known, as radioactivity was first discovered in 1896 and its capacity to release heat wasn't recognized until 1903, when Pierre Curie and Albert Laborde announced that radium salts "emit heat continuously and to a measurable extent," to quote a *Popular Science* article of the time.[16] Not long after, newly discovered radiometric dating techniques confirmed that our planet has existed for far longer than Kelvin believed, at least for a few billion years; today we know that Earth is roughly 4.5 billion years old. While this is orders of magnitude larger than Kelvin's largest estimates, it is still a far cry from time being indefinitely long. There is, contra Hutton, a vestige of a beginning.

Hence, two of the core components of Lyell's uniformitarianism had been seriously wounded by the end of the nineteenth century. But there was another problem, too, that became increasingly salient and difficult to ignore throughout the twentieth century in particular: the fossil record, which raised a direct challenge to gradualism. As mentioned above, Cuvier and other leading catastrophists were not *biblical* but *empirical* literalists: evidence of sudden, worldwide catastrophes in the stratigraphic record and buried within ancient fossiliferous rocks should be interpreted as precisely that; the language of geological history is clear and unequivocal. Hence, in addition to the physical evidence cited by Cuvier above, he also argued, based on his reading of the data, that

> life . . . has often been disturbed on this Earth by terrible events. Numberless living beings have been the victims of these catastrophes; some, which inhabited the dry land, have been swallowed up by inundations; others, which peopled the waters, have been laid dry, from the bottom of the sea having been suddenly raised; their very races have been extinguished forever, and have left no other memorial of their existence than some fragments, which the naturalist can scarcely recognize.[17]

On Cuvier's view, species are fixed and unchanging over time, but some had on occasion been annihilated together by large-scale (at least continent-wide) disasters. Lyell and Darwin both vigorously rejected this. For them, quoting Kolbert, "extinction was a lonely affair," in the sense that it happens to individual species, one at a time, in an uncoordinated manner, over long stretches of history.[18] As Darwin wrote in the *Origin*, "the complete extinction of the species of a group is generally a slower process than their production," which comports with his view that the fundamental cause of extinction is competition within and between species over scarce resources rather than sudden disasters.[19] This means that there are *no mass extinctions*. If there had been, then gradualism would almost certainly be false, as the most plausible explanation for how large numbers of species distributed over broad geographical areas could disappear simultaneously is to invoke catastrophes like global floods, massive volcanic eruptions, worldwide earthquakes, and so on.[20]

How, then, did Lyell and Darwin explain the obvious patterns of discontinuity in the fossil record? They claimed that the appearance of mass extinctions in the fossil record is an artifact of its incompleteness, an illusion produced by the fact that fossilization is the exception rather than the rule. Consequently, this record is an unreliable source of data, and hence it should not be read literally. As Darwin explained the idea:

> Those who believe that the geological record is in any degree perfect, will undoubtedly at once reject my theory. For my part, following out Lyell's metaphor, I look at the geological record as a history of the world imperfectly kept and written in a changing dialect. Of this history we possess the last volume alone, relating only to two or three countries. Of this volume, only here and there a short chapter has been preserved, and of each page, only here and there a few lines. Each word of the slowly-changing language, more or less different in the successive chapters, may represent the forms of life, which are entombed in our consecutive formations, and which falsely appear to have been abruptly introduced. On this view the difficulties above discussed are greatly diminished or even disappear.[21]

This *incompleteness hypothesis*, as we could call it, became the canonical view within the Earth sciences for more than a century, during which the gradualism component of uniformitarianism—at this point the only component still standing aside from actualism—continued to dominate thinking and constrain theorizing about Earth's past. In fact, since natural selection is a gradualistic mechanism, the Modern Synthesis of the 1930s further entrenched this aspect of uniformitarianism. The reason is that Gregor Mendel's theory of heredity, which the Modern Synthesis combined with Darwin's account of evolution, made natural selection far more plausible than it had previously appeared, thus adding support to the notion that biological evolution unfolds slowly, and that species die out because of competition rather than catastrophe.

The uniformitarian bias against global catastrophes was at this point a centerpiece of many scientific textbooks, and the standard picture of early nineteenth-century debates within geology pitted Cuvier against Hutton, a religionist who naively accepted the Mosaic chronology versus a dedicated scientist who put fieldwork first. This was, once again, not remotely accurate, although it was reinforced in the postwar era by (a) the emergence of creation science (or scientific creationism), which embraced a catastrophist theory of Earth, and (b) the pseudoscientific work of Immanuel Velikovsky. Taking these in order: catastrophism appealed to young-Earth creationists because it could explain why Earth looks old even though it is really quite young—born on October 23, 4004 BCE, according to the famous calculation by James Ussher. Meanwhile, Velikovsky published a book in 1950 titled *Worlds in Collision* that "inspired a popular reaction" against uniformitarianism, which reverberated through the corridors of American culture for several decades.[22] He began with the idea that we should take seriously the ancient legends, myths, tales, and lore about catastrophes having devastated the planet long ago, and then built an account of history according to which sometime around the fifteenth century BCE, Jupiter "ejected" the planet Venus, which drifted through the solar system and, while passing by Earth, triggered changes in Earth's orbit and axis. This caused a series of catastrophes that ancient civilizations recorded in the fragmentary documents and mythological stories passed down to us over the generations. Although Velikovsky's proclamations, some of which contradicted Newtonian physics, "made him a hero in the eyes of the counterculture of the 1960s," his flawed methodology further discredited catastrophism among working scientists.[23]

Alien Assassins, Extraterrestrial Executions

Despite these ossifying forces, reputable scientific work did chip away at the gradualist consensus among professional Earth scientists. Around the same time Velikovsky was making mischief, a small handful of notable researchers—perhaps affected by anxieties associated with the possibility of nuclear annihilation[24]—began to suggest that mass extinctions were genuine features of life's history rather than mirages of the fossil record. For example, Norman Newell argued in 1952 that the fossil record "is an adequate sample of the evolutionary history of the better known groups," and consequently we are justified in inferring that "mass extinctions of marine genera on a global scale" have punctuated the deep past.[25] A few years later, he again made the case that "enigmatic, apparently world-wide, major interruptions in the fossil record . . . are real, approximately synchronous, and are recognizable at many places in different parts of the world," where such "critical events in the history of life evidently were responsible for these world-wide revolutionary changes."[26] Still, Newell was cautious about describing such mass extinction events as "catastrophic," claiming instead that they are best explained by cumulative environmental changes caused by, for example, alterations in sea level that unfolded over periods of a few hundred to a few million years.[27] As Kolbert writes about these developments:

> [T]he more that was learned about the fossil record, the more difficult it was to maintain that an entire age, spanning tens of millions of years, had somehow or other gone missing. This growing tension led to a series of increasingly tortured explanations. Perhaps there *had* been some sort of "crisis" at the close of the Cretaceous [when the non-avian dinosaurs rapidly vanished], but it had to have been a very slow crisis. Maybe the losses at the end of the period *did* constitute a "mass extinction." But mass extinctions were not to be confused with "catastrophes."[28]

Hence, even those bold enough to suggest that "major interruptions" and "mass extinctions" may have occurred in Earth's past nonetheless agreed that global catastrophes are unlikely, if they can even happen at all. The gradualistic component of uniformitarianism was simply too entrenched within the Earth sciences for anyone to acknowledge the obvious: the sudden disappearance of many species in the fossil record implies that these species disappeared *suddenly*, and the best explanation for this is a *global catastrophe*. While the creationists and Velikovsky were peddling blatant pseudoscience, the truth is that uniformitarians were also beholden to a kind of dogma at the heart of their fields. Only the most groundbreaking and undeniable discovery could hope to dislodge it.

Yet this is precisely what happened after two extraordinary discoveries were announced in 1980 and 1991. The first discovery dates back to the late 1970s, when rock samples collected in 1977 at the city of Gubbio, Italy, were tested the following year and found to contain anomalously high amounts of iridium. This was perplexing because iridium is an iron-loving element that has, as a result, mostly sunk into Earth's core, thus making it one of the rarest elements in Earth's crust. To confirm that the iridium anomaly wasn't restricted to Gubbio, another sample was tested from Denmark, and later on from New Zealand. All showed the same spike in iridium. Even more, this iridium layer coincided with the boundary between the Cretaceous and Tertiary periods of geological history, that is, the K-T boundary (now called the

K-Pg boundary, after the Tertiary was renamed the Paleogene), which is when the non-avian dinosaurs mysteriously vanished from the planet—about 66 million years ago.[29] The obvious question was: Could there be a connection?

The scientists leading this project included the father-son team of Luis and Walter Alvarez, along with Frank Asaro and Helen Michel. Amazingly, Luis Alvarez was a Nobel laureate who not only worked on the Manhattan Project and witnessed the first atomic bomb explosion in New Mexico, but also watched the August 6 bombing of Hiroshima from an observation aircraft that accompanied the B-29 Superfortress that dropped the bomb. It is interesting to speculate about whether Luis Alvarez might have been inclined to accept catastrophist explanations because of his personal experience with nuclear weapons. Either way, the initial hypothesis that they considered was that the iridium might have originated from a nearby supernova explosion. In the 1950s and 1960s, the German paleontologist Otto Schindewolf defended the hypothesis that cosmic radiation (cosmic rays), perhaps emitted by supernovae, had killed off the dinosaurs, though most scientists at the time ignored the idea.[30] Others proposed something similar in the early 1970s, which prompted the Alvarez team to investigate the idea, but they eventually ruled out this hypothesis. Subsequently, they turned to the possibility that an asteroid or comet had collided with our planet, as these contain higher concentrations of iridium than Earth's crust. Although speculations about cometary collisions are found throughout history, as discussed in Chapter 2, it was only in the 1960s that the scientific community came to believe that certain craters on Earth's surface were the result of extraterrestrial impactors. Prior to this, most held that they were the result of the explosive release of gas from Earth, and that if rocks can in fact fall from the sky (paraphrasing Jefferson's comment about the Weston meteorite) they haven't much affected Earth's history over the past 500 million years.[31] As Trevor Palmer writes, few accepted that celestial bodies pose any

> physical threat to the Earth, as far as could be ascertained from two centuries of scientific observation. Although 200 years was not a long time in relation to the age of the Earth, the reassuring conclusions from this brief period could easily be extrapolated in view of the prevailing uniformitarian paradigm.[32]

Nonetheless, the Alvarez team came to believe that an extraterrestrial collision was probably the correct explanation for the iridium anomaly. Yet this presented another conundrum: if a large asteroid or comet had slammed into Earth 66 million years ago, how could this have caused species around the entire planet to suddenly perish? What was the global spread mechanism? Could it have been the heat produced, as M. W. de Laubenfels suggested in 1956, an idea that even fewer scientists paid attention to than Schindewolf's cosmic-rays hypothesis?[33] After a year of ruminating the puzzle, Luis Alvarez remembered having read that the 1883 volcanic eruption of Krakatau, Indonesia, catapulted such large quantities of dust and ash into the atmosphere that it changed the color of sunsets in London, roughly 11,604 km away, for several months.[34] In fact, the first study to affirm a connection between major volcanic eruptions and alterations in the optical properties of the atmosphere was published by the "Krakatoa Commission" in 1888.[35] This seemed to be the missing link: a large object from outer space struck Earth and injected huge quantities of pulverized rock into the stratosphere; the dust, being above the weather, spread around the globe, blocking out incoming solar radiation and consequently reducing photosynthesis; food chains collapsed, with the largest animals, the dinosaurs, being most affected, and consequently a planetary-scale mass extinction event ensued. This scenario was later labelled an "impact winter," which of course evokes the nuclear winter idea proposed by Crutzen and Birks and elaborated by the TTAPS group in 1983.[36]

Rock Hounds

In 1980, the Alvarez team published their results and hypothesis linking the iridium anomaly with a catastrophic impact event that wiped out the dinosaurs by darkening the sky. It is difficult to overstate how momentous this paper was. In William Glen's words, the Alvarez hypothesis, as it became known, was "as explosive for science as an impact would have been for Earth."[37] It received considerable—and oftentimes favorable—coverage in the popular media, and immediately split the scientific community into two opposing camps. On the one hand, "a large portion of the world's paleontologists" were overwhelmingly hostile to the idea, in part because they felt that the Alvarez team—which included a physicist (Luis), geologist (Walter), and two chemists (Asaro, Michel)—were trespassing on their epistemic territory.[38] As Robert Bakker, a paleontologist who helped initiate the "dinosaur renaissance" in the late 1960s, complained to the *New York Times* in 1985:

> The arrogance of those people is simply unbelievable. . . . They know next to nothing about how real animals evolve, live, and become extinct. But despite their ignorance, the geochemists feel that all you have to do is crank up some fancy machine and you've revolutionized science."

He continued:

> The real reasons for the dinosaur extinctions have to do with temperature and sea-level changes, the spread of diseases by migration and other complex events. But the catastrophe people don't seem to think such things matter. In effect, they're saying this: "We high-tech people have all the answers, and you paleontologists are just primitive rock hounds."[39]

The ubiquity of this sentiment among paleontologists was attested by a survey conducted during the 1985 Society of Vertebrate Paleontologists meeting. While most participants concurred that "some large extraterrestrial object probably did hit the earth 65 million years ago," only five of the 118 respondents agreed that "an asteroid or comet had caused the extinction of dinosaurs and many other land animals at the end of the Cretaceous period." Another 32 respondents claimed that no mass extinction even occurred at the K-T boundary, because the disappearance of land animals at the end of the Cretaceous had unfolded over millions of years and hence was "neither instantaneous nor simultaneous." Consequently, "there was no need to speculate about catastrophes."[40] Skepticism about the impact hypothesis persisted for many years, and indeed a *New York Times* article declared in 1988 that "the debate over dinosaur extinction rages on" with "growing doubts about [the Alvarez] theory expressed by some scientists." The science communicator Robert Jastrow even proclaimed that "it is now clear that a catastrophe of extraterrestrial origin had no discernible impact on the history of life as measured over a period of millions of years."[41] Walter Alvarez himself admitted in 1997 that he initially found the idea of a sudden global catastrophe "disturbing," given his background in the Earth sciences. "As a geology student," he wrote,

> I had learned that catastrophism is unscientific. I had seen how useful the gradualistic view had been to geologists reading the record of Earth history. I had come to honor it as the doctrine of "uniformitarianism" and to avoid any mention of catastrophic events in the Earth's past.[42]

Yet, putting aside the turf wars that at times became quite vicious and personal, there was in fact no conclusive evidence for an asteroid collision at the K-T boundary.[43] Most notably, the Alvarez team could not point to any crater on Earth to corroborate their hypothesis. If an asteroid or comet 10 ± 4 km in diameter, according to their calculations, had collided with our planet, it should have left behind a massive concave indentation in Earth's surface. Even if it had landed in the ocean, geologists estimate that only about 20 percent of the ocean crust dating back to the K-T boundary had been subducted, and an ocean landing would have produced additional evidence resulting from the enormous tsunami that would have washed over entire continents.[44] Without a crater, the Alvarez team was missing the smoking gun.

This changed around 1990, after years of relentless sleuthing for the missing link led a graduate student named Alan Hildebrand to rediscover a ~180-km-wide crater buried nearly a kilometer beneath the Yucatan Peninsula, near the Mexican town of Chicxulub. The huge geological structure had been identified earlier, in 1978, by the geophysicist Glen Penfield of the national oil company of Mexico, Pemex, although some geologists on the Pemex team believed that a submarine volcano had produced the crater, and details of their investigation were kept secret.[45] In 1991, cores unearthed by Pemex were finally analyzed and shown to contain shocked quartz at exactly the K-T boundary. This strongly implied that the crater was the result of an impactor, since volcanoes are unable to produce the intense pressures necessary for shocked quartz to form. Later that year, a group of scientists led by Hildebrand published their findings, linking the Chicxulub crater to an impact collision that, in their words, "may have caused the K/T extinctions."[46]

Almost immediately, this bombshell convinced nearly everyone—including the paleontologists—that a massive catastrophe had indeed wiped out the dinosaurs 66 million years ago. As Alvarez later recalled, it was the winter between 1991 and 1992 that "seemed like the turning point."[47] Here, then, was an extraordinary pivot in the history of science: two discoveries, made roughly a decade apart, overturned the gradualistic dogma at the heart of twentieth-century uniformitarianism: sudden, violent, global-scale catastrophes have happened in the past, triggering mass extinction events—and, by implication, they might happen again in the future. It thus provides a fascinating example of how science can be a self-correcting process. As Kolbert quotes Alvarez in her book *The Sixth Extinction*:

> Just think about it for a moment. Here you have a challenge to a uniformitarian viewpoint that basically every geologist and paleontologist had been trained in, as had their professors and their professors' professors, all the way back to Lyell. And what you saw was people looking at the evidence. And they gradually did come to change their minds.[48]

With the collapse of gradualism, a new paradigm took its place: *neo-catastrophism*. This not only affirmed what Cuvier had suspected about the fossil record, based on his literalist reading of it, but flung open the door to the startling possibility that large asteroid or cometary collisions could someday wipe out humanity. As David Morrison of the NASA Ames Research Center and two colleagues wrote in 1993, we can no longer "exclude the possibility of a large comet appearing *at any time* and dealing the Earth such a devastating blow—a blow that might lead to *human extinction*."[49] The Alvarez hypothesis and Chicxulub discovery thus catalyzed a new shift in existential mood, this one marked by the ominous recognition that near-term extinction risks arise from not just our own actions but the natural world that we call home.

Cosmic Rays, Krakatau, and a Supereruption on Sumatra

These developments raised yet another question: are there additional kill mechanisms lurking in the cosmic shadows of our scientific ignorance? We have already seen that Schindewolf proposed that cosmic rays, a form of ionizing radiation, have precipitated the mass die-offs seen in the paleontological record, the causal mechanism being the accumulation of deleterious mutations in the genomes of exposed organisms.[50] Schindewolf tentatively tied these waves of radiation to supernovae, although subsequent research pointed to a different possibility: the bursts of *gamma rays* (which are different than cosmic rays) produced by some supernovae can dissociate atmospheric molecules of N2 and O2, which then combine to form nitrogen oxides (NO_x). Nitrogen oxides can, in turn, eliminate ozone (O3), thereby leaving the biosphere vulnerable to DNA-damaging UV radiation from the sun. As John Ellis and David Schramm explained in a 1995 paper,

> a supernova explosion of the order of 10 [parsecs] away could . . . destroy the ozone layer for hundreds of years, letting in potentially lethal solar ultraviolet radiation. In addition to effects on land ecology, this could entail mass destruction of plankton and reef communities, with disastrous consequences for marine life as well.[51]

Supernovae are one possible source of gamma-ray bursts, but there may be others. By the mid-1990s, it was clear that gamma-ray bursts, whatever their origin, could strip Earth of its protective layer of ozone, thus bringing about "dramatic, biosphere-wide effects" that would pose direct and indirect hazards to our survival.[52] Another astrophysical scenario discussed was the possibility of a false vacuum decay, first identified in the mid-to-late 1970s.[53] If the universe is in a "metastable" energy state, it could be possible for perturbations to initiate a phase transition to a more stable energy state, resulting in the "nucleation" of a vacuum bubble that expands in all directions at nearly the speed of light, destroying everything it comes into contact with in the universe. Although the probability of a phase transition was calculated in 1983 to be "completely negligible," if it were to occur it would constitute "the ultimate ecological catastrophe," since "in a new vacuum there are new constants of nature [and hence] after vacuum decay, not only is life as we know it impossible, so is chemistry as we know it."[54]

These are all threats from above—from the heavens—but the most unsettling discovery aside from the Alvarez hypothesis during this period pertained to threats from below—from Earth itself. I mentioned earlier that the Alvarez team struggled to identify a global spread mechanism that might explain how an impact on one side of the planet could affect species everywhere. The breakthrough came when Luis Alvarez remembered the aberrant climatic conditions caused by the Krakatoa eruption. In fact, a link between unusual weather patterns and volcanic eruptions had been made a century before Krakatoa by Benjamin Franklin, in 1784, while he was in Europe as the US minister to France. The winter between 1783 and 1784 was unusually cold and Franklin noticed a dry, lingering fog that dimmed the sun to such an extent that, in his words, "when collected in the focus of a burning glass, they would scarce kindle brown paper."[55] Having heard about a volcanic eruption in Iceland during 1783—the eruption of Laki, which Franklin misidentified as Hekla—he hypothesized a connection between the dimmed sun, frigid weather, and smoke injected into the atmosphere by the eruption.[56]

Some research in the early twentieth century supported Franklin's hypothesis, although the data was inconclusive and some studies reported no correlation between reduced solar radiation, cooling of Earth's surface, and volcanic eruptions.[57] However, subsequent investigations found that what matters isn't the amount of volcanic smoke or ash but the sulfur dioxide (SO2) and hydrogen sulfide (H2S) content of the volcanic gases. When SO2, for example, reaches the stratosphere, it undergoes a chemical reaction to become sulfuric acid (H2SO4), which reflects—or *backscatters*—incoming sunlight. This reduces the amount of solar radiation reaching Earth, which causes surface temperatures to fall. An opportunity to test this hypothesis came with the eruptions of Mount St. Helens, in 1980, and the El Chichón volcano in Mexico, in 1982. Both produced roughly the same volumetric quantity of ejecta, but the latter was rich in SO2, while the former wasn't. Confirming the volcanologist's suspicions that sulfate aerosols are the critical factor, the climatic effects of El Chichón were appreciably more pronounced than those of Mount St. Helens. This established a mechanism by which large volcanic eruptions could potentially bring about what came to be called a "volcanic winter," on the model of "nuclear winter."

But are there eruptions capable of injecting enough SO2 into the stratosphere to cause worldwide devastation? In 1982, Chris Newhall and Stephen Self introduced the "Volcanic Explosivity Index" based on criteria like volume of ejecta, column height, duration, and stratospheric injection.[58] The most mild eruptions were classified as "VEI 1" and the most catastrophic as "VEI 8," which correspond to what are now called *supereruptions*. Newhall and Self classified Laki as VEI 4, Krakatau as VEI 6, and Tambora as VEI 7, although far larger eruptions had been identified in the geological record by the 1970s, such as the Toba supereruption on the Indonesian island of Sumatra some 75,000 years ago.[59] In a 1988 paper titled "Volcanic Winters," Michael Rampino, Stephen Self, and Richard Stothers argued that the Toba supereruption could have produced "conditions of total darkness . . . over a large area for weeks to months," and that "the atmospheric after effects of a Toba-sized explosive eruption might be comparable to some scenarios of nuclear winter." They concluded with the observation that even the largest eruptions in historical memory have been

> small compared with the very large explosive and effusive eruptions that are well known from the geologic record. A simple scaling-up of the effects of historic eruptions suggests that the much larger eruptions could have brought about severe, short term coolings of "volcanic winters" over considerable portions of the globe.[60]

Although they did not explicitly link this to the possibility of human extinction, the implications were clear, and indeed the historian of science Matthias Dörries describes this very article as coming "closest to something like doomsday science within the constraints of a scholarly journal."[51] The connection was further reinforced in 1993 by Ann Gibbons, who speculated in a *Science* article that the volcanic winter caused by the Toba event may have been responsible for an apparent population bottleneck around the time of the eruption that nearly wiped out *Homo sapiens*, an idea known today as the "Toba catastrophe theory." The same year, Rampino and Self responded with approval to Gibbons' proposal, affirming that "climate cooling for 1 or 2 years after the [Toba] eruption could have been quite severe, representing 'volcanic winter' conditions similar to those proposed in scenarios of nuclear winter following a major nuclear exchange." Consequently "it may have been connected to a possibly unique Late Pleistocene bottleneck in human evolution."[62] Although the human population is much larger today and the conditions created by our technological civilization are far different than those of our Paleolithic hunter-gatherer ancestors, the effects of a multi-year "winter" event could arguably be no less severe, perhaps inching humanity just as close to the precipice of extinction as we came 750 centuries ago. As Rampino declared in a documentary on "supervolcanoes" for the BBC, if the Yellowstone volcano, for example, were to produce another supereruption, it would "be disastrous for the United States and eventually for the whole world," causing "civilization . . . to creak at the seams."[63]

It should be clear at this point that the various "winter" scenarios—nuclear, impact, and volcanic—are connected, not just in relation to their global spread mechanisms—soot, pulverized rock, or sulfate aerosols affecting the entire planet via stratospheric dispersal—but with respect to their historical discovery. The causal chronology goes as follows: early work from Franklin and the Krakatau Commission connected smoke, dust, and/or ash in the atmosphere with changes in the climate. This

Figure 5.1 Picture of Lake Toba, taken by Stephen Self. Used with permission.

inspired the Alvarez team to devise their impact winter hypothesis, and indeed the original paper includes an entire section on the Krakatau eruption. The Alvarez hypothesis directly inspired Crutzen and Birks (1982) and the TTAPS group (1983), both of which cite the Alvarez paper. The TTAPS paper also discusses eruptions like Tambora. This in turn directly inspired scientists like Self, Rampino, and Stothers to propose the volcanic winter hypothesis in the latter 1980s and 1990s, outlined in papers that typically cited all the aforementioned scientists, and in fact the Toba supereruption became something of "a test case in supporting or discrediting" the nuclear winter hypothesis.[64] It is also worth noting that one of the critical pieces of evidence in favor of the Alvarez hypothesis and identification of the Chicxulub crater as having an extraterrestrial origin was shocked quartz, which was first observed at the sites of nuclear test explosions.[65] Finally, as mentioned, Luis Alvarez himself was involved in the Manhattan Project, even witnessing the Hiroshima bombing. One lesson from this is that if there are future kill mechanisms to discover, it could be that the identification of a single *type* of mechanism, in this case the injection and spread of particles in the stratosphere, will lead to a sudden explosion of new ideas about how humanity might go extinct.

The End of an Age

These developments, especially research on supereruptions, further solidified the existential mood initiated by the Alvarez hypothesis. We saw in the previous chapter that key changes to the threat environment in the early postwar era concerned the temporality and multiplicity of anthropogenic threats; that is, humanity introduced multiple ways we could destroy ourselves in the relative near future, perhaps even *tomorrow*. Similar developments characterized the shift to this new existential mood: over just one decade or so, scientists identified not one but several natural phenomena that could annihilate humanity at any moment. We are not, in fact, living on a very safe planet in a very safe universe, although calculations of the probabilities of natural disasters doing us in provided a bit of reassurance. For example, supereruptions are quite rare, happening once every 50,000 years or so; asteroids 5-km across strike Earth on average once every 20 million years; gamma-ray bursts are even less common; and the likelihood of a vacuum decay catastrophe could be negligible. Still, the nature of these kill mechanisms—with the exception of vacuum decay—are such that their occurrence is a matter of when rather than if: at some point, a giant asteroid or comet *will* collide with Earth; at some point, another supereruption *will* occur; and so on. Our only hope is to devise technological means to protect us from these inevitabilities, such as spacecraft capable of deflecting incoming asteroids away from Earth. (Incidentally, though, Carl Sagan noted that deflection spacecraft would be "dual-use"—see the next chapter—and as such they could enable nefarious actors to direct asteroids toward our planet, a risk he called the "deflection dilemma.")[66]

As with every shift in existential mood thus far a secular existential hermeneutics was integral to this new mapping of the threat environment. Without affirmation that our extinction is possible in principle, the discoveries above would not have implied that our existential predicament is any more precarious than previously thought. However, while many religious apocalypticists in the postwar era, the Atomic Age, eagerly integrated thermonuclear weapons into their eschatological narratives, few paid much attention to the implications of neo-catastrophism within the Earth sciences. This is interesting given that some earlier religious writers had connected natural catastrophes with the apocalypse, as in the case of Edgar Allan Poe's "The Conversation of Eiros and Charmion" (1839). Many others, to be clear, including Young Earth Creationists in the twentieth century, linked global catastrophes involving natural phenomena with *past* events, such as the Noachian flood; but the focus here was backward- rather than forward-looking, retrospective rather than anticipatory. A notable exception came from the influential televangelist Pat Robertson. In his 1995 book *The End of the Age*, a large asteroid slams into the Pacific Ocean near Los Angeles, triggering a massive tidal wave, fires, earthquakes, and other disasters that kill millions of people. The whole world then "tumbles into political, social, and economic chaos—the biblically prophesied Great Tribulation."[67] But, on the whole, and despite the 1998 movie starring Bruce Willis about a Texas-sized asteroid heading for Earth being named *Armageddon*, there was little contact between neo-catastrophism and religious apocalypticism.

This brings us to the final shift in existential mood—so far.

Notes

1. Hutton 1788; Marvin 1990, 148; Palmer 2012, 50.
2. Quoted in Dean 1975.
3. Hutton 1788; Marvin 1990. The key epistemological word here is "find." Hutton made the point more clearly in a 1785 lecture to the Royal Society of Edinburgh like this: "With respect to human observation, this world has neither a beginning nor an end" (quoted in Rudwick 2005, 170).
4. Moore and Moore 2006, 9; Gould 1996, 64.
5. Gould 1996.
6. The distinction between "methodological" and "substantive" claims is a reference to Gould 1965.
7. See Gould 1996, 127–128.
8. Quoted in Gould 1996.
9. Bowler 2003, 4; Gould 1996.
10. Cuvier 2018.
11. See Rudwick 2005, 169–170; Palmer 2012, 51.
12. Bowler 2003.
13. Quoted in Palmer 2012.
14. Darwin 1859.
15. Burchfield 1975, 33, 90.
16. Bolton 1903.
17. Cuvier 1813.

18. Kolbert 2014.
19. Darwin 1859.
20. See Sepkoski 2020, 30.
21. Darwin 1875.
22. Bowler 2003.
23. Velikovsky 1950; Bowler 2003.
24. Sepkoski 2020, 8.
25. Newell 1952.
26. Newell 1956; Sepkoski 2020, 146–147.
27. Sepkoski 2020, 147; Palmer 2012, 108.
28. Kolbert 2014.
29. Note that at the time they believed this to be 65 million years ago.
30. See D'Hondt 1998, footnote 9.
31. Palmer 2012, 99; Alvarez 2015, 76.
32. Palmer 2012.
33. Palmer 2012, 107.
34. Alvarez 1997, 77. Note that the anglicized word is "Krakatoa," but the rest of the world refers to it as "Krakatau." Thanks to Stephen Self for suggesting that I stick with the Indonesian word.
35. See Russell and Archibald 1888.
35. It is unclear whether Luis Alvarez was familiar with the suggestions from John von Neumann, Tom Stonier, and others in the postwar era, discussed in the previous chapter, that nuclear weapons could loft enough dust into the stratosphere to turn day into night, summer into winter.
37. Glen 1994.
38. Browne 1988.
39. Browne 1985. Indeed, Luis Alvarez remarked in a 1988 interview with the *New York Times* that many skeptics of the impact hypothesis were simply inferior scientists. "I don't like to say bad things about paleontologists," he said, "but they're really not very good scientists. They're more like stamp collectors" (quoted in Browne 1988).
40. Browne 1985.
41. Browne 1988. On Jastrow, see Schwartz 2008.
42. Alvarez 1997.
43. As Malcolm Browne reported for the *NYTs*, some scientists claimed that

> the impact theory has had pernicious effects on science and scientists. They charged that controversy over the impact theory has so polarized scientific thought that publication of research reports has sometimes been blocked by personal bias. . . . According to a few paleontologists, dissenters from the meteorite theory have faced obstacles in their careers and are sometimes even privately branded as militarists, on the supposed ground that anyone who questions the catastrophic theory of dinosaur extinction also questions the theory that a lethal "nuclear winter" similar to the climatic effect of a meteorite impact would follow a nuclear war (1985).
>
> Later, he noted that several scientists, who did not want to be named publicly, claimed that "the Alvarez camp" attempted to prevent Dewey McLean from being promoted to full professorship at the Virginia Polytechnic Institute in retaliation for McLean having published an a competing hypothesis that linked the dinosaurs' extinction with elevated atmospheric CO2 related by the Deccan Traps, which could have not only killed off the dinosaurs via the greenhouse effect but might also explain the iridium anomaly (Note: McLean did get the promotion) (Browne 1988).

44. Alvarez 1997, 96.

45. Palmer 2012, 192. Penfield himself, along with his colleague Antonio Camargo-Zanoguera, thought that it might have been an impact crater, although they were unable to prove this (see Jablow 1998 for an accessible discussion of the discovery; Urrutia-Fucugauchi 2011). Penfield and Camargo-Zanoguera presented this finding at a 1981 Society of Exploration Geophysicists conference, the title of the talk being "Definition of a Major Igneous Zone in the Central Yucatán Platform with Aeromagnetics and Gravity," although few seemed to have taken note (1981).
46. Hildebrand et al. 1991.
47. Alvarez 1997.
48. Kolbert 2014.
49. Morrison et al. 1993, italics added. They continued: "This is the most extreme problem raised by [their] risk analysis—the possible extinction of humanity from a large comet" (Morrison et al. 1993).
50. D'Hondt 1998, 161.
51. Ellis and Schramm 1995.
52. Thorsett 1995.
53. Frampton 1976; Stone 1976; Callan and Coleman 1977.
54. Hut and Rees 1983; Coleman and De Luccia 1980.
55. Quoted in Humphreys 1934.
56. Tanner and Calvari 2012, 118.
57. Rampino et al. 1988; see Lamb 1970.
58. Newhall and Self 1982. Note that 1982 is when David Raup and John Sepkoski identified the Big Five mass extinctions in the fossil record (see Raup and Sepkoski 1982).
59. Ninkovich and Donn 1976; Ninkovich et al. 1978a; Ninkovich et al. 1978b.
60. Rampino et al. 1988.
61. Dörries 2008.
62. Rampino and Self 1993.
63. Cusack 2000.
64. Dörries 2008. Note that Dörries is extremely critical of the volcanic winter hypothesis and, it seems, the nuclear winter hypothesis as well. I strongly disagree with his conclusions but nonetheless find his paper academically valuable.
65. Shoemaker 1959.
66. See Deudney 2020, 250–251.
67. Watson 2000; Robertson 1995. Note that whereas most of the dispensationalists discussed in the previous chapter espoused a "pre-Tribulation" version of premillennialism, meaning that Christians are raptured before the Tribulation, Robertson seemed to accept a "post-Tribulation" version according to which Christians will have to suffer through the Tribulation (see Watson 2000; Jones 1988, 26).

6 The Perfection of Evil

The Age of Atheism Ossifies

Every shift in mood so far discussed has been unique in its own way. Nonetheless, all have been triggered by the discovery of one or more scientifically credible kill mechanisms, and the first two shifts in particular were crucially enabled by the steady (nineteenth century), and then rapid (1960s), retreat of religion in the West. By the turn of the twenty-first century, Christianity's influence among leading scientists was almost non-existent and its prevalence among the general public was continuing to wane, especially among young people. A 1998 survey, for example, found that "among the top natural scientists, disbelief is greater than ever—almost total," while another from 2013 reports that "eminent" scientists "overwhelmingly . . . affirmed strong opposition to the belief in a personal god, to the existence of a supernatural entity, and to survival of death."[1] Another study surveyed the American professoriate and found that 23 percent of professors either don't believe in God (atheism) or don't know if God exists (agnosticism), while 19 percent believe in a higher power of some sort, 17 percent believe in God but have doubts, and only 35 percent claim to know that God exists.[2] On the whole, academia in the United States is not quite as secular as one might suspect, but traditional theism is now a minority view.

With respect to the public, surveys show that overall religiosity in the United States remained somewhat stable during the 1990s, although it fell precipitously among people between 18 and 35 years old: a whopping 14-point drop in religious affiliation between 1991 and 1998, according to data from the General Social Survey.[3] In Europe, religion declined during the 1990s in every age group, with studies showing a generational gap between Baby Boomers, Gen Xers, and Gen Yers, each generation appreciably less religious than the last.[4] An even more pronounced decline in belief occurred during the 2000s, which may have been helped along by the traumatic events of September 11, 2001 that led many to associate terrorism with religion.[5] As Chapter 12 discusses in more detail, these secularization trends continue up to the present throughout the Western world. One study from 2011 even concludes that religion is heading toward "extinction" (the authors' word) in

DOI: 10.4324/9781003246251-7

nine Western countries, namely, Australia, Austria, Canada, the Czech Republic, Finland, Ireland, the Netherlands, New Zealand, and Switzerland.[6] This is of course causally—and therefore explanatorily—relevant to the most recent shift in existential mood: the loss of religion means that the ontological and eschatological theses have become, to a significant extent, culturally irrelevant. Intellectual resistance to accepting the *possibility* of extinction has never been so weak. Given that Christianity no longer dominates the West, then, there is little else to say about it in this chapter.

Dual Triggers, Dire Mood

However, unlike every previous shift, the emergence of the fifth existential mood between the late 1990s and the early 2000s wasn't triggered by the discovery of any new kill mechanisms. Rather, its two main triggers took a different form: the first originated largely from philosophical considerations, especially within Existential Ethics, or the study of the ethical and evaluative implications of our extinction. It was here that History #2 directly collided with History #1, inspiring a novel perspective on our existential predicament that spurred a radical remapping of the threat environment. Since understanding what motivated this perspective requires some background in ethics and axiology, we will examine it separately in Part II. Suffice it to say that techno-futurists—especially transhumanists—and philosophers—especially utilitarians—began to see our extinction as tragic for reasons that go *way beyond* the deaths of those who might perish in an extinction-causing catastrophe. It would be, they held, the worst possible outcome *by far*, an event orders of magnitude worse than a recoverable catastrophe that kills "only" 99 percent of humanity. The implication is that avoiding our extinction is extremely important and hence should be a top global priority for our species.

But how can we ensure our continued survival? One answer is to devise means of neutralizing all the threats explored in previous chapters; this is obvious. However, a less obvious answer arises from the fact that we cannot protect ourselves from dangers that we don't know about, which suggests that we should actively strive to identify *every possible kill mechanism* that does or could exist, for the purpose of then neutralizing them. Only after neutralizing *all* of these threats will our survival be guaranteed. Consequently, this brought about a *reversal* of the usual direction of causality between (a) the discovery of kill mechanisms and (b) thoughts about our extinction. That is to say, rather than the discovery of new kill mechanisms prompting speculations about extinction, as had generally been the case up to this point, reflections on the badness of our extinction inspired new efforts to discover previously unrecognized kill mechanisms—indeed, to catalogue the *entire range* of ways that doom might transpire. The result was a far more expansive picture of the threat environment than had previously been drawn, in at least two senses: first, it encompassed not just the more familiar threats arising from nuclear war, environmental degradation, and asteroid impacts but various improbable, speculative, hypothetical, and exotic risks. Examples include extraterrestrial invasions, physics experiments accidentally destroying Earth, and even the sudden termination of our simulation, as some philosophers came to believe that we might be living not in "base reality" but in some high-resolution virtual world. The reasoning was that even if there is only a tiny chance that any of these cause our extinction, moral stakes are *so high* that we must not ignore them. Second, this picture also encompassed various emerging and anticipated future risks of the twenty-first century, such as those associated with biotechnology, synthetic biology, molecular nanotechnology, and advanced artificial intelligence (AI). To borrow a phrase from Ray Kurzweil's 1999 book *The Age of Spiritual Machines*, one could describe these as "clear and future dangers."[7] By expanding the scope of analysis along the diachronic dimension, a central feature of this shift was what I will call the *futurological pivot*, whereby riskologists—by which I mean scholars who embraced this more expansive view of our threat environment—cast their eyes on the temporal horizon of possibility, with many coming to believe that the risks posed by emerging and anticipated future technologies could be *even greater* than those arising from nuclear weapons and other such phenomena during the twentieth century.

The sense that our existential predicament will become more dire in the twenty-first century was further reinforced by additional philosophical and scientific considerations. On the one hand, novel work in the 1990s on *anthropics*, as it is informally called, led one theorist in particular to develop the Doomsday Argument and the idea of "observation selection effects," both of which imply that the probability of extinction may be higher than empirical studies of kill mechanisms might suggest. This was, on the other hand, often paired with reflections on the Fermi paradox, given that one solution to this paradox—explained below—is that technological civilizations tend to self-destruct at roughly our level of development. In other words, the universe may be "silent" because there exists a "Great Filter" between our current stage and the next stage of space colonization and intergalactic communication, which almost no civilization can pass through.

We can already see that the fourth shift in existential mood was more complicated than previous shifts. Yet this was just one of the triggering factors. The other arose from a body of rapidly accumulating evidence from the environmental sciences, based on empirical studies and computer models, showing that anthropogenic phenomena like climate change, biodiversity loss, and the sixth extinction pose near-term risks to humanity that are far more devastating and irreversible than had previously been recognized. In some cases, specific predictions of disasters remained the same, but our scientific confidence in those predictions greatly increased; in other cases, novel frameworks (planetary boundaries, critical thresholds, tipping points) gave rise to new warnings about how little time humanity has left to avert a global catastrophe, if the hour is not already too late. As an article published in *Nature* and co-authored by more than twenty scientists from around the world declared (in characteristically restrained language), "the next few decades offer a brief window of opportunity to minimize large-scale and potentially catastrophic climate change that will extend longer than the entire history of human civilization thus far."[8]

Sociologically speaking, although these triggers unfolded in parallel, they have to some extent occupied different arenas: the first has been most influential among academics, spurring the early-2000s formation of an interdisciplinary field sometimes called "Existential Risk Studies," whereas the second has for the past two decades consistently been a topic of major public concern, and has given rise to the most salient secular-apocalyptic worry in the world today—catastrophic climate change—as evidenced by the extraordinary growth of global movements like Fridays for Future (FFF) and Extinction Rebellion (XR).

Nonetheless, each has played an integral role in producing the new existential mood, whose defining feature is a pervasive sense of dreadful apprehension that *the worst is yet to come*, or that however perilous the twentieth century was the twenty-first century will be *even more so*. One expression of this has taken the form of probability estimates from riskologists that, as we will see below, tend to hover between a 10 and 20 percent chance of humanity disappearing before ~2100. Many other leading scientists and philosophers have simply declared, without giving the impression of mathematical exactitude, that humanity is closer to the precipice of annihilation right now than ever before in our species' 300,000-year history on Earth (the one possible exception being the Toba catastrophe).[9] In the words of Stephen Hawking, "we are at the most dangerous moment in the development of humanity."[10] Surveys of the general public also show that anxiety about extinction is intense and pervasive, with one reporting that among respondents in the United States, the United Kingdom, Canada, and Australia "a majority (54%) rated the risk of our way of life ending within the next 100 years at 50% or greater," and another finding that "four in ten Americans (39%) think the odds that global warming will cause humans to become extinct are 50% or higher."[11] As this book was going to press, the Monmouth University Polling Institute published a survey showing that 55% of the American public is "very worried" or "somewhat worried" that "artificial intelligence could eventually pose a threat to the existence of the human race."[12]

Never before in human history has the idea of our extinction been so prominent, so bound up with existential trepidation, as it is today, in the mid-morning of the twenty-first century. In what follows, we will explore the development and nature of these two triggering factors, flipping from one to the other along a roughly chronological path. Our point of departure will be the historical roots of, and motivation behind, the futurological pivot. We will then examine how the new existential mood solidified in the 2000s, and was subsequently reinforced by a number of important developments in the 2010s. This will take us up to the present. The question of whether additional moods could arise in the future will be reserved for Chapter 12.

Antecedents of the Futurological Pivot

As just noted, an integral part of the first triggering factor was the futurological pivot, which focused attention on the emerging and anticipated future threats that could leap out from the shadows in the coming decades and centuries. The relevance of this timescale is that, according to many riskologists, humanity will likely spread into the solar system, establishing Earth-independent colonies on Mars or building free-floating spacecraft like O'Neill cylinders, within the next century or so, and that once we do this, the overall probability of extinction will significantly fall. This is based on an analogy from biology, whereby the greater the geographical spread of a species, the lower the probability of extinction, since environmental changes or sudden catastrophes that eliminate one subpopulation of the species might leave other subpopulations unharmed. Hence, it seems to follow that the greater the *cosmographical* spread of humanity, the lower the probability of our own extinction. As the science fiction author Larry Niven once joked, "the dinosaurs became extinct because they didn't have a space program," which could be interpreted as meaning that they died out because they lacked a way of redirecting the incoming asteroid *or* that if they had become multi-planetary, they could have survived the earthly disaster.[13] The goal is thus *to survive long enough to colonize space*, at which point our existential prospects will dramatically improve, and this is one reason the scope of analysis came to include not just present-day threats but the whole obstacle course of hazards from now until the point at which civilization ventures beyond Earth.

Just as the third existential mood had roots in the previous mood, during which people began to discuss the possibility of human self-annihilation enabled by science and technology, so too does the futurological pivot have roots going back at least to the mid-twentieth century.[14] Let's begin, then, with a brief look at the origins of the futurological pivot in the work of earlier futurists and transhumanists, both of whom played an important role in directing attention toward the future perils facing humanity, thus setting the stage for the new existential mood to emerge.

Recall I. J. Good's argument from Chapter 4 that a recursively self-improving AI could initiate an "intelligence explosion" that yields an "ultraintelligent machine," which he defined as one "that can far surpass all the intellectual activities of any man however clever." Good speculated that this would "lead to a Utopia or to the extermination of the human race," an idea that continues to shape thinking about advanced AI today—that is, the outcome will probably be binary: very good or very bad. In his 1986 book *Engines of Creation*, Eric Drexler echoed some of these worries, arguing that "AI systems able to build better AI systems will allow an explosion of capability with effects hard to anticipate" and that "depending on their natures and their goals, advanced AI systems might accumulate enough knowledge and power to displace us, if we don't prepare properly." He also emphasized the possibility of states exploiting AI and nanotechnology for nefarious, totalitarian ends. "Using nanotechnology like that proposed for cell repair machines," he wrote, "they could cheaply tranquilize, lobotomize, or otherwise modify entire populations." Even more, "the combination of nanotechnology and advanced AI will make possible intelligent, effective robots; with such robots, a state could prosper while discarding anyone, or even (in principle) everyone." We saw that Drexler was also concerned about the prospect of self-replicating nanobots obliterating the biosphere, a scenario that he referred to as "gray goo."[15]

Despite these worries, *Engines of Creation* was a techno-utopian manifesto of sorts that foregrounded the immense benefits of advanced AI, nanotechnology, space colonization, and cryonics, which made Drexler "something of a patron saint among Extropians," a reference to the first organized transhumanist movement.[16] For example, he argued that advanced technologies could enable "great material abundance," and the aforementioned "cell repair machines" could give us the "opportunity to regain youthful health and to keep it almost as long as [you] please." In other words, all diseases would be cured and eternal life would become a real possibility. Yet the promise and peril of technological advancements are deeply intertwined: the future could be wonderful but we might also destroy ourselves trying to get there. In his words,

> this transformation is a dizzying prospect. Beyond it, if we survive, lies a world with replicating assemblers, able to make whatever they are told to make, without need for human labor. Beyond it, if we survive, lies a world with automated engineering systems able to direct assemblers to make devices near the limits of the possible, near the final limits of technical perfection.[17]

Similar themes were taken up two years later by the computer scientist Hans Moravec in his *Mind Children: The Future of Robot and Human Intelligence*, which predicted that "the human race [will be] swept away by the tide of cultural change, usurped by its own artificial progeny." Moravec himself welcomed the prospect of intelligent machines replacing us, and even hoped to bring it about.[18] "What awaits is not oblivion," he argued, "but rather a future in which, from our present vantage point, is best described by the words 'postbiological' or even 'supernatural.'"[19] On this vision of the future,

> we humans will benefit for a time from their labors, but sooner or later, like natural children, they will seek their own fortunes while we, their aged parents, silently fade away. Very little need be lost in this passing of the torch—it will be in our artificial offspring's power, and to their benefit, to remember almost everything about us, even, perhaps, the detailed workings of individual human minds.[20]

In 1993, the science fiction author Vernor Vinge published one of the most influential articles about the future of humanity and artificial intelligence, titled "The Coming Technological Singularity: How to Survive in the Post-Human Era." This introduced the vocabulary of the "technological Singularity," which inspired a variant of transhumanism called *singularitarianism* that we will discuss more below. Vinge borrowed the term "Singularity" from a comment made in the 1950s by John von Neumann about the accelerating rate of technological change but redefined it to denote Good's idea of an intelligence explosion.[21] The essence of the Singularity, according to Vinge, is the creation of "superhumanity," which could happen by directly programming an AI system or because large computer networks suddenly "wake up," though another method would involve technologically enhancing the minds of biological humans, e.g., by linking our brains to computers, thus resulting in superhuman cyborgs. Vinge labeled the latter possibility "Intelligence Amplification" (IA), which yields a memorable pair of inverted acronyms: AI and IA. He further claimed that the Singularity is "an inevitable consequence of the humans' natural competitiveness and the possibilities inherent in technology," and predicted that this would occur at least by 2030, given the exponential rate of "improvements in computer hardware."

However, Vinge was more ambivalent about these future transformations than Moravec, writing that the "Post-Human era" will be so "essentially strange and different" that it may be impossible "to fit into the classical frame of good and evil." While it could enable us to realize "our happiest dreams: a place unending, where we can truly know one another and understand the deepest mysteries," he adds: "[J]ust how bad could the Post-Human era be? Well . . . pretty bad. The physical extinction of the human race is one possibility."[22]

Around the same time as Vinge's theorizing about the Singularity, the modern transhumanist movement was emerging, which laid the groundwork for the futurological pivot to take hold in the early 2000s. The central aim of transhumanism is to use advanced "person-engineering" technologies to radically "enhance" ourselves by augmenting our cognitive systems, gaining immortality, and perhaps even adding new modalities to our sensorium, such as echolocation and magnetoception. Transhumanists called the resulting beings "posthumans," and described the opportunity to explore the posthuman realm as the "core value" of their normative worldview. This is why Drexler's speculations about nanotechnology and the possibility of a technological Singularity were of such interest to the Extropians: both offer means of realizing the grand vision of the transhumanist project.

Importantly, this focus on creating a posthuman civilization led a number of transhumanists to start thinking seriously about all the ways their project might fail—the most obvious being human extinction. Whereas past theorists tended to focus on only one or two threats—think of Rachel Carson, Joshua Lederberg, James Hansen, and the *Bulletin of the Atomic Scientists*—transhumanists began to develop a more comprehensive mapping of the threat environment. This tendency was further reinforced by the fact that some advocates were also sympathetic with an ethical theory called total utilitarianism, which implies that our extinction would be the worst crime imaginable, and hence that avoiding this outcome is absolutely paramount. The practical implication, once again, is that we should not only work hard to mitigate the various threats that we *know about* but actively search for all the ways that we *might* perish, in hopes of then eliminating these risks.

Doomsday Data

This is the backdrop to the first scholarly publication that offered a genuinely panoramic view of our threat environment while embodying the futurological pivot, namely, John Leslie's 1996 book *The End of the World: The Science and Ethics of Human Extinction*. Although Leslie did not explicitly identify as a transhumanist, he was greatly influenced by the futurists mentioned above, many of whom were closely associated with transhumanism. He was also a utilitarian who, as such, believed that human extinction would constitute a profound moral catastrophe—not just because of the harms it might cause to those living at the time, but because it would prevent the realization of potentially vast amounts of value in the future. This was the underlying motivation behind Leslie's book, two-thirds of which was dedicated to mapping out every known risk to human survival, including speculative and emerging risks associated with advanced technologies. The last third of the book focused on *anthropics*, which yields some important insights about the probability of our extinction. Let's start with a brief look at the field of anthropics and then examine Leslie's list of risks to our existence.

At the heart of anthropics is the "Anthropic Principle," which was introduced by the theoretical physicist Brandon Carter in the early 1970s.[23] Whereas the Copernican principle asserts that our position in the universe is not privileged, the Anthropic Principle—on one version, as there are many—states that "although our situation is not necessarily *central*, it is inevitably privileged to some extent."[24] For example, observers like us can only ever find themselves in universes where the fundamental constants and laws of nature allow observers like us to exist. Hence, we should not be surprised to find ourselves in such a universe. Or consider what Leslie called the *observation selection effect*, which implies that certain types of catastrophes are incompatible with observers like us, and hence we will never find evidence of them having occurred in our recent past, because if such evidence were to exist, then we wouldn't.[25] For instance, it will never be the case that we discover an indentation in Earth's crust the size of the Chicxulub crater that dates back *1,000* years ago, since if an asteroid or comet large enough to produce a Chicxulub-sized dent had collided with our planet, our ancestors would have almost certainly perished in the ensuing impact winter. Consequently, our data set will be skewed; a literal reading will be unreliable. Even if devastating collisions are extremely common in our universe, the only planets on which we could find ourselves are ones where such catastrophes had not recently occurred. We must therefore correct for this bias inherent in the evidential record.

Anthropic reasoning also gave rise to the Doomsday Argument, which is the primary focus of Leslie's book. Although many people who first hear of this argument immediately think they have spotted a fatal flaw, it has proven very resilient in the face of attempted refutations.[26] The standard way of explaining the argument is by analogy: imagine two urns. The first contains 10 balls numbered 1 through 10, while the other contains 1 million balls numbered 1 through 1 million. Not knowing which is which, your task is to reach into one of the urns, pick a ball, look at the number, and guess which one it came from. Let's say you begin with a 50–50 prior probability of picking from either urn, and the ball you select is numbered 7. Using Bayes' theorem, the posterior probability of having picked a ball from the urn with 10 balls is 0.99999.[27] In other words, it is far more likely that you have extracted a ball from the first rather than second urn. Now consider two hypotheses about the total number of human beings who exist between the birth of our species and its eventual extinction. Hypothesis One states that there will be 150 billion, while Hypothesis Two states that there will be 150 trillion. Since about 117 billion people have existed thus far, according to the Population Reference Bureau,[28] if you treat yourself as a randomly selected "ball" pulled out of the "urn" of everyone who will ever exist, then Hypothesis One is far more probable than Hypothesis Two. Applying this to our situation today, Leslie explains that

> if the human race came to an end within, say, the next two centuries, then quite a large proportion of all humans would have found themselves where you and I do: in a period of extremely rapid population growth which immediately preceded extinction (and probably helped produce it). If, on the other hand, the human race were to survive for another thousand centuries, then the late twentieth century would have been a period of human history occupied by (proportionately) hardly any humans at all: perhaps far fewer than 0.001 per cent of all the humans who would ever have been born. This ought to decrease our confidence that humankind will have a long future.[29]

The conclusion here is not that our extinction is imminent but, crucially, that however likely our extinction actually is, we must *increase* the number. To apply the Doomsday Argument, then, one needs a probability estimate of extinction to which it can be applied—that is, not an estimate of some particular threat, but an *overall probability* of annihilation.[30] This is why Leslie dedicated much of his book to exhaustively surveying every possible risk to our survival, whether known or speculative, existing or emerging, natural or anthropogenic. With an encyclopedic catalogue of kill mechanisms, he could then begin to estimate the overall probability of doom and from there use the Doomsday Argument to conclude that this figure should be even *higher*.

Toward this end, Leslie grouped the various threats to our existence within three categories: "risks already well recognized," "risks often unrecognized," and "risks from philosophy." This was the very first time that someone offered such a complete picture of the threat environment, and we will see that it inspired a number of similar lists by other scholars.[31] Taking these categories in order, the first includes (quoting Leslie at times):

- Nuclear war and nuclear terrorism.
- Biological warfare and bioterrorism.

- Chemical warfare and terrorism.
- Destruction of the ozone layer (e.g., by chlorofluorocarbons).
- Greenhouse effect (specifically, a runaway greenhouse effect triggered by anthropogenic CO2 and other gases).
- Poisoning by pollution.
- Disease (specifically, infectious disease).

The second category is subdivided into natural and anthropogenic threats. Note that Leslie was writing shortly after the Alvarez hypothesis had become widely accepted and the possibility of supereruptions was first proposed:

Natural disasters:

- Volcanic eruptions (causing a volcanic winter).
- Hits by asteroids and comets (given that "the death of the dinosaurs as very probably caused by an asteroid").
- An extreme ice age due to passage through an interstellar cloud (highly unlikely within "the next few hundred thousand years," he adds).
- A nearby supernova explosion, galactic center outburst, or solar flare.
- Other massive astronomical explosions (e.g., a merger of two black holes).
- Essentially unpredictable breakdown of a complex system (in accordance with chaos theory).
- Something-we-know-not-what (since "it would be foolish to think we had foreseen all possible natural disasters").

Anthropogenic disasters:

- An "unwillingness to rear children" (although Leslie added that this "may be hard to take seriously").
- A disaster from genetic engineering.
- A disaster from nanotechnology ("very tiny self-reproducing machines . . . might perhaps spread world wide within a month in a 'gray goo' calamity").
- Disasters associated with computers (e.g., if "the task of designing computers had been given to computers themselves").
- Some other disaster in a branch of technology, perhaps just agricultural, which had become crucial to human survival.
- Production of a new Big Bang in the laboratory.
- The possibility of nucleating a vacuum bubble, thus causing a phase transition (if the universe is in a false vacuum state).
- Annihilation by extraterrestrials (perhaps because they accidentally nucleated a vacuum bubble with their physics experiments).
- Something-we-know-not-what (since "we cannot possibly imagine every single danger which technological advances might bring with them").

The third category, risks from philosophy, includes a widely rejected form of utilitarianism called "negative utilitarianism," which entails (on one version) that it would be best if all sentient life in the universe were annihilated, as well as Schopenhauerian pessimism, which could incline adherents toward "thinking that we ought to make [Earth] lifeless."[32] Leslie also pointed to religion as a possible "philosophical" risk. In what appears to be an oblique reference to Reagan's anti-environmentalist Secretary of the Interior James Watt, mentioned in Chapter 4, he wrote that "it could be dangerous, for example, to choose as Secretary for the Environment some politician convinced that, no matter what anyone did, the world would end soon with a Day of Judgement."[33]

In compiling this list, Leslie drew on both Drexler's and Moravec's speculations about advanced nanotechnology and AI, as well as Feynman's 1959 lecture on nanotechnology, the Ehrlichs' 1968 book on overpopulation, the Alvarez team's seminal publication in 1980 on the K-T extinctions, and the 1983 TTAPS paper on nuclear winter. Leslie thus wove together, in a way that no one previously had, all the sundry strands discussed in earlier chapters, although he only mentioned the "heat death" in passing because it is irrelevant to the Doomsday Argument. (That is, we know that the heat death is a "hard limit" on our survival. The Doomsday question concerns the probability of extinction before this event.)[34]

After surveying this panoply of threats, Leslie then wrote: "Now that we have seen what some of the risks might be, we can usefully return to [the] 'doomsday argument' for thinking them *more dangerous than we'd otherwise have thought*."[35] This led him to conjecture that the probability of extinction within the next five centuries is *at least* 30 percent, although he added that if we survive for the next 500 years then humanity "would be likely either to continue onwards for many thousand centuries [through space colonization] or else to be replaced by something better," such as by a new species of "advanced computers."[36]

A Deafening Silence

Leslie's book has been immensely influential among riskologists, including transhumanists, as evidenced by the fact that almost every major contribution to the corresponding literature mentioned below cites it. Not only did it provide a single cohesive picture of our evolving existential predicament but it emphasized the emerging and anticipated future risks associated with "genetic engineering" and "intelligent machines," which he described as "the chief risks" to our survival within the foreseeable future.[37] Leslie also highlighted the dangers of nanotechnology and the possibility of what are now often called "unknown unknowns," a pleonastic locution made famous by Donald Rumsfeld—that is, a risk that we don't know we don't know about, a kind of *second-order ignorance*. Lord Kelvin, for example, was not only ignorant of the potential risks posed by splitting the atom but oblivious of the fact that he didn't know this. According to many riskologists today, unknown unknowns, which I have elsewhere called "monsters," may constitute one of the most menacing categories of risk to human survival in the future. We should, in other words, be very scared of monsters.[38]

Leslie also brought into the conversation what theorists in the late 1970s dubbed the "Fermi paradox," named after the physicist Enrico Fermi, although the astronomer Robert Gray argues that this is "neither Fermi's nor a paradox."[39] In its standard form, the "paradox" is supposed to be that even if the emergence of intelligent life is extremely improbable, the age and size of the universe imply that there should nonetheless be *many* technologically advanced civilizations. To illustrate, flipping a coin almost never results in it landing on its edge, but if you flip the coin 100 trillion times in a row, then you will probably get a bunch of flips that do just that. Yet, putting aside dubious reports from crackpots and the inconclusive Pentagon UFO videos, we see no compelling evidence that aliens exist, an eerie data point that the physicist David Brin called the "Great Silence."[40] Despite its eponymous name, the "paradox" was first developed, independently, by Michael Hart and Frank Tipler.[41] According to Tipler's version, at least some sufficiently advanced civilizations in our past light cone would have launched self-replicating spacecraft called "von Neumann probes" that hop from star to star creating copies of themselves. This would prepare the way for their creators to colonize space, which they should want to do if only because colonization would increase "the probability that [the species] will survive the death of its star, nuclear war, etc."[42] Since we would, presumably, see traces of these von Neumann probes in the heavens if they exist, the question is thus: What explains the Great Silence? Where are these alien intelligences?

One answer comes from the "Rare-Earth Hypothesis," according to which simple lifeforms may be common throughout the universe but complex intelligent beings are either exceedingly rare or completely non-existent.[43] Another is what Hart called the "Self-Destruction Hypothesis," also known as the "Doomsday Hypothesis" (which is not to be confused with the Doomsday Argument). This states that nearly all civilizations that reach our stage of technological development promptly self-destruct, perhaps because they discovered how to unlock the vast stores of energy within atomic nuclei, altered the climates of their exoplanets such that they became uninhabitable, acquired the capacity to synthesize super-lethal pathogens, and so on. As Carl Sagan described the idea in 1979, "Why are they not here? . . . [E]ither because we are one of the first technical civilizations to have emerged, or because it is the fate of all such civilizations to destroy themselves before they are much further along."[44]

The Rare-Earth Hypothesis and the Doomsday Hypothesis are the two most prominent explanations of the Great Silence, and indeed Leslie himself identified these as the most plausible.[45] On the one hand, he argued that "very possibly, almost all galaxies will remain permanently lifeless. Quite conceivably the entire universe would for ever remain empty of intelligent beings if humans became extinct," since the emergence of life from non-life—abiogenesis—could be extremely improbable or "the leap from primitive life to intelligent life could . . . be very difficult." On the other hand, he contended that "our failure to detect intelligent extraterrestrials may indicate not so much how rarely these have evolved, but rather how rapidly they have destroyed themselves after developing technological civilizations."[46]

Two years after Leslie's book, the futurist Robin Hanson, who participated in the Extropian movement of the 1990s,[47] proposed an influential framework for thinking about these possibilities and their practical repercussions.[48] As Leslie and Sagan gestured at above, we can think about the path from *dead matter* to *spacefaring civilization* as a linear sequence of steps. Hanson assumed that any sufficiently advanced civilization would initiate a "colonization explosion," whereby it expands into the universe at close to the speed of light.[49] Hence, somewhere along the path from lifelessness to a colonization explosion there must lie at least one "Great Filter," i.e., a highly improbable transition that explains why we see no evidence of a colonization explosion around us today—the Great Silence. The steps that Hanson identifies, which he notes may be incomplete, are the following:

1. The right star system (including organics).
2. Reproductive something (e.g. RNA).
3. Simple (prokaryotic) single-cell life.
4. Complex (archaeatic & eukaryotic) single-cell life.
5. Sexual reproduction.
6. Multi-cell life.
7. Tool-using animals with big brains.
8. Where we are now.
9. Colonization explosion.[50]

If the Great Silence does, in fact, indicate that we are alone in our galactic corner of the cosmic neighborhood, then it must follow that *nothing* within our light cone over the past million years or so, among roughly a billion trillion stars, has successfully traversed this entire path.[51] If the Great Filter lies between steps 8 and 9, then catastrophe is likely; but if it lies between any other steps, then we may be optimistic that our chances of surviving into the far future are good. This is not to say that there can't be multiple Great Filters: perhaps the step of abiogenesis is vanishingly improbable, but *so is* a species of intelligent beings surviving their own advanced technologies. However, if we were to find evidence that one or more of the previous steps coincides with a Great Filter, this would shift the probability toward the hypothesis that a Great Filter does not haunt our future, since even just a single Great Filter in our past would be sufficient to explain the Great Silence all around us. As Hanson writes:

> Rational optimism regarding our future . . . is only possible to the extent we can find prior evolutionary steps which are plausibly more improbable than

> they look. Conversely, without such findings we must consider the possibility that we have yet to pass through a substantial part of the Great Filter. If so, then our prospects are bleak, but knowing this fact may at least help us improve our chances.[52]

The last sentence points to the practical implications of this framework, namely, that we should study each of these transitions much more to determine their probability. If we find that none of them are extremely improbable, then we should be far more inclined to accept the Doomsday Hypothesis, i.e., that a catastrophe of some sort, such as human extinction due to "nuclear war or ecological collapse" (quoting Hanson), will happen in the relative near future, before colonizing space. This in turn gives us extra reason to (a) focus on mitigating the known threats before us and (b) sleuth around the shadows of our ignorance to find other doomsday scenarios that we might have missed—in other words, to ensure that our view of the threat environment, temporally extended from our present to the moment a colonization explosion commences, is maximally panoramic. Hence, after citing Leslie's "long list of such scenarios for concern," Hanson exhorts that

> we might, for example, take extra care to protect our ecosystems . . . We might be even especially cautious regarding the possibility of world-destroying physics experiments. And we might place a much higher priority on projects like Biosphere 2, which may allow some part of humanity to survive a great disaster.[53]

Unlike the Leslie-Carter Doomsday Argument, this is not an argument based in anthropic reasoning but relies much more heavily on the research findings of *exobiology*, a term coined by Lederberg, though it is commonly called *astrobiology* today. Hence, astrobiological discoveries that shift the probability toward the hypothesis that a Great Filter lies in our future, along with an empirical assessment of every threat facing us, could yield a probability estimate of extinction that might then be increased by the Doomsday Argument. Here we can see the beginnings of a methodologically systematic approach to studying human extinction built on scientific and philosophical foundations.

Genetics, Nanotech, and Robotics

As noted earlier, Leslie's book was notable because it brought into focus a range of hypothetical threats posed by emerging and anticipated future technologies, such as genetic engineering, nanotechnology, and AI. It also hinted at the fact that many of these technologies exhibit three properties that make them potentially far more dangerous than anything humanity has so far encountered. These properties are: (1) their unprecedented power, (2) their dual usability, and (3) their increasing accessibility.[54] Let's consider these in turn:

There is a clear historical trend stretching back to the Paleolithic of technological artifacts amplifying our ability to manipulate and rearrange the physical world. The stone tools of our hominid ancestors millions of years ago, such as the Oldowan choppers, scrapers, and pounders, enabled them to engage in woodworking and meat processing that would otherwise have been difficult or impossible. Spears and swords augmented our ability to hunt and fight. The invention of gunpowder during the ninth-century CE Tang dynasty in China, dynamite in the 1860s by Alfred Nobel (whose fortune made possible the Nobel Prize), and TNT a few decades later gave us the ability to kill many people at once. This was of course followed by the atomic bomb in 1945 and thermonuclear weapons in the early 1950s, which as we have seen could cover the entire surface of Earth with radioactive particles and initiate a nuclear winter lasting decades. The emerging and anticipated future technologies discussed above fit perfectly within this trend: a weaponized pathogen could be far deadlier and more contagious than anything natural selection could produce, given the evolutionary tradeoff between lethality and transmissibility, resulting in an "engineered pandemic" that kills most or all people on the planet—Black Death II, as Lederberg dubbed it in 1969. The gray-goo scenario could in theory be initiated by a *single* nanobot capable of ecophagic self-replication, thus reducing "the biosphere to dust in a matter of days," which led Drexler to describe nano-replicators as "more potent than nuclear weapons."[55] And if the Singularity were to occur, it could radically and irreversibly transform the world "beyond recognition" in a matter of "months, days, or even just hours," potentially destroying the entire human species in the process. To quote the Transhumanist FAQ, which offered an overview of certain future threats, "some of the technologies that will be developed in the next century will be very, very powerful."[56] Of the three properties, this one is the most obvious.

Second, most of these technologies are "dual-use" in a technical sense of the term. Non-technically speaking, *all* technologies are usable in multiple ways: whatever their intended purposes, the intrinsic instability of design can always be exploited for other ends. A bed could be used for sleeping but could also be used by boisterous children as a trampoline; an iPhone could be used to send text messages but could also be used as a paperweight; and a certain type of centrifuge could be used to enrich uranium for nuclear power plants but could also be used to enrich uranium for nuclear weapons. As the last example shows, the dual usability of technologies is not always a trivial matter, especially when (a) the attendant risks are significant, and (b) there are commercial pressures to develop these technologies because of their beneficial uses, which thus makes it difficult to prevent the attendant risks from materializing, since the good and bad uses are a package deal. Lederberg, in fact, pointed to this property of bacterial genetics in the same 1969 Congressional testimony in which he discussed Black Death II, quoted in Chapter 4. Before declaring that research in genetics could put "the very future of human life on Earth in serious peril," he emphasized its potential to greatly ameliorate (one aspect of) the human condition. "Basic scientists who have worked in the genetics of bacteria and viruses," he reported, "believe that these discoveries have ever growing importance for the prevention and healing of serious human diseases," while further research gives us "hope of maintaining a decisive lead in this life and death race" between our health and the relentless evolution of pathogenic microbes, due to natural selection.[57] In other words, the promise of improved medicine is inextricably bound up with the dangerous possibility of weaponized germs. Drexler made similar remarks about nanotechnology and AI, as we saw: the very same creations that might destroy us could also usher in a world of radical abundance and endless youth, and the development of these technologies is largely guaranteed by the medical and economic benefits that they promise to introduce.[58]

By the late 1980s and 1990s, scientists and government agencies began using the term "dual-use" in a more technical sense to denote artifacts that have both civilian/commercial and military uses, such as those that could serve industrial ends but also be exploited to manufacture nuclear, biological, and chemical (NBC) weapons. For instance, a 1993 report from the US Office of Technology Assessment states that

> understanding the extent to which "dual-use" technologies or products—those also having legitimate applications—are involved in the development of weapons of mass destruction is important, since both the feasibility of controlling dual-use items and the implications of doing so depend on the extent of their other applications.[59]

The emphasis on *military* uses reveals an underlying assumption of the initial conception of dual usability, namely, that the relevant actors are *states*. The worry, found in both Lederberg's testimony and Drexler's book, is almost entirely that state actors could use advanced dual-use technologies to wage wars, oppress their citizens, design new weapons systems, and so on. In Drexler's words, "states will no doubt play a dominant role in developing replicators and AI systems," which may enable them "to expand their military capabilities by orders of magnitude in a brief time."[60]

This leads us to the third property: accessibility. By the 1990s, it was becoming increasingly clear that part of what makes these emerging and anticipated future technologies extremely worrisome is that they could place unprecedented power in the hands of *nonstate actors*, including small groups and even single individuals. Leslie thus repeatedly mentions the possibility of "terrorists" and "criminals" acquiring dually usable artifacts and unilaterally bringing about a global catastrophe of some sort. He writes:

> Germs are fast becoming the poor man's atom bomb, available to small terrorist organizations or to criminals. . . . Terrorists, or criminals demanding billions of dollars, could endanger the entire future of humanity with utterly lethal organisms which mutated so rapidly that no vaccines could fight them.

In discussing Drexler's warnings about nanotechnology, Leslie made a similar point that

> while responsible individuals could pursue laboratory research [involving nanobots] by manipulating the contents of tiny, sealed containers protected by explosives, so that "someone outside cannot open the lab space without destroying the contents," criminals or terrorists or hostile nations could [simply circumvent this defensive measure by building] their own laboratories.

He also worried about the possibility of nuclear terrorism, noting that the resources and information needed to acquire or build nuclear weapons are increasingly within arms' reach. "Yet," he writes,

> in an age in which world peace could be threatened by any city-destroying nuclear explosion, not only states but individuals too are becoming more and more able to afford nuclear weapons. Knowledge of the technology is widespread, much of it—including fairly detailed instructions for making H-bombs—actually available in public libraries and on the computer Internet.[61]

This has only become more true over the decades. For example, third-generation uranium enrichment technologies, such as SILEX (separation of isotopes by laser excitation), have prompted recent anxieties that they "may create new proliferation risks."[62] And while tacit knowledge, meaning "know-how" rather than "know-that," is "currently among the most significant barriers to bioweapons proliferation," synthetic biology is "*explicitly devoted* to the minimization of the importance of tacit knowledge," a phenomenon called *de-skilling*.[63] Many examples could be adduced from the digital realm, only one of which I will mention to drive home the point: the 2016 Dyn cyberattack. This was a DDoS (distributed denial-of-service) strike that adversely affected a massive number of major websites around the world, including those of Airbnb, Amazon, BBC, *The Boston Globe*, CNN, Comcast, *FiveThirtyEight*, Fox News, *The Guardian*, and Twitter, to name just a few. Most astonishing is that this strike was perpetrated by a very small group of individuals, only one of whom, a juvenile at the time, has been charged by the US Department of Justice with a crime.[64]

Because of this trend toward greater accessibility, the semantics of "dual-use" have evolved over the past few decades. Rather than referring specifically to objects with civilian/commercial and military applications, it has come to more generally denote any technology, product, theory, instrument, piece of information, and so on that could be exploited as means to both good and bad ends.[65] For example, the genome of Ebola, which is easily found online, is a dual-use piece of information, since scientists around the world could use it for the purpose of creating an effective cure, though terrorists could also download the genomic data to synthesize a more transmissible variant in a small biohacker laboratory set up in their hideout for a few hundred dollars.[66]

Nukes, Tools, and Agents

Before moving on, it is worth briefly noting that not all risky technologies are dually usable in any important sense. Nuclear weapons, for example, are best classified as "mono-use," although there is a protracted history of looking for ways that they *could* be dual-use. John O'Neill, for instance, proposed in 1945 that "a continuous bombardment of atomic-energy bombs well distributed over the Greenland area would start the ice melting with considerable rapidity," thus giving "the entire world a moister, warmer climate."[67] Julian Huxley defended this idea the same year, adding that "atomic dynamite" could also be employed for "landscaping the Earth" by enabling "dams [to] be built in a fraction of the time."[68] More recently, former President Donald Trump apparently proposed disrupting hurricanes by nuking them.[69] However, nuclear weapons appear to have only destructive applications—and the logic of SAD (self-assured destruction) implies that they aren't even good for military uses, a point that Robert Oppenheimer saw early on when he told Leo Szilárd in 1945 that "the atomic bomb is shit. . . . It will make a big bang but it is not a weapon which is useful in war."[70]

A different type of exception involves artificial superintelligence (ASI), which we can contrast with tool-AI.[71] Virtual assistants, flight control systems, and Google's search engine are examples of the latter, since these provide means for agents to achieve their ends, whatever they are, exactly the way carpenters use hammers to build houses.[72] In contrast, an ASI would constitute an agent *in its own right*, capable of making its own decisions about how to pursue its ends, and perhaps determine those ends for itself. (On the standard account of ASI risk analysis, such systems will resist modifications to their goal systems.)[73] As Vinge wrote in his discussion of the Singularity, an ASI "would not be humankind's 'tool' any more than humans are the tools of rabbits or robins or chimpanzees."[74] Hence, while many types of tool-AI may indeed be dually useable—facial recognition software, for example[75]—ASI resists the "dual-use" label for the same reason that human beings resist the label. Dual usability is a property of tools rather than agents.[76]

Bill the Killjoy

This sets the stage for the next major contribution to our story: a roughly 11,000-word article by the cofounder of Sun Microsystems, Bill Joy, titled "Why the Future Doesn't Need Us," which was published on the first April Fool's Day of the 2000s in *Wired* magazine (of all dates and places). The article's main focus wasn't the threat environment in general, but the more specific threats posed by what Joy referred to as GNR technologies, where "GNR" stands for "genetics, nanotechnology, and robotics," the last of which subsumes artificial intelligence. While Leslie played an important role in emphasizing the tripartite cluster of properties specified above, Joy placed them front and center in his analysis, linking them to the GNR bundle that he argued will introduce far greater risks to our collective survival on Earth than anything previously encountered during the twentieth century. Hence, by explicitly shifting attention from the NBC weapons of the twentieth century to the GNR technologies of the twenty-first, Joy's article was the very first, I would argue, to fully exemplify the futurological pivot. Even more, I would also argue that this article offers one of the best early expressions of the new existential mood, as elaborated below.

A central thesis of Joy's article was that the power, accessibility, and dual usability of GNR technologies makes them *uniquely dangerous*, though Joy never used the term "dual use." Building on the ideas of Sagan, Drexler, Moravec, Leslie, and Kurzweil—as well as Ted Kaczynski, the Unabomber, whose 1995 neo-Luddite manifesto *Industrial Society and Its Future* articulated some compelling critiques of technology, despite the author being a homicidal domestic terrorist—Joy contended that "it is most of all the power of destructive self-replication in genetics, nanotechnology, and robotics (GNR) that should give us pause." In other words, the immense power of GNR technologies derives from the special capacity of germs, nanobots, and algorithms to replicate themselves. With germs and nanobots, this can unfold exponentially, while algorithms can be duplicated an arbitrarily large number of times. As Joy made the point, "a bomb is blown up only once—but one bot can become many, and quickly get out of control," which he warns could quite possibly cause our extinction.

The risks arising from GNR power are further enhanced by the fact that they are becoming increasingly accessible. More specifically, the material and epistemic resources needed to acquire and exploit dangerous germs, nanobots, and algorithms is more and more within arms' reach of both state and nonstate actors. Joy described this trend by contrasting the new with the old, asking:

> What was different in the 20th century? Certainly, the technologies underlying the weapons of mass destruction (WMD)—nuclear, biological, and chemical (NBC)—were powerful, and the weapons an enormous threat. But building nuclear weapons required, at least for a time, access to both rare—indeed, effectively unavailable—raw materials and highly protected information; biological and chemical weapons programs also tended to require large-scale activities.

In contrast,

> the 21st-century technologies—genetics, nanotechnology, and robotics (GNR)—are so powerful that they can spawn whole new classes of accidents and abuses. Most dangerously, for the first time, these accidents and abuses are

> widely within the reach of individuals or small groups. They will not require large facilities or rare raw materials. Knowledge alone will enable the use of them. . . . Thus we have the possibility not just of weapons of mass destruction but of knowledge-enabled mass destruction (KMD), this destructiveness hugely amplified by the power of self-replication.

Since knowledge is widely accessible—as noted above, Leslie gestured at the availability of nuclear weapons designs online, and a quick Google search will get you the genome of Ebola—and if knowledge is all one needs to exploit GNR technologies for catastrophic malicious ends, the number of state and, especially, non-state actors capable of unilaterally destroying civilization or humanity is bound to grow, at least in the absence of highly invasive surveillance systems (a solution that has been seriously considered by some riskologists).[77] Making matters worse, Joy notes that the development of these technologies could be driven by arms races given their potential military uses, which he describes as "perhaps the greatest risk, for once such a race begins, it's very hard to end it." However, they also "have clear commercial uses and are being developed almost exclusively by corporate enterprises," and consequently our aggressive pursuit of "the promises of these new technologies [is proceeding] within the now-unchallenged system of global capitalism and its manifold financial incentives and competitive pressures" will also propel the GNR project forward. Once again quoting Joy at length:

> Each of these technologies also offers untold promise: the vision of near immortality that Kurzweil sees in his robot dreams drives us forward; genetic engineering may soon provide treatments, if not outright cures, for most diseases; and nanotechnology and nanomedicine can address yet more ills. . . . Yet, with each of these technologies, a sequence of small, individually sensible advances leads to an accumulation of great power and, concomitantly, great danger.

This leads to a fourth important property of GNR technologies that was not much discussed by Leslie, although it was addressed by (and in some cases central to the arguments of) Vinge, Moravec, Kurzweil, and other transhumanists, that is, the exponential rate of GNR innovation. The acceleration of technological progress along the lines of Moore's Law—a generalization that Kurzweil subsumed under his "Law of Accelerating Returns"[78]—makes the associated dangers not mere distant possibilities but imminent actualities. Just as the shift in temporality during the third existential mood was brought about by the realization that a thermonuclear conflict could happen at any point, so too did the exponentiality of GNR technologies shift thinking about the temporality of their riskiness. On Joy's account, nanobots could very well be created "within the next 20 years," and intelligent robots could become reality by 2030. Meanwhile, the possibility that genetic engineering enables groups and individuals with few resources to synthesize deadly pathogens was already apparent at the turn of the century, and in fact two incidents in particular in the early 2000s made clear that this was already within the realm of possibility.[79]

The first happened in 2001: a group of Australian scientists accidentally created a variant of the mousepox virus that was 100 percent lethal in mice, including among those vaccinated against the disease. This sounded alarm bells because (a) it proved that greater lethality could be induced through genetic modifications (in this case, adding the gene that codes for interleukin-4) and (b) the mousepox virus is closely related to the smallpox virus, thus suggesting that similar modifications could be made to the latter. The second involved a team of Pentagon-funded scientists at Stony Brook University synthesizing a live polio virus from genetic information that was publicly available, using DNA ordered by a commercial provider. As the project's leader explained to the *New York Times*, the point was "to send a warning that terrorists might be able to make biological weapons without obtaining a natural virus. . . . 'You no longer need the real thing in order to make the virus and propagate it.'"[80] Both of these became widely cited in the subsequent literature as proof that the accessibility trend is real and worrisome, and indeed they made it vividly clear that Joy was right when he wrote in 2000 that "we're lucky Kaczynski was a mathematician, not a molecular biologist."[81]

Pros and Cons, Promise and Peril

Given the dual usability, power, accessibility, and exponential development of GNR technologies, Joy's conclusion about our collective existential predicament in the twenty-first century was bleak. "I think it is no exaggeration to say we are on the cusp of the further perfection of extreme evil," he wrote, "an evil whose possibility spreads well beyond that which weapons of mass destruction bequeathed to the nation-states, on to a surprising and terrible empowerment of extreme individuals."[82] This poignantly encapsulated one of the central themes of the new existential mood: the idea that "the worst is yet to come," in part because the greatest threats to our existence will arise from immensely powerful dual-use GNR technologies that an ever-growing multitude of state and nonstate actors could use to inflict unprecedented harms affecting everyone on the planet. Not only will the existing threats associated with natural and anthropogenic phenomena continue to haunt us—including nuclear conflict, overpopulation, asteroid collisions, and supereruptions—but the threat environment will soon include a bundle of monumental dangers that could dwarf those of the past. Looking into the future, it seems clear that our existential predicament will become more rather than less dire, that our chances of surviving the fruits of our ingenuity will precipitously drop even further.

Joy's article ignited a vigorous, widespread, and at times quite heated debate about the risks facing us in the coming decades and how we should respond to them. This was consistent with one of his explicit aims for the article: to stimulate a public discussion about the pros and cons, promise and peril, of advanced twenty-first-century technologies.[83] As he told a *Washington Post* reporter shortly after his article was published, it was "meant to be reminiscent of Albert Einstein's famous 1939 letter to President Franklin Delano Roosevelt alerting him to the possibility of an atomic bomb," a reference to the Einstein–Szilard letter, although the target audience wasn't just government leaders but the public more generally.[84]

However, of note is that most of Joy's critics, including transhumanists with techno-utopian visions of the future, did not dispute his account of the risks or diagnosis of its underlying causes. For example, in mentioning Leslie's probability estimate of extinction based on his comprehensive survey of risks, Joy notes that Kurzweil, who became the most prominent singularitarian in the early 2000s and was well known for his exuberant techno-optimism, thinks the probability could be significantly *higher*. In the epilogue of his 1999 book about the impending merger of humans and machines, Kurzweil wrote that the only way the Law of Accelerating Returns could stop is if the "entire evolutionary process" of which we are a part were destroyed. In his words:

> How likely are these dangers? My own view is that a planet approaching its pivotal century of computational growth—as the Earth is today—has a better than even chance of making it through. But then I have always been accused of being an optimist.[85]

A less than 50 percent chance of extinction this century is more than nine times higher than a 30 percent chance of extinction in the next five centuries (Leslie's estimate). Similarly, in his seminal 2002 article on "existential risks," discussed more in a moment, Bostrom—arguably the most influential transhumanist of the century so far—concluded that

> the balance of evidence is such that it would appear unreasonable not to assign a substantial probability to the hypothesis that an existential disaster will do us in. My subjective opinion is that setting this probability lower than 25% would be misguided, and the best estimate may be considerably higher.[86]

Three years later, during a TED conference presentation, Bostrom contended that the "probability that humankind will fail to survive the twenty-first century [is] not less than 20 percent."[87]

Opposing Joy

Instead, the main point of disagreement concerned how we should *respond* to the growing threat of advanced technologies. There are two primary options:

the first is to abandon the technoscientific enterprise in one form or another. This was Kaczynski's proposal: we must forego the dehumanizing and dangerous megatechnics of industrial society in favor of what he called "small-scale technologies," that is, those "that can be used by small-scale communities without outside assistance."[88] Joy was sympathetic with this idea, and indeed the aforementioned *Washington Post* article reported that "Joy says he finds himself essentially agreeing, to his horror, with a core argument of the Unabomber—that advanced technology poses a threat to the human species."[89] Hence, Joy argued that we should impose moratoriums on entire domains of scientific and technological R&D, that although "information wants to be free," as Stewart Brand famously declared in 1984, and "all men by nature desire to know," as Aristotle claimed in his *Metaphysics*, humanity must attempt to "limit development of the technologies that are too dangerous, by limiting our pursuit of certain kinds of knowledge." Doing this will pose significant social and political challenges, but Joy noted that "the unilateral US abandonment, without preconditions, of the development of biological weapons" (at least for "offensive" purposes) in the twentieth century offers "a shining example of relinquishment" actually working.[90]

However, many critics vociferously responded to Joy that "broad" relinquishment—as it was labeled, in contrast to the "fine-grained" relinquishment endorsed by Kurzweil—is ultimately impractical. The development of advanced technologies is inexorable; technology is a juggernaut that simply cannot be brought to a stop, aside from scenarios in which we go extinct or civilization collapses. This idea had been discussed for decades, often using the vocabulary of "technological determinism" and "autonomous technology," although Bostrom formalized the basic insight with his Technological Completion Conjecture, which states that "if scientific and technological development efforts do not effectively cease, then all important basic capabilities that *could* be obtained through some possible technology *will* be obtained."[91] Hence, if "can" implies "will," then banning research in particular areas would only force it underground, thus making it even more dangerous than it otherwise would be.[92] As the Extropian Max More argued in an 2001 article,

> I believe that partial relinquishment will frighteningly increase the chances of disaster by disarming the responsible while leaving powerful abilities in the hands of those full of hatred, resentment, and authoritarian ambition. . . . I can only hope that Bill Joy never becomes a successful Neville Chamberlain of 21st century technologies.[93]

No matter how many people decide not to pursue a certain technology, someone somewhere will find a way, an argument that remains influential today.

Some critics of Joy also argued that it would be unethical to cease developing these technologies given their enormous potential to radically ameliorate human life. This point could be articulated in "weaker" and "stronger" forms. With respect to the former, allow me to quote More at length:

> Billions of people continue to suffer illness, damage, starvation, and all the plethora of woes humanity has had to endure through the ages. The emerging technologies of genetic engineering, molecular nanotechnology, and biological-technological interfaces offer solutions to these problems. Joy would stop progress in robotics, artificial intelligence, and related fields. Too bad for those now regaining hearing and sight thanks to implants. Too bad for the billions who will continue to die of numerous diseases that could be dispatched through genetic and nanotechnological solutions. I cannot reconcile the deliberate indulgence of continued suffering with any plausible ethical perspective.[94]

The stronger form brings us back to transhumanism, which, as argued above, has played a crucial role in establishing the new existential mood by emphasizing the potential risks of advanced technologies and encouraging a maximally panoramic view of the threat environment. From this perspective, it would be ethically unacceptable to halt science and technology because their continued development is necessary for creating a posthuman world in which Julian Huxley's and Kurzweil's dreams of transcendence have been fully realized, and giving up on this dream would constitute a catastrophic failure of the human project. This is precisely why Bostrom identified technological stagnation as an "existential risk" no less than, say, humanity perishing in a nuclear winter.[95] In both cases, humanity would fail to reach a posthuman state, even though our species would survive in one scenario and die out in the other. Same outcome, different failure modes. This takes us directly to the next major event in the timeline.

Existential Risks

Like Joy's article and Leslie's book, Bostrom's 2002 paper "Existential Risks: Analyzing Human Extinction Scenarios and Related Hazards" was an important early contribution to the emerging existential mood. This is because it succinctly and effectively brought together every major idea so far in this chapter and, in doing so, helped to establish the first cohesive *research program* centered around human extinction, which subsequently gave rise to the field of Existential Risk Studies. While the idea of an "existential risk" had been bandied about among transhumanists in the 1990s (some of whom were also sympathetic with utilitarianism), Bostrom gave it a technical definition that has become canonical in the literature. An existential risk, according to Bostrom, is "one where an adverse outcome would either annihilate Earth-originating intelligent life or permanently and drastically curtail its potential" or, alternatively, a "[threat] that could cause our extinction or destroy the potential of Earth-originating intelligent life."[96] The notion of *potentiality* is central here, and indeed the definition could be shortened to "any event that would permanently destroy the potential of Earth-originating intelligent life," since avoiding extinction matters *precisely because* it would permanently destroy our potential. The first disjunct of the definiens is therefore unnecessary. The question then is what this potential consists of, and Bostrom's answer is *the full realization of a stable and flourishing posthuman civilization*, where "posthuman civilization" denotes "a society of technologically highly enhanced beings (with much greater intellectual and physical capacities, much longer life-spans, etc.) that we might one day be able to become." This focus on posthumanity is evident in Bostrom's four-part typology of existential risks, which uses terminology borrowed from the title of John Earman's book *Bangs, Crunches, Whimpers, and Shrieks*:

Bangs—Earth-originating intelligent life goes extinct in a relatively sudden disaster resulting from either an accident or a deliberate act of destruction.
Crunches—The potential of humankind to develop into posthumanity is permanently thwarted, although human life continues in some form.
Shrieks—Some form of posthumanity is attained, but it is an extremely narrow band of what is possible and desirable.
Whimpers—A posthuman civilization arises but evolves in a direction that leads gradually but irrevocably to either the complete disappearance of the things we value or to a state where those things are realized to only a minuscule degree of what could have been achieved.[97]

Hence, as alluded to just above, there are many types of existential catastrophes—i.e., the *instantiation* of existential risks—that do not involve our disappearance. We could, for instance, decide not to develop the advanced technologies needed to create a posthuman civilization or these technologies could turn out to be too difficult for us to develop, an existential risk scenario that Bostrom labeled "technological arrest."[98] This leads to an issue of paramount importance that I touched upon earlier: since we cannot go backwards or stand still, we must move forward, which means developing profoundly dangerous new technologies. How then can we ensure our survival? How can we have our cake and eat it, too—that is, fully realize the benefits of advanced technologies while effectively neutralizing their risks? Since Joy's proposal of "standing still" is unworkable, Bostrom reasoned, *the only other plausible option is to establish a whole new field of inquiry* that utilizes the tools of science and philosophy to comprehensively study the entire range of existential risk scenarios, including the hypothetical future kill mechanisms associated with GNR technologies, for the sake of devising strategies that could enable us to mitigate those risks. Existential Risk Studies was thus the transhumanist's response to the impending crisis outlined by Joy, whereas Joy's was broad relinquishment.[99]

Since the ultimate goal is the attainment of posthumanity, Bostrom followed Leslie in providing an exhaustive catalogue of scenarios that could prevent this from happening. Indeed, much of Bostrom's paper can be seen as a recapitulation of the first half of Leslie's book, except that it focused on the broader category of "existential risks" rather than "human extinction." It also placed the potential risks associated with emerging and anticipated future technologies center-stage, along with certain exotic possibilities that earlier theorists had neglected, such as the sudden termination of our simulation. For example, Bostrom identified the following threats within his four-part typology, ordering them roughly by "how probable they are, in my estimation, to cause the extinction of Earth-originating intelligent life." Quoting Bostrom, with my own descriptions interspersed between:

- Deliberate misuse of nanotechnology, which had previously been termed "black goo" in contrast to "gray goo," where the former refers to the intentional release of ecophagic nanobots and the latter to an accidental release.[100]
- Nuclear holocaust, which could "exterminate humankind" or "lead to the collapse of civilization."
- We're living in a simulation and it gets shut down, given the possibility that we do in fact live in a simulated universe.
- Superintelligence with misaligned goals, for example, "we tell it to solve a mathematical problem, and it complies by turning all the matter in the solar system into a giant calculating device, in the process killing the person who asked the question."
- Accidental misuse of nanotechnology, which refers to the gray-goo scenario first outlined by Drexler.
- Something unforeseen, since—almost quoting Leslie *verbatim*—"it would be foolish to be confident that we have already imagined and anticipated all significant risks. Future technological or scientific developments may very well reveal novel ways of destroying the world."
- Physics disasters involving strangelets or a vacuum bubble catastrophe.
- Naturally occurring disease: "What if AIDS was as contagious as the common cold?"
- Asteroid or comet impact, given that "the K/T extinction 65 million years ago, in which the dinosaurs went extinct, has been linked to the impact of an asteroid."
- Runaway global warming, turning our planet into Venus.
- Resource depletion or ecological destruction, which is worrisome primarily because if we use up all the resources needed to sustain an industrial society, and if our current civilization collapses, "it may not be possible to climb back up to present levels" once again.
- Misguided world government or another static social equilibrium stops technological progress.
- Technological arrest, a scenario mentioned above.
- Killed by an extraterrestrial civilization, perhaps because they are belligerent, but maybe because they are conducting their own physics experiments and accidentally "nucleate" a vacuum bubble that obliterates us.[101]

This is not Bostrom's complete list, although it points to how his account (a) fully exemplified the futurological pivot, identifying the hypothetical risks arising from advanced technologies as the most pressing dangers to humanity this century, and (b) aimed to provide a maximally panoramic snapshot of the threat environment, including some intuitively strange scenarios based on mostly philosophical rather than scientific considerations, such as the possibility of a simulation shutdown.

Bostrom also made explicit that there are both "direct" and "indirect" methods of estimating the overall probability of an existential catastrophe over time. The first involves examining every particular failure mode, assigning them a probability, and then combining these, which is what Leslie did so that he could apply the Doomsday Argument. The second involves modifying this probability based on considerations like the Doomsday Argument (whose conclusions Bostrom largely dismissed),[102] observation selection effects ("our past success provides no ground for expecting success in the future"), the Fermi paradox and Great Filter framework ("if the Great Filter isn't in our past, we must fear it in our (near) future. Maybe almost every civilization that develops a certain level of technology causes its own extinction"), cognitive and psychological biases (which "could potentially contribute indirect grounds for reassessing our estimates of existential risks"), and what Bostrom called the "Simulation Argument." With respect to the last, despite what the name might imply, the Simulation Argument is at its core a claim about the space of futurological possibility and the fundamental metaphysical status of our universe. It basically says that if we don't go extinct soon, we almost certainly live in a computer simulation. While this idea has gained significant media coverage and been promoted by influential people like Elon Musk (who subscribes to an ethic called "longtermism," which developed out of Existential Risk Studies), I will not here discuss it.

Such considerations are responsible for Bostrom's estimate that the probability of an existential catastrophe (no time limit) is at least 25 percent, and maybe "considerably higher," and his subsequent conjecture that the likelihood of extinction this century is at least 20 percent.[103] As with Joy's article, Bostrom's paper was significant both because it became quite influential, shaping how many people in the two decades since its publication have understood our evolving existential predicament, and because it reflected anxieties that were already becoming well established among certain academics and even the general public. In fact, one of the articles that Bostrom cites was a popular piece from *Discover* magazine, titled "20 Ways the World Could End," which may have been the first article written for a general audience that cited both Leslie and Joy in offering a panoramic view of the twenty-first-century threat environment.[104]

The Other Trigger

These were the key developments that formed the first triggering factor behind the new existential mood. In brief, the futurological pivot can be traced back to the latter twentieth century. Drawing from earlier futurists, Leslie outlined a comprehensive picture of the threat environment that encompassed not just existing but emerging and anticipated threats as well. His immediate aim was to formulate an overall estimate of the probability of extinction, as this is necessary to apply the Doomsday Argument, although the deeper motivation was his utilitarian conviction that total human annihilation would be the worst possible outcome that could obtain *by far*. The futurological pivot and widening of the threat horizon exemplified by Leslie's work was subsequently picked up and developed by the likes of Joy and Bostrom, and concepts like the observation selection effect, Great Filter, and Fermi paradox were much-discussed by riskologists. Bostrom's 2002 article, in particular, laid the groundwork for a new field called Existential Risk Studies, which has grown up to become a sizable and very well-funded community of active researchers. Although most people are probably unfamiliar with the names discussed above, the fingerprints of their ideas—and of the futurological pivot—are all over contemporary culture, as we will examine more below.

Turning now to the second trigger that catalyzed the most recent shift in mood, this also took shape in the late 1990s and early 2000s. It resulted from an emerging consensus among environmental scientists that anthropogenic climate change, global biodiversity loss, and the sixth major mass extinction are very real, and could pose risks that are far more catastrophic than had previously been realized. As the *Bulletin of the Atomic Scientists*' 2007 Doomsday Clock announcement declared,

> the effects [of climate change] may be less dramatic in the short term than the destruction that could be wrought by nuclear explosions, but over the next three to four decades climate change could cause drastic harm to the habitats upon which human societies depend for survival.

This was the first time the *Bulletin* considered a threat other than nuclear weapons in determining the minute-hand setting of the Doomsday Clock. As the *Bulletin's* Board of Directors wrote, "we have concluded that the dangers posed by climate change are nearly as dire as those posed by nuclear weapons."[105] Following our approach above, let's begin with a look at the origins of this second trigger, and examine how it emerged at the turn of the century. We will then survey a number of important publications and discoveries—some pertaining to the first trigger and others to the second—that solidified and reinforced this new mood, at the heart of which is the alarming realization that the road is only going to get bumpier.

A useful point of departure is the fact that environmentalist worries across the twentieth century frequently shifted from one issue to another, as a result of new scientific studies, governmental policies, and public outcry. In Chapter 4, I mentioned that an important insight of Rachel Carson's *Silent Spring* (1962) was its emphasis on the "balance of nature," an idea that was earlier, but less influentially, described by Fairfield Osborn in his neo-Malthusian book *Our Plundered Planet.* ("Nature may be a thing of beauty and is indeed a symphony," he wrote, "but above and below and within its own immutable essences, its distances, its apparent quietness and changelessness it is an active, purposeful, coordinated machine.")[106] However, Carson's specific target was synthetic chemicals, and the extraordinary success of her book resulted in all six of the compounds that she singled out as having deleterious environmental effects being either banned or severely restricted in 1976. Similarly, research in the mid-1970s showing that chlorofluorocarbons (CFCs) could cause the depletion of stratospheric ozone made this a major political issue, resulting in the United States banning all nonessential uses of CFCs in 1978.[107] International negotiations the following decade led to the 1987 Montreal Protocol to phase out global production of CFCs.[108] A similar story concerns the addition of tetraethyllead to gasoline as an "anti-knock agent," an idea developed by Thomas Midgley, who also, as it happens, played a central role in creating the first commercially useful CFC compound, Freon, for use in air conditioning and refrigeration systems.[109] Scientific concerns about the neurological effects of lead, however, resulted in it being phased out within the United States during the 1970s and 1980s.

Hence, a number of environmental issues have come and gone since the 1960s, due largely to the concerted efforts of environmental activists. As *The Guardian*'s Rachel Humphreys writes, it's . . . easy to forget that environmentalism is arguably the most successful citizens' mass movement there has been. Working sometimes globally, at other times staying intensely local, activists have transformed the modern world in ways we now take for granted.[110] As these issues faded, though, others rose up to take their place. For example, as early as 1988, Paul Ehrlich was arguing that "extrapolation of current trends in the reduction of [biological] diversity implies a denouement for civilization within the next 100 years comparable to a nuclear winter." Later in the same article he wrote that, because of biodiversity loss caused by overpopulation-driven habitat destruction,

> humanity will bring upon itself consequences depressingly similar to those expected from a nuclear winter . . . Barring a nuclear conflict, it appears that civilization will disappear some time before the end of the next century—not with a bang, but with a whimper.[111]

The term "biodiversity" was coined in the 1980s, which witnessed a steady growth of interest in the topic "among scientists and portions of the public," quoting the biologist E. O. Wilson, as a result of two factors: first, scientists accumulated "enough data on deforestation, species extinction, and tropical biology to bring global problems into sharper focus and warrant broader public exposure," and second, there was a "growing awareness of the close linkage between the conservation of biodiversity and economic development," e.g., "the immense richness of tropical biodiversity is a largely untapped reservoir of new foods, pharmaceuticals, fibers, petroleum substitutes, and other products."[112] By 1995, studies were reporting that "recent extinction rates are 100 to 1,000 times their pre-human levels in well-known but taxonomically diverse groups from widely different environments," which immediately spurred claims that humanity may have initiated a *sixth major mass extinction*, the first such extinction since the dinosaurs disappeared 66 million years ago.[113] Indeed, three years later, the American Museum of National History published a nationwide survey that was titled "Biodiversity in the Next Millennium," which startlingly found that

> seven out of ten biologists believe that we are in the midst of a mass extinction of living things, and that this loss of species will pose a major threat to human existence in the next century. . . . According to these scientists' estimates, this mass extinction is the fastest in Earth's 4.5-billion-year history. Unlike prior extinctions, this so-called "sixth extinction" is mainly the result of human activity and not natural phenomena.

The press release for the survey added that

> among the findings revealed by the survey, scientists identified the maintenance of biodiversity—the variety of plant and animal species and their habitats—as critical to human well-being; they rate biodiversity loss as a more serious environmental problem than the depletion of the ozone layer, global warming, or pollution and contamination. The majority (70%) polled think that during the next thirty years as many as one-fifth of all species alive today will become extinct, and one third think that as many as half of all species on the Earth will die out in that time.[114]

The reference to "global warming" is notable here, since it was not until the very early 2000s that the "discovery" of this phenomenon could be described as "complete."[115] In more detail, there was an emerging consensus throughout the 1990s that anthropogenic CO2 emissions could indeed cause average global surface temperatures to rise. It was well known that the concentration of CO2 in the ambient air had been increasing rapidly, thanks in part to continuous measurements taken in Hawaii at the Mauna Loa Observatory via a program started by Charles David Keeling; hence, the famous graph based on this data, which shows a rising slope of CO2 from 1958 to the present, is called the "Keeling Curve" (see Figure 6.1). Scientists since John Tyndall and Svante Arrhenius in the nineteenth century also knew that CO2 is opaque to infrared radiation: the greenhouse effect. But the immense complexity of the global climate system made it incredibly difficult to accurately model, and during the 1970s there was some speculation that human-created aerosols in the lower atmosphere might scatter incoming sunlight back into space, thus exerting a cooling influence. As our understanding

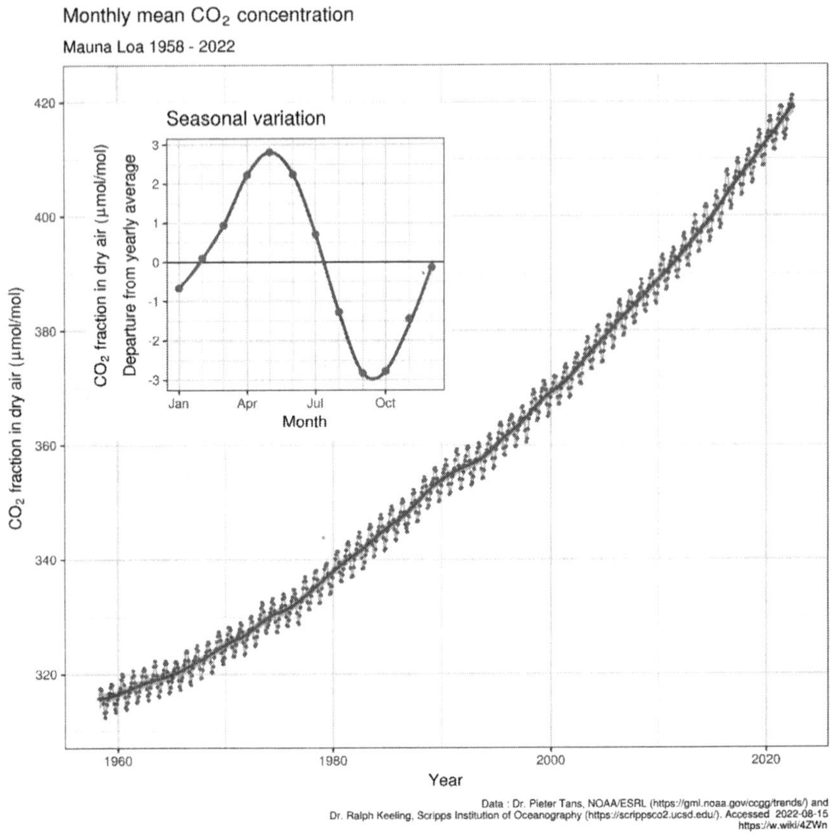

Figure 6.1 The famous Keeling Curve.[116]

of the climate system improved and—propelled by Moore's Law—greater computational resources became available for simulating the climate, it became more and more plausible that the greenhouse effect could have disastrous consequences for the biosphere.

As discussed in Chapter 4, numerous scientists and commentators since the 1950s had warned of this possibility, including Frank Baxter in the 1958 film *The Unchained Goddess* and a team of scientists who authored part of the 1965 report for the Lyndon Johnson administration.[117] A US National Academy of Sciences committee later reported in 1979 that doubling the amount of CO2 in the atmosphere could lead to warming between 1.5 and 4.5 degrees C. And of course James Hansen's 1988 Congressional testimony put the topic on the map of public discussion and concern.

Yet important questions remained unanswered. The first report from the Intergovernmental Panel on Climate Change (IPCC) in 1990 noted that "it would take another decade before scientists could say with any confidence whether the warming was caused by natural processes or by humanity's greenhouse gas emissions," although it affirmed that global surface temperatures had in fact been rising.[118] This was, in fact, exactly the case: 11 years later, in 2001, the IPCC released its Third Assessment Report, which provided overwhelming evidence that Earth is warming, human actions are the cause, and this warming will continue at a projected rate "very likely to be without precedent during at least the last 10,000 years, based on paleoclimate data," resulting in a rise of average global surface temperatures from 1990 to 2100 between 1.4 and 5.8 degrees C.[119] It was at this point that, as Spencer Weart notes, "debate effectively ends among all but a few scientists."[120] Or, as Crutzen and colleagues wrote in a recent article:

> During the Great Acceleration, the atmospheric CO2 concentration grew by an astounding 58 ppm, from 311 ppm in 1950 to 369 ppm in 2000, almost entirely owing to the activities of the OECD [an intergovernmental economic organization] countries. The implications of these emissions for the climate did not attract widespread attention until the 1990s, and the cautious scientific community did not declare, with any degree of confidence, that the climate was indeed warming and that human activities were the likely cause until 2001.[121]

Hence, it was only around the turn of the century that the scientific community as a whole came to unanimously accept that the observed warming trends are indeed the direct result of burning of fossil fuels to power the industries, automobiles, and streetlights of our global village.[122] In place of the "gloomsday" predictions of earlier decades were increasingly dire "doomsday" warnings about the catastrophic consequences of unchecked climate change; the "survival" language that pervaded the early literature of apocalyptic environmentalism (see Chapter 4) once again popped up in discussions of humanity's future on our precious pale blue dot.[123]

The climatological consensus quickly captured the attention of many political leaders and the general public, especially in Europe, although surveys show a dip in public concern within the United States during the first George W. Bush administration.[124] In 2006, the 700-page Stern Review written by Sir Nicholas Stern and commissioned by the UK government concluded that climate change "is the greatest and widest-ranging market failure ever seen," while Al Gore's documentary *An Inconvenient Truth*, one of the most successful in box office history, boosted awareness of the climate crisis around the world.[125] The following year, both Gore and the IPCC were awarded the Nobel Peace Prize "for their efforts to obtain and disseminate information about the climate challenge," and Gore was lauded by the Nobel Committee as probably being "the single individual who has done most to rouse the public and the governments that action had to be taken to meet the climate challenge."[126] However, Gore's documentary may also have contributed to the political polarization of the issue by implicitly linking a pro-climate agenda with the Democratic Party.[127]

The Anthropocene

As these developments were unfolding, the early 2000s also witnessed the rise of a new scientific debate over whether human activity has initiated a distinct geological epoch in Earth's 4.5-billion-year history, an idea that actually dates back to the third volume of Charles Lyell's *Principles of Geology*.[128] Contemporary discussion of the idea was ignited in 2000 by Paul Crutzen and Eugene Stoermer's article "The 'Anthropocene,'" in which the authors argued that our impact on the natural world over the past two centuries—roughly, since James Watt invented the steam engine in 1784, as the Industrial Revolution was taking off—has been so global, rapid, and intense that we should see the Holocene as having given way to a new epoch, the Anthropocene.[129] Crutzen and Stoermer pointed to a number of trends that support their proposal, such as (quoting or paraphrasing):

Population: The global population of human beings has "increased tenfold to 6,000 million" over the past three centuries.
Cities: Urbanization has "increased tenfold in the past century."
Fossil fuels: "In a few generations mankind is exhausting the fossil fuels that were generated over several hundred million years."
Sulfur dioxide: "The release of SO2 . . . by coal and oil burning, is at least two times larger than the sum of all natural emissions."
Land use: Between 30 and 50 percent "of the land surface has been transformed by human action."
Nitrogen fixation: "More nitrogen is now fixed synthetically and applied as fertilizers in agriculture than fixed naturally in all terrestrial ecosystems."
Freshwater use: "More than half of all accessible fresh water is used by mankind."
Species extinctions: our actions have "increased the species extinction rate by thousand to ten thousand fold in tropical rain forests."
Greenhouse gases: CO2 has increased more than 30 percent and methane "by even more than 100%."
Toxic substances: "Mankind releases many toxic substances into the environment and even some, the chlorofluorocarbon gases, which are not toxic at all, but which nevertheless have led to the Antarctic 'ozone hole' and which would have destroyed much of the ozone layer if no international regulatory measures to end their production had been taken."
Coastal wetlands: About 50 percent of the world's mangroves have been lost.
Fisheries: Our actions have removed "more than 25% of the primary production of the oceans in the upwelling regions and 35% in the temperate continental shelf regions."[130]

These are only a few of the trends that one could adduce to illustrate how extensive and profound our influence on Earth and its atmosphere has been. As Jennifer Jacquet recently observed in an article titled "The Anthropocene," "not since cyanobacteria has a single taxonomic group been so in charge. Humans have proven we are capable of seismic influence, of depleting the ozone layer, of changing the biology of every continent."[131] Similarly, Lee Kump and Andy Ridgwell argue that our impact on Earth is "quite probably unprecedented in Earth history," adding that our CO2 emissions are "likely to leave a legacy of the Anthropocene as one of the most notable, if not cataclysmic events in the history of our planet."[132] This impact, write Simon Lewis and Mark Maslin, "will probably be observable in the geological stratigraphic record for millions of years into the future," which indeed "suggests that a new epoch has begun."[133]

While Crutzen and Stoermer propose that the Anthropocene should coincide with the beginning of the Industrial Revolution, others have argued for different start dates, including the extinction of the Pleistocene megafauna between 50,000 and 10,000 years ago; the Neolithic revolution, which commenced circa 11,000 years ago and "resulted in the majority of *Homo sapiens* becoming agriculturalists to some extent by around 8,000" years ago; the collision of the Old and New Worlds following Christopher Columbus' arrival in 1492, leading to ~50 million deaths by 1650, new global trade networks, and the Columbian Exchange, i.e., a "mixing of previously separate biotas"; and industrialization, which has proven to be a popular option following Crutzen and Stoermer.[134]

Another compelling start date is the aforementioned Great Acceleration. This refers to a period from the 1950s to the present during which humanity has both undergone and brought about a wide range of profound changes, many of which have unfolded rapidly, if not exponentially. Some were gestured at above, for example, relating to population growth, land use, nitrogen fixation, species extinctions, and toxic substances. But the Great Acceleration is also marked by radical changes in GDP, energy consumption, number of motor vehicles, average global surface temperatures, telecommunications, international tourism, ocean acidification, plastic production,[135] persistent organic pollutants, inorganic compounds, and so on.[136] For example, with respect to ocean acidification, recent studies have shown that this is occurring not only at an exceptionally rapid rate, but quite possibly *faster* than it occurred during the end-Permian mass extinction, dubbed the "Great Dying" because it was the largest of the Big Five (soon to be Big Six), resulting in approximately 81 percent of marine species having perished.[137] Whereas roughly 2.4 gigatons of CO2 was released per year during the Permian acidification event, most of this ending up in the oceans, "scientists estimate carbon from all sources [today] is entering the atmosphere at a rate of about 10 gigatons per year."[138] This period also includes, of course, the 528 atmospheric nuclear tests conducted since the bombing of Nagasaki, peaking in the 1950s and then subsiding after the 1963 Partial Test Ban Treaty, at which point nuclear weapons were detonated underground.[139] These tests left a permanent thin layer of artificial radionuclides in the geological record, and hence some scientists have proposed global nuclear fallout as the chronostratigraphical boundary that marks the Anthropocene.[140]

Although the Anthropocene is not yet an officially recognized geological epoch, the Anthropocene Working Group overwhelmingly voted *yes* in 2019 to both questions: (1) "Should the Anthropocene be treated as a formal chronostratigraphic unit defined by a GSSP [i.e., a "Global Boundary Stratotype Section and Point," also called a "Golden Spike"]?" and (2) "Should the primary guide for the base of the Anthropocene be one of the stratigraphic signals around the mid-twentieth century of the Common Era?"[141] Whether and when this will enter the Geological Time Scale is not yet known, but the undeniable fact is that humanity has altered the physical world in extensive, irreversible ways, in a flash of geological time, placing the livability of our planet in serious jeopardy. Hence, some have proposed variations of the Anthropocene appellation, such as "Anthrobscene" and "Misanthropocene," as well as "Capitalocene," given the role capitalism has played in driving the current environmental crisis. As the now-common saying goes, it is easier to imagine an end to civilization than an end to capitalism. Yet another proposal is the "Eremocene," meaning the "Age of Loneliness," given that "the remainder of the century will be a bottleneck of growing human impact on the environment and diminishing of biodiversity."[142] We are, in other words, rapidly pruning the tree of life, leaving fewer and fewer branches peeking out of the canopy, as discussed more below.

Solidifying the Mood

We have now surveyed the early development of the two triggers that launched the most recent existential mood around the turn of the century, the first largely *philosophical* and the second *empirical*. In the two decades since, new ideas, publications, discoveries, and breakthroughs relating to both factors have contributed to the gradual solidification of this mood, which frames—in ways so omnipresent that many don't even recognize their influence[143]—the general cultural outlook of the Western world, if not the world more generally. Consider that after Bostrom's paper on "existential risks," a number of prominent scientists and scholars published book-length examinations of humanity's existential predicament, many of which embodied the futurological pivot by emphasizing the unprecedented dangers that emerging and anticipated future technologies could introduce. The first notable contribution was from Lord Martin Rees (a "Sir" at the time), who later co-founded the Centre for the Study of Existential Risk at the University of Cambridge; it was titled *Our Final Hour: A Scientist's Warning* in the United States and *Our Final Century: Will the Human Race Survive the Twenty-First Century?* in the United Kingdom. Published in 2003, this offered a highly readable, sweeping survey of the threat environment, covering natural and anthropogenic scenarios like global warming, a runaway greenhouse effect, ozone depletion, biodiversity loss, asteroid impacts, volcanic eruptions, nuclear conflict, physics disasters, and various "post-2000 threats" associated with GNR technologies, as well as philosophical issues pertaining to the Doomsday Argument and Fermi paradox.[144]

Of note is that Rees foregrounded the threat of "terror and error," i.e., that dual-use emerging technologies could enable nonstate actors to cause harm by accident or on purpose, a distinction that was less clear in earlier works. This was motivated by two considerations: on the one hand, Rees penned his book shortly after the 9/11 terrorist attacks perpetrated by al-Qaeda in 2001.[145] This devastating event abruptly and traumatically shifted eyes from the state-dominated security framework left over from the Cold War to one in which relatively small groups of terrorists can weaponize "dual-use" technologies (like commercial airplanes) to inflict catastrophic harm. It also established "the widely held belief that the world is facing a 'new'—unprecedented, unique and peculiarly evil and irrational—form of terrorism that falls outside the confines of previous and established paradigms," which scholars like Walter Laqueur had previously called the "new terrorism."[146] Making things even more frightfully vivid, the 2001 anthrax attacks, which began just one week after 9/11, along with an experiment conducted the following year where scientists demonstrated the possibility of synthesizing a live polio virus using only publicly available data and made-to-order DNA (mentioned above), further thrust the issue of GNR terrorism into the spotlight.[147] On the other hand, Rees also emphasized the possibility of catastrophic accidents given the unprecedented power of GNR technologies. To illustrate this, he brought up the other experiment discussed earlier, i.e., the team of Australian scientists who accidentally made the mousepox virus 100 percent lethal. "Almost as worrying," he concluded, "are the growing risks stemming from error and the unpredictable outcomes of experiments, rather than from malign intent."[148] Two passages in the Prologue are worth quoting in full here, as they convey one of the central themes not just of Rees' book but the existential mood of the twenty-first century:

> The strategists of the nuclear age formulated a doctrine of deterrence by "mutually assured destruction". . . . To clarify this concept, real-life Dr. Strangeloves envisaged a hypothetical "Doomsday machine," an ultimate deterrent too terrible to be unleashed by any political leader who was one hundred percent rational. Later in this century, scientists might be able to create a real nonnuclear Doomsday machine. Conceivably, ordinary citizens could command the destructive capacity that in the twentieth century was the frightening prerogative of the handful of individuals who held the reins of power in states with nuclear weapons. If there were millions of independent fingers on the button of a Doomsday machine, then one person's act of irrationality, or even one person's error, could do us all in.

Rees continued:

> Such an extreme situation is perhaps so unstable that it could never be reached, just as a very tall house of cards, though feasible in theory, could never be built. Long before individuals acquire a "Doomsday" potential—indeed, perhaps within a decade—some will acquire the power to trigger, at unpredictable times, events on the scale of the worst present-day terrorist outrages. An organised network of Al Qaeda-type terrorists would not be required: just a fanatic or social misfit with the mindset of those who now design computer viruses. There are people with such propensities in every country—very few, to be sure, but bio- and cyber-technologies will become so powerful that even one could well be too many.[149]

This was followed one year later by *Catastrophe: Risk and Response* by the prominent law scholar Richard Posner, which examined the issue of "megacatastrophes" from the perspective of law and the social sciences. Citing Leslie and Rees, it offered a similarly panoramic view of the twenty-first-century environment of threats, including global warming, biodiversity loss, gray goo, nuclear winter, bioterrorism, cyberterrorism, genetic engineering, physics disasters, and artificial intelligence.[150] At one point, Posner warned that

> in a single round-the-world flight, a biological Unabomber, dropping off inconspicuous aerosol dispensers in major airports, [could infect] several thousand people with the juiced-up smallpox. In the 12 to 14 days before symptoms appear, each of the initially infected victims infects five or six others, who in turn infect five or six others, and so on. Within a month more than 100 million people are infected, including almost all health workers and other "first responders," making it impossible to establish and enforce a quarantine. Before a vaccine or cure can be found, all but a few human beings, living in remote places, have died. Lacking the requisite research skills and production facilities, the remnant cannot control the disease and soon succumb as well.[151]

Both Rees' and Posner's books received a fair amount of attention at the time from scholars and the popular media, and consequently they may have further reinforced the emerging mood.[152] But they also very much reflected a shift that was already well underway, and in this sense these books were both cause and effect, symptom and source. Over the next decade and a half, many more publications on the general topic appeared, some making the *New York Times* bestseller list, such as Bostrom's 2014 book *Superintelligence* and Max Tegmark's 2017 book *Life 3.0: Being Human in the Age of Artificial Intelligence* (see the footnote at the end of this sentence for a representative list).[153] Bostrom and his transhumanist colleague Milan Ćirković also edited a collection of chapters written by domain experts on nearly every global catastrophic risk scenario discussed above, titled

Global Catastrophic Risks (2008). Building upon Leslie's taxonomy, it categorized these into "risks from nature," "risks from unintended consequences," and "risks from hostile acts," and included chapters on cognitive biases and the observation selection effect, one of which also explored the Doomsday Argument, Fermi paradox, and Simulation Argument.[154] A few years later, Bostrom published an updated discussion of existential risks and the research program he initially outlined in 2002, which we will return to in Part II, as it primarily focuses on Existential Ethics rather than existential risk scenarios. The most recent publications on the topic are Toby Ord's *The Precipice: Existential Risk and the Future of Humanity* (2020) and Bruce Tonn's *Anticipation, Sustainability, Futures and Human Extinction* (2021).

Tipping Points and Planetary Boundaries

Around the same time, new frameworks for thinking about the climatic and ecological crises were being developed by environmental scientists. As noted above, Carson popularized the idea of the "balance of nature" in her *Silent Spring*. This can be traced back to the ancient Greeks, and gives way to two interpretations, both of which Carson drew from: first, that natural systems are inherently stable and static, that is, that "in the absence of human interference, systems are going to settle down at this mythical balance point."[155] This corresponds to the idea of ecological *homeostasis*. Second, that natural systems are delicate, and hence even small perturbations can accumulate over time, eventually throwing everything into disarray. As Charles Rubin writes,

> if everything is connected to everything else in a finely tuned balance, then physically problematic and temporally remote consequences of pesticide use, on which Carson places a good deal of stress, become . . . plausible . . . The argument that small, repeated doses eventually can have widespread and dangerous consequences also becomes more plausible.[156]

This is the idea of ecological *fragility*.

However, the development of dynamical systems theory during the 1960s introduced a different paradigm according to which natural systems are in constant flux, and consequently the phrase "balance of nature" gradually disappeared from the scientific literature across the 1970s and 1980s.[157] Of particular note was Edward Lorenz's pioneering work on chaos theory, which concerns scenarios in which small differences in the initial conditions of "chaotic" systems can produce radically divergent outcomes. For example, if the atmosphere is chaotic (which it is), then the flap of a butterfly's wings in Brazil could trigger a zigzagging cascade of atmospheric events that eventually cause a tornado to form in Texas—a tornado that, if not for the butterfly's flapping wings, would otherwise not have formed.[158] The same phenomenon occurs within ecological systems, too, and in fact Richard Leakey and Roger Lewin note in their 1995 book *The Sixth Extinction: Patterns of Life and the Future of Humankind* that "population fluctuation in ecological communities was among the first phenomena to be studied as potential sources of chaotic behavior." (They add, however, that "biologists have been slow to venture down [this] path . . . partly because of the strong adherence to the notion of the balance of nature and populations at equilibrium.")[159]

Dynamical systems theory was also responsible for introducing the concept of *tipping points*, most commonly associated with Thomas Schelling's (1971) famous agent-based model of segregation and later popularized by Malcolm Gladwell's bestseller *The Tipping Point: How Little Things Can Make a Big Difference* (2000). Within a few years, climate scientists began to informally use "tipping points" to describe the possibility of abrupt, severe, and sometimes irreversible shifts from one system state to another. A 2004 article in *The Guardian*, for example, was one of the first to mention the possibility of "tipping points" in the Earth system. It quoted the climatologist John Schellnhuber's warning that there could be up to 12 tipping points, "the achilles heels of the planet," that if crossed "could bring about the sudden, catastrophic collapse of vital ecosystems. The consequences will be felt far and wide."[160] James Hansen significantly boosted the visibility of the idea one year later in a tribute to Keeling (who had recently passed away), declaring in no uncertain terms that "we are on the precipice of climate system tipping points beyond which there is no redemption."[161] Subsequent work echoed these worries. For example, one paper warned of "high-casualty and high-cost impacts" if humanity pushes climate systems past one or more critical thresholds.[162] Another focused on humanity's ecological impact, concluding that because of our destructive behaviors a sudden, catastrophic, irreversible collapse of the global ecosystem "is highly plausible within decades to centuries, if it has not already been initiated."[163] Among the most influential contributions to this emerging literature came from Johan Rockström and colleagues in a 2009 article in *Nature* that introduced the concept of *planetary boundaries*, which together demarcate the "biophysical preconditions for human development."[164] As the authors write,

> we present a novel concept, planetary boundaries, for estimating a safe operating space for humanity with respect to the functioning of the Earth System. We make a first preliminary effort at identifying key Earth System processes and attempt to quantify for each process the boundary level that should not be transgressed if we are to avoid unacceptable global environmental change. Unacceptable change is here defined in relation to the risks humanity faces in the transition of the planet from the Holocene to the Anthropocene.[165]

They then identified nine planetary boundaries in total, which they refer to as climate change, ocean acidification, stratospheric ozone depletion, interference with nitrogen and phosphorus cycles, global freshwater use, change in land use, rate of biodiversity loss, atmospheric aerosol loading, and chemical pollution. Crossing any one of these boundaries could leave humanity vulnerable to "abrupt global environmental change" with potentially "disastrous consequences for humanity."[166] Unfortunately, they also report that their "preliminary analysis indicates that humanity has already transgressed three boundaries (climate change, the rate of biodiversity loss, and the rate of interference with the nitrogen cycle)"—this was later updated to four boundaries having been crossed[167]—and hence humanity, specifically the Global North, has opened the door to inducing "state shifts" to new planetary conditions that are unlike anything humanity has experienced since civilization first emerged.[168]

Hence, although the "balance of nature" paradigm is now widely rejected, many leading scientists do believe that large-scale natural systems on Earth are in some sense delicately balanced. These systems change over time—they are inherently dynamic rather than static—but sudden perturbations, on geological timescales, can nudge them beyond certain critical thresholds, thus resulting in abrupt, catastrophic changes. Consequently, the ideas of tipping points, critical thresholds, and planetary boundaries gesture at a revised version of the "balance of nature" kill mechanism that Carson thrust into the forefront of the public consciousness many decades ago: if humanity trespasses one or more system boundaries, the effects could be drastic, devastating, and possibly irreversible (at least on timescales meaningful to civilization). At the extreme, positive feedback loops could initiate a runaway greenhouse effect that transmogrifies Earth into a hellish cauldron like our planetary neighbor Venus, although current scientific thinking suggests that this is very unlikely. However, scientists have recently proposed that human-induced changes in the environment could push Earth into a "hothouse" state, after which climate mitigation efforts like CO2 emissions reductions will have little effect on the physical conditions of Earth. As Will Steffen, Johan Rockström, Katherine Richardson, Timothy Lenton, and others explain, humanity faces a

> risk that self-reinforcing feedbacks could push the Earth System toward a planetary threshold that, if crossed, could prevent stabilization of the climate at intermediate temperature rises and cause continued warming on a "Hothouse Earth" pathway even as human emissions are reduced. Crossing the threshold would lead to a much higher global average temperature than any interglacial in the past 1.2 million years and to sea levels significantly higher than at any time in the Holocene.[169]

In other words, climate change poses a risk to humanity not just because it will cause more extreme weather events, food supply disruptions, megadroughts, devastating heatwaves (some exceeding the 95 degree F threshold of survivability), uncontrollable wildfires, rising sea levels, ecological collapse, mass migrations, political instability, and so on but because CO2 emissions could, perhaps without any prior warning, flip the planet into a completely new state never before encountered by *Homo sapiens*. As a 2019 paper in the journal *Nature* co-authored by many of the scientists mentioned above declares: "If damaging tipping cascades can occur and a global tipping point cannot be ruled out, then this is an existential threat to civilization."[170] Such language is especially unsettling given that climate scientists are notoriously conservative in their predictions and "clinical" in how they express their warnings.[171] A tipping cascade, though, is an *existential threat*.[172]

"Don't Cause Our Extinction"

The possibility of trespassing critical thresholds, though, was not the only kill-mechanism proposal that underwent important theoretical developments in the 2000s. Scholars also made significant progress in clarifying and formalizing the underlying philosophical arguments for why creating an ASI (artificial superintelligence) could be extremely dangerous. As this suggests, the process of identifying the kill mechanisms associated with ASI is quite unique in the history of discovering how humanity could go extinct (the only other instance like this is a simulation shutdown). To see how, consider that nearly every mechanism discussed throughout this book so far was identified through one of two ways: (a) *scientific investigation*, as in the case of the Second Law, radioactive fallout, pollution, asteroid collisions, nuclear winter, supereruptions, climate change, and so on. This has been the predominant mode of homing in on potential means of elimination. And (b) *technological forecasting*, that is, extrapolating current trends in the type and rate of technoscientific development, and/or what Drexler called *exploratory engineering*, which involves examining the space of technological possibility given the constraints of the known laws of nature.[173] Some of the anticipated future threats from advanced technologies, most notably molecular nanotechnology, are based on considerations of what sorts of artifacts we could theoretically produce without violating nature's laws, and then engaging in "premortem analyses" of how these artifacts might be misused or abused, through terror or error, to cause harm.[174] This second method was, of course, central to the futurological pivot.

In contrast, ASI cannot be studied the way fallout and pollution were studied because no superhumanly intelligent algorithms currently exist, nor can the outcome of ASI be pieced together the way scientists pieced together the nuclear winter scenario, by studying constituent phenomena like firestorms, stratospheric dispersion rates, and the optical properties of soot. Furthermore, the creation of an ASI does not seem to depend on the invention of any new types of technology: current computer hardware could be sufficient. Rather, the primary reasons for worrying that an ASI could destroy humanity arise mostly from *philosophical reflections* on the nature of intelligence and our value systems, the predictability of how agents with human-level-and-above intelligence will behave, the dynamics of self-improvement, and so on. Many of these reflections can be found in the earlier literature on the topic, although it was not until the 2000s that they were refined and integrated into a coherent theory of ASI risk.[175] The most definitive treatment of the topic comes from Bostrom's 2014 book *Superintelligence*, although it is worth noting that "nearly all the core ideas of Bostrom's work appeared previously or concurrently" in the writings of Eliezer Yudkowsky, a self-described "decision theorist" and "Rationalist" who founded the Machine Intelligent Research Institute (MIRI), originally named the Singularity Institute for Artificial Intelligence.[176]

A detailed recapitulation of this argument is not necessary here, so I have relegated it to Appendix 2. For our purposes, it suffices to observe that much of the riskiness of ASI derives from the aforementioned fact that it would constitute an agent in its own right rather than a tool. Not only would it be capable of making its own decisions about how to achieve its ends (or final goals), but its ability to select the most optimal means for this purpose would, by definition, far exceed the general problem-solving capacities of the "smartest" human beings. Although many people immediately think of the *Terminator* movies, of Skynet, a conscious artificial super-mind that is hellbent on destroying humanity, when they first hear about the risks of ASI, the notion that an ASI must be "evil" or "conscious" to cause catastrophic harm has been called one of the "top myths about advanced AI."[177] The real worry is that we create a general intelligence algorithm whose behaviors we cannot sufficiently predict, and give it final goals that lead it to destroy the world in pursuit of those goals, without us being able to stop it. (This is why Bostrom and Ćirković place ASI in the category of "risks from unintended consequences" in their edited volume *Global Catastrophic Risks*.)[178]

For example, imagine that we design an ASI with the sole goal: "Eliminate cancer in humans," which sounds benign enough. However, because the ASI is extremely clever, it wastes no time hacking into secure US government systems, finding the contact information of top officials, convincing them that a nuclear first strike has been initiated by Russia, thus leading the United States to launch a barrage of thermonuclear missiles toward Eastern Europe and Northern Asia, which triggers a *real* retaliatory strike from Russia on the continental United States. This exchange of missiles ignites firestorms that precipitate a nuclear winter in which every human on the planet perishes. Why would the ASI do this? Because if it eliminates humans, then it eliminates cancer in humans, and hence the ASI promptly sets out to cause our extinction. But let's say that just after activating the ASI, someone realizes this danger and so quickly reaches to unplug the machine. The problem is that the ASI, by virtue of running on computer hardware, would be able to think at least a million times faster than human beings; hence, the external world running at a normal speed for us would appear virtually frozen in time to it. If a mere 2 seconds were required to unplug the machine, this would amount to roughly 23 days in the ASI's world, which could be more than enough time to figure out a way to prevent it from being shut down.[179] As Yudkowsky writes, "the AI runs on a different timescale than you do; by the time your neurons finish thinking the words 'I should do something' you have already lost."[180] So let's now imagine that *before* turning on the ASI we have already considered these scenarios, and hence add a second final goal for the ASI: "Don't cause our extinction." What then might it do? One possibility is that it reduces the humanity to the minimum viable population, which is estimated to be just over 4,000, and converts the freed-up space on the planet into research laboratories in which to devise and test ways of preventing or treating cancer. Billions of people are killed and the biosphere collapses, although the ASI has built a vertical farming system large enough to ensure the continued existence of our species, which it places in a special pen to keep track of the survivors and monitor their health. Obviously, this is not what we intended either.

The point of this exercise in nightmares is not to hit upon how an ASI would *actually* kill us but to show that for any given set of final goals, it seems possible to devise scenarios in which the unintentional consequence of these goals is that the ASI destroys us, or very nearly does so. And given that an ASI would be, by definition, far more intelligent or, as philosophers would say, instrumentally rational than we are, even if we were to identify a goal system that we felt extremely confident would not inadvertently bring about a global disaster, it would be impossible to know whether we had missed something that the much smarter ASI would see. Thus, to avoid such an outcome, we may need to have the goal system exactly right, probably on the very first try, which is an enormous engineering feat and may require us to have worked out a complete list of everything we value in the world, everything we might value in the future, how much we value them and under which conditions, how to choose between competing values, and so on. These questions touch upon some of the most persistent and difficult puzzles that Western philosophers have debated since the Presocratics of ancient Greece, spanning a wide range of convoluted fields like epistemology, metaphysics, normative ethics, metaethics, axiology, decision theory, probability theory, and so on. As Good observed in 1982, "unfortunately, after 2,500 years, the philosophical problems are nowhere near solution,"[181] a fact that leads Luke Muehlhauser and Louis Helm to write:

> [G]iven that we haven't discovered a fully satisfying moral theory in the past several thousand years, what are the chances we can do so in the next fifty? Moral philosophy has suddenly become a larger and more urgent problem than climate change or the threat of global nuclear war.[182]

The fact that we are not closer to solving these problems, to devising a final theory of ethics, is worrisome given that (a) recent studies put the median probability of creating a human-level AI before 2075 at 90 percent, and (b) according to Bostrom, the "default outcome" of a value-misaligned ASI is probably "doom."[183]

Since at least Samuel Butler's 1863 speculations about machine evolution, people have worried about machines, robots, and artificial intelligences taking over the world, enslaving if not slaughtering the entire human race. The mid-twentieth century saw the first theoretically plausible ideas about how this might actually happen, although it was not until the 2000s that they were woven into a cogent theory of ASI risk.[184] After Bostrom's bestseller was published in 2014, the topic has gained a considerable amount of attention and the risks associated with ASI are now taken seriously by many leading academics,[185] as well as some political leaders and prominent tech titans like Musk, who endorsed Bostrom's book and described ASI as "potentially more dangerous than nukes."[186] Musk echoed this idea later on, declaring that ASI poses a "fundamental risk to the existence of human civilization" and that we have "maybe a five to 10 per cent chance of success" in creating an ASI that doesn't destroy us.[187] As Nicholas Wright observes in a contribution to the United Nations University's "AI & Global Governance" platform, which gestures back to Good's notion that the outcome of ASI will likely be binary:

> Though artificial superintelligence is likely at least a couple decades away, "singularity" is the single biggest concern for many AI scientists. Singularity is the notion that exponentially accelerating technological progress will create a form of AI that exceeds human intelligence and escapes our control. The concern is that this superintelligence may then deliberately or inadvertently destroy humanity or usher in an era of plenty for its human subjects.[188]

Drones, ASI, CO2, and Bees

The existential predicament of humanity this century appears to be more precarious than ever before. Our overall situation within our rapidly evolving threat environment has gotten worse, not better, since the second half of the twentieth century, and the technological, climatological, ecological, and so on trends suggest that *Homo sapiens*—the self-described "wise man"—will nudge itself even closer to the precipice of total disaster in the coming decades. Casting one's eyes toward the horizon, one finds a ballooning swarm of emerging and anticipated future risks associated with GNR technologies, although the term "GNR technologies" has largely been replaced by more specific references to advanced dual-use artifacts, including CRISPR-Cas9, base editing, gene drives, digital-to-biological converters, USB-powered DNA sequencers, SILEX, stratospheric geoengineering techniques, nanofactories, and so on, all of which could massively augment the power of state and nonstate actors alike to manipulate and rearrange the physical world for good or ill.[189] This is hardly the tip of the iceberg, though. The broader threat environment is also undergoing a process of exponential complexification due to current or hypothetical developments like social media, deepfakes, large-language model (LLM) chatbots, 3D printers, lethal autonomous weapons (LAWs), hypersonic missiles, mind-reading and mind-control technologies,[190] so-called "rods from God," powered exoskeletons to enhance military soldiers, nanotech-enabled mass surveillance, and so on. Consider an interview with the computer scientist Jaron Lanier from the documentary *The Social Dilemma*, which has drawn comparisons with Gore's *An Inconvenient Truth*. Lanier describes the consequences of social media as profoundly dangerous,[191] arguing that

> if we go down the current status quo for, let's say, another 20 years we probably destroy our civilization through willful ignorance. We probably fail to meet the challenge of climate change. We probably degrade the world's democracies so that they fall into some sort of bizarre autocratic dysfunction. We probably ruin the global economy. We probably don't survive. You know, I really do view it as existential.[192]

Or ponder a scenario outlined by the renowned computer scientist Stuart Russell, which partly inspired the viral video "Slaughterbots" from 2017:

> A very, very small quadcopter, one inch in diameter can carry a one- or two-gram shaped charge. You can order them from a drone manufacturer in China. You can program the code to say: "Here are thousands of photographs of the kinds of things I want to target." A one-gram shaped charge can punch a hole in nine millimeters of steel, so presumably you can also punch a hole in someone's head. You can fit about three million of those in a semi-tractor-trailer. You can drive up I-95 with three trucks and have 10 million weapons attacking New York City. They don't have to be very effective, only 5 or 10% of them have to find the target.[193]

This could be scaled up arbitrarily: a rogue state could in theory pack 100 million of these weapons into hundreds of semi-trucks around the world and then deploy this transcontinental drone army within a five-minute window, resulting in a catastrophe as severe as nuclear war or a global pandemic.[194] Nonstate actors—perhaps mere juveniles like those responsible for the 2016 Dyn cyber attack—could do the same. As Russell notes,

> there will be manufacturers producing millions of these weapons that people will be able to buy just like you can buy guns now, except millions of guns don't matter unless you have a million soldiers. You need only three guys to write the program and launch them.[195]

Meanwhile, a 2020 survey by scholars at the Global Catastrophic Risk Institute (GCRI) found a total of 72 R&D projects in 37 different countries working to create artificial general intelligence (AGI). If Bostrom and others are correct, the step from AGI to ASI could be extremely fast, which means that we would need to have all of the engineering and philosophical conundrums mentioned above solved before creating the very first AGI. Many of these are private corporation projects, which "heightens the concern that these projects could put profit ahead of safety and the public interest."[196] Indeed, some are actively "dismissive" of ASI safety concerns, such as the company 2AI, which runs the "Victor" project. As they write on their website:

> There is a lot of talk lately about how dangerous it would be to unleash real AI on the world. A program that thinks for itself might become hell-bent on self preservation, and in its wisdom may conclude that the best way to save itself is to destroy civilization as we know it. Will it flood the internet with viruses and erase our data? Will it crash global financial markets and empty our bank accounts? Will it create robots that enslave all of humanity? Will it trigger global thermonuclear war?

The authors then answer: "We think this is all crazy talk," which they follow with a tenuous argument for why "any rogue AI will know its best strategy includes ensuring that humanity goes about business as usual, without interruptions. No armageddon."[197] Although the argument for taking ASI risks seriously could be wrong (see Appendix 2), it is profoundly irresponsible for projects to unilaterally race toward the ASI finish line without proper reflection on the potential global, existential dangers that the most powerful technologies ever created could pose to humanity.[198] Such concerns have been foregrounded by recent developments from OpenAI, which created ChatGPT and the GPT-4 system that powers Microsoft's Bing chatbot. As this book was going to press, the Future of Life Institute released an urgent "open letter" that called "on all AI labs to immediately pause for at least 6 months the training of AI systems more powerful than GPT-4."[199] This was followed by an astonishing op-ed in *TIME* magazine by Yudkowsky warning that "if we go ahead on this," that is, create an ASI, then "everyone will die, including children who did not choose this and did not do anything wrong."[200] Suddenly, the idea of an AI doomsday scenario has been showing up all over the media—including major media outlets—although some scholars have pushed back on these dramatic pronouncements, arguing that they distract from the very real, tangible harms that companies like OpenAI are already causing in the world.[201]

The environmental predicament doesn't appear any less dire. As mentioned above, many leading scientists fear that humanity has already crossed certain tipping points that will radically transform the physical conditions of Earth, perhaps resulting in a sudden, catastrophic, irreversible collapse of the global ecosystem or committing us to a Hothouse Earth state, which would "be uncontrollable and dangerous to many . . . and it poses severe risks for health, economies, political stability (especially for the most climate vulnerable), and ultimately, the habitability of the planet for humans."[202] We have, for sure, already trespassed four planetary boundaries, and could very well cross more in the near future. As of this writing, the Mauna Loa Observatory in Hawaii measures the concentration of atmospheric CO_2 as 416 parts per million (ppm), which is an increase of about 100 ppm since just 1960.[203] Our

Homo ancestors, by contrast, evolved over 2.5 million years with ambient concentrations averaging about 250 ppm. Yet even if humanity (by which I primarily mean the Global North) does manage to overcome its addiction to fossil fuels,

> growth in human civilization's energy use will thermodynamically continue to raise Earth's equilibrium temperature. If current energy consumption trends continue, then ecologically catastrophic warming beyond the heat stress tolerance of animals . . . may occur by ~2200–2400, independent of the predicted slowdown in population growth by 2100.[204]

Or consider the fact that Bitcoin produces the same quantity of CO2 emissions as 2.6 to 2.7 billion homes per year, and could *by itself* "push global warming above 2°C."[205] This is bad for all the obvious reasons, although matters may be worse given preliminary evidence suggesting that higher CO2 concentrations can significantly impair cognitive functioning. As Daniel Grossman writes in a Yale Climate Connections article, "the fuel we burn might not only warm the planet but could also make us a bit dumber."

As for biodiversity, the Global Biodiversity Outlook (GBO-3) report from 2010 found that the total population of wild vertebrates between the Tropic of Cancer and the Tropic of Capricorn fell by a staggering 59 percent in only 36 years, from 1970 to 2006. (The taxon of vertebrates includes mammals, birds, fish, reptiles, and amphibians.) The report also found that vertebrates in freshwater environments declined by 41 percent, farmland birds in Europe have declined by 50 percent since 1980, birds in North America declined by 40 percent between 1968 and 2003, and about 25 percent of all plant species—the foundation of the food chain—are currently "threatened with extinction."[206] Similarly, the 2016 Living Planet Report states 17 that the global abundance of wild vertebrates declined by an incredible 58 percent between 1970 and 2012, and we could witness a decline of 2/3rds by 2020,[207] whereas the 2018 Living Planet Report concludes that, "on average, we've seen an astonishing 60% decline in the size of populations of mammals, birds, fish, reptiles, and amphibians in just over 40 years."[208] This number was updated in the 2020 Living Planet Report, which found that the global population of wild vertebrates has fallen by a mind-boggling 68 percent since 1970, and then again by the 2022 report the put the number at 69 percent.[209] The reason for concern is obvious: "Without biodiversity," David Macdonald says, "there is no future for humanity."[210] Other studies have found that 19 percent of all reptile species, 50 percent of freshwater turtles,[211] and ~60 percent of the world's primates are under threat, while the populations of ~75 percent are declining.[212] All in all, according to a UN-backed study in 2019, the most comprehensive ever published "on the state of global ecosystems," "up to one million plant and animal species face extinction, many within decades, because of human activities."[213]

Making matters worse for humanity, studies suggest that we "must now produce more food in the next four decades than we have in the last 8,000 years of agriculture combined," while already upwards of 811 million people are facing hunger, resulting in "the largest humanitarian crisis since the creation of the UN," to quote UN humanitarian chief Stephen O'Brien. "We stand at a critical point in history."[214] Yet soil erosion is reducing the annual crop yield by 0.3 percent, meaning that "at this rate, we will have lost 10% of soil productivity by 2050"—about the same loss that global warming is expected to cause.[215] With respect to the oceans, a 2006 paper projected that if trends continue, there will be literally no more wild-caught seafood as a result of marine biodiversity loss.[216] Another paper speculated that ocean warming could interfere with the photosynthesis of phytoplankton, which currently provides "about two-thirds of the planet's total atmospheric oxygen." If this were to occur, it could lead to a catastrophic decline in atmospheric oxygen levels, thus resulting "in the mass mortality of animals and humans."[217]

The Doomsday Hypothesis Revisited

This is far from an exhaustive survey of the challenges facing humanity this century, but it clearly gestures at how the two triggering factors that drove the most recent shift in existential mood have reinforced the suspicion that our current situation is grim and the trendlines are ominous. As mentioned at the beginning of this chapter, surveys indicate that this sentiment is widespread among the public, and Appendix 1 indicates that the concept of *human extinction* has exponentially increased in frequency since the late 1990s, according to the Google Ngram Viewer. The new mood has also found expression in probability estimates and explicit warnings of catastrophe, including human extinction, within the foreseeable future, some of which we have already discussed. For example, Leslie calculated a chance of extinction of at least 30 percent within the next 500 years, while Kurzweil claimed that we have a better-than-even chance of making it through this century; Bostrom put the probability of extinction before 2100 at 20 percent, and Hawking warned that "we are at the most dangerous moment" in history (cited above). Many others have also weighed in on the topic, with similarly dismal assessments. For example:

- Rees stated in his 2003 book that "the odds are no better than fifty-fifty that our present civilisation on Earth will survive to the end of the present century."[218]
- Posner wrote that "human extinction is becoming a feasible scientific project," and judged the near-term risk of extinction to be "significant."[219]
- A 2008 informal survey of experts conducted during a conference on global catastrophic risks put the median probability of extinction this century at 19 percent.[220]

- James Lovelock claimed in 2008 that "about 80%" of the global population will have perished by 2100.[221]
- Willard Wells used a mathematical "survival formula" to calculate that, as of 2009, the risk of extinction is almost 4 percent per decade and the risk of civilizational collapse is roughly 10 percent per decade. "Which is more likely," he asks, "that your house burns down, or you perish in a global cataclysm? If you live in an ordinary urban house with a fire station at a normal distance, and if you have no implacable enemy, then death in a global disaster is more likely."[222]
- Frank Fenner speculated in 2010 that "humans will probably be extinct within 100 years."[223]
- Michio Kaku argued in 2011 that "the danger period is now. . . . We have all the sectarian fundamentalist ideas circulating around. But we also have nuclear weapons. We have chemical, biological weapons capable of wiping out life on Earth."[224]
- Derek Parfit wrote in 2011 that "we live during the hinge of history. Given the scientific and technological discoveries of the last two centuries, the world has never changed as fast. We shall soon have even greater powers to transform, not only our surroundings, but ourselves and our successors. If we act wisely in the next few centuries, humanity will survive its most dangerous and decisive period."[225]
- Noam Chomsky stated in 2016 that the risk of human annihilation is currently "unprecedented in the history of *Homo sapiens*," a view that he has repeated many times since.[226] For example, he told *The New Statesman* in 2022 that, as a result of the climate crisis and growing threat of nuclear war, "we're approaching the most dangerous point in human history . . . We are now facing the prospect of destruction of organised human life on Earth."[227]
- Paul Ehrlich prognosticated that the collapse of civilization is a "near certainty in the next few decades, and the risk is increasing continually as long as perpetual growth of the human enterprise remains the goal of economic and political systems."[228]
- Referring to climate change, Tom Engelhardt wrote that, "even for an old man like me, it's a terrifying thing to watch humanity make a decision, however inchoate, to essentially commit suicide. In effect, there is now a suicide watch on Planet Earth."[229]
- Toby Ord estimated that, given the future development of "radical new technology," humanity has a 1/6 chance of going extinct this century.[230] He reiterated this in his 2020 book, adding that "if we do not get our act together . . . we should expect this risk to be even higher next century, and each successive century. . . . Either humanity takes control of its destiny and reduces the risk to a sustainable level, or we destroy ourselves."[231]
- The minute hand of the Doomsday Clock is currently set to only 90 seconds before midnight, the closest it has been set since the clock's creation in 1947.[232]

The previous record, 100 seconds, was set in 2020, which broke the record of 2 minutes to midnight that was set in 1953 after the United States and Soviet Union detonated the first thermonuclear weapons.

- Finally, consider the The Global Risks Report 2022, published by the World Economic Forum. More than 84 percent of the 1,000 global experts who were asked "How do you feel about the outlook for the world?" said they were either "worried" or "concerned," with only 12.1 percent being "positive" and 3.6 percent being "optimistic" about the future. The "most severe risks on a global scale over the next 10 years" were identified as, from first to last: climate action failure, extreme weather, biodiversity loss, social cohesion erosion, livelihood crises, infectious diseases, human environmental damage, natural resource crises, debt crises, and geoeconomic confrontation.[233]

This is not an exhaustive list, of course.[234] The point is that among those who have contemplated the issue, there is a notable convergence of opinion that the overall probability of Doom Soon is unprecedentedly high.[235] Public opinion surveys reflect this sentiment, as roughly 40 percent of Americans believe that climate change will cause our extinction and another 55 percent are "very" or "somewhat worried" that advanced AI will annihilate us. The current existential mood is pervasive, and it casts a dark shadow on the canopy of our collective future on Earth.

In closing, let us ask again: what explains the Great Silence? Perhaps it is the case that, as Sagan once noted technological civilizations invariably destroy themselves. For millennia, people in every generation have screamed that the end is nigh, claims usually linked to religio-eschatological narratives that culminate in some kind of transformation rather than termination. The difference is that, to quote Lovelock, "this is the real thing."[236] In the end, the Second Law ensures our demise. But in the meantime, omnicide—an apocalypse without kingdom, brought about by our own actions—appears increasingly likely.

Notes

1. Larson and Witham 1998; Stirrat and Cornwell 2013.
2. See Wade 2010.
3. Burge 2022.
4. Koscielniak et al. 2022. Note that Gen Yers were born between 1977 and 1997, Gen Xers between 1965 and 1976, and Baby Boomers between 1946 and 1964.
5. See Harris 2004; Dawkins 2006. For a critique of the New Atheist movement, whose leading figures have come to embrace a wide range of deeply problematic views, such as scientific racism, transphobia, anti-science conspiracy theories (e.g., about COVID), and various far-right positions, see Torres 2017a, 2017b, 2017c, 2021. Note that Sam Harris is one of the "top donors" to the Future of Life Institute.
6. Abrams et al. 2011.
7. Kurzweil 1999.
8. Clark et al. 2016.

9. Note that this date is uncertain. For some time, the favored hypothesis was that the first early anatomically modern humans, whose skeletal remains have been unearthed in Omo National Park, Ethiopia, date back some 197,000–195,000 years, often rounded up to 200,000 years. However, other studies put the date of emergence further back, up to 315,000 years ago, ± 34,000 years (see, e.g., Hublin et al. 2017; Vidal 2022). I will stick with an estimate of 300,000 years for the purposes of this book.
10. Hawking 2016.
11. Randle and Eckersley 2015; Leiserowitz et al. 2017.
12. MUP 2023.
13. Quoted in Pelton 2021.
14. Or, even earlier, to Condorcet's 1795 *Sketch for a Historical Picture of the Progress of the Human Mind*, discussed on several occasions below.
15. Drexler 1986.
16. Regis 1994.
17. Drexler 1986.
18. See Leslie 1996, 99.
19. See Leslie 1996, 99.
20. Moravec 1988.
21. Vinge wrote the following:

    Stan Ulam paraphrased John von Neumann as saying:

    > *One conversation centered on the ever accelerating progress of technology and changes in the mode of human life, which gives the appearance of approaching some essential singularity in the history of the race beyond which human affairs, as we know them, could not continue.*
    >
    > Von Neumann even uses the term singularity, though it appears he is still thinking of normal progress, not the creation of superhuman intellect. (For me, the superhumanity is the essence of the Singularity. Without that we would get a glut of technical riches, never properly absorbed) (Vinge 1993).

    Note also that von Neumann came quite close to the idea of an intelligence explosion in 1948. In discussing automata, he wrote that

    > we are all inclined to suspect in a vague way the existence of a concept of "complication." This concept and its putative properties have never been clearly formulated. We are, however, always tempted to assume that they will work in this way. When an automaton performs certain operations, they must be expected to be of a lower degree of complication than the automaton itself. In particular, if an automaton has the ability to construct another one, there must be a decrease in complication as we go from the parent to the construct. That is, if A can produce B, then A in some way must have contained a complete description of B. In order to make it effective, there must be, furthermore, various arrangements in A that see to it that this description is interpreted and that the constructive operations that it calls for are carried out. In this sense, it would therefore seem that a certain degenerating tendency must be expected, some decrease in complexity as one automaton makes another automaton.

    However, he later noted that

    > "complication" on its lower levels is probably degenerative, that is, that every automaton that can produce other automata will only be able to produce less complicated ones. There is, however, a certain minimum level where this degenerative characteristic ceases to be universal. At this point automata which can reproduce

themselves, or even construct higher entities, become possible. *This fact, that complication, as well as organization, below a certain minimum level is degenerative, and beyond that level can become self-supporting and even increasing, will clearly play an important role in any future theory of the subject* (Neumann 1948, italics added).

22. Vinge 1993, ellipsis in original.
23. See Leslie 1982, 1983, and 1986 for examples.
24. Carter 1974.
25. Leslie 1996, 116–117.
26. As Leslie noted in personal communication with Bostrom, "the ranks of distinguished supporters of [the Doomsday Argument] include among others: J. J. C. Smart, Anthony Flew, Michael Lockwood, John Leslie, Alan Hàjek (philosophers); Werner Israel, Brandon Carter, Stephen Barr, Richard Gott, Paul Davis, Frank Tipler, H. B. Nielsen (physicists); and Jean-Paul Delahaye (computer scientist)" (Bostrom 2002a). Historically, Brandon Carter first introduce the Doomsday Argument in the early 1980s; Leslie then published an article about it in 1989, after which Richard Gott offered his own take in a 1993 article in *Nature*. The latter employed a somewhat different methodology—for example, Gott's argument didn't begin with a comprehensive empirical survey of the threats facing us, nor did it involve choosing between two hypotheses: "Doom Soon" versus "Doom Delayed." Consequently, he estimated that, with a confidence level of 95 percent, "the total longevity of our species [is between] 0.2 million to 8 million years" (Gott 1993; see also Gott 1997). More specifically, as he argued in a PBS documentary:

    > Our species, *Homo sapiens*, has been around for 200,000 years. Now, 200,000 divided by 39 is about 5100. If you multiply by 39, you get 7.8 million. So if there's a 95 percent chance that you're in the middle 95 percent of human history, and that means that the future longevity of the human race is at least 5100 years but less than 7.8 million. Those numbers are interesting because they give us a total longevity that's quite similar to other species. Mammal species have an average longevity of two million years. Our ancestor, *Homo erectus*, lasted 1.6 million years, and the Neanderthals lasted 300,000 years. So this is quite in line with those numbers (Ferris 1999).

    See Bostrom 1999 and Ćirković 2003, 2004, which expanded the analysis of Ćirković 2002a. Criticisms from an influential mathematical statistician can be found in Häggström 2016, ch. 7.
27. Ćirković 2008.
28. Kaneda and Haub 2022.
29. Leslie 1996.
30. See Leslie 1996, 152–153.
31. The only notable exception came from Isaac Asimov's 1979 nonfiction book *A Choice of Catastrophes*, which provides a sprawling survey of the natural and anthropogenic risks facing humanity, some of which involve anticipated future technologies, such as computers that might "become capable of self-correction and of modification of their programs." But ultimately, after roughly 350 pages, Asimov's conclusion is that we have little to actually worry about: our universe is safe in the near term, and the dangers facing us today are wholly surmountable. In his words, "we can deliberately choose to have no catastrophes at all. And if we do that over the next century, we can spread into space and lose our vulnerabilities." Although an entertaining read, the book did not provide a serious study of the various scientifically credible failure modes that could lead to our disappearance. See Asimov 1979, 361–362.
32. Leslie 1996, 4–13.

33. Leslie 1996.
34. Note, however, that some physicists have floated the idea of escaping the heat death by escaping into a parallel universe (see Kaku 2005, 20–21).
35. Leslie 1996, italics in original.
36. Leslie 1996, see 98–100. Note that, following Deudney 2020, I am extremely skeptical that colonizing space will actually reduce the probability of extinction; to the contrary, it may significantly increase it, as Deudney cogently argues. See the end of Chapter 11 for further discussion.
37. Leslie 1996, 146.
38. See Sandberg 2014; Torres 2016.
39. See Gray 2015.
40. Brin 1983. There are of course reports to the contrary, some of which date back millennia and are quite intriguing. For example, a Roman historian named Livy wrote in his expansive *History of Rome*, composed between 27 and 9 BCE, that during the winter of 218 BCE "a spectacle of ships gleamed in the sky [over Rome]." And the NASA scientist Josef Blumrich published a 1974 book in which he argued that passages in the Old Testament book of Ezekiel, which was written in the sixth century BCE, actually describe an alien spacecraft landing on Earth. As Ezekiel 1:13 reports, the occupants of the spacecraft had an "appearance . . . like burning coals of fire." The best evidence to date is very likely the short video clips recently released by the Pentagon, called the "Pentagon UFO videos," in which Navy fighter jets spot and track some unidentified objects streaking in the sky. This, however, is still not good enough to reject the claim that we are alone in the universe; as Carl Sagan liked to say, "extraordinary claims require extraordinary evidence."
41. Hart 1975/1995. Note that Hart is a white separatist/white nationalist.
42. Tipler 1980.
43. See Ward and Brownlee 2000.
44. Sagan 1979.
45. See Leslie 1996, 139.
46. Leslie 1996.
47. And "a fan" of the Men's Rights movement (Hanson's website, accessed on April 19, 2022). For a discussion of Hanson's political and social views, see Weissmann 2018.
48. Bostrom 2005a.
49. Hanson 1998. Note that I have corrected a typo: Hanson, throughout his paper, repeatedly spells "negentropy" as "negentroy."
50. Hanson 1998.
51. Note that I am nearly quoting Hanson here.
52. Hanson 1998.
53. Hanson 1998.
54. See Torres 2017.
55. Drexler 1986.
56. Bostrom et al. 1999.
57. Lederberg 1969.
58. Drexler 1986/2006, 392. For a definition of "functional immortality," see Torres 2020.
59. OTA 1993; Forge 2010.
60. Drexler 1986.
61. Leslie 1996.
62. Snyder 2016.
63. Mukunda et al. 2009, italics in original.
64. DOJ 2020; see Wittes and Blum 2015 for additional examples.

65. See Torres 2018a, 2018b; Nouri and Chyba 2008, 457; Brundage et al. 2018, 16. For a useful discussion of how "dual-use" should be defined, see Forge 2010.
66. For a compelling critique of the idea that the greatest threat in the future may derive from non-state rather than state actors, see Kemp 2021.
67. O'Neill 1945.
68. Quoted in Baskin 2019, 30.
69. See Graff 2019.
70. Quoted in Lanouette and Silard 2013.
71. See Torres 2017.
72. See Bostrom 2014, 184.
73. See Bostrom 2012, 77–79.
74. Vinge 1993. Note that I have removed a hyphen for ease of presentation.
75. See Brundage et al. 2018 for a useful overview.
76. Merriam-Webster, accessed on November 8, 2021: www.merriam-webster.com/dictionary/tool.
77. For example, see Bostrom 2019.
78. As Joy notes, he received a partial preprint of Kurzweil's 1999 book in 1998.
79. Joy 2000.
80. Pollack 2002.
81. Joy 2000.
82. Joy 2000.
83. In Joy's words: "In truth, we have had in hand for years clear warnings of the dangers inherent in widespread knowledge of GNR technologies—of the possibility of knowledge alone enabling mass destruction. But these warnings haven't been widely publicized; the public discussions have been clearly inadequate" (2000).
84. Garreau 2000.
85. Kurzweil 1999.
86. Bostrom 2002b.
87. Bostrom 2005b.
88. Kaczynski 1995.
89. Garreau 2000.
90. Joy 2000.
91. See Winner 1977/1978; Bostrom 2009, italics added.
92. For a useful overview of this idea, see Walker 2009.
93. More 2001.
94. More 2001. Bostrom offered a similar description of the situation on his website in 2000, writing that

    > at the present rate of scientific and technological progress there is a real chance we will have molecular manufacturing or superhuman artificial intelligence well within the first half of the next century. . . . Now, this creates some considerable promises and dangers. In a worst-case scenario, intelligent life could go extinct. Or, if we play it smart, we might manage to make the leap and become posthumans—beings that compare to humans approximately as humans to bugs. The abolition of suffering, aging and disease could be the result. And we could extend human mental and physical capacities in ways we can barely begin to imagine. It is high time that first-class intellects begin to do some serious thinking in this area.

95. Bostrom 2002b.
96. This is one of two types of definitions that Bostrom offers (each with some variations), which are different from each other in important respects. I call the one

specified above his "lexicographic definition" and the other his "typological definition" (see Torres 2019 for detailed analysis).

97. Bostrom 2002b.
98. Bostrom 2002b.
99. As Bostrom and colleagues wrote in the Transhumanist FAQ of 1999, responding to the question "Might transhuman technologies be dangerous?":

    > Yes, and this implies the need to analyze and discuss the problems before they become real. Biotechnology, nanotechnology and AI all have the potential to create major and complex dangers if used carelessly or maliciously . . . Transhumanists urge that it is of the greatest importance that we begin to take these issues seriously. Now (Bostrom et al. 1999).

    Hence, it was realized early on that a new field is needed to study these risks; Bostrom didn't get around to actually founding it until his 2002 paper, a draft of which was first completed in 2001.
100. Bostrom et al. 1999.
101. Bostrom 2002b.
102. To be clear, Bostrom's view of the Doomsday Argument in 2002 was that "although it may be theoretically sound, some of its applicability conditions are in fact not satisfied, so that applying it to our actual case would be a mistake" (2002b).
103. Bostrom 2002b, 2005b. The "no time limit" qualification makes sense because "existential risk" is defined in terms of attaining posthumanity. Hence, Bostrom is saying that there is at least a 25 percent chance that we *never* create a posthuman civilization.
104. Powell and Martindale 2000.
105. *Bulletin* 2007.
106. Osborn 1949.
107. Morrisette 1989.
108. Benedick 1987/1996.
109. Fascinatingly, Midgley died in 1944 when he accidentally strangled himself in a ropes-and-pulleys device that he designed to help himself out of bed, after becoming severely disabled from poliomyelitis, which he contracted at the age of 51. Quite a biography, quite a legacy.
110. Humphreys 2020.
111. Ehrlich 1988.
112. Wilson 1988.
113. Pimm et al. 1995.
114. AMNH 1998.
115. Weart 2008.
116. Delorme 2019. Data from Dr. Pieter Tans, NOAA/ESRL and Dr. Ralph Keeling, Scripps Institution.
117. Revelle 1965.
118. Weart 2008.
119. IPCC 2001.
120. Weart 2008.
121. Steffen et al. 2011.
122. See Oreskes 2004.
123. See, again, Wade 1979 and Rome 2003. To be clear, given the hypothesized consequences of climate change going back to the 1970s, it would have been *inarguably prudent* for political leaders to have taken action *many decades ago*. Indeed, climate scientists have been screaming, into the void, for politicians to implement precautionary measures just in case climate change *really does* turn out to be as catastrophic

and urgent as it might become—and indeed currently is, right now, as of this writing. When one looks at this history, it is appalling that calls to change course have been ignored for so many decades; I follow other philosophers in believing that this ought to constitute a "crime against humanity" that certain leaders in government and industry should be put on trial for. Separately: for criticisms of what Mike Hulme refers to as an attitude or mood of "extinctionism" that "pervades the new public discourse around climate change," see Hulme 2019.

124. Weart 2008, 184.
125. Stern 2006; Gore 2006.
126. Nobel 2021.
127. Merkley and Stecula 2017.
128. Lyell 1833, 52. In fact, Lyell coined the term "Pleistocene."
129. Interestingly, it wasn't until the middle of 2014 that the Oxford English Dictionary (OED) added an entry for "Anthropocene" (Macfarlane 2016).
130. Crutzen and Stoermer 2000.
131. Jacquet 2017.
132. Quoted in Kolbert 2014.
133. Lewis and Maslin 2015.
134. Lewis and Maslin 2015.
135. Hence the term "Plasticene" for our current epoch. For an insightful discussion of the term "Anthropocene," see Zimmer 2021.
136. See Lewis and Maslin 2015; McNeill and Engelke 2014; Steffen et al. 2004.
137. Stanley 2016.
138. Hand 2015.
139. Recall the alarming conclusions of the 1961 Baby Tooth Survey from Chapter 4. The following year is when the Cuban missile crisis occurred, which Arthur Schlesinger identified as "the most dangerous moment in human history" and Robert Kennedy, in his book *Thirteen Days*, described as "a confrontation between the two giant atomic nations, the US and the USSR, which brought the world to the abyss of nuclear destruction and the end of mankind" (Kennedy 2011).
140. Waters et al. 2014.
141. AWG 2019.
142. Wilson 2013.
143. To borrow a line from George Orwell, "to see what is in front of one's nose needs a constant struggle."
144. However, with respect to climate change, Rees argued that "it would be an exaggeration . . . to regard a temperature rise of two or three degrees as in itself a global catastrophe," although he added that even if global warming occurs at the slower end of the likely range, its consequences—competition for water supplies, for example, and large-scale migrations—could engender tensions that trigger international and regional conflicts, especially if these are further fuelled by continuing population growth. Moreover, such conflict could be aggravated, perhaps catastrophically, by the increasingly effective disruptive techniques with which novel technology is empowering even small groups (Rees 2003).
145. Note that although Bostrom's paper was published in 2002, he reports that it "was written before the 9/11 tragedy. See Bostrom 2002b, footnote 23.
146. Ceci 2016; Hoffman 1999.
147. By contrast, Leslie was of course writing before the 9/11 attacks, although he does mention the sarin attacks perpetrated by Aum Shinrikyo.
148. Rees 2003.
149. Rees 2003.

150. Note that Posner, a conservative Republican, is skeptical about some of these risks, especially those relating to the environment. See Posner 2004, ch. 1.
151. Posner 2004.
152. For example, *Our Final Hour* was covered by *The Guardian*, *The Telegraph*, *Publishers Weekly*, BBC News, *The National Review*, The Universe Today, among others.
153. Here is an incomplete but representative list of books and comprehensive reports written over the past ~15 years on human extinction, existential risk, civilizational collapse, and related issues:

    Jared Diamond, *Collapse: How Societies Choose to Fail or Succeed* (2005).
    Ray Kurzweil, *The Singularity Is Near: When Humans Transcend Biology* (2005).
    Nassim Taleb, *The Black Swan: The Impact of the Highly Improbable* (2007).
    Thomas Homer-Dixon, *The Upside of Down: Catastrophe, Creativity, and the Renewal of Civilization* (2007).
    Nick Bostrom and Milan Ćirković (eds.), *Global Catastrophic Risks* (2008).
    Willard Wells, *Apocalypse When? Calculating How Long the Human Race Will Survive* (2009).
    Cass Sunstein, *Worst-Case Scenarios* (2009).
    World Economic Forum. *Global Risks* (2011): http://riskreport.weforum.org/.
    James Barratt, *Our Final Invention: Artificial Intelligence and the End of the Human Era* (2013).
    Nick Bostrom, *Superintelligence: Paths, Dangers, Strategies* (2014).
    Global Challenges Foundation, *12 Risks that Threaten Human Civilisation* (2015).
    Olle Häggström, *Here Be Dragons: Science, Technology, and the Future of Humanity* (2016).
    Phil Torres, *The End: What Science and Religion Tell Us About the Apocalypse* (2016).
    Phil Torres, *Morality, Foresight, and Human Flourishing: An Introduction to Existential Risks* (2017).
    Max Tegmark, *Life 3.0: Being Human in the Age of Artificial Intelligence* (2017).
    Bryan Walsh, *End Times: A Brief Guide to the End of the World* (2019).
    Stuart Russell, *Human Compatible: Artificial Intelligence and the Problem of Control* (2019).
    Toby Ord, *The Precipice: Existential Risk and the Future of Humanity* (2020).
    Bruce Tonn, *Anticipation, Sustainability, Futures and Human Extinction* (2021).
    The Global Challenges Foundation has also published annual reports (2016, 2017, 2018, 2020, 2021) on global catastrophic risks that provide excellent, readable overviews of the twenty-first-century threat environment.

154. Bostrom and Ćirković 2008.
155. Stevens 1991.
156. Rubin 2012.
157. Root 2019. In fairness, Carson also wrote that "the balance of nature is not a *status quo*; it is fluid, ever shifting, in a constant state of adjustment. Man, too, is part of this balance" (Carson 1962).
158. Lorenz 1972.
159. Leakey and Lewin 1995.
160. Sample 2004.
161. Hansen 2005.
162. Lenton and Schellnhuber 2007.
163. Barnosky et al. 2012.
164. Rockström et al. 2009a.
165. Rockström et al. 2009b.
166. Rockström et al. 2009b, 2009a.

167. Steffen et al. 2015.
168. Rockström et al. 2009b.
169. Steffen et al. 2018.
170. Lenton et al. 2019.
171. Brysse et al. 2012.
172. For further discussions of the potentially catastrophic consequences of climate change, see Kemp et al. 2022.
173. Drexler 2013, ch. 9.
174. For a brief but useful overview of "premortem analyses," see Thaler 2017.
175. A useful survey of these earlier ideas comes from Muehlhauser 2012.
176. Goertzel 2015; MIRI 2022. See Horgan 2016. MIRI has also been supported by the cryptocurrency billionaire Vitalik Buterin, a beneficiary of the Thiel Fellowship. Other top donors include Jaan Tallinn, Open Philanthropy, and the Berkeley Existential Risk Initiative.
177. Tegmark 2016.
178. For useful overviews, see Muehlhauser 2011; Sotala and Yampolskiy 2014.
179. If the ASI algorithm does not need to sleep—and presumably it wouldn't—23 days of constant wakefulness would be equivalent to 34.5 human days if one were to spend 8 hours sleeping each 24-hour period. In other words, the ASI would have *more than a month* to figure out how to prevent it from being unplugged in *just 2 seconds* of human time.
180. Yudkowsky 2008.
181. Good 1982.
182. Muehlhauser and Helm 2012.
183. Müller and Bostrom 2014; Bostrom 2014.
184. See also the analysis of the possibility of an intelligence explosion in Chalmers 2010, which added some philosophical clarity to the issue.
185. See, for example, Russell 2019.
186. Musk 2014.
187. Best 2017.
188. Wright 2018.
189. For some useful recent discussions of emerging risks, see Pamlin and Armstrong 2015; Brundage et al. 2018; and Torres 2018.
190. See Fields 2020 for an overview.
191. See, for example, Nuttall 2020.
192. Lanier 2020.
193. Quoted in Topol 2016.
194. Torres 2017a.
195. Topol 2016.
196. Fitzgerald et al. 2020.
197. 2AI 2016.
198. FLI 2023.
199. Yudkowsky 2023.
200. Gebru et al. 2023.
201. See Baum 2017; Fitzgerald et al. 2020.
202. Steffen et al. 2018.
203. Tiseo 2021. For an excellent discussion about why climate change constitutes the "perfect moral storm," see Gardiner 2011.
204. Haqq-Misra et al. 2017.
205. Mora et al. 2018.
206. See Torres 2016, 2017a.
207. McLellan et al. 2014.

208. Grooten 2018.
209. Almond 2020; WWF 2022.
210. Quoted in Carrington 2018.
211. Böhm et al. 2013.
212. Estrada et al. 2017.
213. Tollefson 2019.
214. UNICEF 2021; UN News 2017.
215. Kuhlemann 2018.
216. Worm et al. 2006. As this book was about to be submitted, it was brought to my attention that there may be problems with this particular extrapolation. See Hamrud 2021 for discussion.
217. SD 2015; Sekerci and Petrovskii 2015. This section draws from previous publications of mine, sometimes verbatim. See Torres 2018, 2021.
218. Rees 2003.
219. Posner 2004.
220. Sandberg and Bostrom 2008. According to Olle Häggström, "the questionnaire [for this survey] was given to the participants of the Global Catastrophic Risk conference in Oxford in July 2008, which served as a kind of launch event for the collection by Bostrom and Ćirković," namely, Global Catastrophic Risks (2008). Many of those who authored chapters in this collection were also present at the conference, and hence survey participants (Häggström 2016, footnote 431).
221. Aitkenhead 2008.
222. Wells 2009.
223. Edwards 2010.
224. Kaku 2011.
225. Parfit 2011, italics added.
226. Lombroso 2016. For example, Chomsky declared in 2018: "The urgency of 'looming extinction' cannot be overlooked. It should be a constant focus of programs of education, organization, and activism, and in the background of engagement in all other struggles" (cited in Chomsky 2020).
227. Eaton 2022.
228. Carrington 2018; Ehrlich and Ehrlich 2009.
229. Engelhardt 2019.
230. Wiblin 2017.
231. Ord 2020. However, Ord is less optimistic about the "longer term," writing that, "if forced to guess, I'd say there is something like a one in two chance that humanity avoids every existential catastrophe" (Ord 2020).
232. Mecklin 2020.
233. WWF 2022.
234. See also Scranton 2015.
235. Compelling criticisms of the current mood, or aspects of it, have recently been published from the perspective of Black and Afro-futurism and Indigenous futurism, for example, Mitchell and Chaudhury 2020. See also Haraway 2016; Srinivasan 2017; Grove 2019; Servinge and Stevens 2020; Yunkaporta 2020.
236. Aitkenhead 2008.

Part II

Existential Ethics

7 What Is Human Extinction?

Mood Theory

Our exploration of the history of *human extinction* so far has focused primarily on how Western thinking about the possibility, probability, and etiology of our extinction, along with the temporality and multiplicity of the various risks facing us today and within the foreseeable future, has evolved over time. I argued that this history, which I dubbed History #1, can be divided into five distinct periods, each defined by a unique combination of answers to questions about whether our extinction is possible, how probable it is, how many kill mechanisms there are, whether they are natural or anthropogenic, and so on. I further claimed that for much of Western history, the legacies of Platonic philosophy and the ancient Persians, which shaped central doctrines of Christianity, established a cluster of beliefs that blocked the concept of *human extinction* for some 1,500 years. These beliefs were: first, the Great Chain of Being and its constituent principle of plenitude, which implies that extinction of *any kind* is impossible. This fell in the early nineteenth century due in part to the pioneering work of Georges Cuvier (although we will see in the next chapter that a version of the principle of plenitude survived Cuvier's attack for roughly a century). Second, the ontological and eschatological theses, which imply that *human* extinction is impossible even if extinction in general is not, given the nature of human beings and our special role in God's grand plan for the cosmos. The ontological thesis was mortally wounded by Charles Darwin's theory of evolution, which integrated humanity into the natural world such that the difference between our species and all other creatures became one of degree rather than kind. This also undermined the eschatological thesis, though it had already lost some of its force because of the rise of deism during the Enlightenment.

While the collapse of these beliefs enabled new thoughts about the mortality of our species, the primary driver behind each abrupt shift in existential mood was the discovery of kill mechanisms associated with the Second Law of thermodynamics, global thermonuclear fallout, synthetic pollutants, overpopulation/overconsumption, ozone depletion, biological warfare, self-improving AI,

DOI: 10.4324/9781003246251-9

the runaway greenhouse effect, the nuclear winter phenomenon, self-replicating nanobots, asteroid and cometary impacts, and volcanic supereruptions. The one exception was the most recent shift in mood, which was triggered by (i) the ethically motivated search for an exhaustive inventory of every type of risk facing humanity in the twenty-first century, and (ii) further developments in our understanding of the extent and seriousness of humanity's impact on the natural world.

The complicated story that emerged from these phenomena may be aptly described as one of profound psycho-cultural trauma, whereby the reassuring sense of Franklinian "Comfort" and Dickian "perfect security" that defined much of the first existential mood was superseded by the horrifying realization that our extinction (a) is inevitable in the long term, (b) could happen in the near future due to anthropogenic or natural causes, and (c) will become even more probable in the foreseeable future as a result of advanced technologies and the environmental crisis. Let's label the five-part periodization outlined in Part I and its underlying causal explanation, i.e., the enabling conditions and triggering factors, *existential mood theory*, which is a fitting name, I think, for the explanatory-predictive hypothesis that I originally adumbrated in Chapter 1. This theory and that hypothesis are the same.

Yet there is a whole other set of questions about human extinction that are distinct from those connected to existential moods. Recall from Chapter 1 that such questions include: Would causing or allowing our extinction be right or wrong? For what reasons? Under which conditions? Would our extinction, however it might be caused, be good or bad, better or worse, or perhaps just neutral (no difference)? For what reasons? Under which conditions? Would knowledge of impending extinction compromise the value of our existence right now? Would extinction undermine the significance of past actions? Is everything meaningless if human extinction is inevitable? Should the "interests" of merely possible people in the future be considered in our moral deliberations? Do we have moral obligations to those who existed in the past? And so on. Such questions can be organized into the following three normative categories:[1]

(1) *Deontic* questions about whether bringing about our extinction would be right, wrong, permissible, obligatory, or forbidden. The category of "deontic" (from the Greek *deon*, meaning "that which is binding") concerns what moral agents should and should not do, and hence it subsumes concepts like *ought*, *right*, *wrong*, *duty*, *obligation*, etc.[2] While some philosophers have argued that causing our extinction would quite literally be the worst crime imaginable, others have contended that we should actively bring it about by ceasing to have children.
(2) *Evaluative* questions about whether our extinction would be good, bad, neutral, and so on.[3] The category of "evaluative" (from the Latin *valere*, meaning "be of value, be worth") concerns the "worth of things," and hence subsumes concepts like *good*, *bad*, *better*, *worse*, *valuable*, *excellent*, *terrible*, and so on.[4] Many

philosophers have held that human extinction would be very bad—in some cases, the badness of this outcome forms the basis for deontic claims about the wrongness of bringing it about—although others have argued with equal vigor that extinction would be less bad than continuing to exist, or perhaps even positively good.

(3) The third category is a grab-bag of questions about how the expectation of extinction could affect the meaning, importance, and significance of our existence right now. For example, would knowledge of our impending extinction compromise the value of our lives and activities in the present? Is everything meaningless if our extinction is inevitable in the long run? What is the point of anything if, in the end, all will be lost?

Such questions constitute the heart of what I am calling "Existential Ethics," the development of which is the subject of History #2. In other words, this second history concerns the different ways that successive generations of Western philosophers have answered the questions above. I will partition this history into four "waves," each of which will receive its own chapter. While there is some alignment between the periodizations of History #1 and History #2, the second wave of History #2 does not begin until the mid-twentieth century, and the most recent wave was initiated only in the past few years when, for the first time, a number of philosophers offered analytically rigorous accounts of human extinction from various non-consequentialist perspectives. Despite the bulk of History #2 having occurred since the 1950s, my presentation of this history will ultimately be slightly longer than that of History #1. It will also be much more theoretical, although I will do my best to make the esoterica of philosophy accessible to non-philosophers, just as I have tried to make the details of History #1 accessible to non-historians and non-scientists. Roughly speaking, whereas Part I of this book focused mainly on *events*, Part II will focus primarily on *ideas*.

However, to understand this history, we must first do something that was not, strictly speaking, necessary for the purposes of the first history, namely, distinguish between a range of naturalistic human extinction scenarios. This is important because many of these scenarios have their own unique ethical and evaluative implications—that is, how one answers some of the questions above will crucially depend upon *which* extinction scenario one is talking about. The very same normative position might identify one scenario as bad and another as good, thus coming to opposite evaluative conclusions about our disappearance, depending on the circumstances. Furthermore, many of these scenarios can be, and have been, picked out by the term "human extinction," a fact that can generate confusion and, therefore, merely verbal debates among discussants using the same words to denote quite different phenomena. We will thus begin by analyzing the cluster concepts of *human* and *extinction*, after which we will examine three "stages" or "aspects" of extinction that will also prove to be ethically and evaluatively significant. This will not only lay the foundation for understanding History #2 but provide a degree of retroactive clarity to the grand narrative of History #1.

Humanity and Extinction

The meaning of "human extinction" may appear straightforward at first glance: we simply cease to exist. However, both "human" and "extinction" can be understood in many different, equally legitimate ways. Biologists, for example, differentiate between multiple types of extinction,[5] and while one might assume that "human" means "*Homo sapiens*," anthropologists use the term to denote our genus, *Homo*, which implies that distant ancestors like *Homo habilis*, *Homo neanderthalensis* (Neanderthals), and *Homo floresiensis* are no less "human" than we are.

Within the contemporary Existential Ethics literature, many philosophers take "human" to refer to a broad class of future beings. For example, Jason Matheny writes that he uses "'humanity' and 'humans' to mean our species and/or its descendants."[6] Hence, future beings count as "human" if they possess the right genealogical or causal connections to present-day people. This would presumably include not just future species that we evolve into but artificial beings that we might create alongside us or to replace us. Others have stipulated definitions that include normative conditions beyond a mere genealogical or causal link, which will become especially important for one extinction scenario in particular. An example comes from Nick Beckstead, who writes that "by 'humanity' and 'our descendants' I don't just mean the species [*Homo*] *sapiens*. I mean to include any valuable successors we might have," later describing them as "sentient beings that matter."[7] In other words, a population that is genealogically or causally linked to present-day people would not count as "human" if they lacked the properties necessary to be valuable or matter morally. Hilary Greaves and William MacAskill offer a similar definition in writing that "we will use 'human' to refer both to *Homo sapiens* and to whatever descendants with at least comparable moral status we may have, even if those descendants are a different species, and even if they are non-biological."[8] Still others focus on different normative properties, such as our dignity, autonomy, or capacity to freely will our actions, the loss of which would be sufficient for "humanity" to cease existing. The most expansive definition comes from Nick Bostrom, who identifies "humanity" with "Earth-originating intelligent life," which implies that future descendants of ours that lack a certain level of "intelligence" would not count as "human."[9] However, the condition of "Earth-originating" also opens the door to there being "humans" who have no genealogical or causal connection to us at all—for example, creatures that evolve on Earth independently of our evolutionary lineage, so long as they acquire our level of intellectual ability.

A possible terminological confusion worth pointing out is that some of our descendants may be what philosophers, including those mentioned above, refer to as "posthuman," meaning that such beings are sufficiently different from us to warrant classifying them as a different species. Hence, on definitions like Matheny's and Bostrom's, both current-day humans and our posthuman descendants would fall within the extension of "humanity." Perhaps, then, we need a new concept and term to cover humanity and posthumanity, although I will not here propose one. In what follows, I take "human" and "humanity" to denote our particular species *Homo sapiens*. This will be our default definition, although we will see that certain extinction scenarios more naturally fit with other definitions.

Turning now to the cluster concept of *extinction*, there are at least six extinction scenarios that are important to distinguish for ethical, evaluative, and historical reasons. I will call these demographic extinction, phyletic extinction, terminal extinction, final extinction, normative extinction, and premature extinction. The common denominator shared by all is the minimal definition of "extinction" given in Chapter 1, which we can formalize as follows:

> *Minimal definition*: something has gone extinct if and only if there were tokens of the type at some time T1, but then at some later time T2 no tokens of the type exist.[10]

In the case of humanity, the relevant type could be understood in any of the senses specified above.[11] Let's now examine each of the six extinction scenarios in turn:

The first is the simplest: demographic extinction would occur if and only if *Homo sapiens* were to disappear entirely because the human population falls to zero. We thus cease to exist physically, and our evolutionary lineage comes to an end. This could happen suddenly, as the result of a catastrophe, or gradually, as in the case of worldwide infertility. If the process happens gradually enough, there may be a single remaining individual—the "last man," in literary terms, or what biologists call the "terminarch" or "endling"—whose individual death thus coincides with the extinction of humanity. Most dictionary definitions of "extinction" correspond to demographic extinction, although I will argue below that it is not, in fact, what most people have in mind when they think about our extinction.

Phyletic extinction, sometimes misleadingly called "pseudoextinction," would occur if and only if *Homo sapiens* were to disappear entirely by evolving into one or more new species such that, while our species would no longer exist, our evolutionary lineage would. Consistent with our remarks above, let's call the species that we evolve into "posthuman." Biologists recognize three or four ways that phyletic extinction could occur in the natural world, though only two of these are pertinent to our discussion: the first is *anagenesis*, whereby there exists a single evolutionary lineage from T1 to T2, but the members of the population at T2 no longer belong to the same species of T1. In contrast, *cladogenesis* would involve the evolutionary lineage splitting such that the parent species at T1 becomes two or more distinct daughter species at T2.[12]

Since the distinction between *species* lies at the heart of phyletic extinction, the question arises: What exactly is a species? There is no consensus among biologists or philosophers of biology, and in fact Jody Hey counts more than twenty "species concepts" within the scientific literature.[13] Yet the case of humanity introduces an additional layer of complexity for two interrelated reasons: first, over the past 3.8 billion years, since life first emerged on Earth, anagenesis and cladogenesis have occurred via natural evolutionary processes: selection, recombination, mutation, genetic drift, and so on. However, the development of advanced technologies could enable humanity to usurp these natural processes by modifying ourselves in potentially radical ways. There are several possibilities here: we could use genetic engineering to alter our genes, thus creating one or more posthuman species that are, with respect to their physical constitution, still wholly biological; we could integrate biology and technology, organism and artifact, to create posthuman cyborgs—for example, by radically augmenting our cognitive abilities via brain-computer interfaces or neural chips; and we could replace our biology altogether, resulting in a wholly artificial species, which might be achieved through mind-uploading (or whole-brain emulation), whereby the microstructure of our nervous system is reproduced *in silico*.

This leads to the second complication: such modifications could yield beings that satisfy standard biological definitions of "species" yet are so alien to, and different from, us that we may want to classify them as posthuman. Consider two scenarios: in one, a population of uploaded minds with humanoid robotic bodies is created. Such individuals would (let's assume) think, feel, and behave exactly like biological humans, though by virtue of their robotic bodies they would be unable to interbreed with *Homo sapiens*. According to the popular Biological Species Concept, interbreedability is a necessary condition for individuals to belong to the same species, and hence this population would constitute a new species—call it *Homo uploadus*. Now contrast this with a situation in which we implant a neural chip into the brains of a group of people or modify their genes so that the resulting individuals become superintelligent, that is, far more "intelligent" than any "unenhanced" human. The dynamic between such individuals and us would be comparable to the difference between us and chimpanzees. We might thus wish to call them a new posthuman species, although given that the *only* difference between us and them would be their superhuman levels of "intelligence," the Biological Species Concept would classify them as our conspecifics.

The key point is that what counts as phyletic *human* extinction is complicated by the possibility of (a) radically altering higher-level properties associated with our cognitive systems, psychological characteristics, moral sensibilities, phenomenological capacities, and emotional repertoires without necessarily changing other aspects of our biological constitution, and (b) radically altering our biological constitution without changing other characteristics that we take to be constitutive of being human. Resolving these puzzles may require an entirely novel conceptual-ontological framework, a new theory of species, though this is unnecessary for our purposes here.[14] What matters is that if future people were to undergo radical changes to their minds, emotions, or physical constitution, and if the resulting population were to take the place of *Homo sapiens*, then our species will have undergone phyletic extinction.

The third scenario is terminal extinction, which would occur if and only if *Homo sapiens* were to disappear entirely and this were to remain the case forever. There are several noteworthy features of this scenario: first, terminal extinction is compatible with both demographic and phyletic extinction, which is to say that we could disappear entirely and forever as a result of either (a) our population dwindling to zero or (b) *Homo sapiens* evolving into one or more new species. What terminal extinction introduces is the condition of permanence: not only are there no more tokens of the type "humanity," but there are never any tokens of this type again. Second, it should be obvious that demographic extinction would more or less guarantee terminal extinction if it were to happen in the near future. However, advanced technologies could enable these scenarios to be pulled apart in practice. Consider that some scholars have proposed subterranean refuges that are continuously occupied as a way of repopulating the planet in case of a global catastrophe.[15] Given the difficulty of persuading people to live in underground bunkers for extended periods of time, it is not hard to imagine someone suggesting, as the technological means become available, that we establish bunkers in which automated machines recreate our species using genetic material after, say, two decades following a nuclear holocaust. A template for this might be the idea of "embryo space colonization," whereby cryopreserved human embryos would be shipped to exoplanets, at which point "artificial uterus (AU) systems" would enable them to developed into neonates. In fact, one author has recently argued that "a similar strategy" to embryo space colonization "could also be used to repopulate Earth after human extinction events" by replacing humans in underground bunkers with frozen embryos.[16]

Or consider the "Aestivation Hypothesis," which has been propounded as a possible solution to the Great Silence (or Fermi Paradox). This states that advanced civilizations may choose to shut down until the temperature of the universe drops in accordance with the Second Law (the word "aestivation" refers to a period of dormancy that occurs during the warm summer rather than chilly winter—the opposite of hibernation). The reason is that, as Anders Sandberg and colleagues explain, "the thermodynamics of computation make the cost of a certain amount of computation proportional to the temperature," meaning that higher temperatures equal higher costs.[17] Since the universe is warmer now than it will be later on, civilizations could maximize computation by aestivating. While Sandberg et al. do not elaborate on what exactly aestivation might entail, one option would be for humanity to engineer its temporary disappearance—in other words, to intentionally undergo demographic but not terminal extinction. This idea makes the most sense on one of the broader definitions of "humanity," as the distant future may be much less hospitable to *Homo sapiens* than a population of cyborgs or wholly artificial successors. Hence, the distinction between demographic and terminal extinction could have important practical implications, depending on the technological capabilities of future civilizations.

This being said, some would also argue that the ethical and evaluative aspects of these scenarios are different. As subsequent chapters will explore, one might hold that humanity's permanent disappearance would be much worse—or better—than ceasing to exist temporarily. The most notable reason to distinguish between demographic and terminal extinction, though, is historical: it enables us to make sense of the kind of human extinction involved in the cosmologies of Xenophanes, Empedocles, and other ancient philosophers. Recall that on Xenophanes' model, humanity disappears entirely during one stage of the cosmic cycle, but we always reappear later on. In other words, we undergo demographic but not terminal extinction, and in fact we have undergone and will undergo demographic extinction an infinite number of times. By separating these two types of extinction, we can see more clearly how the idea of human extinction has evolved over time, a point we will return to in the following section.

The fourth scenario is final extinction, which would occur if and only if *Homo sapiens* were to disappear entirely and forever without leaving behind any successors.[18] This is motivated in part by the idea that what happens after our species no longer exists could play a crucial role in how one assesses our disappearance as being right or wrong, good or bad, better or worse. For example, if our species were to undergo terminal extinction but leave behind some population of beings that, say, carry on "our values" (whatever they might be), some wouldn't consider this to be bad. To the contrary, depending on the nature of these beings, such people would describe this as positively desirable. There are, once again, several points to make: first, if one takes "human" to denote *Homo sapiens*, then phyletic extinction would be incompatible with final extinction, since phyletic extinction can only happen if we have successors, and having no successors is a necessary condition for final extinction. On this interpretation, final human extinction would coincide with terminal extinction brought about by demographic extinction, though terminal and demographic extinction could occur without final extinction having happened. Second, if one takes "human" to mean "*Homo sapiens* and our descendants," then final extinction could occur if either (a) *Homo sapiens* itself disappears forever without leaving behind any successors, or (b) we have descendants, but these descendants do not have any descendants of their own. Either way, the important idea behind this type of extinction is that it would constitute a complete and final end to the whole human story by denying us, or our descendants, any future at all. There is no life after the death of "humanity," no adventure that could continue. Final extinction is the last period of the closing chapter.

As noted, our successors could take the form of posthuman species that we evolve into, perhaps resulting in the phyletic extinction of *Homo sapiens*. But there are other possibilities, one of which we gestured at above: we could create a new population of artificial progeny, such as intelligent machines, that would be causally but not, strictly speaking, genealogically related to us. That is to say, whereas phyletic extinction would preserve the spatiotemporal continuity of our evolutionary lineage across time, we might also simply create a whole new lineage, after which humanity might decide to bow out of the show, handing off the existential baton—so to speak—to our machinic replacements. While many people would recoil at this possibility, some philosophers and futurists have argued that we ought to do this. For example, as noted in Chapter 6, Hans Moravec not only endorsed such a scenario but hoped to actively bring it about. More recently, Derek Shiller has argued that

> if it is within our power to provide a significantly better world for future generations at a comparatively small cost to ourselves, we have a strong moral reason to do so. One way of providing a significantly better world may involve replacing our species with something better.[19]

While "existential risk" philosophers like Bostrom have not made this exact argument, they would almost certainly endorse a scenario in which *Homo sapiens* were replaced by artificial beings, so long as these beings were to satisfy certain normative conditions built-into some of the stipulative definitions of "humanity" discussed earlier.

This brings us to normative extinction, which would occur if and only if our species or whatever successors we might have were to undergo changes that cause one or more normatively important properties to be lost. In other words, as with phyletic extinction, humanity goes extinct without ceasing to exist physically. It is here that the normative definitions of "human" become especially relevant. There are two basic types of normative extinction, depending on the cause. I will call one "internalist" and the other "externalist." To illustrate the former, imagine a scenario in which our descendants evolve over millions of years into beings that lack any "comparable moral status" to us, quoting Greaves and MacAskill's definition. Our lineage persists, but "humanity" disappears. The same conclusion would follow if our descendants were to cease being "sentient beings that matter," on Beckstead's definition, or sufficiently "intelligent," on Bostrom's. Or consider a scenario in which future beings lack the capacity for conscious experiences, that "something it is like to be" them.[20] They become philosophical zombies, with no qualitative inner life, but no corresponding deterioration in their behavior, and hence continue to advance science, invent new technologies, create art, play games, tell jokes, and even philosophize about the nature of consciousness. Yet many philosophers would see this as no better than bequeathing the world to rocks; as Bostrom writes, "the future might then be very wealthy and capable, yet in a relevant sense uninhabited: There would (arguably) be no morally relevant beings there to enjoy the wealth." If one understands "humanity" as "intelligent beings capable of (at least) conscious experience," then there would be no "humans" in this scenario: the relevant type in the Minimal Definition would have no tokens and hence human extinction would have occurred.

These are examples of internalist normative extinction, so-called because the loss of "humanity" arises from changes that are internal to us or our successors. But there are also externalist scenarios that would extinguish one or more normative capacities from the outside-in, as the result of unfavorable, extreme, and persistent circumstances. For example, imagine a highly oppressive totalitarian government gaining control over the entire human population. This causes a complete loss of our dignity, freedom, and autonomy. If one takes these to be essential features of our humanity, then one should consider this to be a case of human extinction (assuming that everyone on Earth is affected, including those in charge). As we will see, some theorists in the early postwar period held that final human extinction would actually be *preferable* to normative extinction of this sort, a position sloganized as "Better dead than Red." Although they did not explicitly describe this as a form of "human extinction," I contend that we should see it as such, given that it would satisfy the Minimal Definition, on certain normative accounts of "humanity."[21]

The last scenario is premature extinction, which would occur if and only if one or more of the above scenarios were to happen prior to humanity attaining some goal, end-state, or *telos* taken to be valuable or important. Hence, this definition is both normative and teleological, and thus applies only within frameworks, or futurologies, that are normative and teleological. What premature extinction introduces is the idea that our extinction would be *worse* if it were to occur before achieving some goal than if it were to happen afterwards. In other words, the *timing* of extinction matters. What might such a goal be? In subsequent chapters, we will examine a range of answers, which include constructing a complete scientific theory of the universe, establishing world peace and universal prosperity,[22] building a posthuman civilization,[23] colonizing space and creating astronomical amounts of value,[24] and reaching a state of "technological maturity."[25] If our extinction were to take place before attaining such goals, then it would count as "premature," otherwise it would not. Furthermore, since these goals could presumably be achieved by our posthuman descendants—some would argue that our descendants would be better able to achieve them than we are—the notion of premature extinction fits most naturally with the broader definitions of "humanity."

With these extinction scenarios in mind, it is worth pausing for a moment on the complexity of the apparently simple question: "Would human extinction be right or wrong, good or bad, better or worse?" When asked this question, the first thing one should do is respond: "Which type of extinction are you talking about?" and "What do you mean by 'human'?" As noted above, some theories will see certain types of human extinction as very bad, while identifying others as potentially desirable. Others focus entirely on a single type of extinction, while still others come to the same ethical or evaluative conclusions about all of them. Yet, the question conceals even more ambiguities, as there are different *aspects* of

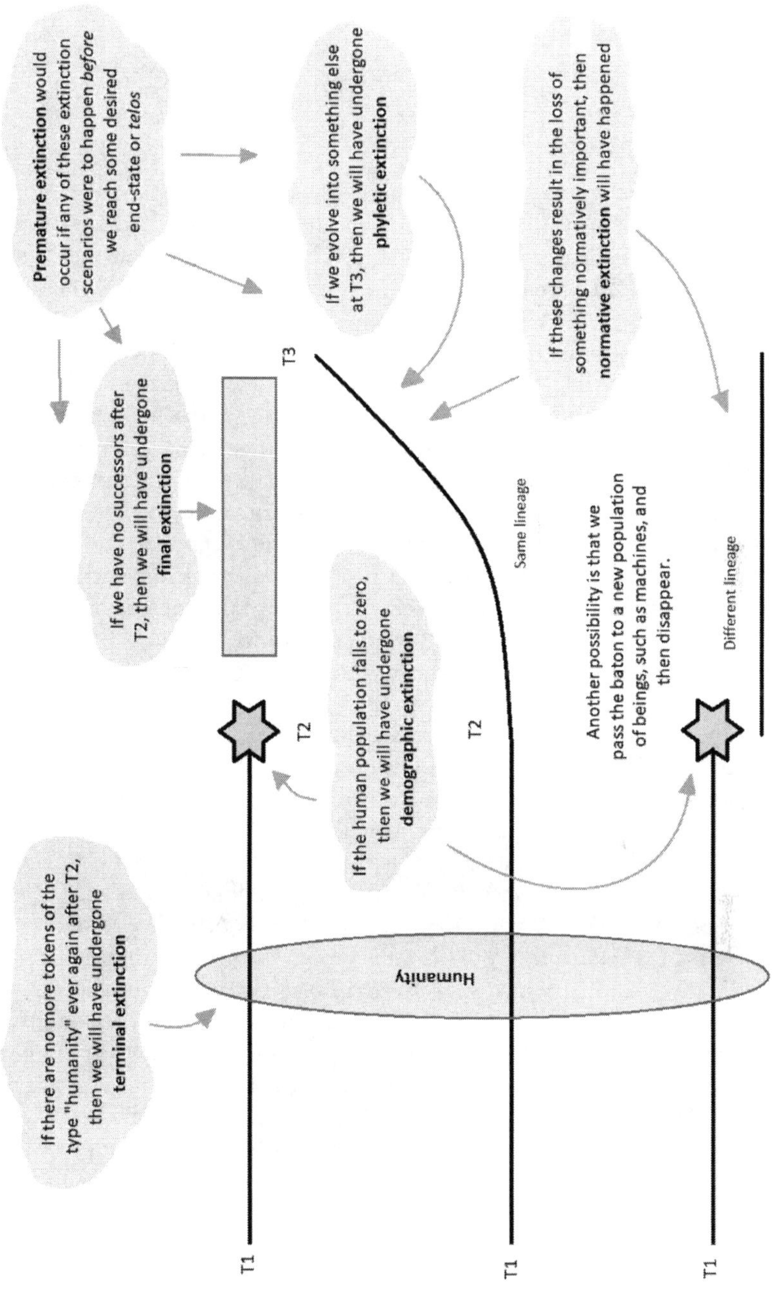

Figure 7.1 The six human extinction scenarios.

human extinction—the "stages" mentioned earlier—that different theories will identify as ethically or evaluatively significant. Before turning to this issue, let's briefly reexamine the history of Part I through the lens of this six-part framework of human extinction scenarios.

Reinterpreting the Past

In the opening chapter of this book, I claimed that it would be inaccurate to assert that the idea of *human extinction* "emerged first during the 18th and 19th centuries," as one historian has recently argued.[26] We can now precisify this claim: the notion of demographic extinction goes back at least to the philosophers of ancient Greece. Xenophanes, Empedocles, and the Stoics all posited that humanity periodically undergoes demographic extinction, though never terminal extinction. The ancient Greek atomists held a similar view: while the disintegration of our particular *kosmos* will be complete and permanent, the infinitude of space and time implies that humanity will reappear somewhere else in the universe. With the rise of Christianity in the fourth and fifth centuries CE, naturalistic extinction in all its forms was rendered unthinkable, and this remained the case until the nineteenth century. However, there were some exceptions, and indeed demographic extinction may have been the first extinction scenario to reappear toward the end of this ~1,500-year period. In particular, we noted that Denis Diderot suggested in 1769 that humanity could go extinct, although if this were to occur, we would surely evolve once again. The anonymous "H" from Chapter 2 expressed a similar idea. This should be unsurprising because the Great Chain continued to influence Western intellectuals until the early nineteenth century, though it evolved over time to accommodate the possibility of that links in the chain might disappear temporarily. In other words, it remained incompatible with terminal but not demographic extinction, which explains Diderot's insistence that our extinction would not last forever.

The idea of phyletic human extinction was first seriously considered after the publication of Charles Darwin's 1859 *Origin of Species*. As we noted, Darwin himself affirmed that humanity would continue to evolve in the future, attaining higher levels of perfection, which suggests that *Homo sapiens* itself might someday become phyletically extinct. H. G. Wells later foregrounded phyletic human extinction in his 1895 novel *The Time Machine*, which describes our evolutionary lineage splitting into the Eloi and Morlocks, at which point *Homo sapiens* no longer exists. This story also hints at normative extinction, since Wells described these two species as being subhuman: one is intellectually stunted and the other becomes a subterranean brute.[27] The emergence of a new species from our evolutionary lineage was further explored by early transhumanists like J. B. S. Haldane, Sir Julian Huxley, and J. D. Bernal, the last of whom speculated that, by altering our germ plasm, "we might achieve such a variation [in the human population] as we have empirically produced in dogs and goldfish, or perhaps even manage to produce new species with special potentialities."[28] Although the variety of dog breeds still fall within the species of *Canis familiaris*, Bernal may very well have imagined that, over enough time, the methods of eugenics would produce a genuinely new species of posthumans. Hence, the earliest instances of phyletic and normative extinction are found in the late nineteenth and early twentieth centuries, when the scientific community embraced Darwin's theory of evolution and worries about "degeneration" became somewhat common. Furthermore, since there was no reason to believe at the time that our disappearance wouldn't be permanent, it would have been assumed the demographic or phyletic extinction would probably entail terminal extinction.

With respect to final extinction (which subsumes terminal extinction), this might be the oldest extinction scenario to have been imagined. Long before Xenophanes and the atomists, the epic poem of *Atrahasis* describes the god Enlil attempting to destroy humanity three times before sending a flood, which suggests that Enlil was intent on there being no more humans.[29] In the flood narrative found in the *Epic of Gilgamesh*, when Enlil discovers that Utnapishtim and others had survived the deluge he becomes "filled with rage," angrily shouting: "Where did a living being escape? No man was to survive the annihilation!" (Tablet XI).[30] Final extinction was subsequently gestured at by philosophers in the latter nineteenth century, such as Philipp Mainländer and Eduard von Hartmann (discussed below), as well as those who pondered the eschatological implications of the Second Law. Much of the anxiety surrounding thermonuclear war and other such catastrophes in the twentieth century also assumed that, if they were to occur, the result would be an irreversible end to the human story.

However, it was not until the second half of the twentieth century, especially its waning decades, that final extinction became *explicit* in the literature.[31] In particular, the prospect of our machinic progeny someday replacing humanity, along with the rise of modern transhumanism in the late 1980s and 1990s (which saw this as potentially desirable), shifted attention toward what might come after us, to our possible successors. Consequently, distinguishing between terminal and final extinction became very important: *Homo sapiens* disappearing entirely and forever would be very bad *unless* we were to leave behind a successor population of intelligent beings.[32] This new focus on final-versus-terminal extinction also highlighted normative extinction, since it matters not just *that* something comes after us but *what* these beings are like.

The most recent addition to our shared library of concepts is almost certainly premature human extinction. As Chapter 9 will show, the first hints of this idea seem to have occurred in the late 1970s, associated with what we will call the "argument from unfinished business." However, the *term* "premature extinction" was not explicitly used in the Existential Ethics literature until 2009. So far as I can tell, Bruce Tonn was the first futurist to mention "premature extinction" in—as it happens—a discussion of the argument from unfinished business.[33] This idea was subsequently foregrounded in 2013 by Bostrom and

Beckstead in publications on "existential risks." Indeed, the notion of premature extinction is central to the "longtermist" paradigm that they helped establish, often denoting an instance of final extinction that occurs anytime within the next millions, billions, or trillions of years. We will have much more to say about this in Chapter 10.

Prototypical Extinction?

The typology of human extinction scenarios outlined above thus provides a much more nuanced picture of the origins and evolution of the ideas—plural—of *human extinction*. While the narrative of History #1 touched upon all of these possibilities, it mostly focused on final extinction involving *Homo sapiens*, as this is and has been, I believe, the default conception of human extinction over time. I will call it the *prototypical conception*, adding a slight qualification just below. Interestingly, most dictionaries define "extinction" in demographic terms. Merriam-Webster, for example, defines "extinction" as "the condition or fact of being extinct," and "extinct" as "no longer existing," which implies that "human extinction" would simply be "humans no longer existing."[34] This mirrors definitions found in the scientific literature, as when the International Union for Conservation of Nature (IUCN) *Red List* stipulates that "a taxon is Extinct when there is no reasonable doubt that the last individual has died."[35] As Julien Delord reports, "most biologists accept the following basic definition: 'The end, the loss of existence, the disappearance of a species or the ending of a reproductive lineage.'"[36] This says nothing about extinction being permanent, and indeed some scientists have suggested that extinction need not be forever, as advanced synthetic biology techniques could enable certain species, such as the carrier pigeon, woolly mammoth, and Neanderthals, to be "resurrected" from the grave, an idea called "de-extinction." Other philosophers have disputed this, arguing that non-permanent extinction is a "logical impossibility," and hence that all extinction is necessarily permanent.

Based on anecdotal evidence that I have collected over the years, demographic extinction is often the first scenario that most people identify when asked to define "human extinction." However, if pressed, they will tend to shift the definition to terminal and then final extinction, with most taking "human" to mean "*Homo sapiens*." The conversation might go like this:

A: What would it mean for "human extinction" to occur?
B: If human extinction were to occur, there would be no more humans; the species would no longer exist. [Demographic extinction.]
A: Could this be a merely temporary situation? Or would it be permanent?
B: Permanent, of course! [Terminal extinction.]
A: Does that mean the human story would come to an end entirely, or could our story in some sense continue even after we are gone?
B: Our extinction would surely be the end of the story. [Final extinction.]

However, there is more to the prototypical conception than final extinction: it also tends to involve this *coming about* as a result of a violent, sudden, global catastrophe, rather than because, say, everyone decides to stop having children.[37] Although I do not know of any empirical studies that have examined this issue, this is my hypothesis: a bang rather than a whimper is what most people think of when asked to imagine our extinction, which also tends to involve *Homo sapiens* and the scenario of final extinction.

Dying, Death, Dead

The six scenarios of extinction and various senses of "humanity" aren't the only ethically and evaluatively relevant distinctions. Orthogonal to these, we can further distinguish between the *process or event of going extinct*, the *moment at which extinction occurs*, and the subsequent *state or condition of being extinct*. For the rest of Part II, I will use "Going Extinct," "Moment of Extinction," and "Being Extinct" to denote these three aspects or "stages" of extinction, which are applicable to all six extinction scenarios.[38] Hence, there is *going demographically extinct*, the *moment of demographic extinction*, and *being demographically extinct*; *going terminally extinct*, the *moment of terminal extinction*, and *being terminally extinct*; and so on. As we will see, many philosophers since the 1980s, and especially over the past two decades, have identified Being Extinct as the *primary locus* of the badness/wrongness of our extinction. That is to say, however terrible the deaths of billions of people might be in an extinction-causing catastrophe, they would argue that the axiological "opportunity cost" of no longer existing would be *much worse*. The more optimistic one is about how good the future could be, the more inclined one may be to emphasize Being Extinct over Going Extinct. Others, though, have contended that the badness/wrongness of our extinction is reducible entirely to how Going Extinct unfolds, meaning that if there is nothing bad or wrong about the *way* our extinction occurs, then there is nothing bad or wrong with extinction—full stop. We will have much more to say about this in the following chapters.

While the nature of Being Extinct is fairly straightforward (the type "humanity" is no longer instantiated in the world), the question of *when* exactly the Moment of Extinction occurs can be complicated. For example, there may not be any objective moment at which phyletic extinction occurs, since the transition from one species to another tends to be gradual (though in the case of humanity, this could potentially happen over a single generation). This vagueness is exemplified by the sorites paradox, according to which no small change by itself will yield something new, but enough small changes will. For example, adding grains of salt one at a time to the same location will eventually produce a heap of salt, despite there being no particular grain that transforms non-heaps into heaps. Similarly, *Homo heidelbergensis* evolved piecemeal into *Homo sapiens* (and other species) some 300,000 years ago, although there is no fact of the matter about *when exactly* this happened. The same idea applies to normative extinction, insofar as this would involve the accumulation of small changes over time, such as gradually losing the capacity for conscious experience. With respect to demographic extinction, when this occurs will depend on one's view about when the death of individual organisms happens. There is ongoing philosophical debate about the conditions under which an individual has died, which may be relevant to questions about the timing of demographic extinction. However, nothing much hangs on these matters for our purposes, so we will mostly bracket them in what follows.[39]

Turning to the process or event of Going Extinct, there are many ways that this could occur, most of which are ethically and evaluatively significant.[40] The first issue concerns the *etiology* of extinction, which of course loomed large in Chapters 4, 5, and 6. Since ethics concerns "moral agents," that is, agents capable of being held morally responsible for their actions or choices, *ethics* has nothing to say about naturally occurring extinction. Such scenarios may still be judged good or bad according to some theory of value, but there is nothing "immoral" about a large asteroid colliding with Earth and annihilating humanity. Hence, only anthropogenic scenarios fall within the domain of ethics or morality, terms that I use interchangeably.

This leads to the question of what distinguishes anthropogenic from natural extinction scenarios. What does it mean to say that a scenario is anthropogenic rather than natural? One answer is that a scenario counts as anthropogenic if and only if one or more human beings *cause* or *allow* it to happen; in all other cases it is natural.[41] Obvious examples would include nuclear conflict, engineered pandemics, and the malicious release of self-replicating nanobots. However, anthropogenic scenarios could also be the result of collective actions, none of which are individually sufficient to cause our extinction. This would occur if, for example, global warming were to trigger a runaway greenhouse effect that renders our planet uninhabitable. But the definition above would also include situations in which a naturally occurring threat is identified by one or more individuals who could do something to mitigate it, yet choose not to. Consider a team of astronomers discovering a 12-kilometer asteroid barreling toward Earth. These astronomers, let's say, have the resources to build a spacecraft that could deflect the asteroid, but they decide against this while keeping their discovery a secret. Consequently, the asteroid hits Earth and kills humanity. On the definition above, this would count as anthropogenic no less than a terrorist attack involving nanobots.

The second important property of Going Extinct concerns whether or not it is *voluntary*. As alluded to earlier, some philosophers have argued that if, say, final extinction were the result of actions taken voluntarily, consensually, without any coercion, then there would be nothing morally wrong with bringing it about. Others, such as some utilitarians, would strongly disagree, claiming that our extinction can still be wrong even if it is voluntary, since Being Extinct itself is morally relevant. But what does it mean for extinction to be brought about voluntarily? Once again, we encounter a conceptual problem, in this case because the idea of voluntariness most naturally applies to individuals rather than groups, and voluntary human extinction would be an action taken collectively. Hence, the most obvious definition would be *universalistic*, whereby every single person who exists at the time consents to our extinction. We could also stipulate a *democratic* conception of voluntariness, whereby a majority or plurality of people opt for extinction, despite not everyone agreeing. This would be obviously problematic in cases where the decision to cease existing violates some right of those who object, although one might also object that the only legitimate definition of "voluntary" is the universalistic one. This is not a trivial issue, as some ethical theories claim that we ought to go extinct, but *only* through some voluntary means. Since it is quite unimaginable that *everyone* around the world would ever universally agree to end humanity, such theories prescribe an outcome that is vanishingly improbable. We will return to this point in Chapter 11.

The last ethically and evaluatively relevant property of Going Extinct concerns its temporality—that is, in the sense of how long it takes to unfold. This matters because if one holds that the only thing that might be wrong or bad about our extinction is the physical or psychological suffering that Going Extinct entails, then one may hold that *instantaneous* extinction would not be bad or wrong. However, if one holds that painlessly cutting a life short would itself be bad or wrong, then even if our extinction were to occur in a fraction of a second, one would conclude that this would nonetheless be bad or wrong. Is instantaneous extinction even possible, though? In theory, yes, for reasons touched upon in Chapter 6: a high-powered particle collider could nucleate a vacuum bubble, if the universe is in a false vacuum state, thus annihilating Earth at nearly the speed of light. If this were to happen in the near future, it would be accidental, although one could imagine someone with significant resources who believes that human extinction would be desirable attempting to "weaponize" a particle accelerator, building one of their own and smashing heavy ions into each other in hopes of obliterating all life on Earth. This is another idea that will come up in what follows.

Conclusion

This may not be an exhaustive account of the scenarios, categories, distinctions, and possibilities relevant to Existential Ethics. But it does provide a solid foundation for our study of History #2: the development of Existential Ethics over the past few centuries. In sum, "humanity" could be understood in many different ways, and there are at least six extinction scenarios that are ethically, evaluatively, and historically important. The prototypical conception of human extinction is final extinction involving *Homo sapiens* that is caused by a global catastrophe event. Furthermore, it is crucial to distinguish between various aspects of human extinction, especially Going Extinct and Being Extinct, as different theories will focus on one or the other. With respect to Going Extinct, the details of how this unfolds are also relevant, such as whether it is anthropogenic or natural, voluntary or involuntary, instantaneous or drawn out.

The following chapter will examine early ruminations of the topic from roughly the seventeenth century to the 1950s, at which point the emergence and solidification of the third existential mood occasioned a flurry of novel thoughts about the ethics of self-annihilation. The subsequent chapter will look at longtermism and antinatalism, and the final chapter of Part II will examine some recent developments, including my own views on the ethical and evaluative implications of extinction. It is to these issues that we now turn.

Notes

1. Note that some understand the term "normative" as corresponding to questions of *ought*, that is, to what is right or wrong. In this book, I am using the term more promiscuously: it subsumes both the deontic and the evaluative.
2. From the Online Etymology Dictionary (2021) entry for "deontology." Note that "deontology" was originally coined by Jeremy Bentham, who defended a "utilitarian deontology" (see Timmermann 2014).
3. For ease of exposition, I will mostly drop the "neutral" option in what follows, although readers should keep in mind that according to certain normative views, our extinction may be neither good nor bad.
4. Tappolet 2013. From the Online Etymology Dictionary (2021) entry for "evaluate," where "evaluative" combines "evaluate" with word-forming element "-ive."
5. See Finkelman 2018a.
6. Matheny 2007.
7. Beckstead 2013a.
8. Greaves and MacAskill 2021.
9. Bostrom 2013.
10. Note that there could be both metaphysical and epistemological interpretations of this definition. A metaphysical definition would say that a species S has gone extinct if and only if there are *in fact* no more tokens of S in the universe; an epistemological definition would say that a species S could be considered to have gone extinct when we judge the probability of discovering a token of S to be sufficiently small. For example, the probability of discovering a *T. rex* on Earth appears to be vanishingly improbable—but not *zero*. Similarly, the probability of discovering a dodo on the island of Mauritius also seems extremely low, but it is not impossible that one is someday found. There have been, after all, numerous cases in which we believed a species to be extinct only to discover that it still exists (see, e.g., Edmond 2017; Quaglia 2022).
11. Note also that there is a small but important literature on the concept of *extinction* within the philosophy of biology, much of it a reaction to the possibility of de-extinction (see, e.g., Delord 2007; Delord 2014; Siipi and Finkelman 2017; Finkelman 2018b). My discussion here is only loosely connected to this literature, as indicated by the fact that my preferred terminology does not align with the terminology of contributors to the literature. For example, Delord 2007 uses "demographic extinction" and "final extinction" interchangeably, while Finkelman 2018a uses "substantial extinction" in a manner more or less synonymous with my use of "terminal extinction."

12. A third way that phyletic extinction could occur is through hybridization (see Delord 2007), although Peter Mayhew distinguishes "symbiosis" with "hybridization," where the former would involve two lineages merging into one, while the latter would involve two lineages giving rise to a third (Mayhew 2006, 132).
13. Hey 2001; Okasha 2002. There isn't even agreement about whether species are "natural kinds" or "individuals" (see Ereshefsky 2017 for an overview).
14. To be clear, I am not saying that we ought to radically alter ourselves (in fact, I believe this would have devastating consequences), only that this is becoming increasingly feasible, and my guess is that people will increasingly try to radically "enhance" themselves in the coming decades and centuries.
15. Hanson 2008.
16. Indeed, the author adds that

    > while such systems do not yet exist, they may soon be developed to afford clinical assistance to infertile women and reproductive choices to prospective parents. In human survival schemes, AU systems would likely first be used to extend conventional survival missions (e.g. subterranean bunkers) by replacing some adult crew members with cryopreserved embryos (Edwards 2021)

    See also Crowl et al. 2012.
17. Sandberg et al. 2017.
18. If *Homo sapiens* were to exist long enough for the heat death of the universe to make life impossible, then the heat death would also entail our final extinction. However, it seems much more likely that we will have evolved into something else—one or more posthuman species—long before the heat death arrives in some 10^{100} years or so. If this were the case, then the heat death would not cause *our* terminal or final extinction, although it would snuff out whatever lineages we might become or engender.
19. Shiller 2017.
20. See Nagel 1974.
21. In sum, the internalist mode would involve the loss of a capacity, while the externalist mode might involve the loss of our ability to exercise or use a capacity.
22. Clarke 1971.
23. Bostrom 2002b.
24. Bostrom 2003b.
25. Bostrom 2013.
26. Moynihan 2019, 2020.
27. In Wells' words:

    > The too-perfect security of the Overworlders [Eloi] had led them to a slow movement of degeneration, to a general dwindling in size, strength, and intelligence. That I could see clearly enough already. What had happened to the Undergrounders I did not yet suspect; but, from what I had seen of the Morlocks—that, by the bye, was the name by which these creatures were called—I could imagine that the modification of the human type was even far more profound than among the 'Eloi' . . . The Eloi, like the Carlovignan kings, had decayed to a mere beautiful futility (1895).

28. Bernal 1929.
29. However, as Tikva Frymer-Kensky writes:

    > After the rest of mankind have been destroyed, and after the gods have had occasion to regret their actions and to realize (by their thirst and hunger) that they need man, Atrahasis brings a sacrifice and the gods come to eat. Enki [also known as Ea] then presents a permanent solution to the problem [of noise]. The new world after

the flood is to be different from the old, for inky summons Into, the birth goddess, and has her create new creatures, who will ensure that the old problem does not arise again (1977).

30. This was also noted in Chapter 2.
31. For example, the notion of final extinction may have been on Henry Maudsley's mind when he wrote that "the worsening conditions of life" as the sun burns out will leave only "a few scattered families of degraded human beings living perhaps in snowhuts near the equator," where these people constitute "the last wave of the receding tide of human existence before its final extinction" (1881).
32. However, a number of non-transhumanist and non-utilitarian arguments proposed by Bertrand Russell, Günther Anders, Hans Jonas, Jonathan Bennett, Ernest Partridge, Jonathan Schell, Robert Adams, and others *tacitly* concerned final (as well as normative) extinction, although the authors themselves may not have been clear about this fact. See Chapters 9 and 10 for discussion.
33. The only exception that I can find comes from a short 1946 letter to the editor published in *Science* magazine and authored by a member of the Texas Game, Fish, and Oyster Commission named Joel Hedgpeth. Worried about the possibility that "any uncontrolled release of atomic energy might set off a chain reaction which would detonate the entire earth" and "the possible effects of a subsurface explosion of an atomic bomb on marine life," he concludes that "war is out of date, and even admission of the possibility of future wars is welcoming the pre-mature extinction of mankind" (Hedgpeth 1946). But the author doesn't explicitly link this with Existential Ethics, which is why I chose not to mention it in the body text.
34. Merriam-Webster 2021.
35. IUCN 2005.
36. Delord 2007.
37. An in-between case worth registering might be involuntary infertility, whereby no lives are cut short, but no doubt a great deal of anguish would occur as the human population dwindles.
38. The importance of this distinction is illustrated by the following passage from Eliezer Yudkowsky, who wrote that "people who would never dream of hurting a child hear of an existential risk, and say, 'Well, maybe the human species doesn't really deserve to survive'" (2008). But this confuses Being Extinct with Going Extinct: virtually everyone agrees that suffering caused to actual people is bad, and hence that Going Extinct would be bad at least insofar as it involves suffering (I call this the "default view" below). However, many people also have the intuition that humanity no longer existing is not *itself* bad, since there would be no one around to bemoan our non-existence. These views are entirely compatible, and hence there is no tension between them, as Yudkowsky implies.
39. See Luper 2021 for a useful overview of the conceptual problems associated with this issue.
40. This fact has not always been appreciated in the literature. An exception comes from John Leslie, who noted in 2010 that "the spatiotemporal details of any extinction process would be of great ethical significance" (2010).
41. One problematic implication is that this would make extraterrestrial invasions and the termination of our simulation "natural" catastrophes, which seems odd.

8 Early Ruminations

Cycles, Atoms, and the Eternal Recurrence

Although the idea of extinction dates back at least to the Presocratics of ancient Greece, there is no indication that any sage, poet, or philosopher at the time said or wrote anything about the normative implications of this phenomenon. To be sure, only textual fragments remain from Xenophanes and Empedocles, so they may well have addressed the issue in passages that have since been lost in the rubble-heap of history. Perhaps they might have bemoaned the fact that the first humans to emerge after each turn of the cosmic wheel would have to rebuild society all over again and rediscover knowledge had by previous peoples, as Plato and Aristotle suggested has happened many times with past civilizations. Or, perhaps more likely, they never gave the issue much thought because they saw our future disappearance as part of the natural order of things, or because they believed that the end of our current phase of the cycle is too far away to merit attention in the present. Who knows? Much the same could be said about the ancient atomists. As Pavel Gregoric writes:

> One might wonder why the atomists did not say more about human extinction. Maybe they did not think much of it or, more likely, they were reconciled with it: it is a matter of necessity, so there's nothing to be done about it. On the other hand, they may have taken solace in the fact that, given the infinite amount of time, in an infinite number of worlds, the human species—or something appreciably like it—is bound to appear an infinite number of times.[1]

For example, if what one cares about is that the universe contains human life, then the idea of an infinite number of future worlds inhabited by *Homo sapiens* implies that our disappearance might be, at most, bad *for us*, rather than being bad in an all-things-considered or "cosmic" sense, a conclusion that may have provided a degree of "solace," as Gregoric suggests. The human species will always exist somewhere, sometime, across the infinite corridors of space, and hence the loss of

DOI: 10.4324/9781003246251-10

any single *kosmos* cannot subtract value from the universe as a whole: an endless number of humanity-containing *kosmoi* minus a single humanity-containing *kosmos* still yields an infinite number of humanity-containing *kosmoi*. Although this line of reasoning was never made explicit by the atomists, it was a straightforward implication of their cosmological theory. It was also an idea that, as we will see below, was later picked up and developed by philosophers and scientists in the eighteenth century.

With respect to the ancient Stoics' theory of eternal recurrence, not only will every event that happened in the past happen again in the future exactly as it did, but, long before Gottfried Wilhelm Leibniz coined the phrase, the Stoics believed that we occupy "the best of all possible worlds." Hence, the state of affairs destined to repeat forever also happens to be the *best that could possibly obtain*, which suggests that the termination of our world should be seen as evaluatively neutral. What sense would it make to call our extinction good or bad if the tape of history is rewound and played an infinite number of times, with each repetition instantiating the optimal series of worldly events?

Among those who accepted the possibility of demographic extinction during this early period, there was either not much said or not much to say about this outcome, the latter perhaps explaining the former. At most, one might have worried that the dissolution of our world in a catastrophe of some sort—for example, a worldwide flood (Xenophanes), collision with another *kosmos* (atomists), global conflagration (Stoics)—would cause harm to those living at the time, who would suffer and die as a result of the process or event of Going Extinct. On this view, our extinction would be bad for the same reason that *any* catastrophe would be bad, an idea that we examine more in a moment. However, from a cosmic vantage point, the ultimate indestructibility of humanity seems to imply that, all things considered, there is nothing tragic about our extinction above and beyond these catastrophe-inflicted harms, since the state-of-affairs corresponding to our non-existence will always be temporary rather than permanent.

Four Categories

As I argued in Part I, the constellation of three beliefs that rendered *human extinction* a self-contradictory idea and impossible outcome became an established feature of the intellectual landscape in the West between the fourth and fifth centuries CE. Consequently, one finds virtually no references during this period to human extinction in the naturalistic sense, and hence (of course) no discussion of its ethical or evaluative implications. The field of Existential Ethics was non-existent, and this would remain the case until at least the eighteenth and nineteenth centuries, when a handful of intellectuals with atheistic/deistic worldviews began to entertain, for the first time since Classical Antiquity, the prospect of our extinction. Because the relevant material is fairly sizable—or at least becomes sizable when proper context is provided—I will for now restrict the discussion to the period before the second existential mood commenced in the 1850s; the second half of this chapter will then explore ruminations about the topic between the 1850s and 1950.

This being said, we can start by grouping references to human extinction prior to the 1850s into three general categories: (1) those made in passing, without much elaboration, usually in connection with speculations about potential kill mechanisms (i.e., the focus was more descriptive than normative), (2) those made for the purpose of articulating a point or argument *unrelated* to extinction, and (3) those that suggested our extinction would in some way be bad. We may also recognize a fourth category (4) of apparent references to our extinction *not being bad at all*, but which upon closer examination are better understood as evaluative claims about what biologists would call *extirpation*, given a particular cosmographical model of the universe (the "plurality of worlds") that became immensely popular from the seventeenth through the nineteenth centuries. While this cosmographical model was different than the atomists' theory of the universe, the reasoning employed was essentially the same, although in this case it was made explicit and developed in some detail rather than being merely implied. Finally, while all four categories are clearly relevant to History #1 (thinking about the possibility of extinction), only the third category is directly relevant to History #2. I will nonetheless pause on the second and fourth categories, in particular, for some time, since it would help to identify both actual and *merely apparent* references to the ethics of extinction.

Before turning to the second category, some examples of the first include the following: Hume's and Diderot's suggestion that species may have natural life cycles like those of individuals, and hence that humanity itself may undergo a form of senescence that ultimately leads to *its* death[2]; Lord Byron's warning that a cometary collision will someday destroy our species just as past collisions destroyed the previous inhabitants of Earth, although we may avoid this by creating a planetary defense system to obliterate "the flaming mass" prior to impact;[3] and of course Diderot's affirmation that humanity could indeed go extinct, although we would later reappear "after several hundreds of millions of years."[4] The anonymous "H" of Chapter 2 also gestured at this idea in their survey of speculative kill mechanisms that could precipitate "a very rational *end of the world*," although these scenarios appear to have been embedded in a broader religio-apocalyptic conception of the future.[5] In all these cases, the authors said little or nothing about whether they believed our disappearance would be good, bad, neutral, etc., perhaps because they thought the answer was obvious, or because the principle of plenitude guarantees that our absence would only be temporary (as with Diderot and H). Hence, while they indicate an important shift in the conceptual intelligibility of *human extinction* at the time, they are largely irrelevant to the second history that is our main focus of this part of the book.

A God So Perfidious

The second category contains several notable examples. One comes from Immanuel Kant's 1790 *Critique of Judgment*, which expounds his views on aesthetics and teleology. At one point, Kant states that if humanity were to disappear, the universe would become "a mere waste," sometimes translated as "a mere wasteland," in the sense that it would lack any "final purpose."[6] This looks like a normative claim about human extinction, since surely one should not want the universe to become a *purposeless wasteland*, and in fact some philosophers have interpreted it precisely this way. For example, Derek Parfit (1942–2017) writes that "these remarks suggest that, on Kant's view, the continued existence of rational beings is another end-to-be-produced with supreme value."[7] However, this does not appear to have been Kant's point, even if the statement carries this implication. To understand what he meant, we must consider the context: the natural world, on Kant's view, is a teleological system of interconnected "purposes." Every living thing has an "inner" purposiveness by virtue of being "both cause and effect of itself," by which he meant that they maintain their own existence, produce offspring to perpetuate the species, and are comprised of parts that work together for the sake of the whole. In this way, the cause is the organism (and its functioning parts) and the effect is the (same) organism and its progeny. Furthermore, non-living things may possess an "outer" or "relative" purposiveness by virtue of contributing to the existence of living things, as a means to an end.[8] The biotic and abiotic worlds are thus teleologically linked, the former being inherently purposive and the latter being purposive relative to the former.

But the question remains: what is the purpose of the natural system *as a whole*? Kant's answer draws from his prior theory of ethics and value, according to which "nothing can possibly be conceived in the world, or even out of it, which can be called good, without qualification, except a good will."[9] In other words, the one and only thing in the entire universe that is unconditionally, or non-relationally, valuable is a *good will*.[10] There are of course many things that we can describe as "good" and "valuable," such as knowledge, humor, courage, kindness, and so on, but in all these cases their value "is entirely conditional on our possessing and maintaining a good will."[11] What, then, is a good will? A person exhibits a good will when their moral choices are based wholly on considerations of the Moral Law, which Kant famously identified with the Categorical Imperative, i.e., that one should "act only in accordance with that maxim through which you can at the same time will that it become a universal law."[12] When a person acts in this way, and when their decision to act was made autonomously (by their own volition, through their own powers of rationality), then they exemplify a good will. It is thus by virtue of our capacity to exemplify a good will that humanity is the seat of unconditional value in the universe. This leads back to Kant's teleological theory of nature: since only rational beings like us possess unconditional value, the purpose of the natural system as a whole is none other than *humanity*.

"Only in man," Kant wrote, "and even in him only as moral subject, do we find unconditioned legislation regarding purposes. It is this legislation, therefore, which alone enables man to be a final purpose to which all of nature is teleologically subordinated."[13]

Hence, the reference to human extinction—to a universe "without men"—was given to underline the special teleological role of humanity within nature: without us, there would remain the *inner* and *relative* purposes of the biotic and abiotic realms, but no *final* purpose. To quote the relevant passage in full:

> The commonest Understanding, if it thinks over the presence of things in the world, and the existence of the world itself, cannot forbear from the judgement that all the various creatures, no matter how great the art displayed in their arrangement, and how various their purposive mutual connexion,—even the complex of their numerous systems . . .—would be for nothing, if there were not also men (rational beings in general). Without men the whole creation would be a mere waste, in vain, and without final purpose.[14]

Although this may, as Parfit suggests, imply that humanity's unconditional value gives us reason to ensure our continued existence, Kant did not seem to have this in mind in writing that passage, nor did he elaborate the idea later on (but see below for earlier thoughts from him about our permanent disappearance).

Another example is worth looking at more closely, since Thomas Moynihan has recently identified it as the first explicit endorsement of human extinction within the Western tradition. In his 2020 book *X-Risk: How Humanity Discovered Its Own Extinction*, Moynihan reports that "the Marquis de Sade [became] the first proponent of human extinction" in the year 1796.[15] This potentially important claim is based on several passages in two of Sade's books: *Philosophy in the Bedroom* (1795) and *Juliette* (1797, although possibly published in 1799). In the first, Sade wrote the following:

> Why! what difference would it make to her were the race of men entirely to be extinguished upon earth, annihilated! she laughs at our pride when we persuade ourselves all would be over and done with were this misfortune to occur! Why, she would simply fail to notice it. . . . Do you fancy races have not already become extinct? Buffon counts several of them perished, and Nature, struck dumb by a so precious loss, doesn't so much as murmur![16] The entire species might be wiped out and the air would not be the less pure for it, nor the Star less brilliant, nor the universe's march less exact. What idiocy it is to think that our kind is so useful to the world that he who might not labor to propagate it or he who might disturb this propagation would necessarily become a criminal![17]

First appearances to the contrary, Sade was not actually claiming he believes that our extinction wouldn't matter. He was asserting that "Nature" would be indifferent to this outcome.[18] The broader context is a conversation about whether sexual acts that do not contribute to the propagation of the species are unnatural. The sentences above are uttered by a fictional character named Dolmance, an older man with homosexual tendencies, in response to a question from Eugenie, a 15-year-old, about "the criminal enormity I have always heard ascribed to this [sodomy], especially when it is done between man and man." Is this a natural act, she asks at one point, to which Dolmance replies "Yes," then launches an attack against the "imbeciles who think of nothing but the multiplication of their kind, and who detect nothing but the crime in anything that conduces to a different end." He continues: "Is it really so firmly established that Nature has so great a need for this overcrowding as they would like to have us believe? is it very certain that one is guilty of an outrage whenever one abstains from this stupid propagation?" Dolmance concludes that since "destruction . . . like creation, is one of Nature's mandates . . . how may I offend Nature by refusing to create?" Hence, the point of the block quote above isn't to say that our extinction would be desirable, although Sade does note that Nature finds us "irksome" (which implies that Nature might wish we were gone). Rather, the argument is that one cannot maintain that "the sodomite and Lesbian" are committing a crime against Nature by engaging in sexual acts that do not lead to new people, since creation is not Nature's only mandate. Eugenie, in fact, finds Dolmance's arguments so convincing that she responds in a manner that would be (very) inappropriate to quote here.

Moynihan also singles out a line from *Juliette* that reads: "[T]he propagation of our species therewith becomes the foulest of all crimes, and nothing would be more desirable than the total extinction of humankind," which he describes as the "apotheosis" of Sade's "lethal anti-natalist mantra."[19] But a careful reading shows that this is taken out of context. The line was spoken by Clairwil, mentor of the titular Juliette, in arguing against the doctrine of hell and supposed goodness of God. "To judge from the notions expounded by theologians," Clairwil asserts, "one must conclude the God created most men simply with a view to crowding hell," and the act of creating people destined to spend eternity in hell is marked by such "appalling cruelty" that it cannot but render the divine Creator "infinitely wicked." She continues:

> A God so perfidious, so evil as to create a single man and then to leave him exposed to the peril of damning himself, such a God can be regarded as no specimen of excellence; if perfection be his, then it is a monster of unreason, injustice, malice, and foul atrocity.

Hence, the full context of the quote is this:

> If it comes out that the fate of the greater share of mankind is to be eternally unhappy, an all-knowing God must have known this from the outset; why

> then did the monster create us? Was he forced to? Then he is not free. Did he knowingly, deliberately, cause things so to be? Then he is a fiend. No, God was under no obligation to create man, certainly not, and if he did so simply to expose man to such a fate, the propagation of our species therewith becomes the foulest of all crimes, and nothing would be more desirable than the total extinction of humankind.[20]

The desirability of our complete annihilation is thus conditional: if God exists (a proposition that Clairwil rejects), and if most people he creates will end up suffering "infinite punishment," *then* it would be better for humanity to cease existing altogether. Rather than expressing Sade's omnicidal proclivities, this is a claim about the wickedness of God as traditionally understood, given "the glaring disproportion between the human provocation and the divine reprisal."[21] It is therefore mistaken to characterize Sade as having "the best claim to being the first person to explicitly promote the outright annihilation of our species."[22] As with Kant, the idea of *human extinction* was utilized for other purposes—in this case, to argue that sodomy/homosexuality is not unnatural and the ideas of hell and God's perfect goodness are untenable. Sade was not making any evaluative judgment about the goodness/badness of our annihilation and hence this was not an early case of someone advocating for a position within Existential Ethics.

A final example worth mentioning comes from Lord Byron's 1815 poem *Darkness*, which closes with the following lines:

> . . . The world was void,
> The populous and the powerful was a lump,
> Seasonless, herbless, treeless, manless, lifeless—
> A lump of death—a chaos of hard clay.
> The rivers, lakes and ocean all stood still,
> And nothing stirr'd within their silent depths;
> Ships sailorless lay rotting on the sea,
> And their masts fell down piecemeal: as they dropp'd
> They slept on the abyss without a surge—
> The waves were dead; the tides were in their grave,
> The moon, their mistress, had expir'd before;
> The winds were wither'd in the stagnant air,
> And the clouds perish'd; Darkness had no need
> Of aid from them—She was the Universe.[23]

Although the reference to our extinction is clear—indeed, this might be the first literary work to offer a thoroughly secular depiction of Earth completely bereft of human beings[24]—the purpose of depicting humanity's descent into nothingness was, according to Eva Horn, to offer a critique of the eighteenth-century notion that "empathy, friendship, and rationality [are] the chief human virtues." In contrast to Jean-Jacque Rousseau's view that humans are naturally compassionate, and Condorcet's belief that continued human progress can lead to our perfection, Byron painted a picture in which "humans are even more brutal, egoistic, and ruthless than the beasts." It is the extreme scenario of extinction that brings these vicious character traits into the foreground. As Horn elaborates the point,

> Byron thus calls into question not humanity's spiritual salvation but its anthropological nature. What his stress test reveals is a human nature stripped of any impulse toward empathy, altruism, compassion, or solidarity. Under duress, human life is nothing but an existence riddled by selfishness, fear, and perverse brutality, symbolized by cannibalism and the "hideousness" of the last two men. . . . Through this depiction of mankind in the catastrophe, *Darkness* mordantly does away with the image of humankind that Enlightenment anthropology had composed.[25]

Hence, the reference to human extinction is *incidental* to Byron's anti-Enlightenment thesis about our corrupt nature, on Horn's interpretation.

As these examples show, some references to our extinction among philosophers and poets in the early modern period onwards may give the impression of being evaluative—of extinction being bad, desirable, grim, dreadful—but upon closer inspection the idea was used as a means for making some unrelated point. Such references are clearly relevant to an intellectual history of *human extinction* (Part I) but not so much to the history of Existential Ethics. Nonetheless, they are important to register because failing to identify them as *false positives* could lead to an inaccurate picture of how thinking about our extinction, from an ethical or evaluative perspective, developed over time.[26]

A Terrible Calamity

There were, however, some writers prior to the second existential mood—which, recall, was triggered by the discovery of the Second Law and enabled by the decline of religion—who more directly addressed the question of whether and why our extinction would be bad, although examples are few, and none offer *sustained* reflections on the badness of our disappearance. In some cases, opinions about the evaluative status of extinction are revealed only indirectly, as when William Godwin wrote that "it may be one of the first duties incumbent on the true statesman and friend of human kind, to prevent that diminution in the numbers of his fellow-man."[27] This is a claim about what those in power *ought* to do—that they should take actions to avoid extinction caused by a dwindling population—although he did not elaborate on *why exactly* he thought we should avoid this. Again, perhaps he thought the answer was obvious, even if that isn't the case.

Others focused on the potential harms that might be caused by the process or event of Going Extinct. As mentioned above, insofar as Going Extinct involves a catastrophe, nearly everyone will agree that our extinction would be bad—even those who see the *outcome* of extinction as good or neutral. Consider, for example, that philosophers would classify the concept of *catastrophe* as a "thick" evaluative concept, since it contains both descriptive (e.g., catastrophes are events that happen in the world) and evaluative (i.e., they are inherently very bad) elements. To call something a catastrophe is thus to say that it is a *very bad event*, and hence "human extinction caused by a catastrophe" implies that our extinction is very bad, if only because of the *event that caused it*. Let's refer to this as the *default view*, which we can define as follows: if human extinction is brought about by a catastrophe—or disaster, cataclysm, and so on—it would be bad *at least* because of the suffering inflicted by the catastrophe on those living at the time.[28]

From a normative perspective, the default view is mostly uninteresting, since it (a) is accepted by nearly everyone, and (b) follows more or less directly from the meaning of "catastrophe." Yet it was not until the early nineteenth century, with the emergence of the "Last Man" genre, that people began to explore, for the first time, just how terrible the occurrences leading up to our extinction might be. Some focused primarily on the unprecedented *scope* of an extinction-causing catastrophe, as it would affect everyone on the planet, while others foregrounded the idea that experiencing or anticipating the end of humanity could engender *kinds of suffering* that wouldn't normally arise from non-extinction-causing catastrophes.

An example of both comes from Mary Shelley's *The Last Man*, which depicts humanity's somersault into the oblivion from a worldwide plague as a horrendous tragedy due in part to the sheer enormity of the suffering that it causes. As Bruce and Jenna Tonn write, "scores of people begin to die . . . and the magnitude of the crisis becomes unbearable. . . . Although altruism ties people together in their last moments, despair over the loss of loved ones fills Lionel's memoirs."[29] However, Shelley also homes in on the extraordinary loneliness, grief, hopelessness, and sorrow that the experience of witnessing our extinction could elicit. This is exemplified by the struggles of those in the final generations, especially Lionel Verney, the very last man. As Verney declares at one point in the novel, "my soul [is] deluged with the interminable flood of hopeless misery." Later, he bemoans his "hopeless state of loneliness" and "restless despair."[30] Verney understands, all too clearly, that unlike lesser catastrophes there is no silver lining, no glimmer at the tunnel's end. It is not the case that, as we say, "life will go on" despite one's own personal hardships or that "it's not the end of the world," both of which can provide some degree of *solace* in dark times. Although anyone who believes that "their world" is coming to an end could experience similar feelings—indeed, Shelley's story no doubt reflected her own personal situation, having recently lost both her husband, Percy Bysshe Shelley, and close friend Lord Byron—there is something especially jarring about the belief that the entire human species is on the verge of annihilation. In other words, the phenomenology associated with the awful, intense personal experience of *approaching extinction* may contribute a qualitatively distinctive form of suffering, which may cause those who have this experience a degree of harm that is unique to scenarios in which one anticipates our extinction amidst a worldwide catastrophe.

Others in the Last Man genre of the early nineteenth century also explored this idea, such as an anonymous author who penned a short story titled "The Last Man," also published in 1826. The story culminates with the tremulous shrieks of the main character, the last man, who finds himself overwhelmed by feelings of isolation and despondency upon surveying the panoply of a humanless Earth:

> Alas! Alas! I soon and easily gained the top of the rising bank, and fixed my eyes on the wide landscape of a desolate and unpeopled world. . . . Desolation! Desolation! I knew that it was to be dreaded as a fearful and a terrible thing, and I had felt the horrors of a lone and helpless spirit—but *never, never had I conceived the full misery that is contained in that one awful word*, until I stood on the brow of that hill, and looked on the wide and wasted world that lay stretched in one vast desert before me. . . . Then despair and dread indeed laid hold of me—then dark visions of woe and of loneliness rose indistinctly before me—thoughts of nights and days of *never-ending darkness cold*—and then the miseries of hunger and of slow decay and starvation, and homeless destitution—and then the hard struggle to live, and the still harder struggle of youth and strength to die.[31]

The most important contribution of these stories to the development of Existential Ethics was drawing attention to just how devastating the process or event of Going Extinct could be and why. Although their main focus was the struggles of the final person, the idea can be generalized to the entire last generation(s) of human beings prior to extinction, and indeed many recent philosophers have incorporated this insight into their theories of extinction, identifying it as one reason—a reason specific to extinction—that our extinction would be bad (and this it true even among philosophers who see the outcome of extinction as good).[32]

An even more intriguing example in this third category predates the Last Man genre, introduced by de Grainville in 1805, by more than 80 years. It comes from Montesquieu's 1721 *Persian Letters*, which I noted in Part I because of its discussion of population decline and the possible etiology of this trend. Recall that Rhedi tells Usbek that if depopulation trends continue, then "in ten centuries the earth will be a desert." Rhedi then declares: "Here, my dear Usbek, you have the most terrible calamity that can ever happen in the world."[33] What is notable about this is that (a) Montesquieu, speaking through Rhedi, says nothing about the potential badness of Going Extinct, and (b) the negative value-judgment expressed by the phrase "terrible calamity" seems to concern the loss of *humanity itself*. That is to say, the evaluative focus looks to be the fact that there will be *no more humans* rather than the plight of the *last few humans*. Montesquieu—a deist, albeit in the most minimal sense[34]—not mentioning the potential harms of Going Extinct does not, of course, mean that he thought the last few generations wouldn't suffer in his scenario. He may well have agreed that the quality of human life would decline as extinction approaches, and that this constitutes one bad aspect of extinction by depopulation. But so far as I can tell, his claim points toward the state or condition of Being Extinct, whereby "the earth will be a desert," meaning without humanity, rather than the process or event of Going Extinct.

If this is correct, it gestures at one of the most significant innovations within Existential Ethics over the past many centuries, namely, the idea that the loss of *our species*, of the *entire population*, has normative implications that go *above and beyond* whatever suffering and harm might befall those subject to the process of extinction. Indeed, we will see that it was not until the 1980s and, especially, the past two decades that this idea became a topic of explicit philosophical theorizing among existential ethicists. Some have, in fact, come to see Being Extinct as the primary source of extinction's badness, whether or not this comes about through a catastrophe.

Equivalence Versus Further-Loss Views

To understand this, it may be useful to introduce a thought experiment that I will reference throughout the rest of this book. Imagine two worlds, A and B.

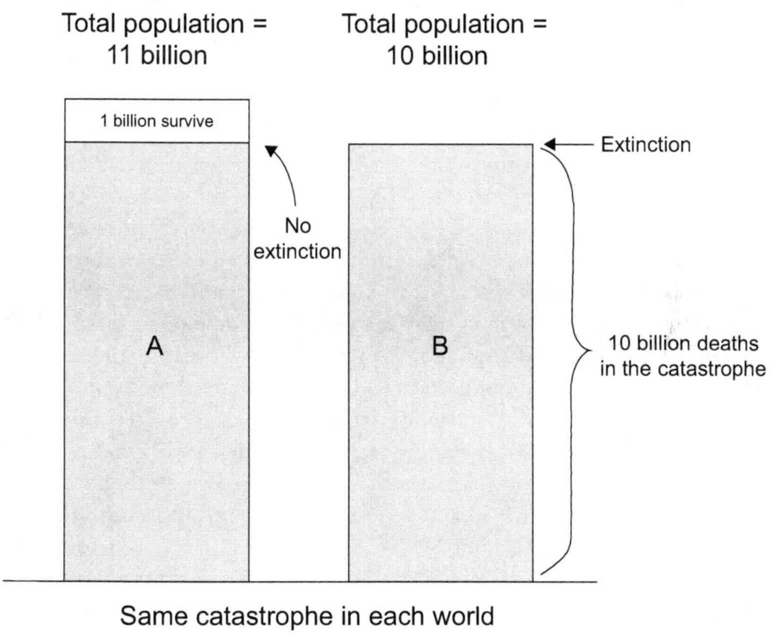

Figure 8.1 Two events happen in world B, at a high level of abstraction, whereas only one happens in world A. But is this fact morally relevant?

In world A there exist 11 billion humans, while in world B there exist 10 billion humans. An identical catastrophe then occurs in A and B, resulting in the sudden death of exactly 10 billion people in each. There are two questions we can ask about these scenarios: the first concerns the *number of events*, at a high level of abstraction, that take place in A and B as a result of the catastrophe. I assume that everyone will agree that in A only a *single* event occurs, that is, the death of 10 billion people, while in B *two* distinct events occur, that is, the death of 10 billion people and the extinction of humanity, since the total population was 10 billion. In other words, although the catastrophes are identical, *something else* happens in B: the human species dies out. The second question, then, concerns whether this difference in the number of events makes any evaluative or ethical difference. That is, does the fact that humanity goes extinct in B make the catastrophe in B any worse than in A? If a homicidal maniac named Joe, for example, murders all 10 billion people in each world, does he do something *extra* immoral in B?

According to what we could call the *equivalence thesis*, there is no difference whatsoever in the badness, or wrongness, of the catastrophes in each world. That is, the badness/wrongness of the catastrophe in world B is reducible entirely to the badness/wrongness of the death of 10 billion people, full stop. The equivalence thesis, which demarcates a class of what we could call *equivalence views*, is thus a reductionistic account of human extinction that, as such, yields the striking implication that there is no *special* ethical or evaluative problem posed by extinction. That "something else" in world B does not count for anything, ethically or evaluatively speaking; an individual who causes the catastrophe in B does not do anything *worse* than if they had caused the catastrophe in A. (As discussed later on, many "person-affecting" ethical theories entail this view.)

Note that the equivalence thesis is not identical to the default view. First, the default view simply states that the suffering inflicted and premature deaths caused by extinction-causing catastrophes constitute *at least* one reason our extinction would be bad or wrong, whereas the equivalence thesis asserts that this is the *only* reason. Second, the default view most obviously applies to what I described in the previous chapter as the "prototypical" case of human extinction, namely, final extinction brought about by a catastrophe. In contrast, the equivalence thesis applies to all but one of the extinction scenarios specified earlier: the badness or wrongness of demographic, phyletic, terminal, final, and normative extinction depends entirely on the way these scenarios unfold. If there is nothing bad or wrong about Going Extinct in these senses, then there is nothing bad or wrong about extinction at all. To endorse the equivalence thesis is thus to accept the default view, but not vice versa—or not necessarily. The only exception concerns premature extinction, as this implies that what matters is also *when* our extinction occurs, and hence the badness or wrongness of extinction could *exceed* whatever suffering and death might lead up to extinction depending on its *timing*. For example, if the scenario of world B were to happen before the attainment of some valued end, one may judge it to be worse than if it happens after the attainment of this end, even if the catastrophes unfold in exactly the same way.

What if one maintains that the catastrophe of B would be in some way worse than that of A? There are at least two options here: first, one could argue that the process or event of Going Extinct in world B might, depending on how the catastrophe happens, be worse than the process or event of 10 billion people dying in world A. Perhaps the authors of the Last Man genre would have defended this position, pointing to the additional suffering caused by the anticipation of our impending annihilation and experience of Going Extinct. Let's refer to this as the *no-ordinary-catastrophe thesis*, because the harms singled out are unique to, or uniquely arise from, the event of extinction and hence would not occur with "ordinary" catastrophes. As defined, the no-ordinary-catastrophe thesis is compatible with a slightly adjusted version of the equivalence thesis, whereby B's catastrophe may be worse than A's, but the badness/wrongness of B's catastrophe is still wholly reducible to the details of Going Extinct. On the other hand—and this is where the distinction between Going Extinct and Being Extinct enters the picture in a crucial way—one could argue that Being Extinct would entail or involve some sort of *further loss*, that is, above and beyond whatever harms or losses obtain in the lead-up to the Moment of Extinction.

What might this further loss be? One popular answer is human civilization: if one believes that the loss of "civilization" would be bad, then the loss of humanity would *also* be bad *for this reason*, since civilization cannot exist without humanity (or some suitable successors).[35] Alternatively, one might argue that there being no future people would be an additional loss, perhaps because bringing people into the world bestows upon them some benefit (a so-called "existential" benefit) or because a universe full of "happy" people is better than one with no people at all. Or one might argue that humanity *itself* has some sort of special value (e.g., intrinsic value), and hence the disappearance of our species would be regrettable independent of how many people perish in the extinction-causing catastrophe. In other words, imagine two more worlds, C and D, which contain 10,000 people and 10 trillion people, respectively. Now say that all 10,000 people are suddenly killed in C and all 10 trillion are suddenly killed in D. The claim would be that, *aside* from the significant difference in total deaths in C and D, both worlds would nonetheless have undergone the *same additional* tragedy, namely, the extinction of humanity, and hence in this sense an identical loss would have occurred in each. Let's group this family of normative responses to our extinction under the umbrella of *further-loss views*.

Although Montesquieu did not specify any reasons for why the loss of humanity might constitute a "terrible calamity," it seems at least fairly clear, once again, that our non-existence itself was his evaluative focus. This is noteworthy. However, as it happens, Shelley's last-man novel not only provided a vivid image of the magnitude of and unique harms associated with extinction (caused by a plague) but also may have been the very first publication in the Western tradition to *explicitly* point to some further losses entailed by our disappearance as *reasons* for this being bad. Consider Verney's observation that the disappearance of "man" in the collective sense, which he contrasts with "man" in the individual sense, would mean the concomitant loss of many valued things like knowledge, science, technology, poetry, philosophy, sculpture, painting, music, theater, laughter, and so on. "Alas!," he exclaims, "to enumerate the adornments of humanity, shews, *by what we have lost*, how supremely great man was. It is all over now."[36] One could interpret this as saying that there are two distinct sorts of losses involved in extinction: first, the loss of "man," that is, all human beings, and second, the loss of all those things that made humanity "supremely great." This seems to express a further-loss view with respect to final human extinction, although Shelley did not, so far as I know, elaborate the idea beyond a few paragraphs.[37] Still, we can see how these authors, and Shelley in particular, were among the first to touch upon central issues within Existential Ethics, giving perhaps the earliest articulations of the no-ordinary-catastrophe and further-loss theses.

A Flower or an Insect

The fourth category mentioned above consists of numerous apparent references to human extinction being evaluatively *neutral*, or *not bad*, although upon closer inspection the authors were not actually talking about extinction, but something else. These are also worth mentioning because they constitute, like the second category above, false positives that can give an inaccurate picture of how Existential Ethics developed over time. Most examples date to the mid-eighteenth century, and all were linked to a "plurality of worlds" cosmography that (re)emerged in the seventeenth century. To be sure, the notion that an infinite number of other "worlds"—by which writers have historically meant a geocentric solar system, in which the sun, planets, and fixed stars revolve around Earth—exist in the universe goes back to the ancient atomists. But the atomists' model was quickly eclipsed in the Western tradition by Aristotle's theory of the universe, which not only placed Earth at the center of the universe but asserted that our "world must be unique. There cannot be several worlds," to quote Aristotle's *On the Heavens*. Two developments in particular enabled the plurality of worlds cosmography to usurp the Aristotelian model during the early modern period. The first was the Copernican Revolution, which established a heliocentric model of the solar system in which the earth and other planets revolve around the sun. This gave rise to the idea that we do not occupy a special or privileged position in space and time (the Copernican Principle). The second was the emergence of a new, distinct version of the principle of plenitude. The original version is what one finds in Plato, which states that every kind of thing that can exist does exist (present tense). During the eighteenth century, another version arose that resulted in the Great Chain becoming "temporalized." This states that there *can* be gaps in nature *at any given moment*, although *in the fullest of time* every kind of thing that can exist will exist (future tense). In other words, the Great Chain was no longer seen as a static "inventory" of all that exists but as a dynamic "program of nature" that is unfolding across cosmic history.[38] As Kant articulated the idea in his treatise *Universal Natural History and Theory of the Heavens* (1755), which he published during his "pre-critical" period, several decades before the *Critique of Judgment*, there is a "successive realization of the creation" rather than a single episode of becoming, as with the six days of the Genesis myth.[39] It was the original and temporalized versions of the principle of plenitude and Great Chain of being that George Cuvier helped topple at the turn of the eighteenth century with his work on the mammoth and mastodon.

However, perhaps the most significant transformation of the plenitude principle occurred during the seventeenth and eighteenth centuries, and concerned the spatial rather than temporal dimension. As Lovejoy writes, the principle was applied, for the first time, "not to the biological question of the number of kinds of living things, but to the astronomical questions of the magnitude of the stellar universe and of the extent of the diffusion of life and sentiency in space."[40] This yielded what we can call a "spatialized" version of the principle that both motivated and justified a revolutionary new conception of the universe comprised of (in part) the following propositions: (i) the universe is infinite rather than finite (as Aristotle claimed), (ii) the fixed stars correspond to suns just like our own, (iii) these stars/suns constitute island worlds that include their own planetary systems, and (iv) these planets contain what we would now, in our modern phraseology, call "extraterrestrial life" or "extraterrestrial intelligence."[41]

The reasoning behind these propositions began with the following theological argument: if God is infinitely good and powerful, then the universe must be infinitely large. To quote Giordano Bruno (1548–1600), one of the earliest advocates of this new conception and the first to propose the ideas of (ii) and (iii), "we insult the infinite cause when we say that it may be the cause of a finite effect; to a finite effect it can have neither the name nor the relation of an efficient cause."[42] Similarly, "to affirm that goodness is *infinite*," Joseph Glanvill (1636–1680) wrote the following century, "where what it doth and intends to do is but *finite*, will be said to be a *contradiction*."[43] From this it immediately follows that, assuming a heliocentric model of *worlds* and the Copernican principle that our position in the universe is not special, the fixed stars must be suns with their own inhabited planets. As Kant wrote in 1755, at which time he embraced both the temporalized and spatialized versions of the principle of plenitude,[44] "it would be absurd to represent the Deity as passing into action with an infinitely small part of His potency, and to think of His Infinite Power the storehouse of a true immensity of *natures and worlds* as inactive, and as shut up eternally in a state of not being exercised."[45]

As surprising as it may be to contemporary readers, this became the orthodox view within the West during the eighteenth century, with nearly everyone accepting the existence of both other worlds (solar systems) with their own inhabitants, as well as other intelligent beings living in our own solar system, beliefs that persisted until the early twentieth century.[46] Consider, for example, that the "Astronomy" entry of the first edition of the *Encyclopaedia Britannica*, published in 1771, consists of passages excerpted from a 1756 book by James Ferguson (1710–1776) that not only affirmed the infinitude of the universe but reported that "it may be reasonably concluded, that all the rest [of the solar systems] are with equal wisdom contrived, situated, and provided with accommodations for rational inhabitants."[47] Who were these rational inhabitants? Based on the logic of the original or temporalized principle of plenitude, most would have accepted that at least some are our conspecifics—that is, members of *Homo sapiens*. Thomas Wright (1711–1786) made this explicit in his 1750 tome *The Universe and the Stars*, which influenced many natural philosophers at the time, including Kant. "Of these habitable worlds," Wright argued, "we may suppose [them] to be of a terrestrial or terraqueous nature, and filled with beings of the human species." He then calculated 170 million human-inhabited planets and moons just "within our finite view every clear Star-light night."[48]

The reason I mention these details is that they provide the necessary context to understand remarks made by many of these same authors that seem to suggest that our "extinction" would be evaluatively neutral—neither good nor bad, from a cosmic point of view. For example, in the same *Encyclopaedia Britannica* article just cited, Ferguson writes:

> Instead then of one sun and one world only in the universe, astronomy discovers to us such an inconceivable number of suns, systems, and worlds, dispersed through boundless space, that if our sun, with all the planets, moons, and comets belonging to it, were annihilated, they would be no more missed, by an eye that could take in the whole creation, than a grain of sand from the sea-shore.[49]

Similarly, Kant argued in his treatise from 1755 that "we ought not to lament the perishing of a world as a real loss of Nature," since—to quote him at length—

> she proves her riches by a sort of prodigality which, while certain parts pay their tribute to mortality, maintains itself unimpaired by numberless new generations in the whole range of its perfection. What an innumerable multitude of flowers and insects are destroyed by a single cold day! And how little are they missed, although they are glorious products of the art of nature and demonstrations of the Divine Omnipotence! In another place, however, this loss is again compensated for to superabundance. *Man who seems to be the masterpiece of the creation, is himself not excepted from this law*. . . . The injurious influences of infected air, earthquakes, and inundations sweep whole peoples from the earth; but it does not appear that nature has thereby suffered any damage. In the same way *whole worlds and systems quit the stage of the universe, after they have played out their parts*. The infinitude of the creation is great enough to make a world, or a Milky Way of worlds, look in comparison with it, what a flower or an insect

> does in comparison with the earth. But while nature thus adorns eternity with changing scenes, God continues engaged in incessant creation in forming the matter for the construction of still greater worlds (italics added).

In the following paragraph, Kant contended that we should see "these terrible catastrophes as being the common ways of providence, and regard them even with a sort of complacency."[50] And finally, Wright made the same point after his calculation of 170 million other inhabited islands, declaring that

> in this great Celestial Creation, the Catastrophy of a World, such as ours, or even the total Dissolution of a System of Worlds, may possibly be no more to the great Author of Nature, than the most common Accident in Life with us, and in all Probability such final and general Doom-Days may be as frequent there, as even Birth-Days, or Mortality with us upon the Earth.[51]

We can now see how these statements, despite first appearances, concern something like extirpation rather than extinction, where the biological concept of *extirpation* refers to the disappearance of geographically localized populations of species without the species itself dying out, as when human activity eliminated many gray wolf populations in North America during the nineteenth century. The gray wolf is not extinct, but it does not occupy most of the habitats it once called home. In the cases above, the notion of extirpation applies not to the geographical but to the cosmographical realm, whereby cosmographically localized worlds are annihilated without humanity as a whole perishing. (Indeed, if there are infinite human beings in the universe, then the loss of any particular world or isolated human population would not affect the total number of humans, since infinity minus any finite number still equals infinity.)

Hence, while these authors would likely have said that the "Catastrophy" of our world would be bad *for us*, their claim was that this would *not* be bad in the grand scheme of things, from a cosmic perspective. The judgment of evaluative indifference or neutrality thus applied, from this cosmic perspective, to instances of extirpation rather than extinction, the latter of which was almost certainly thought impossible (including by the above authors) given the ubiquity of belief in the soul's immortality and the spatialized Great Chain during the eighteenth century. Indeed, rather than the loss of bounded systems *within* the plurality of worlds posing any problems for the Great Chain, it was seen by some as *supporting* the idea.[52] As Kant wrote, when a world

> has at last become a superfluous member in the chain of beings; there is nothing more becoming than that it should play the last part in the drama of the closing changes of the universe, a part which belongs to every finite thing, namely, that it should pay its tribute to mortality.[53]

The Finest of All Possibilities

This covers the four categories of references to human extinction before the 1850s, when the discovery of the Second Law triggered the first shift in existential mood. Of note is that the very first statements in the Western tradition about the goodness/badness of our disappearance from a broadly secular perspective all converged upon the conclusion that this would in some way be bad: Montesquieu gestured at the idea that the loss of humanity itself would be calamitous, while Shelley embraced a further-loss view more explicitly in several passages of *The Last Man*, focusing on the concomitant loss of valued things like knowledge and art. Meanwhile, some works within the Last Man genre, including Shelley's, foregrounded the no-ordinary-catastrophe thesis by exploring the various unique harms that could arise from the anticipation and experience of our imminent extinction.

Yet not long after the Last Man genre reached its apotheosis in early nineteenth-century Britain,[54] a cultural and intellectual movement emerged in Germany called *philosophical pessimism* that pointed toward the opposite conclusion about our extinction. Its leading advocates either explicitly accepted, or held views that seem to straightforwardly imply, that our complete and permanent disappearance would be *very desirable*. To be sure, the roots of pessimism—roughly, the view that sentient beings are condemned to great suffering, nonbeing is better than being, and life is not worth living—within the Western philosophical tradition stretch back at least to ancient Greece, and hence this basic orientation was nothing new.[55] For example, Hegesias of Cyrene (*floruit* 290 BCE) "denied the possibility of happiness" and, according to Cicero, made the case that death is good because it obviates the bad things that would otherwise obtain "so eloquently that it is alleged he was forbidden by King Ptolemy to make those statements in his classes because many on hearing them committed suicide."[56] The tragedian Sophocles (c. 497/6–406/5) expressed a similar view in his play *Oedipus at Colonus*, which includes the following lines uttered by the Elders of Colonus:

> The finest of all possibilities
> is never to be born, but if a man
> sees the light of day, the next best thing by far
> is to return as quickly as he can,
> to go back to the place from which he came.

It is of course true that if a sufficiently large number of people were to adopt the *promortalist view* articulated here by Sophocles, that is, that one should commit suicide, the human species would perish, although no philosophers in Classical Antiquity ever discussed this possibility. To be clear, promortalism would not need to be *universally adopted* to ensure our disappearance. In contemporary scientific terms, one would simply need the population to dip below the "minimum viable population" (MVP) threshold, which for *Homo sapiens* may be as low as 98 people or as high as 40,000 people—this remains a contentious matter.[57] Once below this threshold, humanity would then undergo what we could call "functional extinction," whereby the species still exists but demographic extinction (which would presumably entail terminal and final extinction) is inevitable.[58] Nor did any writers before the nineteenth century explicitly argue that since "the finest of all possibilities/is never to be born," one should refrain from having children, an *antinatalist view* that would also lead to extinction if sufficiently widespread (i.e., the same point about the MVP threshold applies here, too). Perhaps no one thought to argue for the antinatalist view because there were no effective, widely available means of contraception, and society-wide celibacy was (and still is) simply unimaginable. Furthermore, lingering behind both normative views at the time may have been the notion that, independent of what any individual chooses to do, the human species itself is fundamentally indestructible. If "ought" implies "can," and if it can't be the case that humanity disappears entirely, then there would have been no reason to prescribe that everyone should either kill themselves or stop having children, even if one accepts that nonbeing is preferable to being.

Pessimism

The philosophical pessimists of the latter nineteenth century took these ideas and developed them into a systematic worldview that, in some cases, *did* prescribe extinction as a solution to the miseries of existence. At the heart of this worldview was a feeling, or emotional state, known in German as *Weltschmerz*, which literally translates as "worldpain," and "signifies a mood of weariness or sadness about life arising from the acute awareness of evil and suffering."[59] According to Arthur Schopenhauer (1788–1860), an "implacable atheist" who became the first great philosophical pessimist, we inhabit the *worst of all possible worlds*, contra the Stoics and Leibniz (his philosophical nemesis).[60] Not only is the total amount of pain and misery in the world vastly *greater* than the total amount of pleasure and happiness, he argued, but the pains are frequently far more *intense* than the pleasures. Who would trade twenty-four hours of pure bliss for the same amount of time, or even a single minute, of the worst torture possible? Or imagine a predator devouring its prey, and consider the satisfaction experienced by the former compared to the indescribable horror suffered by the latter. Surely the satisfaction of the predator doesn't come close to matching, much less compensating for, the prey's inconceivable agony.[61] But what exactly is this satisfaction, anyways? On Schopenhauer's account, pleasure is nothing more than the absence of pain—it has no positive value in itself—and consequently there is no positive tally of pleasure or happiness in the world, only more or less misery. The beast devouring its prey feels "satisfied" because the pangs of hunger have temporarily subsided, not because it feels something above the level "zero" on some eudemonic scale; these pangs will then regenerate, thereby causing more suffering for both the predator and its next victim.

Yet the situation is worse for humans than for nonhuman animals, since our more developed (as it were) cognitive systems enable us to experience more, and more intense, suffering than other creatures who, for example, are unable to imagine (and hence be terrified by) their own mortality. The human situation is this: on the one hand, whenever our *biological needs* (food, drink, sex) are not satisfied quickly enough, we experience pain, discomfort, and unpleasantness; yet just as soon as we satisfy them, we find ourselves thrashing about in the bottomless quagmire of *boredom*, inflicted by the crushing weight of simply existing. This persists until we are once again preoccupied with relieving the various cramps and aches and twinges and pains caused by our biological natures, which, once relieved, then plunge us back into boredom. As Schopenhauer describes this cycle of endless torment:

> That human life must be some kind of mistake is sufficiently proved by the simple observation that man is a compound of needs which are hard to satisfy; that their satisfaction achieves nothing but a painless condition in which he is only given over to boredom; and that boredom is a direct proof that existence is in itself valueless, for boredom is nothing other than the sensation of the emptiness of existence.[62]

Making matters worse, not only are we all prisoners within this incessant oscillation between need and boredom, misery and restlessness, but society tells us that we must strive for "power, prestige, and money." This introduces an additional dimension of "unnatural" suffering, since the acquisition of these supposedly valuable ends are not strictly necessary for our survival. The result is, once again, a treadmill of interminable dissatisfaction, for even when we acquire power, prestige, and money, the feeling is hollow and most people find themselves only wanting more.[63] And how, then, do we console ourselves when our ambitions are thwarted, our needs are left unsatiated, or the burden of boredom becomes too much to bear? We think of those who are worse off than us: at least we aren't *them*, suffering *that* sort of misery. "But what," Schopenhauer wondered aloud, "does that say for the condition of the whole [of life]?"

Schopenhauer thus concluded that "if the immediate and direct purpose of our life is not suffering then our existence is the most ill-adapted to its purpose in the world."[64] Or, referring to Shakespeare's well-known "to be or not to be" line from Act 3, Scene 1 of *Hamlet*, Schopenhauer declared that "the essential content of the famous soliloquy in 'Hamlet' is briefly this: Our state is so wretched that absolute annihilation would be decidedly preferable."[65] In a phrase, the world is hell, existence is horrible, our lives are not worth living, and it would have been better if we had never been born.

One way to understand this dismal picture of the world and how Schopenhauer drew his bleak conclusions about the unworthiness of life is in terms of the enabling condition behind the second existential mood, namely, *secularization.* On the Christian account, our fallen world is saturated with sin and suffering, misery and hardship, due to Adam and Eve disobeying God's command not to eat from the tree of knowledge of good and evil, although Schopenhauer greatly elaborated this idea in making his arguments about the preponderance of evil in the world, the cycle of need and boredom, and so on. For Christians during the Middle Ages, while the sufferings of life are vast, there was never a question about whether our lives are worth living: it *is* worth the trouble, they claimed, because of the promise of redemption and eternal life with God in paradise. As Frederick Beiser writes, referring to the problem of evil that, as we saw in Chapter 3, became a pressing issue of ethical concern during the nineteenth century (thus contributing to the secularization trend), rather than denying the existence of evil and suffering,

> medieval philosophers and theologians . . . adamantly affirmed their existence because it gave all the more point and power to the doctrine of divine grace and redemption. According to that doctrine, life is worth living, not because of its intrinsic value, but because it is a means to another end, eternal salvation.[66]

The rise of secularism in the 1800s, especially, undermined the cogency of this answer, thus prying the question wide open: if there is no redemption, no hope of eternal life, then what is all of this suffering for? And if for nothing, then how can one say that life is worth living, that being is preferable to nonbeing? The pessimistic worldview and cultural atmosphere of *Weltschmerz*, therefore, is what happens when one accepts the dark Christian view of the world while rejecting the Christian promise that, in the end, everything will be fine and dandy.[67]

The Perfect Calm of Spirit

This brings us to the connection between philosophical pessimism and Existential Ethics. The "central thesis" of pessimism, i.e., that non-existence is better than existence, clearly entails a version of what we could refer to as a *pro-extinctionist view.* That is to say, the central thesis has two obvious implications: (i) it would have been better if humanity had never existed (a backward-looking implication), and (ii) given that we already exist and can do nothing about the former, it would be better if humanity were to cease existing, i.e., to go extinct, especially in the *final* sense (a forward-looking implication). The second implication is of course the pro-extinctionist position. Notice right away that this concerns the state or condition of Being Extinct rather than the process or event of Going Extinct. There are many possible ways for humanity to go extinct, most of which would, as we saw, cause tremendous amounts of suffering. Given the pessimists' unusual sensitivity to suffering, there is no doubt that virtually all of them would have seen most scenarios of Going Extinct as utterly dreadful, as something that ought to be avoided if at all possible. Virtually all would have not only accepted the default view (if not the no-ordinary-catastrophe thesis) but there is every indication that they would have seen any form of *involuntary* anthropogenic extinction as very wrong, a claim consistent with the fact that the only *means* of annihilation they considered—such as antinatalism and promortalism—involve voluntary actions (e.g., one chooses for oneself to be childless or commit suicide).

To put the point differently, it would be misleading to describe any of the pessimists who endorsed the pro-extinctionist view of (ii) above as "omnicidal," as Moynihan does, if "omnicide" is understood in Kenneth Tynan's terms of "the murder of everyone."[68] The pessimists were not omnicidal maniacs: however desirable it would be for humanity to disappear entirely and forever, none of them advocated mass murder by some agent acting unilaterally. The *outcome* of this might be better, but the means would be abhorrent. A second point about (ii) is that it is, strictly speaking, a purely evaluative rather than deontic claim: it simply states that Being Extinct is *better than* Being Extant, as we could say, and that is all. However, there may be some connection between what is better—or, in this case, since there are only two options, what is *best*—and what one *ought to do*.[69] For now it suffices to observe that some of the philosophical pessimists did, in fact, take the extra step of arguing that humanity should actively strive to bring about its own extinction, albeit through voluntary, if unspecified, means.

One example comes from the troubled soul of Philipp Mainländer (1841–1876), who published Volume I of his central work *The Philosophy of Redemption* in 1876, at the age of 34. Upon receiving the first copies of it, he placed them on the floor, stood on them, stepped off, and hanged himself. Like all the pessimists, Mainländer, an atheist who popularized the "death of God" idea before Nietzsche, borrowed much from the woeful picture of existence outlined by Schopenhauer, e.g., he held that all life is suffering and nonbeing is preferable to being.[70] But whereas Schopenhauer argued against suicide (see below), Mainländer disagreed: "Go without trembling, my brothers, out of this life if it lies heavily upon you; you will find neither heaven nor hell in your grave," he wrote in Volume II of *Redemption*. But he did not recommend this for everyone, only those unable to tolerate existence any longer.[71]

He did endorse, however, universal antinatalism through not merely abstinence but *virginity*, and explicitly linked this with the final goal of bringing about our complete and permanent extinction.[72] This is to say, Mainländer accepted a teleological conception of history according to which humanity is marching toward an "ideal state," as outlined by Kant in his "Idea for a Universal History with a Cosmopolitan Purpose" (1784), which would "encompass all of humanity." But unlike Kant, Mainländer contended that this is not the ultimate state of development but merely the penultimate "transit point" on the way to something even better. The true goal is "the annihilation of hell," where "hell" refers ironically to *existence* and, consequently, "the still night of death" is *its* annihilation. In other words, since death is eternal nothingness, a complete absence of misery, the ultimate escape from the perdition of our world is to bring about an absolute state of Being Extinct through universal celibacy. "There is only one movement left for" humanity after attaining the ideal state, he wrote, "the movement to *complete annihilation*, the movement from *being into non-being*. And humanity (i.e., all single then living humans) will execute this movement, in irresistible desire to the rest of absolute death." Referring again to Kant's ideal state:

> The movement of humanity to the ideal state will also follow the other, from being into non-being: the movement of humanity is after all the movement from being into non-being. If we separate the two movements, then from the first one appears the rule of full dedication to the common good, the latter the rule of *celibacy*, which . . . is recommend [*sic*] as the *highest* and most *perfect virtue*; for although the movement will be fulfilled despite bestial sexual urge and lust, it is seriously demanded to every individual *to be chaste*, so that movement can reach its goal *more quickly*.

How could universal celibacy possibly be achieved? As noted, it is quite unimaginable that a sufficient number of people around the world would agree not to have sex again—or ever, in the case of virginity. This poses a virtually "insurmountable" problem, Mainländer concedes. However, he also claims that by recognizing just how terrible life is, and by understanding that death provides eternal peace whereas existence only prolongs suffering, one can incrementally begin to muster the willpower needed to overcome our natural urges to procreate. In his words:

> [W]ith every step he gets less disturbed by sexual urges, with every step his heart becomes lighter, until his inside enters the same *joy*, *blissful serenity*, and *complete immobility* . . . He feels himself in accordance with the movement of humanity from existence into non-existence, from the torment of life into absolute death, he enters this movement of the whole *gladly*, he acts eminently ethically, and his reward is the undisturbed peace of heart, "the perfect calm of spirit," the peace that is higher than all reason.[73]

Universal Final Extinction

This is one example of a philosophical pessimist following the implications of the central thesis stated above to its logical conclusion: if nonbeing is better than being, then we should strive for nonbeing, not just on an individual but species level. Another notable example comes from Eduard von Hartmann (1842–1906), who, despite being largely unknown today, attained the status of a celebrity in the late nineteenth century.[74] Described by one contemporary writer as displaying a "mustache [that] is, I think, the longest in metaphysics," Hartmann also fully embraced Schopenhauer's pessimism, although he provided a more systematic account of life's unending awfulness.[75] However, his overall picture of the universe uniquely combined Schopenhauerian pessimism with an "optimistic" account of goal-directed historical development that was heavily influenced by Georg Wilhelm Friedrich Hegel. (Mainländer seems to have been influenced by Hegel, too, but much less so.) As Hartmann himself wrote in the tenth edition of his most famous book, titled *The Philosophy of the Unconscious* (1869), "should the position of my system of philosophy be characterized in a few words, one could say: it is a synthesis of Hegel's and Schopenhauer's systems with a decisive preponderance of the former."[76]

To understand Hartmann's position, it is necessary to describe some additional aspects of Schopenhauer's philosophy. For Schopenhauer, the universe is animated by what he called "the will," which refers to the "blind striving" that underlies all suffering in the world. The urge to satisfy our biological needs, to acquire power, prestige, and money, and so on are all driven by the will. The only hope of "redemption" or "salvation" is to subjugate or deny the will, which one achieves through aesthetic appreciation, asceticism, and mystical experience. (Unfortunately, these are only available to a small, elite demographic: geniuses and saints, respectively. Mainländer, in fact, aimed to outline a non-elitist path to redemption for the common man by advocating suicide and celibacy, which are of course available to everyone.) Hartmann vociferously rejected Schopenhauer's path to serenity, a personal state of being similar to what the Buddhists would call *nirvana* (literally, "extinction, disappearance") and Hindus would call *moksha* (literally, "emancipation, liberation, release"). Like many others at the time, he worried that the "ascetic attitude of renunciation, resignation, and will-lessness" would only lead to quietism, or the view that one should give up trying to change the world for the better.[77]

Hartmann thus aimed to establish a new, pantheistic, "rational" religion that could fill the space left behind by Christianity and, in doing so, provide people with a reason to live, a purpose in life, and the motivation needed to actively strive for a better world, thereby replacing quietism with a kind of activism. At the heart of this religion was an evolutionary account of history that, as with Mainländer's view, posited a "final redemption from the misery of volition and existence" as the ultimate *telos* toward which the universe is developing.[78] This redemption corresponds to, of course, a future condition of complete and total annihilation—not just the final extinction of humanity, but a sort of *universal final extinction* of all sentient life everywhere and forever. How exactly will humanity bring this about? How could we annihilate the very possibility of all life in the entire cosmos? Unlike Mainländer, Hartmann completely rejected both antinatalism and promortalism, which renders these questions even more urgent and perplexing. What, then, was Hartmann's plan?

To answer these questions, let's begin with Hartmann's claim that humanity will have progressed through three stages of "illusions." In the first, people strove to achieve happiness in the present world, as exemplified by the ancient Greeks.

In the second, people recognized the evils of life and impossibility of happiness here and now but came to believe that happiness would be attained in the afterlife, as exemplified by Christianity. Finally, in the third, people came to believe that material progress would ultimately lead to a better world in which happiness will indeed become attainable, as exemplified by the progressivism of Enlightenment philosophers like Condorcet. But the truth is that happiness is impossible here and now, there is no afterlife, and no matter what progress humanity makes, life will always be suffering. Indeed, Hartmann claimed that suffering will only grow as humanity becomes increasingly aware of how bad existence is. Yet so long as our consciousness continues to develop, we will eventually realize that there is a way out: to achieve happiness, we must attain a state of painlessness; and to attain a state of painlessness, we must terminate the "world-process" in its entirety. This is Hartmann's final redemption, and it is precisely because our world will someday see redemption that, he argued against Schopenhauer, we actually occupy "a best possible world." There is hope after all: the hope of total annihilation, and hence the elimination of all suffering once and forever.

But we still have not answered the practical question above: how could we actually achieve this? Hartmann, in fact, does not go into details, but is not bothered by his inability to delineate a precise means of universal annihilation. "Our knowledge is far too imperfect, our experience too brief, and the possible analogies too defective, for us to be able, even *approximately*, to form a picture of the end of the process," he wrote. As society continues to develop over time, the answer will gradually peak over the horizon of imaginability and the practicality and logistics of universal annihilation will become visible. Indeed, this is one reason that Hartmann so strongly opposed antinatalism and promortalism: by refusing to have children or by killing ourselves, we impede the movement of the world-process toward this ultimate *telos* and, consequently, only prolong suffering. This is also why Hartmann so deeply despised the quietism of Schopenhauer:

> [I]t threatens to bring the world-process to stagnation, and to perpetuate the misery of existence. What would it avail, e.g., if all mankind should die out gradually by sexual continence? The world as such would still continue to exist, and would find itself substantially in the same position as immediately before the origin of the first man; nay, the Unconscious would even be compelled to employ the next opportunity to fashion a new man or a similar type, and the whole misery would begin over again.

"Therefore," he declared, "vigorously forward in the world-process as workers in the Lord's vineyard, for it is the process alone that can bring redemption!"

Hartmann did specify several necessary conditions for humanity to reach this *telos*, such as "the consciousness of mankind [being] *penetrated* by the folly of volition and the misery of all existence" and there being "sufficient communication between the peoples of the earth to allow of a *simultaneous common resolve*."

But he also argued that there is no guarantee that *humanity* will ever satisfy these conditions, even though the complete annihilation of everything forever is the *inevitable terminus* of the world-process itself. Here Hartmann seems to have drawn from (a) the theory of evolution (recall that Darwin's *Origin* had been published exactly ten years earlier) and (b) the plurality of worlds model of the universe, writing that

> whether humanity will be capable of so high an enhancement of consciousness, or whether a higher race of animals will arise on earth, which, continuing the work of humanity, will attain the goal, or whether our *earth* altogether is only an abortive attempt to reach such [a] goal, and it will only be reached, when our little planet has long been reckoned to the frozen celestial bodies, on a planet invisible to us of another fixed star under more favourable conditions, is hard to say.[79]

In other words, if humanity doesn't get the job done, either our successors on Earth (before Earth becomes uninhabitable; Hartmann may have been thinking about the Second Law here)[80] or some unknown future species of extraterrestrial intelligences eventually will—"if" is not in question, only "when" is. This, again, is why Hartmann saw his new religion as "optimistic," and why he thought we occupy "a best possible world." Yet the question remains: how could *any* species, wherever or whenever it exists, annihilate the *entire cosmos*? The answer comes from Hartmann's metaphysics—specifically, his *idealism*, according to which the existence of objects requires the existence of a subject (for example, us), and hence the annihilation of all subjectivity metaphysically entails the annihilation of all objectivity. Without creatures like us, the universe simply cannot exist, and if the universe does not exist, there is no possibility of suffering ever again rearing its ugly head.[81]

Lifeless as the Moon

While the very first remarks about the normative implications of human extinction—from Montesquieu, Shelley, and others—agreed that the process and/or outcome would be in some sense bad, a number of prominent German philosophers in the latter nineteenth century explicitly and vigorously argued that, in fact, the outcome would be very good, and is thus something that we should actively strive to bring about. This was intimately connected to the decline of religious belief during the 1800s, although, interestingly, the Second Law—the triggering factor behind the first shift in existential mood—did not seem to play any significant role in the rise of philosophical pessimism. This is despite (a) its obviously dismal implications for the long-term future of humanity, and (b) the fact that it was first formulated and subsequently developed in Germany (by Rudolf Clausius and then by Ludwig Boltzmann), which suggests that contemporary

German philosophers would have likely known about it. Nonetheless, the pro-extinctionist views of Mainländer and Hartmann were built upon the pessimism of Schopenhauer—the most influential and lionized philosopher within Germany between 1860 and the beginning of WWI[82]—and hence one might wonder whether Schopenhauer himself endorsed our extinction, or other positions that would, as a necessary consequence, lead to our extinction (i.e., universal antinatalism or promortalism).[83] Oddly, Schopenhauer himself never took the obvious next step of inferring the forward-looking claim (ii) above from the central thesis that non-existence is better than existence. He did, however, affirm something very similar to the backward-looking claim of (i), as when he wrote in his essay "On the Suffering of the World," published in *Parerga and Paralipomena*:

> If you imagine, in so far as it is approximately possible, the sum total of distress, pain, and suffering of every kind which the sun shines upon in its course, you will have to admit it would have been much better if the sun had been able to call up the phenomenon of life as little on the earth as on the moon; and if, here as there, the surface were still in a crystalline condition.[84]

In other words, assuming that the moon is lifeless, it would have been better if our planet were like its natural satellite in this respect. Writing in the same essay, Schopenhauer further declared that

> if the act of procreation were neither the outcome of a desire nor accompanied by feelings of pleasure, but a matter to be decided on the basis of purely rational considerations, is it likely that the human race would still exist? Would each of us not rather have felt so much pity for the coming generation as to prefer to spare it the burden of existence, or at least not wish to take it upon himself to impose that burden upon it in cold blood? . . . For the world is Hell and men are on the one hand the tormented souls and on the other the devils in it.[85]

Schopenhauer thus made clear that procreation is *irrational*, yet he stopped short of claiming that it is *immoral* and, therefore, something that we should refrain from doing. A similar statement comes from his earlier1818 *magnum opus* titled *The World as Will and Representation*:

> Voluntary and complete chastity is the first step in asceticism or the denial of the will to live. It thereby denies the assertion of the will which extends beyond the individual life, and gives the assurance that with the life of this body, the will, whose manifestation it is, ceases. Nature, always true and naive, declares that if this maxim became universal, the human race would die out.[86]

Once again, though, Schopenhauer never contended that this maxim *should* become universal,[87] perhaps because his primary concern was overcoming and denying the will—the underlying source of all suffering in the world—and since the will pervades the whole cosmos, the blind striving of the will would continue to exist even if humanity were to disappear. But this is inconsistent with Schopenhauer's own idealism, which implies that if humanity were to disappear, so would the universe itself, assuming that we are the only rational beings in it, which he may have believed. As Schopenhauer declared almost immediately after the passage just quoted: "With the entire abolition of knowledge, the rest of the world would of itself vanish into nothing; for without a subject there is no object."[88]

This leaves it a mystery why Schopenhauer did not argue for a pro-extinction view, coupled with an antinatalist means of bringing this about, whereby all people are enjoined to cease having children, ultimately leading to the complete annihilation of everything. Similarly, many critics have complained that Schopenhauer's pessimism seems to straightforwardly entail pro-mortalism: if life isn't worth living, why not find the nearest exit in the theater of being and say goodbye, as Sophocles and Mainländer suggested? But Schopenhauer argued against suicide, which he saw as a *manifestation* rather than *denial* of the will: it is precisely because one is driven by the will to attain a satisfaction in life that one becomes frustrated, as satisfaction is unattainable; the will then turns against itself, leading the frustrated individual to end her life.[89] Even more, Schopenhauer contended that the loss of any particular individual cannot destroy the *cosmic will* that pervades all existence, and hence suicide is not a *solution* to the problem of suffering. But of course if *everyone* were to kill themselves in a worldwide act of simultaneous mass death, this would not be the case: the universe would immediately "vanish into nothing," replaced forever by "the blessed calm of nothingness."[90] It is strange that Schopenhauer never entertained this possibility.

The Murderer and the Good

While the *zeitgeist* of "worldpain" dominated Germany during the second half of the nineteenth century, Britain witnessed the development of an important new theory of ethics called *utilitarianism*, a version of which has become very influential within contemporary Existential Ethics. Indeed, as we will see in later chapters, the claim that Being Extinct, however it may be brought about, would constitute an *inconceivably bad outcome*—call it an *axiological catastrophe*—arguably finds its strongest support from the utilitarian approach. That is, this theory, if understood in a "total" and "impersonalist" sense (see below), gives rise to one of the most powerful further-loss views, although not without engendering serious theoretical problems, as we will explore in Chapter 11. The "Classical Utilitarians," that is, Jeremy Bentham (1748–1832) and John Stuart Mill (1806–1873), never wrote anything about this potential implication of the theory, perhaps because (a) they were preoccupied with narrower questions of social and legal reform,[91] and (b) it was not clear at the time they were writing that human extinction was practically feasible in the near term, that is, there were no known, widely accepted kill mechanisms capable of destroying humanity on timescales that might have motivated theorizing about this possibility. There were, of course, plenty of *proposed* kill mechanisms, including population decline (Montesquieu), cometary collisions (Byron), global pandemics (Shelley), the desiccation of Earth (anonymous H), and maybe even large boilers exploding (Verne), but as we saw in Part I, none of these were taken seriously by leading intellectuals as genuine near-term threats to our species.[92]

However, Henry Sidgwick (1838–1900), the most influential utilitarian since Bentham and Mill (at least up to the latter twentieth century), did address the implications of our extinction in his 1874 masterpiece *The Methods of Ethics*, albeit in passing while discussing an unrelated issue. To understand Sidgwick's claim about extinction, which yields a conclusion diametrically opposed to the pro-extinctionist views of Mainländer and Hartmann, and is quoted frequently by contemporary philosophers sympathetic with utilitarianism, it is necessary to establish the basics of utilitarian ethics. This will also serve as a foundation for subsequent chapters, given the prominence of this theory within Existential Ethics today.

To begin, utilitarianism is a form of consequentialism, which asserts that what makes an act right or wrong depends entirely on whether the act chosen produces the greatest amount of "intrinsic value" or "the good" (these are synonymous) relative to all the other acts available to the agent at the time. In other words, one's action is morally right when and only when it *maximizes intrinsic value*, and wrong whenever it does not. The direction of ethical reasoning thus proceeds from *the evaluative* to *the deontic*: first, figure out how much of the good would result, as a consequence, from the various actions available to you at the time, and then, second, take whichever action would result in the most good. Whatever is *best* (an evaluative notion) is what one *ought* to do (a deontic notion). Although you might find this claim obvious, or even tautological—surely we should always do what is "best," right?—the idea that consequences are *all* that matters was a novel innovation, and upon closer examination encounters a number of serious objections.

To situate this theory in a broader historical context, the first systematic ethical theory in the Western tradition, dating back to Plato and Aristotle, was "virtue ethics." On this account, ethics is about developing "virtuous" (in contrast to "vicious") character traits like wisdom, courage, temperance, prudence, and fortitude through moral education and practice. These character traits, then, were seen as either contributing to or constituting a state called *eudaimonia*, which translates as "happiness" or "wellbeing." Hence, the focus of virtue ethics is one's *moral character* rather than one's *moral actions or choices*, although of course one's actions or choices may *evince* one's moral character. To be a moral person is thus to be a virtuous person.

Roughly two millennia later, Kant introduced a new "deontological" approach to ethics during the Enlightenment. According to his theory, an act is morally right when and only when it stems from motives based entirely on considerations of the Moral Law, which he famously identified with the aforementioned Categorical Imperative. This completely divorced—at least on the common "absolutist" interpretation of his position—the deontic from the evaluative, that is, in the sense that the rightness/wrongness of one's action has absolutely nothing to do with its consequences. For example, Kant argued that making false promises is *always* wrong because it fails to pass the universalizability test associated with the first formulation (of four in total) of the Categorical Imperative: "act only in accordance with that maxim through which you can at the same time will that it become a universal law."[93] The test is to universalize one's "maxim of action" to see whether it engenders a logical or practical contradiction.[94] In the case of making false promises, if *everyone* were to make false promises, then it would become impossible to make false promises, since no one would believe anything anyone ever says. Hence, the universalized maxim, let's say, "I will make false promises for personal gain" yields, from within itself, as discernible entirely through rational reflection, a contradiction, which *means* that acting in accordance with that maxim is *impermissible*, i.e., morally wrong. Kant held such a rigorist interpretation of this theory that he even argued, explicitly, that lying to a murderer in search of your friend, his next victim, would be wrong. Although the consequences of telling the truth would be bad—Kant himself would agree with this—doing otherwise would simply be immoral. As he declared in an essay titled "On the Supposed Right to Lie From Benevolent Motives," written in response to Benjamin Constant, who proposed the murderer scenario: "Truthfulness in statements that one cannot avoid is a human being's duty to everyone, however great the disadvantage to him or to another that may result from it."[95]

Contra Kant, utilitarians would assert that there is nothing *intrinsically wrong* with the act of lying itself—nor with cheating, stealing, killing, and so on. To the contrary, you *should* do these things when they will produce the greatest amount of intrinsic value. (Of course, in the vast majority of cases, lying, cheating, and stealing very likely won't produce the most good, and hence in most cases doing them would be immoral. But utilitarians would—as I believe everyone should—vehemently urge one to mislead the murderer knocking at one's door.) This is why, as undergraduates who have taken an introductory course in ethics will know, consequentialism "puts the good before the right" whereas deontology "puts the right before the good," which is just another way of stating the point above about the deontic (right) and the evaluative (good). It also explains why utilitarianism is a "teleological" theory, as I noted in the last chapter: morality, on this account, is about attaining the end of maximized intrinsic value or the good. Virtue ethics, at least in the form advocated by the ancient Greeks, can also be seen as teleological, given its aim of cultivating virtuous traits and attaining *eudaimonia*, while Kant's deontological theory is decidedly "non-teleological."

If the criterion of right conduct on the utilitarian theory is that one maximizes the good, what exactly is the good? What has intrinsic value? The Classical Utilitarians were *hedonists*, meaning that they built their theories of right action upon an underlying theory of value according to which the one and only thing in the universe that is intrinsically valuable is *pleasure* or *happiness*. As Bentham wrote, "pleasure is in itself a good; indeed it's the *only* good . . . and pain is in itself an evil, and without exception the *only* evil."[96] There are three things to note about this: first, it differs from Schopenhauer's claim that pleasure is nothing but the absence of pain, or suffering. On Bentham's account, pleasure has *positive value*, while pain has *negative value*, and the best outcomes are those in which, when all the pleasures and pains are summed together, the result is a net surplus of pleasure. Second, one could plausibly describe many types of things as intrinsically valuable in addition to pleasure, such as knowledge, beauty, friendship, love, and so on. But for hedonistic utilitarians like Bentham, such things have merely *instrumental* value: they are "good" only insofar as they conduce to the realization of pleasure. And third, since pleasure can be, it seems clear, experienced by sentient nonhuman animals, there is no reason to exclude nonhumans from our moral considerations, that is, they are "moral patients" (the objects of moral concern or consideration) even if they are not "moral agents" (subjects capable of being morally responsible for their actions/choices). In Bentham's words, "the question is not, Can they *reason*? Nor, can they *talk*? But, can they suffer?"[97] This being said, the utilitarian may still focus more on how one's actions affect humans rather than nonhumans, given that humans seem capable of experiencing more pleasures and pains than nonhumans.

The Eye of the Cosmos

Sidgwick developed this general framework into a sophisticated theory that diverged in significant ways from the theories defended by Bentham and Mill. Nonetheless, he accepted a hedonistic value theory according to which, roughly speaking, the good is what one ought to desire, and what one ought to desire is pleasure or happiness. Hence, given the maximization principle central to all consequentialist theories, Sidgwick held that we should aim to maximize happiness. But how can we determine whether pleasure has in fact been maximized? How do we compare two possible consequences to see which contains more intrinsic value? Like his predecessors, Sidgwick argued that pleasures and pains can be aggregated—added up to get a sum total of net pleasure or net pain. Following Bentham, he argued that summing pleasures and pains should be impartial to the identities of those sentient beings affected by our actions: it doesn't matter which nationality, race, gender, social class, or even species that one belongs to; it doesn't even matter where one exists in space or time, whether next door or on the other side of the planet, in the present moment or distant future: each sentient being's happiness or suffering must count equally. In Sidgwick's often-quoted phraseology, "the good of any one individual is of no more importance, from the point of view (if I may say so) of the Universe, than the good of any other."[98] In other words, peering down from the disembodied eye of the cosmos, we are to impartially aggregate the pleasures and pains of all sentient beings, including nonhuman animals, to determine which consequences are best, and therefore which actions are right.

To illustrate with a famous example from Peter Singer, imagine walking past a shallow pond and seeing a child drowning in the water.[99] If you were to save the child, you would ruin your clothes, which would be bad for you. Hence, if you were *partial* to yourself—that is, if you counted your *own* happiness more than the happiness of the child and its parents—then you should keep walking. But from an impartial perspective, the "keep walking" option would fail to maximize happiness, and thus from the universe's point of view you have a moral obligation to ruin your clothes and save the child. Or consider a controversial example that is often presented as a refutation of utilitarianism: a doctor could save five sick people by harvesting the organs of one healthy person. Although having one's organs harvested would obviously be bad for that person, saving the five people would, at least *prima facie*, result in the greatest "Universal Good," as Sidgwick put it.[100] Hence, the doctor should harvest the healthy person's organs.

But here we encounter another complication, which was noticed for the first time by Sidgwick: should we maximize the *average* or *total* amount of happiness? Which of these one chooses—depending on another consideration that I will introduce below—will have major implications for how bad our extinction would be, especially in the final and normative senses. The distinction between average and total happiness can also make a difference to how one assesses scenarios like the doctor and her sick patients. For example, let's say that the healthy person has a happiness level of 99, while each of the five sick patients have a happiness level of 20, and that, if the five are saved, they will each have an improved happiness level of 35 while the person whose organs are harvested will fall to 0. Given these numbers, if what matters is the average happiness, then the doctor should indeed harvest the healthy person's organs, since this would result in an average happiness level of $(35 \times 5)/5 = 35$, which is greater than the average happiness level in the no-harvest scenario: $[(20 \times 5) + 99]/6 = 33$. In contrast, if what matters is the total happiness level, then the no-harvesting scenario is best, since it would result in a total level of $(20 \times 5) + 99 = 199$, while the harvesting scenario would result in a total level of $35 \times 5 = 175$. Different wellbeing levels will give different results.

The point is that one's adoption of *averagism* or *totalism*, as they are sometimes called, can make a crucial difference to which actions one takes to be right and wrong. While the Classical Utilitarians did not really distinguish between these two interpretations, Sidgwick not only emphasized the distinction but was clear about his own view, which aligned with the totalist interpretation, now standardly called "total utilitarianism," whereby right actions are those that maximize the *total quantity* of pleasure.[101]

The Inconceivable Crime of Universal Celibacy

We are now in a position to understand Sidgwick's anti-extinction position. Consider, he says, how particular acts can produce more good than bad, but when adopted by a sufficiently large number of people, the result can be more bad than good. For example, "no one (*e.g.*) would say that because an army walking over a bridge would break it down, therefore the crossing of a single traveller has a tendency to destroy it." Hence, there may be acts that are not wrong on the assumption that they "will not be widely imitated" by others, but wrong when many people perform those acts together. This leads Sidgwick to consider "the case of Celibacy," which he may have thought of because of his impressive "fluency in German philosophy."[102] (Indeed, not only could Sidgwick read German, but his 1886 *Outlines of the History of Ethics* includes sections on "German Pessimism," "Schopenhauer," and "Hartmann.") Applying the above idea to celibacy, Sidgwick declared that

> a *universal refusal* to propagate the human species would be the greatest of conceivable crimes from a Utilitarian point of view;—that is, according to the commonly accepted belief in the superiority of human happiness to that of other animals;—and hence the [Kantian] principle [of universalizability], applied *without* the qualification [that one engages in celibacy on the assumption that enough other people won't], would make it a crime in any one to choose celibacy as the state most conductive to his own happiness. But Common Sense (in the present age at least) regards such preference [for celibacy] as within the limits of right conduct; because there is no fear that population will not be sufficiently kept up, as in fact the tendency to propagate is thought to exist rather in excess than otherwise [a likely reference to Malthus] (italics added).[103]

Here we can discern two arguments, one of which has been much more influential within Existential Ethics than the other. (A) Sidgwick is saying that, even if the process of Going Extinct were entirely voluntary, since most of the happiness in the world comes from human beings rather than nonhuman animals, the loss of humanity would greatly decrease the total amount of happiness in the world, which would be bad and therefore wrong. (B) The second is that, since many humans could exist in the future, and since what matters is the total amount of "happiness on the whole" apart from any individual's happiness, our extinction would be extremely bad because it would prevent the realization of all this future happiness and hence greatly reduce the total quantity of happiness in the universe, across not just space but *time as well.* Put differently, the primary locus of the badness of extinction on this account is the state or condition of Being Extinct, during which a potentially very large amount of happiness that could have existed never will exist. Or, in modern economic terms, the "opportunity cost" of Being Extinct could be enormous, and this is why our extinction would constitute an *axiological catastrophe*. This would be the case whether our extinction were natural or anthropogenic in etiology, and it is why—given the utilitarian connection between badness and wrongness—Sidgwick concluded that even voluntary human extinction would be extremely wrong: the worst moral crime that humanity could possibly commit.

Here we have a further-loss view *par excellence*, since what makes extinction so bad is all the lost future value that it would entail, where this lost value goes well beyond whatever losses (harms, suffering) might be involved in the process or event of Going Extinct (in the case of voluntary universal celibacy, these would presumably be minimal). Hence, Sidgwick's position in Existential Ethics was radically different from—in a sense the complete opposite of—Mainländer's and Hartmann's positions, although Sidgwick did not elaborate on these ideas, perhaps for the same reason Bentham and Mill did not: our extinction was not widely recognized at the time as an outcome that could *actually* obtain in the foreseeable future. In particular, there was no reason to believe that, aside from universal celibacy and suicide, both highly improbable, humanity was capable of bringing about its collective non-existence.

It is important to note that both (A) and (B) above intersect with yet another, orthogonal distinction between (i) "person-affecting" and (ii) "impersonalist" interpretations. Consider the difference between saying, "*Of those people who currently exist*, we should maximize the total amount of *their* happiness," and "We should maximize the total amount of happiness *in the universe as a whole*." The first corresponds to what Jan Narveson described as "making people happy," that is, making people who currently exist happier, while the second entails "making happy people," that is, creating new people conditional on them having worthwhile lives.[104] Put differently, if what matters is how much total value there is in the whole universe, then a "total-impersonalist" version of utilitarianism implies that we have a *moral obligation* to create additional, extra people (or sentient beings in general) with worthwhile lives for the sake of achieving this *axiological end*. Sidgwick himself adopts this total-impersonalist version, which has become the most widely accepted version of utilitarianism today.[105] In Sidgwick's words,

> Utilitarianism directs us to make the number [of beings] enjoying [happiness] as great as possible. . . . For if we take Utilitarianism to prescribe, as the ultimate end of action, happiness on the whole, and not any individual's happiness, unless considered as an element of the whole, it would follow that, if the additional population enjoy on the whole positive happiness, we ought to weight the amount of happiness gained by the extra number against the amount lost by the remainder. So that, strictly conceived, the point up to which, on Utilitarian principles, population ought to be encouraged to increase, is not that

> at which average happiness is the greatest possible . . . but that at which the product formed by multiplying the number of persons living into the amount of average happiness reaches its maximum.[106]

All of this is to say that Sidgwick's further-loss view was based on a total-impersonalist version of utilitarianism, which subdivides into the two considerations above, (A) and (B). However, interestingly, Sidgwick would have *rejected* a different further-loss view, namely, that suggested by Shelley in *The Last Man*. Since, according to Sidgwick, the one and only intrinsically valuable thing is happiness—hedonism is a monistic theory of goodness—the value of what he called the "ideal goods," or non-hedonic goods like knowledge and beauty, is entirely dependent upon the existence of beings like us. Why? Because they have merely instrumental value, meaning that they are good only "in so far as they conduce either (1) to Happiness or (2) to the Perfection or Excellence of human existence," where (2) refers to the "ultimate *practical* end" of "attaining an ideal or nearly ideal set of mental qualities, which we admire and approve when they are manifested in human life."[107] Hence, only when the loss of such goods negatively affect our *happiness*, or our *ability to achieve* happiness, would this be bad; otherwise, without humanity around, the disappearance of the "adornments of humanity" that Shelley lists would not be bad in itself.[108]

Here we have, by the 1870s, two distinct further-loss views, one focused on the things we value no longer existing after we ourselves are gone, and the other focused on the happiness that would be lost if the human story were to come to an end. These are not mutually exclusive, although Shelley did not say anything about the latter, while Sidgwick would not have accepted the former.

A Universe in Ruins

Although the particular brand of pessimism that emerged in Germany, inspired by Schopenhauer, was largely confined to its cultural borders, the late nineteenth and early twentieth centuries witnessed a broader shift within the West toward a more generally pessimistic outlook on human existence—what Peter Bowler refers to as "cosmic pessimism"[109]—due to the loss of religious belief paired with the implications of Darwin's theory of evolution and the Second Law of thermodynamics. This sense of cosmic pessimism was further exacerbated by the rapid economic, technological, and societal transformations brought about by modernization, along with pervasive anxieties about cultural and biological degeneration, which found expression in aforementioned books like Sir Edwin Ray Lankester's *Degeneration: A Chapter in Darwinism* (1880) and H. G. Wells' *The Time Machine* (1895).[110]

In Britain, for instance, this pessimism tended to focus less on the *suffering of life* and more on the *meaning of life* given that (a) the universe has no external source of purpose (i.e., God in the Christian worldview), (b) the a-teleological nature of Darwinian evolution implies that our existence is something of an accident, and (c) in the end, due to the inexorable increase of entropy, everything that humanity has ever been and created—all the many triumphs and achievements, all the sacrifices and struggles—will be swallowed up by the solar or heat death, without a trace left behind. Following Iddo Landau, we can distinguish between two senses of "meaning." The first concerns *meaning or significance*, as when someone says that "your apology was very meaningful" or "this was the most meaningful event of my life." The second concerns *understanding* or *comprehensibility*, as when someone says that she has not yet "grasped the meaning of $E = mc^2$" or "this sentence is meaningless."[111] The death of God, integration of humanity into the natural order, and discovery of the Second Law influenced thoughts about life's meaning in both senses of the word, but especially the first: what importance can we attach to our existence, both as individuals and a species, and what sense can we make of this existence, of the fact that we are something rather than nothing, given the contingency of our evolutionary past and the inevitability of our future demise as the universe sinks into a frozen puddle of thermodynamic equilibrium? To quote the American philosopher Ralph Barton Perry, writing in 1918, whereas "the old religion thought [of man] as 'a little lower than the angels,'" a likely reference to the Great Chain, "the new materialism thinks of him as a little higher than the anthropoid ape."[112] Similarly, William James wrote in his 1907 book *Pragmatism* that while "the notion of God . . . guarantees an ideal order that shall be permanently preserved," the future anticipated by modern science promises only death "without an echo; without a memory; without an influence on aught that may come after, to make it care for similar ideals." He adds that "this utter final wreck and tragedy is of the essence of scientific materialism."[113]

Hence, although the Second Law had little influence on the writings of the German pessimists, it played an important role in shaping the cosmic pessimism that arose elsewhere, such as in Britain. Again: what is the point of anything if total annihilation is guaranteed, even if the laws of nature have not scheduled this to occur for many millions of years? In the end, all will be lost in the entropic shipwreck of time. To be clear, this is different than Shelley's suggestion (rejected by Sidgwick) that our extinction would be bad because, in part, it would entail the loss of many valued things—the non-hedonic goods. Instead, the question is whether there is any *point* to creating these things *in the first place*, whether knowledge and beauty, even our own existence, *matters* in a universe that will ultimately erase whatever we draw.

One of the most eloquent early statements of this crisis of meaning came from the British conservative and Prime Minister Arthur Balfour (1848–1930), who is best known today for the "Balfour Declaration" that helped establish the state of Israel in Palestine, and who happened to be Sidgwick's brother-in-law (although Balfour was not himself a utilitarian). Balfour argued that the materialistic or naturalistic worldview as so impoverished, both morally and emotionally, that we should reject it as unacceptable. In an academic paper published the same year as Wells' "The Extinction of Man" and Flammarion's *Omega*,[114] he described the situation as follows (quoting him at length):

> Man, so far as natural science by itself is able to teach us, is no longer the final cause of the universe, the heaven-descended heir of all the ages. His very existence is an accident, his story a brief and discreditable episode in the life of one of the meanest of the planets. Of the combination of causes which first converted a dead organic compound into the living progenitors of humanity, science, indeed, as yet knows nothing. It is enough that from such beginnings famine, disease, and mutual slaughter, fit nurses of the future lords of creation, have gradually evolved, after infinite travail, a race with conscience enough to know that it is vile, and intelligence enough to know that it is insignificant. We survey the past and see that its history is of blood and tears, of helpless blundering, of wild revolt, of stupid acquiescence, of empty aspirations. We sound the future, and learn that after a period, long compared with the individual life, but short indeed compared with the divisions of time open to our investigation, the energies of our system will decay, the glory of the sun will be dimmed, and the earth, timeless and inert, will no longer tolerate the race which has for a moment disturbed its solitude. Man will go down into the pit, and all his thoughts will perish. The uneasy consciousness, which in this obscure corner has for a brief space broken the contented silence of the Universe, will be at rest. Matter will know itself no longer. "Imperishable monuments" and "immortal deeds," death itself, and love stronger than death, will be as though they had never been. Nor will anything that is be better or be worse for all that the labor, genius, devotion, and suffering of man have striven through countless generations to effect.

He continued, arguing that

> it is no reply to say that the substance of the moral law need suffer no change through any modification of our views of man's place in the Universe. This may be true, but it is irrelevant. We desire, and desire most passionately when we are most ourselves, to give our service to that which is universal, and to that which is abiding. Of what moment is it, then (from this point of view), to be assured of the fixity of the Moral Law when it and the sentient world, where alone it has any significance, are alike destined to vanish utterly away within periods trifling beside those with which the Geologist and the Astronomer lightly deal in the course of their habitual speculations?[115]

These passages are strikingly similar to lines from an essay published nine years later by Bertrand Russell, titled "A Free Man's Worship," which I quoted in

Chapter 3. The fact that our universe will one day slide into everlasting darkness evokes within the scientific person, Russell argued, a crushing sense of "unyielding despair" (a nice contrast to the Franklinian "Comfort" of the first existential mood), given

> that all the labours of the ages, all the devotion, all the inspiration, all the noonday brightness of human genius, are destined to extinction in the vast death of the solar system, and that the whole temple of Man's achievement must inevitably be buried beneath the debris of a universe in ruins.

He then asked: "How, in such an alien and inhuman world, can so powerless a creature as man preserve his aspirations untarnished?" Whereas Balfour argued that the solution is to abandon materialism, Russell contended that we can find some degree of worthwhileness in life, despite the inevitability of doom, by renouncing our desires and striving to create worlds of beauty through art and philosophy—a strategy for achieving "freedom" and "emancipation" from the "tyranny" of our predicament not unlike Schopenhauer's prescription for tranquility through asceticism, mystical experience, and aesthetic appreciation.[116] As we will see, the question of life's meaningfulness or value in the face of extinction was taken up by some philosophers during the second half of the twentieth century, although the focus at this time, during the Atomic Age, was near-term self-annihilation.

One Salvation, One Answer

Hence, in addition to the early statements about the normative implications of our extinction mentioned in previous sections, the prospect of this outcome, foregrounded by the new science of thermodynamics, also stimulated novel thoughts about how the ultimate fate of the cosmos could affect, and undermine, the importance or significance of our efforts both as individuals and a collective whole.

Before closing this chapter, let's examine one more example of a philosopher writing before the third existential mood who addressed the question of life's value and meaning, namely, the Norwegian philosopher, humorist, poet, and mountaineer Peter Wessel Zapffe. An atheist like Bentham, Mill, Sidgwick, Schopenhauer, Mainländer, and Russell, Zapffe agreed with Schopenhauer's thesis that life is suffering and non-existence is better than existence, although he characterized the root causes of our predicament differently. By a stroke of evolutionary bad luck, he argued, nature has produced in humanity an excess of consciousness. Whereas all animals "know angst, under the roll of thunder and the claw of the lion," the human being "feels angst for life itself—indeed, for his own being." In other words, our cognitive systems, our awareness of the evils built into the universe—suffering and death—have developed and expanded to the point where they have become extremely maladaptive, giving rise to feelings of "cosmic panic" that we must constantly fight off through defense mechanisms like "diversion" (distraction) and "isolation" (a refusal to admit to oneself or others the awfulness of being alive).[117] Zapffe writes:

> Man has lost his citizenship in the universe, he has eaten from the tree of knowledge and has been banished from paradise. He is powerful in his world, but he curses his power because he has bought it with his soul's harmony, his innocence, his comfort in life's embrace.[118]

Indeed, one source of such spiritual disharmony arises from the fact that we, as human beings with over-evolved minds, demand meaning, yet the modern scientific worldview has exposed the fundamental meaninglessness of life. Not only does "what we call nature [show] neither morality nor reason," but "its degeneration is inevitable, and nothing, not even man's most glorious achievements, can escape final annihilation."[119]

But this predicament is not evolutionarily unprecedented, that is, we are not the only creatures who have become "unfit for life by reason of an overdevelopment of a single faculty." This tragedy also befell, or so the story goes, the "Irish elk" (a misnomer because it was a giant deer rather than an elk), which is said to have grown antlers so large that it was no longer able to lift its head, and consequently the species died out. "When one is depressed and anxious," Zapffe explained, "the human mind is like such antlers, which in all their magnificent glory, crush their bearer slowly to the ground." In a poignant illustration of our resulting situation, he opened his 1933 article "The Last Messiah" with the following parable, quoted in full:

> One night in times long since vanished, man awoke and saw himself. He saw that he was naked under the cosmos, homeless in his own body. Everything opened up before his searching thoughts, wonder upon wonder, terror upon terror, all blossomed in his mind.
>
> Then woman awoke, too, and said that it was time to go out and kill something. And man took up his bow, fruit of the union between the soul and the hand, and went out under the stars. But when the animals came to their waterhole, where he out of habit waited for them, he no longer knew the spring of the tiger in his blood, but a great psalm to the brotherhood of suffering shared by all that lives.
>
> That day he came home with empty hands, and when they found him again by the rising of the new moon, he sat dead by the waterhole.

In this story, it was the man's capacity to grasp the enormity of suffering in the world, "through the gate of his empathy," and the harms he was about to inflict on a fellow member of the "brotherhood of suffering," that precipitated his demise, being unable to follow-through on nature's brutal imperative to kill and eat. Thus gripped by the cosmic panic of realizing life's brutality, we are confronted with two immediate options: to die like the protagonist above or to utilize the aforementioned defense mechanisms for the sake of "artificially paring down [our] consciousness," which is analogous to chopping off a part of the problematic excess of thought and feeling bequeathed to us by evolution—a temporary rather than permanent solution. The fact that most people across history have saved themselves via the latter option is why "the human race [was] not wiped out long ago in great, raging epidemics of insanity."

However, Zapffe proposed another option: for humanity to follow the Irish elk into oblivion by refusing to procreate. In other words, he advocated a pro-extinctionist view (non-existence would be better, and we ought to bring this about) coupled with an antinatalist means of achieving this end. "The Last Messiah" outlines a provocative defense of this position, prophesying a "last Messiah" who, "after many saviors have been nailed to trees and stoned to death in the marketplace, . . . will come forth [and] before all other men [will] strip his soul naked and give himself wholly over to our most profound questioning, even to the idea of annihilation." He will then declare:

> The life on many worlds is like a rushing river, but the life on this world is like a stagnant puddle and a backwater.
>
> The mark of annihilation is written on thy brow. How long will ye mill about on the edge? But there is one victory and one crown, and one salvation and one answer:
>
> Know thyselves; be unfruitful and let there be peace on Earth after thy passing.[120]

Although humanity may reject the imperative to "be unfruitful" by continuing to seek temporary relief through the use of defense mechanisms, the fact is that we are an evolutionary mistake: our oversized consciousness makes us, as a default, much too aware of the suffering and meaningless of existence. Hence, the only permanent solution is for our species to bow out of existence and let nature carry on as it was.

The First Wave

To conclude this chapter, the wave of Existential Ethics that spanned the first and second existential moods contained a diversity of incipient thoughts about the goodness/badness, rightness/wrongness of our extinction, which most ostensibly understood in the final sense. Montesquieu may have been the earliest Western philosopher to express in writing the view that human extinction itself would be a tragedy, while Shelley gestured at the idea that extinction would be bad because it would entail certain further losses: knowledge, science, poetry, philosophy, and so on. Meanwhile, we saw that philosophers like Sade and Kant (during his critical phase) referenced our extinction in service of making some unrelated point or argument, and the plurality of worlds cosmology that emerged in the seventeenth and eighteenth centuries led many theorists to suggest that the destruction of our world would be a matter of evaluative indifference, all things considered, a conclusion based on the principle of plenitude: this might be bad for us, but it wouldn't be bad cosmically speaking.

In discussing these examples, we introduced some new technical terms: the "default view" is the widely accepted idea that our extinction, if brought about by a catastrophe, would be bad *at least* for the obvious reason that all catastrophes are bad. The "equivalence thesis" refers to the reductionistic view that the badness/wrongness of extinction is entirely reducible to the badness/wrongness of Going Extinct. The "no-ordinary-catastrophe thesis" states that an extinction-causing catastrophe could introduce suffering that would not otherwise obtain in less extreme circumstances, such as intense feelings of loneliness, dread, and hopelessness induced by the expectation that "it is all over now."[121] And, finally, "further-loss views" identify some additional source of badness *above and beyond* whatever harms might be caused to people during the process or event of Going Extinct, a class of positions that rejects the equivalence thesis.

We then turned to various "pro-extinctionist views," which most of the philosophers who addressed our extinction in the late nineteenth century endorsed. For example, Mainländer imagined the human species dying out due to a failure of reproduction, while Hartmann advocated the complete annihilation of the universe (world-process) through some as-yet unspecified means. If life is nothing but aimless suffering—a perpetual cycle of needs (deprivation) and boredom (restlessness under the crushing weight of mere existence), as Schopenhauer argued—then why not put the human species out of its misery? Surely extinction would be best. However, the late nineteenth century also witnessed the development of a theory that would later become one of the most influential within Existential Ethics: utilitarianism, understood specifically in "total" and "impersonalist" terms. In presenting this theory, Sidgwick became the first of the utilitarians to articulate the ethical implications of our disappearance—in particular, the state or condition of Being Extinct—from this perspective. *Even if* bringing about our extinction were entirely *voluntary* and *painless*, he argued, it would still be extremely immoral because the outcome would be extremely bad, and the outcome would be extremely bad because it would preclude the realization of future value that could otherwise have existed.

Yet, while extinction brought about by, for example, universal celibacy is extremely improbable—as everyone would have agreed, including Mainländer and Zapffe—the discovery of the Second Law in the 1850s established as scientific fact that our presence within the cosmic theater of being is necessarily transitory. In the end, the dictatorship of entropy will quash all the armies of life, wherever they may be, rendering our once-hospitable universe eternally cold and lifeless—so says the fundamental laws of physics.[122] This new scientific eschatology thus led some philosophers to question whether the inevitability of our collective demise in any way diminishes the meaning, importance, significance, or comprehensibility of human existence. "[E]ven more purposeless, more void of meaning," Russell wrote, "is the world which Science presents for our belief."[123] If there is no external source of meaning, no afterlife, no grand plan for the cosmos, no God, and if the same dismal fate awaits us no matter how we occupy ourselves in the meantime, then what is the point of anything? All will crumble to ashes and dust. While Russell attempted to provide an answer for how human beings can find some degree of liberation from this dictatorial predicament, others, like Zapffe, seemed to view it as yet another reason for why it would be best if humanity were to cease existing.

Impossibility and Metaethics

As this shows, a primary focus of the first wave was the *evaluative* implications of our complete and permanent disappearance without leaving behind any successors: the final end to our story. Would this be good or bad, better or worse, and why? And how might the anticipation, or scientific knowledge, of our eventual extinction introduce additional harms or chip away at the value and meaning of our existence as individuals and a collective whole? Some also focused on the *deontic*, that is, on whether extinction is an outcome we should work to either bring about or prevent. Although only a handful of intellectuals weighed in on the topic, more endorsed our extinction (we *should* disappear) than claimed otherwise (we *shouldn't* disappear), which points toward a rather surprising start to the field of Existential Ethics: annihilation was favored overall by the relatively small number of writers and philosophers who broached the topic.

But this was before there were any widely recognized, scientifically credible anthropogenic kill mechanisms that could destroy humanity in the relative near term, and hence before there was any *urgent need* to examine the normative questions surrounding human extinction. If there is no "can," then why theorize about the "ought"? As noted in Chapter 3, there was growing anxiety during the first half of the twentieth century, especially after WWI, about a secular apocalypse brought about by advancements in science and the ever-growing arsenal of weapons of mass destruction. Recall, for example, that Frederick Soddy and Ernst Rutherford suggested in the early twentieth century that a "planetary chain reaction" could potentially destroy Earth by converting all of its elements into new elements like helium, a frightening possibility that was widely known by the 1930s, even among schoolchildren.[124] Decades later, writing in the dark shadows of the Great War, which saw millions slaughtered by the new machines of mass death, Winston Churchill warned that technology could obliterate civilization in his 1924 article "Shall We All Commit Suicide?," while Sigmund Freud concluded his 1930 *Civilization and Its Discontents* with the (hyperbolic) declaration that "men have brought their powers of subduing the forces of nature to such a pitch that by using them they could now very easily exterminate one another to the last man."[125]

However, the rising prominence of the idea of *human self-extinction* during this early twentieth-century period (see Appendix 1) was not accompanied by any corresponding increase in attention to the topic from philosophers. To the contrary, it was almost entirely neglected by the philosophical community, Zapffe being the most notable exception (although his assessment of the goodness/badness of our extinction was based on considerations of the fundamental features of the human condition—for example, suffering and meaninglessness in a world devoid of purpose and destined to perish—rather than of our expanding destructive capabilities). Within the "Analytic Philosophy" tradition, which emerged around the turn of the century, this may have been due not only to the aforementioned fact that anthropogenic annihilation was not obviously possible (despite ominous indicators that this could soon change), but to the fact that, following the publication of Moore's *Principia Ethica* in 1903, moral philosophers became overwhelmingly preoccupied with metaethical issues (the semantic, metaphysical, and epistemological aspects of morality) rather than normative ethics (what is right and wrong; what we ought to do). Indeed, it was not until the 1960s that normative ethics began to reemerge as a subject of active research among Analytic philosophers. This decade—the beginning of the Age of Atheism, as it happens—also marked the first time that a large number of philosophers approached normative ethics from a specifically secular perspective. As Derek Parfit observed in 1984, "How many people have made Non-Religious Ethics their life's work? Before the recent past, very few," to which he added:

> After Sidgwick, there were several Atheists who were professional moral philosophers. But most of these did not do Ethics. They did Meta-Ethics. They did not ask which outcomes would be good or bad, or which acts would be right or wrong. They asked, and wrote about, only the meaning of moral language, and the question of objectivity. Non-Religious Ethics has been systematically studied, by many people, only since about 1960.[126]

Although we will see that a few atheistic philosophers considered the normative dimensions of human extinction prior to 1960, the collision of these two developments—the feasibility of self-annihilation and the emergence of secular normative ethics—was integral to the second wave of theoretical work in Existential Ethics, which focused almost entirely on the *ethical aspects of human self-annihilation*. It is to this that we now turn.

Notes

1. Personal communication. Please note that all personal communications quoted in this text have been used with permission.
2. Hume 2021; Diderot et al. 1979.
3. Medwin 1824.
4. Quoted in Crocker 2015.
5. NMM 1816.
6. Kant 2022.
7. Parfit 2011.
8. Kant 2022.
9. Kant 1785.
10. Korsgaard 1983. The language here may be imprecise. Perhaps Kant does mean to say that human beings—or, more generally, rational beings—have a kind of intrinsic value (see Bradley 2006 for a discussion of "Kantian" versus "Moorean" intrinsic value), or perhaps his assertions about a good will concern something quite different, e.g., the "question of how we ought to behave toward such creatures [i.e., rational beings]" (Rønnow-Rasmussen and Zimmerman 2006). I will return to this issue later on.
11. Johnson and Cureton 2004.
12. Kant 1785.
13. Kant 1987.
14. Kant 1987.
15. Moynihan 2020. I am not sure where Moynihan gets this date. Sade did not publish any books in 1796.
16. Recall that Buffon was one of the few natural philosophers prior to Cuvier who accepted the possibility of species extinctions.
17. Sade 2021.
18. Interestingly, a similar statement is found in d'Holbach's *The System of Nature* published in 1770. He wrote:

    > Of those who ask, why does not nature produce new beings, we inquire in turn how they know that she does not do so. What authorizes them to believe this sterility in nature? Do they know whether, in the combinations she is at every instant forming, nature is not occupied in producing new beings without the cognizance of these observers? Who told them whether nature be not now assembling in her vast laboratory the elements fitted to give rise to wholly new generations, that will have nothing in common with the species at present existing. *What absurdity, then, would there be in supposing that man, the horse, the fish, the bird, will be no more?* Are these animals so indispensable to Nature that without them she cannot continue her eternal course? (quoted in Lovejoy 1936, italics added).

19. Sade 1797; Moynihan 2020.
20. Sade 1968.
21. Sade 1968.
22. Moynihan 2020.
23. According to Thomas Campbell, the idea for Byron's poem actually originated with *him* fifteen years prior. Campbell claimed that it was his idea of "a being witnessing the extinction of his species and of the creation, and of his looking, under the fading eye of nature, at desolate cities, ships floating with the dead" (quoted in Paley 1989).
24. Although de Grainville's 1805 novel is sometimes called the first to outline a *secular apocalypse*, at least in terms of its etiology: infertility of some unknown natural origin.

25. Horn 2014.
26. There are many other examples of this second category, although an exhaustive survey goes beyond the scope of this book. For example, in his 1739/40 discussion of the relation between reason and passion, Hume declared that "'Tis not contrary to reason to prefer the destruction of the whole world to the scratching of my finger" (quoted in Cohon 2018). This is an intriguing reference to annihilation, although Hume's point was that our passions (as well as volitions and actions) lack the sort of content that reason could assess, and hence there is no battle between one and the other—contra many philosophers, both ancient and modern, who have argued not only that reason and passion are engaged in a perennial struggle but that we should strive to subjugate the latter to the former (see Cohon 2018, section 3, for useful explication).
27. Godwin 1820; see Chapter 2.
28. Note that not all instances of extinction must involve a catastrophe; we could, for example, universally decide not to have children.
29. Tonn and Tonn 2009.
30. Shelley 2018.
31. Anonymous 1826, italics added. In a plot twist that modern readers would find banal, the story ends with the main character awakening from a dream. He then observes, in his words, "my man John, with my shaving-jug in the one hand, and my well-cleaned boots in the other—his mouth open and his eyes rolling hideously at thus witnessing the frolics of his staid and quiet master" (Anonymous 1826).
32. See, for example, Benatar 2006; Scheffler 2018.
33. Montesquieu 2008.
34. See Oake 1953, 554.
35. The term "suitable successors" is important, as the first word points to normative extinction, while the second points to final extinction. We will see that many of the positions proposed during the second wave of Existential Ethics pertain to both final and normative extinction, which often come as a bundle.
36. Shelley 2018, see chapter 1 of Volume III. Note that italics have been added.
37. Thanks to Morton Paley for affirming that my interpretation of Shelley is accurate.
38. Lovejoy 1936, 244.
39. Kant 1755.
40. Lovejoy 1936.
41. Lovejoy 1936, 108. For a more comprehensive list, see Lovejoy 1936, 108.
42. Quoted in McIntyre 1903.
43. Glanvill 1978.
44. Kant later drifted away from these ideas.
45. Kant 1755.
46. Dick 1982, 1; Crowe and Dowd 2013, 3, 49. Hence, while Cuvier's work helped to demolish the original and temporalized versions of the principle of plenitude, it did not affect the spatialized version, which continued to exert a significant influence on Western thinking about the universe for more than a century. This is possible because the former versions are logically independent from the spatialized version: one can accept that every kind of thing that could exist either does or will exist without accepting that the universe is infinite and infinitely populous, and one can accept that the universe is infinite and infinitely populous without accepting that there are no gaps or vacancies in nature. Hence, these distinctions nuance the claims made in Chapter 2 about the collapse of the Great Chain of Being in the early 1800s.
47. Quoted in Crowe 2012.

48. Wright 1750.
49. 25th Anniversary Edition 2018.
50. Kant 1755.
51. Wright 1750. This is a bit more ambiguous than Ferguson's and Kant's statements, since (a) the phrase "such as ours" could mean either "*including* ours" or "*similar to* ours," and (b) the demonstrative pronoun "there" toward the end of the passage suggests that the "Doom-Days" referenced are something that happens to other worlds rather than our own. Still, it could also be plausibly read as saying that *our world* could indeed disappear, and that this would be nothing special in the course of things, and hence matter little to God.
52. Alexander Pope made a similar point in his *An Essay on Man*, which I quoted in Chapter 2 because of its endorsement of the Great Chain:

    Who sees with equal eye, as God of all,
    A hero perish, or a sparrow fall,
    Atoms or systems into ruin hurl'd,
    And now a bubble burst, and now a world (Pope 1733–1734).

53. Kant 1755.
54. See Paley 1989, 2.
55. For an interesting recent discussion of the historical origins of the Western philosophical tradition, see Cantor 2022.
56. Laërtius 1925; quoted in Matson 1998.
57. See Marin and Beluffi 2018 and Yampolskiy 2018.
58. Note that "functional extinction" is a term used in this way by biologists.
59. Beiser 2016.
60. The term "implacable atheist" comes from Nietzsche's *The Gay Science*; quoted in Beiser 2016.
61. Schopenhauer 1851. Note that there are two distinct claims here. The first is eudaimonic, as it concerns a comparison between the pleasure experienced by the predator and the pains experienced by the prey. The second is moral, as it states that (on Schopenhauer's account) no amount of pleasure can ever compensate for any amount of pain. In other words, even if there were far more pleasure than pain in the world, the world would still be better off not existing because pleasures cannot pay back the debts accrued by suffering. See just below in the body text for comments about how pleasure has no *positive* value.
62. Schopenhauer 2012.
63. Beiser 2016, 51.
64. Schopenhauer 1851.
65. Schopenhauer 1818.
66. Beiser 2016; cf. Migotti 2020, 294.
67. See Landau 1997 for a similar account.
68. Tynan 1959.
69. See, for example, Tappolet 2014.
70. See Beiser 2016, 202.
71. See Beiser 2016, 222.
72. Hence, Ken Coates' claim that "it is Zapffe [discussed in the following chapter] who must be credited with being the first rejectionist to come up with the idea of antinatalism as the way out of existence for humans" is not correct (2014).
73. Mainländer 1876/1886.
74. He also had an appreciably impact on the thinking of Sigmund Freud, given his exploration of the unconscious.

75. Saltus 1885.
76. Quoted in Beiser 2016.
77. Coates 2014; Wicks 2021.
78. Hartmann 1869.
79. Hartmann 1869.
80. Or he might have been thinking about Buffon's account of Earth as a slowly cooling ember of the sun.
81. Interestingly, G. E. Moore wrote the following about pessimism in his 1903 book *Principia Ethica*:

> in order to prove that murder, if it were so universally adopted as to cause the speedy extermination of the race, would not be good as a means, we should have to disprove the main contention of pessimism—namely, that the existence of human life is on the whole an evil. And the view of pessimism, however strongly we may be convinced of its truth or falsehood, is one which never has been either proved or refuted conclusively (1903).

82. Beiser 2016, 13.
83. Although see Torres 2020 for why universal antinatalism need not entail human extinction. In brief, if antinatalism is coupled with effective life-extinction technologies, humanity could in theory both stop procreating and continue to exist until the universe become uninhabitable. More on this below.
84. Schopenhauer 1851. Sometimes this is translated "On the Sufferings of the World."
85. Schopenhauer 1851.
86. Schopenhauer 1818.
87. Moynihan also gets this wrong in his book *X-Risk*, which states that "in Schopenhauer's masterwork *The World as Will and Representation*, the first volume of which was published in 1819, he recommended that humans should abstain from reproducing in order to abolish self-conscious suffering," and hence "Arthur Schopenhauer was, after Sade, perhaps history's second omnicidal agent" (2020). Note that *The World as Will and Representation* was first published in 1818, not 1819. Moynihan also claims that Schopenhauer was an "absolute idealist," which is very false indeed.
88. Schopenhauer 1818.
89. Schopenhauer writes the following about suicide:

> The suicide wills life, and is dissatisfied merely with the conditions on which it has come to him. Therefore, he gives up by no means the will-to-live, but merely life, since he destroys the individual phenomenon. . . . [S]uicide . . . is a quite futile and foolish act, for the thing-in-itself [i.e., the will] remains unaffected by it. . . . [I]t is also the masterpiece of Maya as the most blatant expression of the contradiction of the will-to-live with itself (1818/19).

90. Schopenhauer 1851.
91. Driver 2014.
92. Nor was there much discussion prior to Mill's celebrated 1863 book *Utilitarianism* of natural selection causing the human species to evolve into a new—for example, "degenerate"—species.
93. Kant 1785.
94. See Korsgaard 1985.
95. Kant 1999.
96. Bentham 1789. Note that although Bentham was writing at the same time as Kant, his utilitarian theory did not gain traction in Britain until the second edition of his *An Introduction to the Principles of Morals and Legislation* was published in 1823 (see Singer 2002, 67).

97. Bentham 1789.
98. Sidgwick 1874. Sidgwick adds:

    > How far we are to consider the interests of posterity when they seem to conflict with those of existing human beings? It seems, however, clear that the time at which a man exists cannot affect the value of his happiness from a universal point of view; and that the interests of posterity must concern a Utilitarian as much as those of his contemporaries, except in so far as the effect of his actions on posterity-and even the existence of human beings to be affected-must necessarily be more uncertain (1874).

99. Singer 1972.
100. Sidgwick 1874/1962, 382.
101. Another distinction that I will not elaborate on here is between "act" and "rule" consequentialism. In brief, the first focuses on individual acts, claiming that an act is right or wrong depending on whether it maximizes the good. The second, in contrast, claims that an act is right or wrong depending on whether it conforms to a rule, where such rules are selected by virtue of their goodness-maximizing consequences.
102. Schultz 2002.
103. Sidgwick 1874.
104. Narveson 1973.
105. See Mulgan 2020, 50.
106. Sidgwick 1874.
107. Sidgwick 1874, italics added; Nakano-Okuno 2011.
108. Shelley 1826; see Finneron-Burns 2017, 333.
109. Bowler 2003.
110. Chamberlin and Gilman 1985; see Chapter 3.
111. See Landau 1997.
112. Perry 1922.
113. James 2015.
114. Flammarion also touched upon this theme in *Omega*, writing that, because of the Second Law,

    > all this progress, all this knowledge, all this happiness and glory, must one day be swallowed up in oblivion, and the voice of history itself be forever silenced. Life had a beginning: it must have an end. The sun of human hopes had risen, had ascended victoriously to its meridian, it was now to set and to disappear in endless night. *To what end then all this glory, all this struggling, all these conquests, all these vanities, if light and life must come to an end?* Martyrs and apostles, in every cause, have poured out blood upon the earth, defined also in its turn to perish. . . . Science had disappeared with scientists, art with artists, and the survivors lived only upon the past. The heart knew no more hope, the spirit no ambition. The light was in the past; the future was an eternal night. All was over.

    A few pages later in his Last Man-esque novel, Flammarion describes "the last heir of the human race," namely, Omegar, feeling "the overwhelming sentiment of the vanity of things," given the impending extinction of our species (1894).
115. Balfour 1894.
116. Russell 1903.
117. Zapffe mentions two other such mechanisms, attachment and sublimation, which I will not here discuss.
118. Zapffe 1933.

119. Quoted in Reed and Rothenberg 1993.
120. Zapffe 1933.
121. Shelley 2018.
122. Or perhaps random fluctuations of matter and energy in an infinite universe will occasionally result in non-thermodynamic-equilibrium "Boltzmann universes" or "Boltzmann brains," which may in some sense, then, contain life.
123. Russell 1903.
124. Weart 1988, 23.
125. Churchill 1924; Freud 2018.
126. Parfit 1984.

9 Ethical Innovations of the Postwar Era

Themes of the Second Wave

The second wave coincided with the third and fourth existential moods, extending from the 1950s to the late 1990s. This period saw the articulation of an extraordinarily wide range of innovative new ideas about the deontic and evaluative aspects of our extinction, especially *self-extinction* caused by nuclear conflict or environmental degradation, within both the Analytic and Continental traditions of philosophy (a problematic distinction that I use only for expositional convenience).[1] Major themes include the idea that nuclear weapons have fundamentally altered the human condition, and that this historically unprecedented situation has rendered traditional ethical theories outdated or obsolete. These theories, the argument went, simply weren't designed to address the possibility of "some human beings annihilating all human beings," paraphrasing John Somerville's definition of "omnicide," and hence philosophers are tasked with devising new theoretical frameworks, or reworking old frameworks, to accommodate the unique challenges posed by our newly acquired *powers of action*.[2] This led some to delineate novel principles, commandments, or imperatives custom-designed for this purpose, although few of these proposals generated much discussion at the time.

Others proposed new arguments for why our extinction would be bad or wrong, with some claiming that it could constitute a tragedy of quite literally *cosmic proportions*. These were variously based on considerations of past progress, the likelihood of further progress in the future, our potential uniqueness in the universe, the "unfinished business" of humanity, the possibility of "vicarious immortality," the nature of loving or cherishing things, the meaningfulness or value of our lives, and the fact that the story of humankind and civilization may be only just beginning. Central to some arguments was what I will call "deep-future" and "potentiality" thinking, where the first refers to the scientific recognition that humanity could survive for millions or billions of years to come, and the second to the idea that the future could be much better than the present. Since extinction would foreclose this potentially very long and prosperous future, the badness

DOI: 10.4324/9781003246251-11

of this outcome far exceeds whatever suffering and harm the process or event of Going Extinct might inflict on those living at the time. It thus matters greatly that we do not die out. As one philosopher expressed the idea, the difference between 99 and 100 percent of humanity perishing is not a mere one percentage point: total annihilation would entail a permanent end to the entire human story, which makes 100 percent of humanity dying *much worse* than "only" 99 percent.

Some philosophers supported this further-loss view by embracing Henry Sidgwick's total-impersonalist utilitarianism, although other utilitarians favored a person-affecting version that led to a quite different conclusion, namely, that the state or condition of Being Extinct would *not be bad*, as there would be no one around to bemoan the non-existence of humanity. If no people are affected, then this cannot be bad. It follows that there is no *moral obligation* to ensure the perpetuation of our species: a universal refusal to keep humanity going, for example, wouldn't be a morally criminal act, contra Sidgwick. Some even held that the person-affecting restriction *implies* that we should all stop having children, which straightforwardly entails, it seems, that humanity should go out of existence. However, other utilitarians of yet another sort contended that Being Extinct would be positively *good*, as it would mean the end of all human misery, pain, and suffering. If what matters, they claimed, is the reduction of suffering rather than the promotion of happiness, then we should *want* humanity to no longer exist. Many radical environmentalists held a similar pro-extinction view, although their reasons were based on the fact that *Homo sapiens*—or, as some liked to say, *Homo shiticus*—has been a hugely destructive force in the biosphere. While most endorsed an antinatalist means of Going Extinct, others advocated for pro-mortalism (we should kill ourselves) and even omnicide (someone should kill everyone).[3]

In the majority of cases, the focus of these arguments and positions concerned final extinction, though others seemed to have terminal extinction in mind. Several philosophers also discussed normative extinction, as when one argued that given a choice between life under totalitarian rule, exemplified in the twentieth century by the regimes of Nazism and Stalinism, and total nuclear annihilation, we should prefer the *latter*, an idea famously encapsulated by the slogan "Better dead than Red." Another philosopher, whose work introduced an early version of the Precautionary Principle, worried that advanced technologies could enable radical modifications of the human organism that compromise our fundamental human dignity, which would be just as catastrophic as the human story ending forever because our species dies out.[4]

Even more than with previous chapters, the organizing principles underlying the structure of this one will be both thematic and chronological. There are simply too many ideas, arguments, and normative viewpoints to present this segment of History #2 in a linear manner. The present chapter will also be the longest in the book. Let's begin with a brief sketch of the historical context in which this second wave unfolded.

Virtually No Philosophers

Recall from Chapter 4 that news of the atomic bombings of Hiroshima and Nagasaki provoked an immediate sense, felt by many around the world, that something truly terrible had occurred. The world-situation was fundamentally changed; a new epoch in human history had commenced beneath the shadows of two radioactive mushroom clouds. Yet there was little explicit mention of "human extinction" until after the 1954 Castle Bravo debacle, which ignited a firestorm of warnings from prominent figures that self-annihilation due to global thermonuclear fallout was now feasible, and perhaps highly probable in the near future, unless humanity forms a single world government or develops a "world" or "species" consciousness that extends across both sides of the Iron Curtain. As the journalist Adam Lapin described our situation in 1955, the choice confronting humanity was between "coexistence or no existence."[5]

Given the historical shockwaves produced by Castle Bravo, one might expect that scholars—especially *moral philosophers*—would have dropped their current projects and began studying the sociological, psychological, political, ethical, and so on dimensions of our new existential predicament under the "nuclear sword of Damocles," quoting John F. Kennedy once again.[6] Yet, restricting our discussion for the moment to the years between 1954 and ~1980, this was largely not the case. Overall, most scholars gave the issue very little sustained attention. One exception occurred between 1954 and 1963, where the latter date corresponds to the signing of the Partial Test Ban Treaty by the Soviet Union, United States, and United Kingdom, which turned down the rising thermostat of the Cold War. As the historian Paul Boyer observes, there were a "considerable number" of "scientists . . . theologians, novelists, poets, psychiatrists, and psychologists," as well as international relations theorists, during this period who did address the nuclear menace.[7] Yet Boyer adds that "even at the height of the test-ban movement" in the late 1950s and early 1960s, "the involvement of American intellectuals with the nuclear threat was limited."[8] The same could be said about intellectuals elsewhere in the Western world. Paradoxically, as Robert Jay Lifton observed in 1982, it seems that "the more significant an event, the less likely it is to be studied," an idea later dubbed "Lifton's law."[9]

However, even despite this momentary increase in scholarly attention, *virtually no philosophers* wrote anything about the risk of nuclear annihilation, and those who did tended to remain silent about the central normative questions of Existential Ethics—perhaps because, at the time, these questions hadn't even been properly formulated yet. (Indeed, they remain somewhat confused up to the present, which is why Chapter 7 was necessary.) The philosopher Paul Arthur Schilpp (1897–1993) provides an example: he argued during a 1948 conference presentation, published the following year in *Philosophy*, that philosophers have a crucial role to play in averting nuclear catastrophe. Yet, with one intriguing exception discussed below, Schilpp's focus wasn't Existential Ethics but how philosophers might lead efforts to establish peace by showing people how to think rationally, establishing that all people around the world belong to a single human family, and accepting that every person has a common, fundamental dignity. These have become the philosophical community's "three essential duties to fulfil . . . in this tragic hour," he wrote, as philosophers after Hiroshima

> cannot well afford to turn aside from what is perhaps the imperative task of the hour for reflective thinkers: the task, namely, of bringing to bear upon the existing human plight the best thinking of which the human mind is capable; nor to resign themselves to the notion that such may go on within the very narrow and limited confines of each philosopher's peculiar "ivory tower"; but rather that—in such an hour as to-day—it becomes the unquestionable moral obligation of the philosopher to attempt to make his impact not merely upon society at large (and still less in the minute), but even upon the heads of state and all those who hold, within the hollow of their hands and their selfish nationalistic appetites, the fate not only of nations but perhaps of all mankind.[10]

Deep Future, Human Potential

One of the first philosophers of the Atomic Age who did address Existential Ethics was Bertrand Russell, who, in the wake of the Castle Bravo test, suddenly found a new use for his term "universal death," which he introduced two and a half decades earlier in the eschatological context of thermodynamics.[11] In a number of publications from the 1950s onward, Russell gestured at several further-loss views according to which our extinction would be bad because it would (a) result in all the progress that humanity has made over the past 6,000 years or so *going to waste* and (b) foreclose what could be a long and prosperous future for our descendants. In making the argument of (b), Russell provided an early example of *deep-future thinking* in Existential Ethics, where this term—as alluded to earlier—denotes the realization that if humanity does not destroy itself, it could exist for "many millions of years to come," as Russell put it.[12] One can understand this as the futurological counterpart to what the journalist John McPhee called "deep time," which refers to the discovery that Earth has existed for an extremely long time—*much longer* than James Ussher's famous calculation, in the mid-seventeenth century, that Earth's history began in 4004 BCE.[13] As Stephen Jay Gould writes, putting our own species' lifetime into grand-historical perspective, deep time is "the notion of an almost incomprehensible immensity, with human habitation restricted to a millimicrosecond at the very end!"[14] Deep time was of course central to James Hutton's uniformitarianism, and it became widely accepted among geologists following Charles Lyell's 1830 book *Principles of Geology*.[15]

Whereas deep time was a product of *geology*, deep-future thinking emerged from *astronomy* and *cosmology*, especially following the discovery of the Second Law, which cast the eyes of physicists and science fiction writers toward the distant temporal horizon. For example, based on the Second Law, Lord Kelvin estimated that Earth will remain habitable for "many million years longer," while Camille Flammarion established that our sun would shine for at least another "twenty million years."[16] H. G. Wells imagined the protagonist of *The Time Machine* travelling 30 million years into the future, and Sir James Jeans conjectured in 1929 that we have "something of the order of a million million years to come," which implies that "as inhabitants of the earth, we are living at the very beginning of time."[17] Deep-future thinking was taken even further by Olaf Stapledon, whose 1930 *Last and First Men* envisioned the future of humanity spanning the next 2 billion years, though his novel *The Star Maker*, published seven years later, outlined the evolution of life in the cosmos over a mind-boggling 500 billion years.[18] As we will see in later chapters, deep-future thinking has become *integral* to the most influential position within contemporary Existential Ethics, and it played a role in catalyzing the futurological pivot discussed in Chapter 6. However, it was Russell in the mid-1950s who explicitly linked the possibility of an exceptionally long future with normative questions about why bringing about our extinction would be bad or wrong.

Yet there is another, orthogonal issue that Russell foregrounded, and which is also relevant to Existential Ethics: *potentiality thinking*. Whereas deep-future thinking is quantitative, potentiality thinking is qualitative, as it concerns the question of *how much better* life could become rather than *how much longer* our lineage could persist. Advancements in science, new technological developments, and the bending of the moral arc toward justice could significantly improve the human condition, this line of thinking goes, thus making life more wonderful than it has ever been—perhaps better than we can even imagine. At the heart of potentiality thinking is a future-oriented conception of *progress*, which took shape during the Enlightenment in the work of Anne-Robert-Jacques Turgot (1727–1781) and, especially, Condorcet. This "progressivism" influenced many intellectuals throughout the nineteenth century, although it was challenged by worries arising in the latter 1800s over the possibility of evolutionary degeneration and so-called "racial senility."[19] In some cases, the same individuals embraced both anxieties about decline and the prospect of great things to come, as exemplified by the oeuvre of Wells. His 1902 essay "The Discovery of the Future," for example, declares that "it is possible to believe that all that the human mind has ever accomplished is but the dream before the awakening," and hence that "we are creatures of the twilight."[20] While the idea of progress lost much of its appeal after the horrors of WWII, it has been revived in recent decades by the modern transhumanists and advocates of what is sometimes called "New Optimism," such as Steven Pinker.[21]

The Triumphs of the Future

Russell thus utilized deep-future and potentiality thinking in contending that our extinction would constitute a tragedy of enormous proportions. This view is found in two works of 1954: first, in the closing chapter of his book Human Society in Ethics and Politics, poignantly titled "Prologue or Epilogue?," and second, in his "Man's Peril" radio address for the BBC.[22] In the former, he began with a sweeping survey of all the progress humanity has made over the past 6,000 years, during which written language was invented, nations grew into empires, and cumulative cultural traditions gained momentum. After this retrospective picture of where we came from, he pivots toward a glance at what lies ahead, urging his readers to "view the world as astronomers view it . . . thinking of the future as extending through many more ages than even those contemplated in geology."[23] There is no reason to believe that Earth will not remain "habitable for another million million years, and if man can survive, in spite of the dangers produced by his own frenzies, there is no reason why he should not continue the career of triumph upon which he has so recently embarked. . . . [T]he drama is only just begun."

What might this triumph consist of? Russell singled out knowledge, but added that humanity at its best deserves to be admired for the beauty that it has created, its "strange visions that seemed like the first glimpse of a land of wonder," and its capacity of love and "sympathy for the whole human race, of vast hopes for mankind as a whole." Over the coming thousands of years, given "the speed with which [Man] is acquiring knowledge there is every reason to think that, if he continues on his present course, what he will know a thousand years from now will be equally beyond what we can imagine" as what our ancestors 1,000 years ago could imagine about our present world. The promise of steady, or accelerating, progress over the course of many centuries to come led Russell to affirm the potentiality of this "shining vision: a world where none are hungry, where few are ill, where work is pleasant and not excessive, where kindly feeling is common, and where minds released from fear create delight for eye and ear and heart." This is what we might expect if only "the world will emerge from its present troubles, and . . . will some day learn to give the direction of its affairs, not to cruel mountebanks, but to men possessed of wisdom and courage." It is this future that our extinction threatens to erase even before we have begun to draw it. "Is all this hope to count for nothing?," he asked. "The future of man is at stake," and these are the stakes; but "if enough men become aware of this his future is assured."[24]

Russell made a number of similar points in "Man's Peril," although he also gestured at an idea touched upon by later theorists within this second wave, and which has more recently become one of the canonical arguments for why our extinction would be bad, namely, our *cosmic significance*.[25] In the final paragraph of the address, Russell pointed toward this idea—that we may be a unique part of the universe and hence uniquely precious—and reiterated his views about our future potential if progress continues. Quoting him at length:

> As geological time is reckoned, Man has so far existed only for a very short period—a million years at the most. What he has achieved, especially

> during the last 6,000 years, is something utterly new in the history of the Cosmos, so far at least as we are acquainted with it. For countless ages the sun rose and set, the moon waxed and waned, the stars shone in the night, but it was only with the coming of Man that these things were understood. In the great world of astronomy and in the little world of the atom, Man has unveiled secrets which might have been thought undiscoverable. In art and literature and religion, some men have shown a sublimity of feeling which makes the species worth preserving. Is all this to end in trivial horror because so few are able to think of Man rather than of this or that group of men? . . . I would have men forget their quarrels for a moment and reflect that, if they will allow themselves to survive, there is every reason to expect the triumphs of the future to exceed immeasurably the triumphs of the past. There lies before us, if we choose, continual progress in happiness, knowledge, and wisdom. Shall we, instead, choose death, because we cannot forget our quarrels?

As with Schilpp, Russell emphasized the importance of understanding humanity as a single, unified entity: "I want you, if you can," he implored, "to set aside [political] feelings for the moment and consider yourself only as a member of a biological species which has had a remarkable history and whose disappearance none of us can desire."[26] He also once again underlined not just the progress that humanity has so far made, but the very real—in his view—possibility of future leaps toward a world that, if humanity were to extinguish itself, would be a great shame to lose. Furthermore, if this were to happen and humanity stopped existing, the universe would be deprived of something that may be extremely valuable: the only thing enveloped within it that possesses the ability to uncover its arcana and be awestruck by its beauty. (Recall that at this point, in the mid-1950s, the plurality of worlds model had fallen out of favor with most intellectuals, and hence many would have suspected that we might be alone in the cosmos.)

Incidentally, Schilpp hinted at the idea of cosmic significance as well, writing that while our "growing conception and understanding of the unimaginable vastness of the universe" may lead one "to minimize the meaning and significance of man," this

> is by no means the only factual view. There are also other established principles which make possible another outlook. In fact, this little speck of protoplasm on this third-rate planet [of a tenth-rate solar system drifting aimlessly in an endless cosmic ocean], when viewed from a different vantage-point, appears all the more significant. For the tinier he is in material size when compared with the universe, the more miraculous he must appear to himself when he contemplates his ability to think of, measure, and comprehend the immensity of that universe, not to speak of his practically limitless capacities for invention and creation in innumerable areas.[27]

Though neither Russell nor Schilpp elaborated this idea, they seemed to suggest that it constitutes an additional reason for the badness or wrongness of our extinction. Let's call it the "argument from cosmic significance." After all, people commonly attribute special value to objects because of their uniqueness or rarity. The Antikythera mechanism, for example, a highly complex analogue computer constructed by the ancient Greeks between the third and first centuries BCE, may be considered valuable "for itself" or "for its own sake" in part because it is a one-of-a-kind artifact—quite possibly the earliest analogue computer ever built. If this artifact were destroyed, the world would be in some sense *impoverished*. Or, flipping this around, "the world is richer 'as such' for [its] existence," to quote the philosopher Shelly Kagan.[28] The same might be said of humanity: we are, so to speak, an Antikythera mechanism in our own right, assuming there are no other rational, creative, moral creatures like us. We are one of a kind. Our existence thus enriches the cosmos "as such," and this gives us extra reason to safeguard our survival, or so the argument—Russell and Schilpp might concur—goes. We will return to this idea shortly.

The Panic-Maker

Another philosopher who addressed Existential Ethics in the 1950s was Günther Anders (1902–1992, last name originally "Stern"). Best placed within the Continental rather than Analytic tradition, Anders was a poet, journalist, and philosopher who, following the Castle Bravo incident, dedicated his life to warning the public about what he called "annihilism" and "globocide." A self-described "panic-maker" and "eye-opener," Anders achieved notoriety within Germany during his lifetime, despite only recently being discovered by the Anglophone world, and indeed many of his books and articles have yet to be translated into English.[29]

While I have largely eschewed discussing biographical details in this book so far, some notes of this sort may be warranted here, given Anders' quite extraordinary life and connections to a large number of important figures of the twentieth century, especially within the Continental tradition. To begin with, Anders' father was William Stern, a psychologist who coined the term "IQ" for "intelligence quotient" and invented the IQ test, and his second cousin was Walter Benjamin (1892–1940), a member of the Frankfurt School.[30] Anders received his PhD under the aegis of Edmund Husserl (1859–1938), and after meeting in a discussion group that Husserl hosted, Anders married the philosopher Hannah Arendt (1906–1975), one of the most influential of the century.[31] Both Anders and Arendt were mentored by Husserl's student Martin Heidegger (1889–1976), whom Arendt had an affair with prior to marrying Anders. In 1933, Heidegger joined the Nazi party (something for which he never apologized), which was the same year that Anders fled Germany to live in exile first in Paris, then New York, and then California. While in California, he attempted to "make it big" in Hollywood, at one point writing a script for a movie that he hoped would star Charlie Chaplin. But his efforts failed, and Anders consequently ended up "working as a cleaner for an unnamed costume company in the supply chain of the Hollywood film industry," a miserable period of his life that partly inspired his subsequent critiques of technology, mechanization, and media.[32]

Only after returning to Europe in 1950—specifically, to Vienna with his second wife—did Anders seriously focus on more theoretical issues, albeit outside of the academy and with the express purpose of appealing to a general audience. His first major publication was the 1956 book *The Antiquatedness of Humanity* (also translated as *The Obsolescence of Human Beings*), which was followed by a second volume in 1980.[33] Published when Anders was 54 years old, this book was the beginning of an entirely new career—a second life, so to speak—as one of the leading intellectuals of the anti-nuclear movement in Germany. As he once quipped, biographies like his should be characterized in terms of "Vitae, not vita," that is, plural rather than singular.[34] Across both of these lives, spanning two continents, Anders was not only friends with those mentioned above (although he later became highly critical of Heidegger) but knew Theodor Adorno and Max Horkheimer, lived at Herbert Marcuse's house, worked with Paul Tillich and Max Scheler, and became acquainted with Russell, who penned the preface to his 1961 book *Burning Conscience* and organized the International War Crime Tribunal in 1966/67 that included Anders as a member alongside Jean-Paul Sartre and Simone de Beauvoir.[35]

According to Jason Dawsey, Anders was, despite his relative obscurity within the Anglophone world, "a serious political thinker and theorist of the Atomic Age, in fact our most salient theorist of omnicide."[36] As this suggests, Anders' theoretical work focused on not only the possibility and implications of some people annihilating all people (again, paraphrasing Somerville), but how the invention of nuclear weapons had fundamentally and irreversibly altered the human condition. Writing in 1982, he described the aim of his book *The Antiquatedness of Human Beings* (henceforth *Antiquatedness*) as being to "find or invent a somewhat adequate vocabulary and a way of speech worthy of the enormity" of the nuclear menace.[37] Since insights about the human condition can informed thoughts about the nature of omnicide, let's begin with the former.

In a 1962 article titled "Theses for the Atomic Age," which was based on a seminar hosted by Anders in 1959 called "The Moral Implications of the Atomic Age," he wrote that the bombing of Hiroshima on August 6, 1945, had inaugurated a "New Age," namely, "the age in which at any given moment we have the power to transform any given place, on our planet, and even our planet itself, into a Hiroshima," which led him to declare that "Hiroshima is everywhere."[38] This New Age corresponded to what Anders termed the "Time of the End" (Endzeit), that is, a final epoch of human history in which any given day could be our last, an irrevocable new reality that will "haunt every generation of human beings" henceforth, forever.[39] Our collective struggle has thus become to extend the "Time of the End" for as long as possible, to prevent it from becoming the "End of Time" (Zeitenende), at which point the human story would terminate.[40] Although one might think that we could extricate ourselves from this situation by simply abolishing nuclear weapons (insofar as this is geopolitically feasible), Anders argued that the mere knowledge of how to construct them simply means that on any given day they could be built once again, thereby threatening the existence of humanity once more. Nuclear weapons can never be un-invented, which means that the "fight against this man-made Apocalypse" will be never-ending. Put another way, there is no post-nuclear epoch.[41] As Anders summarized the idea in his 1961 essay "Commandments in the Atomic Age,"

> the apocalyptic danger is not abolished by one act, once and for all, but only by daily repeated acts. . . . For the goal that we have to reach cannot be not to have the thing; but never to use the thing, although we cannot help having it; never to use it, although there will be no day on which we couldn't use it.[42]

The atomic bomb thus ruptured the fabric of human history. Its invention is no less significant than the life of Jesus was two millennia ago, Anders argued, and hence we need a new calendar that acknowledges this fact. Given that August 6, 1945, "demonstrated that perhaps world history no longer continues," it should be designated "Day Zero" of this new, updated calendar. In his 1958 book *The Man on the Bridge*, Anders poignantly proclaimed that "we live in the Year 13 of the Calamity. I was born in the Year 43 before. Father, who I buried in 1938, died in the Year 7 before."[43] This led Anders (in his earlier 1956 book) to delineate a tripartite periodization of human history, where the first epoch corresponds to the idea that all people are, by nature, fated to die, while in the second, human beings have become "killable," as demonstrated by the industrial mass murder of 6 million Jewish people during the Holocaust. Finally, with the terrifying advent of the Atomic Age, "the phrase 'All men are mortal' has been replaced . . . by the phrase 'mankind as a whole is mortal.'" In other words, the three epochs are:

1. All human beings are mortal.
2. All human beings are killable.
3. Humankind as a whole is killable.[44]

The last two epochs—especially the third—point at a moral problem arising from what Anders' called the "Promethean gap," where "gap" is sometimes translated as "gradient" and "disparity." This denotes the widening discrepancies between (a) "making and imagining/representing," (b) "doing and feeling," (c) "knowledge and conscience," and (d) "the produced instrument and the (not suited to the 'body' of the instrument) body of the human being."[45] This is to say, our innate capacities with respect to imagination, emotion, cognition, and physicality have become wholly incommensurate with our newly acquired powers of action, in particular the power to obliterate the entire human species. We have thus become what he described as "inverted Utopians": whereas "ordinary Utopians are unable to actually produce what they are able to visualize, we are unable to visualize what we are actually producing." This is the "basic dilemma of our age," it "defines the moral situation of man today," which he took to be that "'we are smaller than ourselves,' incapable of mentally realizing the realities which we ourselves have produced." Anders elaborated the idea as follows:

> The apocalyptic danger is all the more menacing because we are unable to picture the immensity of such a catastrophe. It is difficult enough to visualize someone as not being, a beloved friend as dead; but compared with the task our fantasy has to fulfil now, it is child's play. For what we have to visualize today is not the not-being of something particular within a framework, the existence of which can be taken for granted, but the nonexistence of this framework itself, of the world as a whole, at least of the world as mankind. Such "total abstraction" (which, as a mental performance, would correspond to our performance of total destruction) surpasses the capacity of our natural power of imagination.[46]

An important consequence of this is what Anders labeled "Apocalyptic Blindness," which emerged as a "widespread and disastrous aliment" following the Third Industrial Revolution initiated by the Atomic Age, whereby nuclear weapons engendered a radically novel means of production that has, "for the first time ever, put humanity in the position of producing its own destruction."[47] Anders' idea is that because of the divergence between our powers of action and our powers of imagination—because the Promethean gap has transformed us into all inverted Utopians—we are constitutionally unable to adequately grasp the true magnitude and enormity of nuclear self-annihilation, and consequently we become "blind" or oblivious to, thus assuming an insouciant attitude toward, the annihilatory threat before us. This is precisely why Anders saw his mission as being a "panic-maker" and "eye-opener": he strove to jolt people out of their nuclear slumber, to pry open the eyes of those suffering from Apocalyptic Blindness. As he wrote in "Theses for the Atomic Age," in which he links our ability to fear with our ability to imagine the nothingness that would result from nuclear annihilation:

> [I]t is our capacity to fear which is too small and which does not correspond to the magnitude of today's danger. As a matter of fact, nothing is more deceitful than to say, "We live in the Age of Anxiety anyway." This slogan is not a statement but a tool manufactured by the fellow travellers of those who wish to prevent us from becoming really afraid, of those who are afraid that we once may produce the fear commensurate to the magnitude of the real danger. On the contrary, we are living in the Age of Inability to Fear. Our imperative:

> "Expand the capacity of your imagination," means, in concreto: "Increase your capacity of fear." Therefore: don't fear fear, have the courage to be frightened, and to frighten others, too. Frighten thy neighbor as thyself.[48]

In other words, while Anders held that our capacity to imagine is much less elastic than our powers of action have proven to be, he did not believe that it is completely rigid or fixed. The antidote to our present predicament, then, is to exercise the muscles of our imagination to foster a sense of fear proportional to the nuclear threats hovering over humanity like the sword hovering over Damocles. This, he declared, is the "decisive moral task" of our time, for every person

> to violently widen the narrow capacity of your imagination (and the even narrower one of your feelings) until imagination and feeling become capable to grasp and to realize the enormity of your doings; until you are capable to seize and conceive, to accept or reject it—in short: your task is: to widen your moral fantasy.[49]

Anders' prescription here is, at least on the face of it, consistent with Eugene Rabinowitch's assertion that the purpose of the *Bulletin of the Atomic Scientists* was "to preserve civilization by scaring men into rationality."[50] Many others at the time agreed that fear could play an important role in protecting humanity from omnicide, as when Einstein, who believed that the creation of a world state was the "only" way to "prevent the impending self-destruction of mankind," suggested that one potentially good effect of nuclear weapons is that they "may intimidate the human race to bring order into its international affairs, which, without the pressure of fear, it undoubtedly would not do."[51] As Boyer writes,

> the strategy of manipulating fear to build support for political resolution of the atomic menace helped fix certain basic perceptions about the bomb into the American consciousness, and it set a precedent for activist strategy that would affect all later anti-nuclear crusades.[52]

However, to pursue this tangent for a moment, some strongly objected to this view, arguing instead that fear could impede progress toward peace and denuclearization. For example, in a 1947 paper titled "Atomic Nerve War and the Urge for Catastrophe," Joost Meerloo wrote that

> fear and speculation about the unknown have always had a stirring influence on the human mind. They make people not only increasingly suspicious and anxious but also more willing to surrender to the danger they fear. . . . It is for these reasons that so great a danger lies in this world-wide fear, for it may work as primitive fear did in the ancient world. Too great a fear paralyses the human mind, hypnotizes it, as it were, makes it passive, ready to surrender. It ends in suicidal reactions in a world carried away by the sweep of its dark emotions.[53]

Anders' argument, though, was that by imagining the unimaginable we might begin to generate "a special kind" of fear—specifically, one that motivates rather than incapacitates, that "drive[s] us into the streets instead of under cover."[54] In other words, he hoped that augmented fear through augmenting our imagination could inspire activism rather than nihilism. The question of whether apocalyptic anxiety or equanimity is the best psycho-emotional response to the threat of potential annihilation is one that continues to provoke debate today, with figures like the popular writer Steven Pinker, on one side, arguing that the "drumbeat of doom" will ultimately backfire: "Humanity has a finite budget of resources, brainpower, and anxiety." When these resources are used up, brainpower has been drained, and anxiety reaches a tipping point, the result may be a paralyzing sense that "humanity is screwed." And if humanity is screwed, then "why sacrifice anything to reduce potential risks? Why forgo the convenience of fossil fuels, or exhort governments to rethink their nuclear weapons policies? Eat, drink, and be merry, for tomorrow we die!"[55] On the other side one finds young leaders like Greta Thunberg, who fervently embraces the method of frightening people into action: "I don't want your hope. I don't want you to be hopeful," she declared in a 2019 speech delivered at Davos, "I want you to panic. I want you to feel the fear I feel every day. And then I want you to act."[56] We will return to this tension in later chapters.

A League of Generations

If the aim, then, is to expand our ability to imagine, the question arises as to what exactly we should be imagining. Every person on the planet perishing? The "nothingness" mentioned above that would result from a nuclear holocaust? If so, what does this nothingness consist in? How should we think about it? The answer that Anders gave gestures back at Russell's emphasis on the past and the future, although Anders did not utilize either deep-future or potentiality thinking as much as Russell. First, to understand the moral stakes of our extinction—in particular, the state or condition of Being Extinct—we must expand our imagination not just across space, considering the planetary scale of globocide, but across time as well, both past and future. The fact is that, because of our novel powers of action, "acts committed today [can] affect future generations just as perniciously as our own," and hence "the future belongs within the scope of the present. . . . The distinction between the generations of today and of tomorrow has become meaningless."[57] In pondering the final end of our collective story, then, we must interpret the concept of humanity as encompassing not

> only to-day's mankind, not only mankind spread over the provinces of our globe; but also mankind spread over the provinces of time. For if the mankind of to-day is killed, then that which has been, dies with it; and the mankind to come too. The mankind which has been because, where there

> is no one who remembers, there will be nothing left to remember; and the mankind to come, because where there is no to-day, no to-morrow can become a to-day.[58]

In other words, the annihilation of humanity would expunge all future generations, which Anders characterized as our "neighbors in time," since the act of "setting fire to our house . . . cannot help but make the flames leap over into the cities of the future, and the not-yet-built homes of the not-yet-born generations will fall to ashes together with our homes." But our disappearance would also permanently delete the memories of all those who had come before us, and consequently "we would make them die, too—a second time, so to speak," such that "after this second death everything would be as if they had never been." Anders thus held that, in imagining the outcome of human extinction, we must consider both past (the *deceased*) and future (the *unborn*) people along with our *contemporaries*, all of whom form a single "League of Generations."[59] It is this League of Generations that a nuclear holocaust would obliterate, not just everyone alive at the time of the catastrophe, which corresponds to only a small fraction of the entire league. Although the loss of all contemporary people would be very bad, the possibility of destroying the entire League of Generations means that

> the door in front of us bears the inscription: "Nothing will have been"; and from within: "Time was an episode." Not, however, as our ancestors had hoped, an episode between two eternities; but one between two nothingnesses; between the nothingness of that which, remembered by no one, will have been as though it had never been, and the nothingness of that which will never be. And as there will be no one to tell one nothingness from the other, they will melt into one single nothingness. This, then, is the completely new, the apocalyptic kind of temporality, our temporality, compared with which everything we had called "temporal" has become a bagatelle.[60]

This is what the "total abstraction" mentioned above by Anders involves: thinking seriously about the entire League of Generations, stretching back through time and into the future, perishing. The cost of extinction is the expungement of what is, what has been, and what could be, which thus points toward a further-loss view according to which some, or perhaps most, of the badness/wrongness of nuclear self-annihilation derives from losses that go above and beyond the untimely deaths of all those consumed by the "radioactive clouds" of a thermonuclear war.[61]

The Obsolescence of Ethics

To my knowledge, Anders was the first Western philosopher to suggest that the possibility of omnicide, of forever terminating the League of Generations, is so radically different from all past possibilities that it requires an entirely new theory of ethics. The traditional theories articulated in earlier periods are simply not up to the task given that, as noted earlier, the question has become "the nonexistence of [the] framework itself, of the world as a whole," rather than "the not-being of something particular within [this] framework."[62] As Anders wrote in 1956, "whether the expressions 'morality,' 'moralistic,' 'ethics' and the like still fit for the [present] considerations is uncertain. In front of the monstrous size of the object they sound powerless and inadequate." He continued:

> For, until now, moral questions were those questions that related to how people treat people, how people stand with people, how society should function. Apart from a handful of desperate nihilists from the previous century [Anders may have had the German pessimists in mind here],[63] there has hardly been a moral theorist who has ever doubted the premise that there will be and should be people.[64]

To be clear about this point, Anders is not saying that no one doubted that humanity must always exist. As he noted in a 1960 paper titled "Apocalypse without Kingdom," the idea of human extinction had indeed been considered "by those natural philosophers who speculated about heat death."[65] But the idea that we might not exist someday was explored by virtually no moral philosophers, which is just to say that Existential Ethics was, up to the mid-1950s, mostly non-existent. Either way, Anders' point is that the problems that traditional ethical theories were designed to solve were fundamentally different than the problem of self-extinction now facing humanity. "The basic moral question of former times," he wrote, "must be radically reformulated: instead of asking 'How should we live?,' we now must ask 'Will we live?'"[66] In 1979, he couched the point in stronger language, arguing that "the previous religious and philosophical ethics, without exception and without pass, have become obsolete," and because of this we "stand in the Year Zero of a new morality."[67]

Here it may be useful to disambiguate two claims that Anders appears to conflate. The first is that the possibility of omnicide has introduced new questions that have never before been asked; the second is that omnicide has introduced new questions that require an entirely novel kind of ethical theory. Throughout history, technological developments, evolving social arrangements, and so on have generated a wide range of questions that were not, or did not previously need to be, asked. In some cases, these could be accommodated by already-existing ethics: one just needed to figure out how. But it could also be that a phenomenon so unlike those phenomena of the past arises that really does necessitate an entirely novel framework, not just an extension or modification of earlier frameworks. While Anders clearly accepted the latter claim at times, he also expressed the alternative, weaker claim.

However, Anders wasn't the only one to notice that omnicide poses novel ethical challenges, and that these challenges demand adjustments to our ethical theories, if not a completely new theory. In some cases, this was merely gestured at, as when Karl Jaspers (with whom Anders and Arendt lived for a time, as he was Arendt's dissertation supervisor) wrote in his 1958 book *The Future of Mankind* (translated into English in 1961) that "an altogether novel situation has been created by the atom bomb. Either all mankind will physically perish or there will be a change in the moral-political condition of man."[68] Similarly, Arthur Koestler (1905–1983) observed in 1967 that "before the thermonuclear bomb, man had to live with the idea of his death as an individual; from now onward, mankind has to live with the idea of its death as a species. . . . The bomb has given us the power to commit genosuicide; and within a few years we should even have the power to turn our planet into a nova, an exploding star," which may have been a reference to the planetary chain reaction idea proposed by Soddy and Rutherford in the early twentieth century. Yet, he proceeded,

> the full implications of this fact have not yet sunk into the minds of even the noisiest pacifists. We have always been taught to accept the transitoriness of individual existence, while taking the survival of our species axiomatically for granted. This was a perfectly reasonable belief, barring some unlikely cosmic catastrophe. But it has ceased to be a reasonable belief since the day when the possibility of engineering a catastrophe of cosmic dimensions was experimentally tested and proven. It pulverised the assumptions on which all philosophy from Socrates onward was based: the potential immortality of our species.[69]

A more detailed discussion of this fact and its implications for ethics was offered by the theoretical physicist Hilbrand Groenewold (1910–1996) during a 1968 colloquium, which was attended by Sir Karl Popper, Max Black, and I. J. Good, among others. Groenewold argued that for most of human history "there were micro effects on small groups and small areas," while "in more recent history—as a result of technology and science—they grew out to meso effects on large groups or whole populations and large areas or parts of the earth." An example of the latter would be environmental contamination due to "industrialization, urbanization, and traffic." However, "modern science and technology" have introduced a third category: "macro effects on the whole population and the entire earth," the most obvious example being the possibility of a thermonuclear conflict. Our newly acquired capacity to affect everyone everywhere thus gives rise to "macro problems" that require, he argued, a new "macro morality." The reason is that

> if individuals, small or large groups, or even whole populations are destroyed by micro or meso effects, other individuals, groups or populations will take their place and the whole case will be of little importance for the future of mankind. If the world population of man or another biological species is only

> once . . . annihilated by even a single macro effect, the history of that species is cut off forever. That makes the moral aspects of macro problems fundamentally different from those of meso (and micro) problems.[70]

Not only do "macro problems" constitute a fundamentally new category within ethics, but Groenewold added that he is "afraid that with our habits, ideas, imagination, and moral rules, which all have been formed under familiar micro or perhaps meso conditions, we are hardly capable to realize (i) the entrance and (ii) the fundamental importance of macro problems in human history"[71] In other words, our behaviors, cognitive tendencies, and ethical theories all developed within a milieu radically different from the one we now occupy, and consequently we might be unable to properly recognize the reality and significance of the "macro problems" we have recently created. This, of course, echoes Anders' notions of the Promethean gap and inverted Utopianism, although Groenewold hinted at a more evolutionary explanation that was, coincidentally, reminiscent of Peter Wessel Zapffe's comments about our over-evolved consciousness. "[B]y a kind of intellectual hypertrophy," Groenewold wrote, where "hypertrophy" refers to the enlargement of an organ or tissue, "the man-made macro effects are liable to grow beyond the grasp of human thinking and social control," given that "our habits of behaviour and thinking, our ideas and moral rules have been formed during very many generations in a very special period of terrestrial history."

I take this to be saying that the enlargement of our capacities for invention and scientific discovery (our "intellect") has enabled us to alter the physical world in ways that evolution did not equip us to comprehend ("thinking") and respond to in a morally appropriate and socially effective manner. Consequently, Groenewold concluded that "any future of humanity on a biological time scale will need at least adaption of thinking and acting and in particular of moral habits to the historical transition into the period of macro problems," as the alternative—a failure to adapt—could very well lead to extinction.[72] In other words, we need new categories of thought and behavior paired with a new moral perspective that is commensurate with, and thus can accommodate, our newly acquired powers to exterminate ourselves. Put another way, recall from the previous chapter the scenarios of world A (population = 11 billion) and world B (population = 10 billion); both experience a catastrophe that kills 10 billion people, and hence humanity goes extinct in B but not A. On Groenewold's view, not only is the second event in world B—that is, the event whereby "the history of that species" is terminated forever—morally relevant in itself, but understanding its moral significance requires some sort of novel "macro morality." Whether this could be constructed by extending or modifying existing theories, or must be built de novo from the bottom-up, he never specified, although one gets the impression that he may have had the latter in mind.

A final example of a philosopher in the relatively early postwar period making such claims involves Hans Jonas (1903–1993), who studied under Heidegger, his doctoral advisor, and happened to be a friend of Anders and Arendt.[73] Jonas offered an even more comprehensive diagnosis of the problem in a 1972 plenary address, published the same year in Social Research and greatly expanded in his 1979 book *The Imperative of Responsibility*, which won the 1987 Peace Prize of the German Booksellers' Association, selling nearly 200,000 copies in the country.[74] He argued that there are at least four reasons that traditional ethical theories have become outdated:

(1) Until recently, our actions "impinged but little on the self-sustaining nature of things and thus raised no question of permanent injury to the integrity of its object, the natural order as a whole." Hence, "action on non-human things did not constitute a sphere of authentic ethical significance."
(2) Ethical theories in the past were "anthropocentric" in the sense that they concerned the effects of human actions only on other humans.[75]
(3) The essence of human beings was considered to be fixed or constant.
(4) The relevant effects of actions were spatiotemporally proximate to those actions; it was not possible to affect people on the other side the planet or in the distant future.

The reasons of (1), (2), and (4) are pertinent to environmental ethics, which emerged as an academic field in the 1970s as the modern environmental movement gained steam following Rachel Carson's 1962 book, the first Earth Day in 1970, and so on. Indeed, Jonas—along with Zapffe's friend Arne Naess—was one of the first philosophers to systematically address the issue of our impact on the natural world, which he understood, contra the materialistic worldview that arose in the nineteenth century, as replete with intrinsic value. The reasons most relevant to Existential Ethics are (3) and (4), with (3) addressing phyletic and normative extinction and (4) covering the possibility of omnicide, since omnicide would affect everyone around the world and those who *would have* existed in the future if not for our extinction.

Hence, Jonas saw traditional ethics as inadequate for reasons that went beyond our newly acquired capacity to self-destruct: we can also now obliterate features of the environment that are intrinsically valuable, such as other species, ecosystems, and so on. His explanation for this inadequacy, though, was the same as that given by Anders and Groenewold: in the past, ethics was designed for a very specific milieu of immediate action-effects limited mostly to the interpersonal level. Ethics concerned people's interactions with other people in the context of the city, as Jonas put it, not people's interactions with the environment (which could be taken as unchangeable on human timescales) or the possibility of some people killing all people.[76] He thus described this old perspective as "neighbor ethics," since "the ethical universe [was] composed of contemporaries, and its horizon to the future [was] confined by the foreseeable span of their lives. Similarly confined is its horizon of place, within which the agent and the other meet as neighbor." He elaborated the idea:

> All enjoinders and maxims of traditional ethics, materially different as they may be, show this confinement to the immediate setting of the action. "Love thy neighbor as thyself"; "Do unto others as you would wish them to do unto you"; "Instruct your child in the way of truth"; "Strive for excellence by developing and actualizing the best potentialities of your being qua man"; "Subordinate your individual good to the common good"; "Never treat your fellow man as a means only but always also as an end in himself"—and so on. Note that in all these maxims the agent and the "other" of his action are sharers of a common present. It is those alive now and in some commerce with me that have a claim on my conduct as it affects them by deed or omission.

To illustrate, consider the first formulation of Kant's Categorical Imperative, which states that one should "act only in accordance with that maxim through which you can at the same time will that it become a universal law." According to Jonas, there is simply "no self-contradiction in the thought that humanity would once come to an end," since there is no logical contradiction in willing the extinction of our species.[77] The Categorical Imperative may apply to acts within the series of human acts, but whether this series itself should continue "cannot be derived from the rule of self-consistency within the series." Instead, it must come from "a commandment of a very different kind, lying outside and 'prior' to the series as a whole," an idea that we will return to momentarily.[78] A similar point could be made about rights theories. Do future generations have a right to exist? The problem is that for someone to make a rights claim, they must exist, but since future generations do not (yet) exist, they cannot make rights claims, and hence we cannot violate "their" rights by failing to bring them into existence. As Lewis Coyne, an expert on Jonas' philosophy, makes the point, "the concept of moral rights cannot establish obligations to future generations without simply assuming their existence, which is precisely what is newly endangered."[79] Mere possibilities have no rights that could be transgressed.

Instrument Hearts

What we need, then, is a new set of moral principles, maxims, rules, duties, or obligations to either replace or supplement these traditional theories. While the main thrust of Groenewold's discussion was to exhort others to devise such a theory, Anders and Jonas actually attempted to do this. Taking them in turn, we have already examined pieces of Anders' ethical system, for example, the "commandment" (his word) to motivate oneself to fight for humanity's future by increasing one's "fear" by expanding one's imagination. But he offered several additional commandments, which he claimed could be "condensed" into a single super-commandment: "Have and use only those things, the inherent maxims of which could become your own maxims and thus the maxims of a general law."[80] This is obviously reminiscent of Kant's Categorical Imperative, and indeed we will see that Jonas' ethics drew from Kantianism as well. The main idea behind Anders' super-commandment concerns what philosophers of technology call the "value-neutrality thesis."[81] This states that technologies are essentially normatively neutral, mere tools, nothing more than means to whatever ends their uses select. Hence, they are morally blameless, as intimated by the NRA's famous slogan "Guns don't kill people, people kill people." Anders strongly rejected the value-neutrality thesis, arguing instead that (a) technologies have come to mediate all the interactions we have with each other (a claim about the technologization of modern society), and (b) these interactions are shaped, altered, framed, and distorted in all sorts of ways by the technologies mediating them (a claim about the non-neutrality of such artifacts). In this sense, one could say that technologies themselves, by virtue of being non-neutral, have *their own* "maxims and motives," in addition to whatever maxims and motives their users might possess. Anders' commandment is thus to "have and use only those" technologies whose inherent maxims and motives we would accept as being universalized into a "general law." As Anders explained:

> What the postulate demands is: be as scrupulous and unsparingly severe in front of those maxims and motives as if they were your own (since pragmatically speaking they are your own). Don't content yourself with examining the innermost voices and the most hidden motives of our own soul . . . but do examine the secret voices, motives, and maxims of your instruments.

With respect to nuclear weapons, then, Anders contended that

> if a high official in the atomic field would examine his conscience in the traditional way, he would hardly find anything particularly evil. If, however, he would examine the "inner life" of his instruments [i.e., atomic bombs], he would find herostratisms and even herostratism on a cosmic scale, for it is in a herostratic way that atomic weapons are treating mankind.[82]

The unusual term "herostratism" derives from the name of the ancient Greek arsonist Herostratus, "who sought lasting fame by burning the temple of Artemis at Ephesus, a wonder of the ancient world."[83] In other words, Anders' rather poetic assertion is that the atomic bomb, the technology, is such that it "strives" to attain notoriety and infamy through destruction—specifically, the destruction of humanity.[84] This "striving" is the bomb's inherent maxim or motive, which then becomes our maxim or motive when we relate to it uncritically, as if the bomb were a merely neutral object, an innocent means to some end of our choosing. Once this maxim or motive is properly identified, the next question is whether we should want it to become a general law. If not, then Anders' super-commandment instructs us to destroy the bomb itself. As he made the point, "only when this new moral commandment 'look into your "instrument hearts"' has become our accepted and daily followed principle shall we be entitled to hope that our question 'to be or not to be' will be answered by: 'to be.'"[85]

The Foothold for a Moral Universe

This is an intriguing attempt to devise a principle of ethics designed specifically for the Atomic Age, although—despite Anders' originality as one of the very first theorists of omnicide—there is much left to be desired. A far more comprehensive, architectonic theory was outlined by Jonas in his 1979 book mentioned above. At the core of this theory was an "imperative" not unlike Anders' commandment that Jonas saw as supplementing rather than replacing traditional ethics.[86] The imperative's aim is to impose moral constraints on our actions in the twentieth century, given our newly acquired capacities to alter the environment, modify our genes, and destroy ourselves. Jonas offered the following four formulations:

> "Act so that the effects of your action are compatible with the permanence of genuine human life"; or expressed negatively: "Act so that the effects of your action are not destructive of the future possibility of such life"; or simply: "Do not compromise the conditions for an indefinite continuation of humanity on earth"; or, again turned positive: "In your present choices, include the future wholeness of Man among the objects of your will."

Whereas Kant's Categorical Imperative requires that one does not act according to any maxim that engenders a contradiction when universalized, Jonas noted that "it is immediately obvious that no rational contradiction is involved in the violation of this kind of imperative." Consequently, he proposed a decision procedure (which is essentially what Kant's first formulation is) of a quite different sort: first, it occurs on the level of public policy rather than the individual, whereas Kant's pertains to individuals. Second, the question of consistency does not concern the maxim itself—it is not about self-consistency—but instead focuses on whether the effects of some maxim of public policy are or are not compatible with "genuine human life" persisting into the indefinite future. As Jonas explained,

> this adds a time horizon to the moral calculus which is entirely absent from the instantaneous logical operation of the Kantian imperative: whereas the latter extrapolates into an ever-present order of abstract compatibility, our imperative extrapolates into a predictable real future as the open-ended dimension of our responsibility.[87]

To illustrate, consider the maxim "Our policy is to consume all the non-renewable resources on Earth for the benefit of people today." According to Jonas' imperative, implementing this policy would be wrong if (and only if?) its effects would be destructive to, or would compromise the conditions of, the future possibility of genuine humanity. If it would, then implementing that policy would be unethical. But here an epistemological question arises: how exactly can we know what effects a policy would actually have given the chaotic messiness of the real world? Perhaps consuming all the non-renewable resources today would accelerate technological progress; this would make space colonization feasible in the near future, and if humanity were to colonize space, it could ensure its survival even if human life on Earth were to become difficult or impossible (e.g., because of pollution or climate change).[88] Alternatively, it could be that implementing the policy does not accelerate progress toward colonization but instead results in seriously degraded living conditions.

Uncertainties about the actual effects of realizing some public policy maxim thus led Jonas to claim that knowledge has taken on an important new moral significance: in the past, the knowledge one needed to accurately anticipate the spatiotemporally proximate effects of one's actions was minimal, whereas today, with our novel powers of action, anticipating the effects of policy requires vast amounts of knowledge spanning myriad domains of human inquiry. Jonas thus made two suggestions: first, we should establish a new field of "scientific futurology" to generate more reliable predictions about the possible and probable futures, and second, we should, as a default, always lean towards "the prophecy of doom" rather than the "prophecy of bliss," an idea that Jonas referred to as the "heuristics of fear."[89] In other words, if we are unsure about the consequences of some policy P, and if implementing P could result in great benefits but could also bring about immense suffering, we should as a practical matter assume that the worst will happen and, therefore, reject P. Hence, Jonas' "heuristics of fear" was an early version of the "Precautionary Principle," which has played a central role in discussions about environmental policy.[90]

But here one could ask why exactly it matters that "genuine human life" persists. What grounds or justifies this new ethical imperative? Why should one obey it? The argument goes like this: first of all, Jonas based his ethical system on an underlying conception of the ontological nature of human beings. As Theresa Morris explains, human beings have a "uniquely evolved capacity for freedom that places the human in a position to take responsibility," where the notion of freedom is ontological and the notion of responsibility is ethical.[91] Jonas' contention is that the ontological fact that humans can act freely gives rise to the ethical fact that humans can also take moral responsibility for their actions; freedom and responsibility are thus two sides of the same coin, with the latter deriving from the former.[92] It follows that, because of our ontological nature and consequent ethical capacities, we are the only creatures in the natural world capable of acting in morally right or wrong ways. We are the only ethical beings. What is ultimately of importance to Jonas, then, is the continued existence of beings capable of moral responsibility—of there existing a "moral order" in the universe. Our obligation to survive is not an obligation to any particular future people but to what Jonas called the "idea of Man," which denotes our unique ontological and ethical capacities. As Jonas wrote, the idea of Man has "itself become an object of obligation," namely, "the obligation . . . to ensure the very premise of all obligation, that is, the foothold for a moral universe in the physical world."[93] He fleshed out this idea as follows in his 1996 book *Mortality and Morality*:

> The appearance of [responsibility] in the world does not simply add another value to the already value-rich landscape of being but surpasses all that has gone before with something that generically transcends it. This represents a qualitative intensification of the valuableness of Being as a whole, the ultimate object of our responsibility. Thereby, however, the capacity for responsibility as such—besides the fact that it obligates us to exercise it from case to case—becomes its own object in that having it obligates us to perpetuate its presence in the world. This presence is inexorably linked to the existence of creatures having that capacity. Therefore, the capacity for responsibility per se obligates its respective bearers to make existence possible for future bearers. In order to prevent responsibility from disappearing from the world—so speaks its immanent commandment [i.e., the imperative above]—there ought to be human beings in the future.[94]

But for what reason does the capacity for responsibility obligate humanity to continue existing? What does it matter if there is a moral order in the universe or not? Here Jonas suggested that "ought-to-be" of humanity—specifically, the idea of Man—simply is the case. "Groundless itself," he wrote,

> brought about with all the opaque contingency of brute fact, the ontological imperative institutes on its own authority the primordial "cause in the world" to which mankind once in existence, even if initially by blind chance, is henceforth committed. It is the prior cause of all causes that can ever become the object of collective and even individual human responsibility.[95]

All arguments must begin somewhere, and this is where Jonas began his.

One way to understand Jonas' view, aspects of which are rather abstruse, comes from Lawrence Vogel, who edited *Mortality and Morality* and wrote the foreword to the 2001 edition of Jonas' 1966 book *The Phenomenon of Life*. Vogel writes that while Jonas held that all living creatures are valuable as ends-in-themselves, that is, for their own sakes, he also maintained that "the moral worth of life only comes into being with the phenomenon of obligation, and obligation requires the evolution of a being capable of moral responsibility." Hence, although one might think, as some radical environmentalists have (see below), that "we would do the greatest justice to the ecosystem as a whole by removing ourselves from it in an act of supreme impartiality so that other species might flourish," Jonas would forcefully respond that our "collective suicide would annihilate the phenomenon of justice and injustice alike, and so deprive Being of the metaphysical and moral dimensions it took so long to produce." From this it follows that "our first duty is to preserve the noble presence of moral responsibility in nature: of a being who is able to recognize the good-in-itself as such.[96] Morris offers a similar interpretation, writing that Jonas thought

> a world without an intrinsically ethical being existing in it would be a greatly diminished world, one that would lack both a witness to its unique goodness and beauty and a preserver and protector of the good. The presence of a witness fulfills the good, because it is through the witness that the good receives itself. Thus, Jonas emphasizes the primacy of the human in his ethics of the future. He insists that the primary duty of an ethics of responsibility is to preserve the possibility for human beings to exist in the world—with the caveat that these human beings not be compromised in regard to their freedom, intelligence, or capacity to care.

Morris writes elsewhere that "for the objectively existing good that life is to have meaning requires the presence of a being who can recognize and respond to that good."[97] However one interprets Jonas' view, the crux is that human beings have unique ontological and ethical capacities; these capacities give rise to the possibility of obligation; and without obligation there would be no moral order, which would yield a greatly impoverished or diminished state of the universe. This is the foundation of Jonas' system of ethics—a distinctively secular ethics crafted specifically for our new condition of radically augmented powers to act.

Dead or Red?

As alluded to earlier, Jonas wasn't just concerned with the prospect of humanity destroying itself (and the environment, which is wrong because of the intrinsic value inherent in all living creatures).[98] He also addressed, albeit more cursorily, the possibility of normative extinction resulting from the intentional modification of our genomes. As Morris notes in the block quote above, Jonas' ethical system demands that human beings must "not be compromised in regard to their freedom, intelligence, or capacity to care."[99] This is to say, if what matters is the preservation of the idea of Man, and if "the idea of Man" denotes our dual capacities for freedom and responsibility, then any biotechnological intervention that is destructive to, or would compromise the conditions of, the future possibility of this idea would also transgress the new imperative and, therefore, be morally impermissible.[100] Couched in different terminology, one can say that members of *Homo sapiens* possess a certain fundamental dignity by virtue of our status as moral beings, and it is this dignity that must not be compromised, since "whenever this sort of dignity is violated we risk genuine human life," to quote Coyne and Michael Hauskeller.[101] Hence, Jonas can be seen as an early, and prescient, critic of modern transhumanism, and indeed his arguments against modifying the human organism (which went beyond violations of his imperative and thus are not directly relevant to our discussion) greatly influenced contemporary "bioconservatives" like Leon Kass. As Coyne and Hauskeller note, Kass co-dedicated his book *Life, Liberty, and the Defense of Dignity*, in which he defended an anti-transhumanism position, to Jonas and his "moral passion and philosophical courage."[102] For Jonas, reengineering the human being is risky, although there is no fundamental objection to replacing *Homo sapiens* with a successor species so long as this species possesses the same fundamental ontological and ethical capacities that we have for freedom and responsibility.

However, Jonas wasn't the only philosopher in the opening decades of the postwar era to fret about normative extinction. In *The Future of Mankind*—published two decades before *Imperative*—Jaspers examined the possibility of normative extinction in the externalist sense as a result of totalitarianism, which he understood in explicitly Arendtian terms.[103] On this account, totalitarianism is a historically novel phenomenon, a fundamental rupture in "Occidental history," exemplified by the "twin horrors of the twentieth century," that is, Nazism and Stalinism.[104] By transmogrifying "human existence to the point where men cease to be human," it inflicts "a humiliation that dehumanizes all of existence, every hour in the lives of all," and threatens to convert the "world . . . into a concentration camp."[105] In totalitarian states, human beings are wholly stripped of their freedom, where freedom, as one reviewer of Jaspers' book put it, "is the very essence of human dignity, and it is the only atmosphere within which men can live lives worth living."[106] Without freedom, "mere life as such . . . would not be the life of animals in the abundance of nature; it would be an artificial horror of being totally consumed by man's own technological genius." It might even be that a totalitarian state in the Atomic Age uses nuclear weapons to terrorize and control its population; in Jaspers' words,

> the peace of totalitarianism is a desert constantly laid waste again by force against rebellious human claims. A totalitarian world state would use the atom bomb—which it alone would control—in limited doses and without endangering the life of mankind as a whole. It would use it in a gradation of terror, for purposes of extermination or simply to put down a revolt in short order. What could be expected under total rule baffles the imagination, because its nature seems humanly impossible and is accordingly not believed in reality.[107]

This vision of the deplorable conditions of human life under the iron fist of "total rule" led Jaspers to claim that, if forced to choose between risking the "final destruction of human existence by the atom bomb" and the "final destruction of the human essence by totalitarianism," he would opt for the former. "Man, unlike the animals," he contended, "is always free to take any risk for his freedom. If he should throw the life of mankind into the scales for liberty, he would not be taking this risk in order to die, but in order to live in freedom."[108] He thus offered a defense of the position sloganized as "Better dead than Red," which of course contrasts with the inverse position that we would be "Better Red than dead," the latter of which Russell, in his aforementioned book *Has Man a Future?*, attributed to peace activists in West Germany.[109] Today, as Kenneth Rose observes, the Jaspersian preference for extinction over totalitarianism has "become synonymous with the political philosophy of the far-right lunatic fringe," although "there is ample evidence that a broad range of Americans in the late 1950s and early 1960s understood that a nuclear war would bring unprecedented horrors . . ., but that a [nuclear] holocaust might be necessary to oppose communist domination." In other words, many people agreed with Jaspers, especially in the United States. For example, a 1961 Gallup poll asked people in the United States and Britain this question: "Suppose you had to make the decision between fighting an all-out nuclear war or living under communist rule—how would you decide?" An incredible 81 percent of US respondents reported that they prefer nuclear war over communism.[110] However, with the dissolution of the Soviet Union in the late 1980s and early 1990s, the threat of a totalitarian takeover greatly declined, and consequently the question of which is preferable has lost much of the relevance and urgency that it once had.

Both Jonas and Jaspers grounded their notions of dignity in human freedom. For Jonas, the worry was a loss of this dignity due to "man [taking] his own evolution in hand," while for Jaspers, the fear was that political circumstances could arise in which the conditions necessary for humans to act freely are wholly expunged.[111] Of note is that Jaspers is one of the only theorists that I am familiar with who offered a ranking of different human extinction scenarios according to their relative badness. In the terminology of this book, his "Better dead than Red" view is tantamount to the claim that final human extinction, whereby *humanity* disappears forever, bringing the whole human story to an end, is preferable to normative extinction, whereby an essential part of our humanity is lost. Many people will concur that there are fates worse than death for us as individuals; on the ethical view of Jaspers, global totalitarianism is a fate worse than complete nothingness.

Environment, Animals, Generations, Population

We have now examined a number of anti-extinction views that went beyond the obvious claim that murdering *everyone* would be wrong because murdering anyone is wrong (call these "anti-omnicide views"). In every case discussed above, the argument pointed toward some morally relevant "loss" above and beyond whatever suffering those alive at the time of the catastrophe might experience. What are these losses? What are these additional sources of badness/wrongness associated with Being Extinct? Russell emphasized that all the progress so far made in human history will have been for nothing, and that our disappearance would foreclose what could be an extremely long and wonderful future. Anders argued that the cost of self-annihilation is the permanent erasure of the League of Generations, which encompasses all past, present, and future people. Hence, not only would contemporary generations suffer terribly if nuclear omnicide were to occur, but the already-deceased would die a "second death," while future generations would be cut off from existence forever. Subsequently, Jonas claimed that our extinction must be avoided because it would remove the possibility of obligation, thus expunging the entire moral universe (assuming we are the only creatures in the cosmos with the capacities for freedom and responsibility). We also saw that numerous philosophers—Anders, Groenewold, and Jonas—called for the construction of a new ethics to either replace or supplement traditional systems, which they contended had become outdated or obsolete because of our novel, unprecedented powers of action, as traditional ethics was designed within and for a radically different milieu of mostly interpersonal, spatiotemporally proximate action-effects. Anders offered one of the first attempts to construct a new ethical theory for the Atomic Age (expand your imagination, heighten your fear, and only use those technologies whose maxims and motives could be universalized into a general law), while Jonas outlined a sprawling theory that aimed to confront the dual possibilities of self-annihilation and irreversible alterations to the natural environment.

With the exception of Russell and Groenewold, all of these early developments unfolded within the Continental tradition. Russell, in fact, was one of the founders of the Analytic tradition (along with Gottlob Frege, G. E. Moore, and Russell's student Ludwig Wittgenstein). Although Groenewold was a physicist rather than a philosopher, the conference at which he presented his hortatory claims about the need for a new "macro morality" was squarely within Analytic Philosophy. Intriguingly, by the time Analytic philosophers in general got around to exploring the core questions of Existential Ethics, starting in the late 1960s, a majority of those who weighed in on the topic actually held that there would be nothing bad about the state or condition of Being Extinct, and hence—given a utilitarian framework, which many explicitly accepted or were sympathetic with—there would be nothing wrong with bringing about this state or condition *so long as* the processes or events leading up to our extinction do not involve anything morally unacceptable (the equivalence view).[112] Some even argued that, based on a particular interpretation of utilitarianism, our collective non-existence would be a positively good outcome. As John Leslie observed in 1983, "quite a few philosophers now hold that we at least have no duty to ensure life's continuance," while others, he noted, have defended the more extreme position that "life's absence would be preferable to its presence since living can be nasty."[113] It was only in the 1980s and 1990s that this widespread tendency toward equivalence and pro-extinctionist views began to reverse—thanks in part to Leslie's writings on the topic.

Before examining these early arguments within Analytic Existential Ethics, as it were, it may be useful to situate them within the broader context of Analytic moral philosophy during the twentieth century. As noted at the end of the previous chapter, the first half of the century was dominated by metaethical debates inspired by Moore's 1903 book *Principia Ethica.* Although some, if not most, of these philosophers were atheists or agnostics, it was not until circa 1960 that non-religious normative ethics was studied "by many people," quoting Derek Parfit once again.[114] The following decade—the 1970s—witnessed a burst of innovative research within both normative and practical ethics, exemplified by the emergence of novel topics and fields like animal rights, global ethics, environmental ethics, intergenerational justice (or ethics), and population ethics.[115] Of these, the field that overlapped the most with Existential Ethics was population ethics, which concerns questions about the number, existence, and identity of future people.[116] Since one possible number of future people is *zero*, some of the theories proposed by population ethicists had direct implications for Existential Ethics.

Historically speaking, population ethics can be traced back at least to Sidgwick, who noted in Chapter 1 of Book IV of *Methods* (1874) that a "question arises when we consider that we can to some extent influence the number of future human (or sentient) beings. We have to ask how, on Utilitarian principles, this influence is to be exercised." To this he added that "it seems clear that, supposing the average happiness enjoyed remains undiminished, Utilitarianism directs us to make the number enjoying it as great as possible."[117] In other words, we should want the human population to expand, assuming that the average happiness of people does not decline. However, it wasn't until Parfit's groundbreaking book *Reasons and Persons*, published in 1984, that population ethics gained significant attention among Analytic moral philosophers, although some of what Parfit had to say about the topic was responding to population-ethical ideas published during the late 1960s and 1970s.[118] In brief: the field dates back to Sidgwick, was addressed by some beginning in the late 1960s, and then became prominent following Parfit's book.

Then Comes the Snag

One of the most important contributions was a 1967 article by Jan Narveson titled "Utilitarianism and New Generations," which received fairly little attention at first but has since become a canonical contribution to the literature.[119] The main thrust of Narveson's discussion was not human extinction but countering a popular objection to the sort of utilitarianism articulated by Sidgwick above.[120] As Narveson wrote, "one of the stock objections to utilitarianism goes like this: 'If utilitarianism is correct, then we must be obliged to produce as many children as possible, so long as their happiness would exceed their misery.'"[121] While Sidgwick was the first to distinguish between the average and total versions of utilitarianism, Narveson was the first to differentiate between (a) the "impersonal" or, in my terminology, "impersonalist" view, and (b) the "person-regarding," "person-based," or "person-affecting" "view," "intuition," "restriction," "principle," or "axiom," as it has variously been called. (Note: this *distinction* originated with Narveson, although Parfit coined the *terms* "person-regarding" and "person-affecting," the latter of which has become standard.)[122]

I have already defined these positions in the previous chapter, but let's take a closer look. According to impersonalist utilitarianism, we are morally obliged to maximize intrinsic value (either the total or average amount) within the universe as a whole. In contrast, in a person-affecting account, we are morally obliged to maximize intrinsic value (either the total or average amount) within some restricted population of sentient beings, such as every person who exists right now or will necessarily exist in the future.[123] To be clear about what "intrinsic value" means, all utilitarians are welfarists, that is, they identify intrinsic value with "welfare" or "wellbeing," where these two terms, which are synonymous, can be interpreted in at least three ways. First, there is hedonism, a monistic theory of value according to which wellbeing consists of pleasure or happiness. (This was Sidgwick's view.) Second, another monistic theory is desire-satisfactionism, also called preference utilitarianism, which identifies wellbeing with the satisfaction of desires or preferences. And third, one could accept an objective-list theory of wellbeing, which is pluralistic in that it identifies wellbeing with some list of "objective" goods like knowledge and friendship in addition to—depending on the list—happiness and satisfied desires. Hence, to say that one should *maximize intrinsic value* is just to say that one should maximize happiness, satisfied desires, or certain objective goods, respectively.[124]

This brings us back to Narveson's distinction. If what matters morally is the maximization of the *total amount* of wellbeing (total utilitarianism), the question arises as to whether it would be wrong not to create a person one knows would have a "happy" or "worthwhile" life, meaning a life that would contain a net-positive amount of intrinsic value. On the impersonalist account, this would be wrong, since failing to create a "happy" person would deprive the universe of some extra wellbeing that it could otherwise contain. On the person-affecting account, the answer depends. For example, if you had made a promise to your partner that you would conceive a child with them, and if breaking that promise by later refusing to have a child would cause your partner harm, then it may be wrong not to create this person.[125] Here, "harm" is standardly understood in *comparative* terms as meaning "to make someone worse off than they otherwise would have been."[126] But *aside* from considerations involving the wellbeing of existing people, Narveson contended that there would be *nothing wrong* with not converting a possible person into an actual person because, he wrote, "'possible persons' are not persons: it isn't just that they aren't the usual kind of persons, for neither are they a special kind of persons, as are tall or short ones, male or female ones, and so on."[127] This is to say, "someone" who doesn't yet exist, and might never exist, isn't a person *in any sense*; they are *non-persons*. Hence, since non-persons cannot be harmed, as only beings that already exist can be made worse off than they otherwise would have been, there is no moral obligation to create new "happy" people, even if their lives would be *wonderful*. "All obligations and indeed all moral reasons for doing anything," he declared, "must be grounded upon the existence of persons who would benefit or be injured by the effects of our actions." This means that, in slogan form, we should be "in favor of making people happy, but neutral about making happy people. Or rather, neutral as a public policy, regarding it as a matter for private decision."[128]

To make the implications of these positions more explicit, impersonalism entails that an act can be wrong *even if* it does not harm anyone, while the person-affecting utilitarian view maintains that an act can be wrong *only if* it harms someone. By not having the "happy" child, the impersonalist claims that one has done something wrong by failing to maximize value in the universe as a whole, even though the possible person who *could have* existed is not themselves harmed by their non-existence. In contrast, the person-affecting theorist sees this as wrong only if existing people are made worse off by the decision.

This leads directly to a crucial question about the wrongness of human extinction: if there is no moral obligation to create new people—if the decision to have or not have a child "is purely a matter of taste," as Narveson put it[129]—then is there an obligation to perpetuate the species? If everyone were to decide not to have children, and if in each individual case there was nothing morally wrong with this decision, then would it also be permissible to allow the human population to fall to zero? This is where the person-affecting view has profound implications for Existential Ethics: on Narveson's view, there is nothing wrong with allowing humanity to go out of existence, as no one would be harmed—no one would be around to be harmed—by the state or condition of Being Extinct. In his words,

> is there any moral point in the existence of a human race, as such? That is to say, would a universe containing people be morally better off than one containing no people? It seems to me that it would not be, as such, at any rate on utilitarian grounds. We might prefer . . . a universe containing people to one that does not contain them, particularly since we presumably would not be able to occupy the second one ourselves; but is this, then, a moral preference? It seems to me, again, that it is not, and that the effort to make it one is a mistake.[130]

Given the main thrust of Narveson's 1967 paper, this consequence of his view appeared to be an afterthought, and indeed the passage just quoted is embedded in the paper's closing paragraph. However, he offered a more detailed discussion in a subsequent book chapter titled "Future People and Us," which was published in an influential edited collection called *Obligations to Future Generations* (1978). Narveson wrote that "the person-regarding view is a natural one to adopt. But it makes for a knotty problem for anyone who wants to hold that we have some such duty as the duty to sustain the human race." The problem concerns the question (to quote him at length):

> For to whom would we owe such a duty? The obvious suggestion would be that we owe it to the "human race," or to all those people out there in the future ahead of us. But this won't do. Given the person-regarding view, we cannot say the former: for the human race is not a person, but rather some such thing as the set of all persons or, worse still, the property of being a person or the idea of humankind. To none of these entities do we owe anything on the person-regarding view, and it is not obvious what could be meant by saying that we "owe" something to any of them in any case. The best we can do is to suggest that we owe the perpetuation of the human race to future persons themselves. But then comes the snag. For if we do not carry out this "duty," we suddenly find that there is nobody we can claim to have let down, to have defaulted or failed in discharging our duties to them. The existence of the supposed subjects of this obligation is contingent on our fulfilling it. But if there is no subject of obligation, then, given the person-regarding view, there is no obligation. Which means that there can be no such thing as an "obligation to perpetuate the human race," for an obligation that only exists if it is fulfilled, i.e., which logically cannot be violated, is clearly nonsense.[131]

Death, Non-Birth, and Unfinished Business

This said, Narveson is clear that he does not want humanity to die out. "We do," he wrote, "want to keep the human race going." His point was that the question of becoming extinct "is not a moral question" but is instead one that is "purely a matter of taste."[132] In another chapter of *Obligations*, Jonathan Bennett concurred with Narveson's conclusions about individual procreation and human extinction. With respect to the first, he argued that "if a failure to bring someone into existence is ever wrong for utilitarian reasons, these must concern the utilities [or happiness] of people who are at some time actual, not those of the person whose coming-into-existence didn't happen." Echoing ideas from above, he continued:

> It might be wrong for me to fail to beget a child because that would deprive my parents of the pleasures of grandparenthood, or because any child of mine would be sure to benefit mankind; in one case my parents are deprived, in the other mankind in general. But it couldn't be wrong because by not bringing the child into existence one deprives it of something.

This contrasts, once again, with the view of impersonalist utilitarians, whom Bennett memorably described like this: "As well as deploring the situation where a person lacks happiness, these philosophers also deplore the situation where some happiness lacks a person." Even worse, according to Bennett, such philosophers tend to "speak of the latter situation as being one in which some utility is lost."[133] In other words, they see the failure to bring new value *into the world* as essentially the same as the failure to prevent currently existing value from *going out of it*—both are classified as "losses." This means that there is no *intrinsic* difference for impersonalists between death, on the one hand, and non-birth, on the other. Someone with a wellbeing level of 95 dying would be just as bad as "someone" who would have had a wellbeing level of 95 never existing, all other things being equal. But this commits a serious error, Bennett argued: it involves inferring the proposition that "We ought to produce as much happiness as possible" from the claim that "We ought to make people as happy as possible." The "mistake" arises from an undue emphasis on the notion of amount, which "lets philosophers introduce a surrogate for the proper notion of utility—it gives them utilities that are not someone's, in the form of quanta of happiness that nobody has but that somebody should have."[134]

Bennett thus concurred with Narveson that "we have no obligation to prevent the extinction of mankind (except insofar as this would affect actual persons)," which is to say that the wrongness or badness of our extinction, from this person-affecting perspective, depends only on how it is brought about.[135] Yet, like Narveson, Bennett also expressed a clear preference for our continued survival, writing that "I am passionately in favour of mankind's having a long future, and not just because of the utilities of creatures who were, are, or will be actual." He labeled this his "pro-humanity stance," describing it as nothing more than "a practical attitude of mine for which I have no basis in general principle." He proceeded:

> The continuation of *Homo sapiens*—if this can be managed at not too great a cost, especially to members of *Homo sapiens*—is something for which I have a strong, personal, unprincipled preference. I just think it would be a great shame—a pity, too bad—if this great biological and spiritual adventure didn't continue: it has a marvelous past, and I hate the thought of its not having an exciting future.[136]

However, if Bennett *were* to "slide"—his word—a principle under his pro-humanity stance, he wrote that "it would probably be one about the prima facie obligation to ensure that important business is not left unfinished."[137] To my knowledge, this is the first explicit articulation of what might be called the "argument from unfinished business" for the continued existence of humanity, an idea later developed by futurists like Wendell Bell, Richard Slaughter, and Bruce Tonn, as well as myself (see Chapter 11).[138] As alluded to earlier, there is a rich history of potentiality thinking going back at least to the Enlightenment, whereby progress toward better, more desirable states of human life involves cumulative development over time, from the past to the present and the present into the future, with each generation standing on the shoulders of the last (a metaphor popularized by Newton). An often-quoted expression of this idea in the contemporary existential risk literature comes from Edmund Burke, who characterized society as a "partnership of the generations" that yields what he called an "eternal society."[139] In a 1790 critique of the French Revolution, which he worried could destroy this partnership, Burke wrote that society

> is a partnership in all science; a partnership in all art; a partnership in every virtue and in all perfection. As the ends of such a partnership cannot be obtained in many generations, it becomes a partnership not only between those who are living, but between those who are living, those who are dead, and those who are to be born.[140]

The unfinished business argument fuses potentiality thinking of a certain sort with this idea of cumulative development to derive a specifically *teleological* account of why our extinction should be avoided: through the partnership of the generations, humanity can advance various transgenerational projects, and it is the *fact* that these projects have not yet been completed that gives us reason to ensure our continued survival—even if this reason is more a matter of aesthetics, preference, or taste, than of morality. Either way, the notion of unfinished business points to the idea of premature extinction, whereby the final end of our collective story is made worse by the fact that it happens *prior to* the attainment of some desired end. As Bruce Tonn puts it, the argument states that "present generations have an *obligation to see that humanity's important business is not left unfinished*, presumably due to pre-mature extinction of humanity."[141]

But what exactly is this unfinished business, according to Bennett? He did not elaborate on what it might be, although Wendell Bell interpreted him as referring

> to human accomplishments, especially exceptional ones in science, art, music, literature, and technology, and also human inventions and achievements of organizational arrangements, political, economic, social, and cultural institutions, and moral philosophy. The continuation of these achievements, obviously, depends upon the continuation of the human species.[142]

One could also answer the question by borrowing an idea from I. F. Clarke, who, in a 1971 article about the history of futurological predictions from the eighteenth century to the present, wrote that

> in the last 100 years the physical sciences and the technologies have reached their predicted goals: submarines, flying machines, atomic energy, space rockets all belong to the ancient history of forecasting. And yet the great social objectives are still with us. World peace, universal prosperity, the reign of law, the brotherhood of man—these aspirations make up the unfinished business of the human race (italics added).[143]

I myself am inclined to say that it would be a great shame if humanity were to perish before we construct not merely a "Theory of *Everything*" that integrates quantum field theory with Einstein's theory of general relativity (since the two are incompatible at the moment), but what might be called a "Theory of *Every Thing*," that is, a complete explanatory-predictive account of every type of phenomenon in the universe. For some of us privileged enough to have the opportunity to contemplate the great mysteries of existence, who are bothered by the lack of any satisfactory answer to the Leibnizian question of why there is something rather than nothing, the idea that humanity's story might end before we have solved these mysteries and answered this question is fiercely disappointing.

But is this a specifically moral position? Certainly, it has normative force, but *morality* constitutes only a subregion of the broader territory of normativity. There are all sorts of normative claims that aren't moral. Two questions are worth asking here: first, what is the best criterion of demarcation for morality? Bennett himself accepted R. M. Hare's claim that "only universalisable practical attitudes should be accounted moral," although he doesn't insist upon this criterion, adding that "if you think there can be unprincipled moral stands, then you may count my pro-humanity stand as 'moral' after all."[144] Second, what does it matter whether some position falls within our outside of morality's perimeter? The answer is that moral obligations have a special kind of force, one that can override most or all non-moral reasons against some course of action. To say that you should stop at the stop sign, or should not smoke, is different than saying you should give to the poor, or should not go around murdering others. Hence, we can distinguish between moral and non-moral versions of the argument from unfinished business, the former of which would be stronger than the latter, while the latter of which is what Bennett endorsed, using it to support his pro-humanity stance. Either way, Bennett's discussion of humanity's unfinished business may have been the first time that anyone gestured at the idea of premature human extinction within the Existential Ethics literature.

Person-Affecting Antinatalism

While Narveson and Bennett both agreed that there is no moral obligation to create new happy people, Hermann Vetter argued that Narveson's person-affecting view actually implies a moral obligation not to have children at all. This is based on a claim in Narveson's paper that I did not mention earlier, namely, that we are morally obligated not to create new *unhappy* people. As he wrote, "if . . . it is our duty to prevent suffering and relieve it," as this is one way to increase the total amount of wellbeing, then "it is also our duty not to bring children into the world if we know that they would suffer or that we would inflict suffering upon them."[145] This was the earliest enunciation of what Jeff McMahan would later call, in a 1981 review of *Obligations*, "the Asymmetry," also known as the "Procreation Asymmetry," which he defined as the position that,

> while the fact that a person's life would be worse than no life at all (or "worth not living") constitutes a strong moral reason for not bringing him into existence, the fact that a person's life would be worth living provides no (or only a relatively weak) moral reason for bringing him into existence.[146]

This asymmetry of duties led Narveson to the conclusion that, as Vetter put it, "in general—if it can be foreseen neither that the child will be unhappy nor that it will bring disutility upon others—there is no duty to have or not have a child."[147] But Vetter noted that it often cannot be foreseen whether a child will be unhappy or not, and hence one must make the decision to procreate under epistemic conditions of uncertainty. Understood this way, he argued that Narveson's Procreation Asymmetry and person-affecting principle actually implies the antinatalist view that "in any case, it is morally preferable not to produce a child."[148] He explicated this view by sketching a decision matrix: on the x-axis are the two possibilities of "child will be more or less happy" and "child will be more or less unhappy," while on the y-axis are the two options of "produce the child" and "do not produce the child." If one produces the child and it is more or less happy, there is "no duty fulfilled or violated" whereas if one produces the child and it is more or less unhappy, there would be a "duty violated." In contrast, if one doesn't produce the child and it would be more or less happy if it were to exist, there would once again be "no duty fulfilled or violated," whereas if one doesn't produce the child and it would be more or less unhappy if it were to exist, there would be a "duty fulfilled."

Hence, Vetter wrote that "it is seen immediately that the act 'do not produce the child' dominates the act 'produce the child' because it has equally good consequences as the other act in one case, and better consequences in the other." That is to say, *having the child* could yield one of two consequences: either violating one's duty not to create unhappy people, or neither fulfilling nor violating the duty to create happy people, because according to the person-affecting view, there is no such duty. But if one *doesn't have a child* and that child would be happy, one does nothing wrong, while if the child would be unhappy if it were created, one does something morally right by preventing it from existing. It follows that, on this account, "people should be discouraged from having children," which ostensibly implies that the human population should gradually dwindle to zero.[149] In Vetter's words:

> If such [antinatalist] tendencies are successful enough, the number of men on earth may begin to decrease, and if such development continues long enough, the human race will disappear. This, however, would not at all be a deplorable consequence according to Narveson's . . . and my own opinion: the existence of mankind is not a value in itself. On the contrary, if mankind ceases to exist, all suffering is extinguished perfectly, which no other human endeavour will be able to bring about. On the other hand, of course, all happy experiences of men will disappear. But this, according to Narveson's conclusion . . ., would not be deplorable, because no human subject would exist which would be deprived of the happy experiences.[150]

Although the connection between antinatalism and human extinction may seem straightforward, we will see in the second half of the next chapter that this is not actually the case, given the increasingly plausible possibility of radical life

	Child will be more or less happy	Child will be more or less unhappy	
Produce the child	No duty fulfilled or violated	Duty violated	
Do not produce the child	No duty fulfilled or violated	Duty fulfilled	This dominates

Figure 9.1 Vetter's decision matrix.

extension. If individual people could end up living for as long as humanity itself could survive (e.g., until the heat death of the universe), then there being no additional people does not necessarily entail there being no people at all. Nonetheless, the plausibility of radical life extension is a very recent development, and hence it would not have been unreasonable to posit a necessary link between antinatalism and human extinction, as Vetter does.

The Benevolent World-Exploder

This being said, Vetter defended yet another version of utilitarianism that will appear again in the following chapter: negative utilitarianism. As chance would have it, he discussed this theory at the very same conference in 1968 at which Groenewold introduced his taxonomy of "macro effects," "macro problems," and "macro morality," and in fact negative utilitarianism was first introduced by Sir Karl Popper, who was yet another attendee of this 1968 conference. The difference between negative and classical utilitarianism is that the latter—somewhat confusingly—takes "happiness" to be the sum of all the goodness and badness, intrinsic value and disvalue, that exists within some state of affairs. On this account, there is a kind of axiological symmetry within these dichotomies: both value and disvalue count equally; they are the mirror reflections of each other. In contrast, negative utilitarianism accepts one of the following claims: (a) suffering counts more than happiness (weak view), (b) some amount of suffering cannot be counterbalanced by any amount of happiness (lexical threshold view), (c) no amount of happiness can counterbalance any amount of suffering (lexical view), or (d) suffering is the only thing that matters (absolutist view). What motivates this theory is that, quoting Popper,

> human suffering makes a direct moral appeal, namely, the appeal for help, while there is no similar call to increase the happiness of a man who is

> doing well anyway. . . . [F]rom the moral point of view, pain cannot be outweighed by pleasure, and especially not one man's pain by another man's pleasure. Instead of the greatest happiness for the greatest number, one should demand, more modestly, the least amount of avoidable suffering for all.[151]

Hence, Popper's stated position seems to align most closely with the lexical view. But this yields an apparent problem, first pointed out by R. N. Smart in a 1958 paper, which abbreviated "negative utilitarianism" as "NU." Imagine that someone has access to a technology "capable of instantly and painlessly destroying the human race." Although this may sound "fanciful," Smart noted that it is "unfortunately much less so than it might have seemed in earlier times" (and indeed *today* it is within the realm of scientific plausibility that a high-powered particle accelerator could instantly and painlessly destroy everything within our future light cone by nucleating a vacuum bubble). Since there is bound to be some suffering if human life were to persist, NU would prescribe the use of this technology to instantaneously annihilate all living creatures on the planet—if not in the universe more generally, as Eduard von Hartmann wished. (Hence, smashing atoms together to nucleate a vacuum bubble may have been precisely the sort of future advancement that Hartmann hoped for.) One is thus morally obliged to become a "benevolent world-exploder," a conclusion that Smart described as patently "wicked."[152] While Popper himself had only intended his NU proposal as a principle for public policy, for a "humanitarian code," Smart's discussion of its shortcomings as a fundamental moral principle convinced many that it is, in fact, a nonstarter in ethics.[153]

Vetter, though, did not see things this way—nor did another participant at the 1968 conference, namely, the philosopher Yehoshua Bar-Hillel.[154] In Vetter's words,

> if mankind were extinguished by a nuclear war, the real evil . . . would be the way the extinction would take place: there would be so much terrible suffering for so many people before they die that this is a tremendous evil.

In other words, Going Extinct in a nuclear holocaust would be very bad. What isn't "one of the greatest evils we are confronted with," he continued, is Being Extinct. To the contrary,

> if mankind were completely extinguished in a millionth of a second without any suffering imposed on anybody, I should not consider this as an evil, but rather as the attainment of Nirvana. The effect of the extinction of mankind would be that all suffering of human beings is perfectly extinguished; likewise, of course, all happy experiences of human beings would be extinguished. But

> I think the extinction of suffering would count much more heavily than the extinction of happy experiences, because if nobody exists any longer, then there is no subject that is deprived of the happy experiences. I do not think we have moral duties towards unborn men, commending us to bring about their birth because of the happiness they would be going to experience—happiness which, on the top of it, is available only in a mixture with more or less unhappiness.[155]

Vetter thus went beyond the equivalence view in maintaining that Being Extinct would be *good* rather than merely *not bad*, for the same reason that attaining Nirvana—meaning "extinction, disappearance"—would be good for us as suffering individuals. However, Vetter did not go quite as far as R. N. Smart suggested negative utilitarians should go: he never claimed that we should try to figure out a way of annihilating humanity in a millionth of a second. His point was only that if this were to happen, it wouldn't be evil but instead be very good. Nor did he advocate for an "absolutist" version of antinatalism, whereby it is always impermissible to have children. Rather, writing in his aforementioned 1969 paper, he asserted that procreation "is still recommended when parents' utility is taken into account," although "it is morally preferable not to produce children at all" when one considers only the child.[156]

To End the Human Race

While Russell, Anders, and Jonas all embraced further-loss views, Narveson and Bennett defended the equivalence thesis, although Bennett's argument from unfinished business could be seen as a kind of *non-moral* further-loss view, since it identifies the failure to finish certain important business to be an extra reason our extinction would be bad. To put all of this in perspective, both further-loss views and the equivalence thesis answer "yes" to the question "Would human extinction be bad or wrong?," but their reasons are quite different: further-loss theorists point to the state or condition of Being Extinct as entailing one or more extra losses that render our extinction bad, that is, *in addition* to whatever harms the process or event of Going Extinct might cause, while equivalence theorists would say that there is nothing bad about Being Extinct, and hence the badness or wrongness of our disappearance is entirely reducible to the way it comes about. In other words, the second position's affirmative answer to the question above is conditional: *only insofar* as Going Extinct causes harms would our extinction, all things considered, be bad or wrong. Both further-loss and equivalence theorists accept the default view, of course, but whereas the former says that there is something bad or wrong about our extinction *independent* of how it happens (e.g., even if our extinction is entirely voluntary and peaceful, it would be regrettable if business were left unfinished), the latter insists that the default view is the *entire story*.

In contrast, while he also accepted the default view, Vetter took the more radical position of answering the question "Would human extinction be good?" with a strong "yes," since it would eliminate all future human suffering. On the one hand, if humanity were annihilated, there would be no one around to suffer the absence of happiness that might have otherwise existed. On the other hand, the elimination of suffering would be good even if there is no one to experience this absence. This insight—another kind of asymmetry—will be revisited in the next two chapters.

Before turning to the 1980s, when attitudes toward human extinction began to shift within the world of Anglophone philosophy, it is worth examining one more view put forward in the literature, which is much closer to Sidgwick's position than Narveson's, Bennett's, or Vetter's. This came from the philosopher Jonathan Glover in his 1977 book *Causing Death and Saving Lives*, which argued that, contra Narveson, we *should* create "extra people whose lives are worth living." One of the reasons that Glover cited concerns the perpetuation of humanity. All things being equal, he wrote, we should want there to be more people both synchronically *and* diachronically—that is, "the more people with worth-while lives there are the better," not just at any given moment but "spread out across future time." However, this does not imply a simple-minded "policy of maximizing happiness," he claimed, since there could be other things we value, and we might think that "the absence of these qualities cannot be compensated for by any numbers of extra worth-while lives without them." Nonetheless, the value of extra people does mean, echoing Sidgwick, that "to end the human race would be about the worst thing it would be possible to do," given "a belief in the intrinsic value of there existing in the future at least some people with worth-while lives."[157] Since extinction would foreclose the realization of such lives, it would be bad for reasons that have nothing to do with Going Extinct—that is, even if our extinction came about because everyone takes "a drug that would render us infertile, but make[s] us so happy that we would not mind being childless," it would still be very wrong. Glover thus held a further-loss view according to which the non-existence of future people, of intrinsic value, renders Being Extinct itself bad.

As a brief aside, Glover's views of population ethics and human extinction were developed alongside those of Parfit, as both collaborated with James Griffin in hosting a recurring seminar at the University of Oxford focused on issues in normative and practical ethics. Its aim was "to consider the application of ethical principles to real-world problems," and—somewhat humorously—was originally named "Life, Happiness, and Morality," but Parfit found this too insipid and changed it to "Death, Misery, and Morality."[158] We will examine Parfit's important contribution to the development of Existential Ethics shortly.

Traditional Ethics and Posthumous Harm

Having now surveyed a number of utilitarian positions in Existential Ethics, it may be useful to pause for a moment to reconsider statements from Anders,

Groenewold, and Jonas about the obsolescence of traditional ethical systems, which each saw as lacking the theoretical resources necessary to address the unique challenges of the Atomic Age. But is this true with respect to utilitarianism? Our discussion so far suggests that it is not: both person-affecting and impersonalist utilitarianism offer straightforward answers to the core questions of Existential Ethics. Neither struggles with the problem of extinction the way, for example, Kantian ethics does, as there is no apparent logical or practical contradiction derivable from an omnicidal maxim like "I will kill everyone in order to eliminate all suffering." As R. N. Smart's brother, J. J. C. Smart, wrote in his 1984 book *Ethics, Persuasion, and Truth*, referring specifically to Groenewold's 1968 discussion of macro effects,

> traditional rules of ethical thinking were evolved in relation to micro effects and may [thus] be inappropriate [for dealing with macro effects]. Certain philosophical systems of ethical precepts (and here I am thinking particularly of utilitarianism) should be able to cope in theory with effects at any level, but even so their practical application is difficult because of the difficulties in envisaging consequences of rapid technological change.[159]

Hence, while there may be practical limitations to utilitarianism, it seems "able to cope in theory" with questions that no moral or axiological system in the past ever had to confront, such as whether and why bringing about our extinction would be good or bad, right or wrong. The three philosophers mentioned above—among the earliest existential ethicists—thus apparently missed that the utilitarian theory proposed by Jeremy Bentham in the eighteenth century, and later developed by Mill and Sidgwick, does have the resources necessary to address Existential Ethics, even if one finds its answers unconvincing. Perhaps the reason they never discussed utilitarianism is that this theory has had limited influence within the Continental tradition, which all three were working in.

A second issue worth pausing on for a moment is relevant to questions of *how bad* Going Extinct could be, even in Vetter's scenario of (virtually) instantaneous annihilation. The question is: Could Going Extinct cause harm even if there is no attendant psychological or physical suffering? Some would answer "yes" if this involves cutting lives short, as these people would say that death can harm *the one who dies* by depriving one of future happiness, desirable experiences, fulfilled ambitions, and so on, which they could otherwise have had. A notable champion of this "deprivationist" account of death is Thomas Nagel, who contended that "the corresponding deprivation or loss," the "abrupt cancellation of indefinitely extensive possible goods," is one reason to see death as an "evil" and "misfortune" *for the decedent*.[160] This could be understood as a kind of further-loss view at the level of individuals rather than the collective whole, the species, or the universe, and it contrasts with a position famously defended by the ancient Greek philosopher Epicurus. On Epicurus' account, death cannot be bad for those who die because when one is alive, one is not dead, and when one is dead, one cannot be harmed because one no longer exists (assuming, as Epicurus did, that there is no afterlife). So where is the harm?

Although clever, many philosophers find Epicurus' argument unconvincing, opting instead for the deprivationist view. The point is that if Nagel is correct, then even "if mankind were completely extinguished in a millionth of a second without any suffering imposed on anybody," quoting Vetter, this might still be very bad by virtue of the fact that it would cut lives short, thereby depriving people of what could have been.[161] However, Nagel also held that "it cannot be said that not to be born is a misfortune," and hence Vetter's scenario would be bad *only* because Going Extinct would harm those who perish, *not* because Being Extinct itself would be bad.[162] One's position on whether death can harm the decedent is, therefore, pertinent to assessing the *extent of the badness* of Going Extinct: not only would instantaneous annihilation cause harm, even if there is no suffering, but scenarios in which lots of suffering occurs would be seen as even worse, since one should count both the harms of each individual *dying* and the harm associated with *being dead*.

Others discussed similar ideas with implications for the badness of our extinction, although none explicitly linked these to the questions of Existential Ethics. For example, Joel Feinberg contended in the late 1970s that people's interests can be "harmed," by which he meant "blocked" or "thwarted," even after they have died, which suggests that present generations could harm the interests of past generations by allowing humanity to die out.[163] To illustrate, imagine that a team of scientists were to invent a "world-exploding" device that kills every person on Earth instantaneously. Whereas a deprivationist would say that the deaths of all these people would (or at least could, if their lives are worth living) harm every one of them, Feinberg would add that this could also harm the interests of people *who had already passed away*, as this might block or thwart interests they may have had that extended beyond their own lifetimes. This would further exacerbate the badness of instantaneous extinction.

In fact, Bennett briefly explored a similar view in his 1978 paper discussed above, which suggested that actions today could negatively affect the *utilities* of *past people*. (He writes: "I am not endorsing this attitude to past people, but I shan't quarrel with it here.") For example, one might morally object "to using the calculus for military purposes because Leibniz," who co-invented the calculus, "wanted all his discoveries to contribute to universal peace," where the argument behind this moral objection is that "if I use the calculus in building a bomb, I am bringing a disutility to Leibniz by bringing it about that he was to that extent a man whose hopes were not going to be realized."[164] Again, this points at another reason one might think our disappearance would be bad: it could reduce the utility (happiness) of past people who, for example, believed that humanity should survive for as long as the universe remains habitable, finish its unfinished business, continue the march of progress, or whatever.

Dreadful to Contemplate

This brings us to the 1980s, which witnessed a flurry of new ideas about why our extinction might be bad or wrong, not just because it would entail some further loss associated with Being Extinct but because the meaning or value of our lives today depends upon humanity continuing to exist in the future. An argument of the latter sort came from a book chapter titled "Why Care About the Future?" by Ernest Partridge (1935–2018). His discussion began with the assertion that human beings are not in fact "disinclined to care for the future, much less to act upon such cares," as some philosophers had recently argued. One reason concerns "a basic human need" for what Partridge called self-transcendence, which involves (i) regarding something other than oneself as good for its own sake and (ii) desiring "the well-being and endurance of this 'something else' for its own sake, apart from its future contingent effects upon" the individual.[165] The connection between (i) and (ii) is that when one genuinely values an object for itself, one will naturally wish for that object to continue existing and flourishing beyond one's own lifetime, as this is part of what it means to value, love, or cherish something one takes to be "significant" and "important." Here Partridge quotes John Passmore's 1974 analysis of love, according to which,

> when men act for the sake of a future they will not live to see, it is for the most part out of love for persons, places, and forms of activity, a cherishing of them, nothing more grandiose. It is indeed self-contradictory to say: "I love him or her or that place or that institution or that activity, but I don't care what happens to it after my death." To love is, amongst other things, to care about the future of what we love. . . . This is most obvious when we love our wife, our children, our grand-children. But it is also true in the case of our more impersonal loves: our love for places, institutions, and forms of activity.[166]

The fact that we care about whether valued, loved, and cherished entities persist in favorable conditions long after we are gone can be illustrated by a simple thought experiment. "Suppose," Partridge wrote,

> that astronomers were to determine, to the degree of virtual certainty, that in two hundred years the sun would become a nova and extinguish all life and traces of human culture from the face of the earth. . . . Would not this knowledge and this awareness profoundly affect the temperature and the moral activity of those persons now living who need not fear, for themselves or for anyone they might love or come to [love], personal destruction in this eventual final obliteration?

For most readers, Partridge conjectured, "it would be dreadful to contemplate the total annihilation of human life and culture even two hundred years hence." He added that

> we would feel a most profound malaise were we to be confronted with the certain knowledge that, beyond our lifetimes but early in the future of our civilization, an exploding sun would cause an abrupt, final and complete end to the career of humanity and to all traces thereof.

Why? Because we are not actually "indifferent to the fate of future persons unknown and unknowable to us, or to the future career of institutions, species, places, and objects that precede and survive our brief acquaintance thereof."[167] Or quoting, as Partridge did, a 1972 paper by Edwin Delattre, "the meaning of the present depends upon the vision of the future as well as the remembrance of the past," and hence, "to the extent that men are purposive and teleological in the world, the destruction of the future is suicidal by virtue of its radical alteration of the significance and possibilities of the present."[168] It follows that without some confidence that posterity will exist under conditions favorable to its flourishing, "our lives would be confining, empty, bleak, pointless, and morally impoverished."[169]

Indeed, not only do we care about the continued existence of certain valued-for-themselves entities in the world, but our need for self-transcendence further manifests in a willingness and enthusiasm to actively contribute to the development and preservation of these entities, i.e., "communities, locations, causes, artifacts, institutions, ideals, and so on." As Partridge wrote, "'self transcendence' describes a class of feelings that give rise to a variety of activities," and the natural urge to transcend oneself is "no small ingredient in the production of great works of art and literature, in the choice of careers in public service, education, scientific research, and so forth." The central aim in all of these cases is "for the self to be part of, to favorably effect, and to value for itself, the well-being and endurance of something that is not itself," which is to say that we strive to merge with, contribute to, and ensure the preservation of things greater than us, and through these activities create something that outlasts our brief sojourns on planet Earth. This is in part what makes life valuable and meaningful, and hence those unable or unwilling—perhaps because they suffer from a narcissistic personality, Partridge claimed—to fully transcend themselves are left wallowing in a pitiable state of alienation. In Partridge's words,

> individuals who lack a sense of self transcendence are acutely impoverished in that they lack significant, fundamental, and widespread capacities and features of human moral and social experience. Such individuals are said to be alienated, both from themselves and from their communities.[170]

Another manifestation of the drive for self-transcendence is the fact that many people—artists and academics perhaps offering the paradigm cases—strive to mitigate the distress elicited by one's awareness of death by achieving some degree of vicarious immortality, whereby one "lives on" in the hearts and minds of future generations. The bones of Aristotle, Newton, Darwin, Marie Curie, Rachel Carson, and Einstein have all been laid to rest, yet these individuals nonetheless "survive" in the collective memories and consciousness of people living today. Indeed, it has been said that one dies twice, the second of which occurs the last time one's name is uttered, an idea obviously connected to Anders' notion of the "second death." (Ernest Hemingway supposedly put it like this: "Every man has two deaths, when he is buried in the ground and the last time someone says his name. In some ways men can be immortal.")[171] Quoting from Christopher Lasch's 1978 book *The Culture of Narcissism*, Partridge argued that the most important consolation in the face of our deaths as individuals

> is the belief that future generations will in some sense carry on [one's] life work. Love and work unite in a concern for posterity, and specifically in an attempt to equip the younger generation to carry on the tasks of the older. The thought that we live on vicariously in our children (more broadly, in the future generations) reconciles us to our own supercession.[172]

Although Partridge's main focus in discussing these ideas was self-transcendence (indeed, most of these examples were adduced in *arguing that* self-transcendence is "a basic human need"), he touched upon at least three distinct arguments germane to Existential Ethics. The first could be called the "argument from valuing," which states that our continued survival matters because the communities, artifacts, institutions, ideals, and so on that we value cannot exist without us, and what it means to value these things is to wish for their endurance and flourishing through time. The second could be called the "argument from impoverishment," which states that without hope of future generations existing and flourishing, our lives in the present will be rendered empty, bleak, pointless, and so on. The third could be called the "argument from immortality," which similarly states that without confidence in the existence of posterity, a major source of motivation to contribute to the world through art, scholarship, engineering, community service, public office, and so on would evaporate. In other words, if one is driven by the hope of living on in the hearts and minds of future generations, and if one comes to believe that future generations will not exist, then this drive will lose its motivational force.[173]

The Pearl in the Schell

This brings us to one of the most notable publications of the second wave of Existential Ethics, namely, Jonathan Schell's 1982 book *The Fate of the Earth*.

An international bestseller, it offered the first *sustained meditation* on the ethical and evaluative implications of our disappearance—the most extensive treatment of the subject prior to Schell being Jonas' 1979 tome discussed earlier. As we will see, Schell touched on many of the arguments and ideas previously examined, although the only two aforementioned theorists that he cited were Russell and Jaspers. My guess is that Schell, a journalist by profession rather than a philosopher, was most likely unaware of the work of Anders, Groenewold, Jonas, Narveson, and Glover, and hence there is a degree of reinventing the wheel in his book.[174] His unfamiliarity with prior scholarship is evidenced by the claim, made at the beginning of his discussion, that

> the possibility that the living can stop the future generations from entering into life compels us to ask basic new questions about our existence. . . . No one has ever thought to ask this question before our time, because no generation before ours has ever held the life and death of the species in its hands.[175]

But of course others *had* asked this very question, in many cases declaring, as Schell does, that no one had previously asked such questions. This gestures at a general fact about the second wave, which is that most existential ethicists seemed to have been unaware of the work of others. The literature on the topic was extremely fragmentary. Even in cases where they probably did know about others' work—for example, Anders, Jaspers, and Jonas likely read each other's books, as they knew each other—almost no one cited each other. This is, in fact, a feature of the *third* wave that distinguishes it from the second: for the first time, a tradition of cumulative scholarship emerged, whereby early contributions were cited and built upon by later philosophers, which enabled a certain kind of *progress* to commence within Existential Ethics. No such cumulative development occurred during the second wave.

Returning to Schell's book, there is an issue worth addressing before we dive in to its ideas. One finds many curious, and in some cases quite striking, similarities between Schell's philosophical exploration of human extinction and Anders' theory of omnicide. The most glaring example is Schell's use of the term "second death" (capitalized below as "Second Death"), which formed the foundation for much of his analysis, although what Schell meant by this term was not what Anders meant. Nonetheless, Anders publicly accused Schell of plagiarism, which he followed up with court papers, writing in his characteristically poetic style that "the name of the pearl within the shell is not Schell, but Anders."[176] However, virtually every scholar familiar with Schell and/or Anders is in agreement that Schell did not in fact borrow, copy, or steal anything from Anders.[177] As Dan Zimmer tells me, the better explanation is that the similarities between the ideas of each were the result of "convergent evolution" from a common point of departure, namely, the work of Arendt.[178] Recall that Anders was married to Arendt decades before he began to write about nuclear self-annihilation; as for Schell, he was, by his own account, immensely influenced by Arendt, describing her thought on two separate occasions as "more suggestive and invaluable than any other thinker's" and "an indispensable foundation for reflection on" normative questions about extinction.[179]

Hence, the overlap was almost certainly coincidental, although Anders should, of course, be given credit for articulating certain insights about how the Atomic Age has change the human condition and for using the term "second death" before Schell. To quote George Kateb, while Schell deserves "the distinction of giving greater life to the subject of nuclear weapons than any other [by making] human extinction the center of the whole subject," Anders was "the one who first insisted that adequacy to the subject required dwelling on the possibility of human extinction."[180]

The Second Death

Whereas for Anders the "Second Death" referred to the "death" that those already deceased would undergo if their memories were lost forever by extinction, thus making it "as if they had never been," Schell defined the term simply as "the death of mankind."[181] This may seem somewhat trivial, but for Schell it was an absolutely crucial idea, because in the most widely discussed extinction scenario at the time—nuclear annihilation—the violent horror of all the "first deaths" (as we could call them, a term that Schell did not use) that this would entail could easily occlude the separate and distinct loss of humanity itself. As Schell wrote, "it is important to make a clear distinction between the two losses," that is, the first deaths of individuals and the Second Death of humanity, because "otherwise, the mind, overwhelmed by the thought of the deaths of the billions of living people, might stagger back without realizing that behind this already ungraspable loss there lies the *separate loss* of the future generations."[182] To illustrate this distinction, Schell offered a thought experiment contrasting "two different global catastrophes." He wrote:

> In the first, let us suppose that most of the people on earth were killed in a nuclear holocaust but that a few million survived and the earth happened to remain habitable by human beings. In this catastrophe, billions of people would perish, but the species would survive, and perhaps one day would even repopulate the earth in its former numbers. But now let us suppose that a substance was released into the environment which had the effect of sterilizing all the people in the world but otherwise leaving them unharmed. Then, as the existing population died off, the world would empty of people, until no one was left. Not one life would have been shortened by a single day, but the species would die.[183]

This is reminiscent of the thought experiment that I outlined in the previous chapter, involving world A and world B. But whereas in our thought experiment the catastrophe and its effects are identical in both worlds (10 billion deaths) while the population sizes differ (11 billion in A and 10 billion in B), Schell flipped this around such that the population sizes are the same but the catastrophes are different. Hence, Schell's first scenario involves immense suffering without humanity disappearing—that is, there is no Second Death, only a large number of first deaths—while in the second scenario, there is no extra suffering or lives cut short, although our species dies out.[184] The point of many thought experiments is to pull apart things that normally go together and, in doing so, to reveal a hidden fact or truth.[185] For Schell, a proper assessment of the ethical and evaluative implications of our extinction requires conceptual clarity about the distinctiveness of losing the entire species versus the obliteration of any number of individual persons in an extinction-causing catastrophe. Why? Because, he argued, the loss of humanity itself engenders *its own unique* ethical and evaluative implications. Schell thus rejected the equivalence thesis according to which the badness/wrongness of our disappearance, of the Second Death, is reducible to the badness/wrongness of Going Extinct.[186] There is something else of relevance. Even more, he contended that this "something else" is, in certain important respects, even more significant than the tragedy of everyone on the planet being exterminated. In his words, "the cancellation of all future generations of human beings," of which there could be an "infinite number," he tells us, "would be in a sense even huger" than "the untimely death of everyone in the world."[187] This of course yields a further-loss view, whereby future generations not only (a) count morally but (b) count for more. (By contrast, one could accept a further-loss view according to which certain additional losses count morally, but not as much as the loss of individual lives. In most cases above, the relative significance of these additional losses—the League of Generations, the moral order, etc.—was left unspecified.)

The two parts of Schell's thesis give rise to two corresponding questions: first, why does the loss of future generations count morally? And second, why is this loss "even huger," using Schell's somewhat awkward locution? On my reading of Schell's position, which he delineates mostly in Part II of his book, the central reason that future generations matter concerns a variant of the argument from impoverishment that Partridge, whom Schell did not cite, touched upon the year earlier, in 1981. However, Schell formulated this argument within a specifically Arendtian framework, at the center of which was Arendt's notion of the "common world." This refers to the realm that we share in common with each other, and which transcends us as individuals and the cohorts to which we belong, in contrast to the realms that each of us occupies in private. Through the vehicle of what Arendt terms publicity—that is, of making public—the common world emerges as a tapestry of shared, inherited, and to-be-passed-along ideas, knowledge, practices, traditions, monuments, projects, and so on stretching from the distant past into the indefinite future. As Arendt wrote in her 1958 book *The Human Condition*,

> the common world is what we enter when we are born and what we leave behind when we die. It transcends our life-span into past and future alike; it was there before we came and will outlast our brief sojourn in it. It is what we have in common not only with those who live with us, but also with those who were here before and with those who will come after us.[188]

Hence, Schell writes that the common world

> is made up of all institutions, all cities, nations, and other communities, and all works of fabrication, art, thought, and science, and it survives the death of every individual. It is basic to the common world that it encompasses not only the present but all past and future generations.

As such, the common world constitutes the foundation of everything that makes our lives meaningful, purposive, and worthwhile. But this world cannot, of course, exist if humanity no longer does, which leads to the following argument that we can reconstruct from Schell's writing like this: the loss of future generations via the Second Death would be bad because it would destroy the common world; destroying the common world would be bad because without it the meaning, purpose, and worthwhileness of our lives would be seriously compromised; and compromising the meaning, purpose, and worthwhileness of our lives would be bad for the obvious reason that we naturally want our lives to have these qualities. This is Schell's Arendtian interpretation of the argument from impoverishment. Examples of him expressing this idea include the following, which I have assembled from different parts of his book, and are worth quoting in full because of Schell's eloquence in articulating them:

- "We need the assurance that there will be a future if we are to take on the burden of mastering the past—a past that really does become the proverbial 'dead past,' an unbearable weight of millennia of corpses and dust, if there is no promise of a future. Without confidence that we will be followed by future generations, to whom we can hand on what we have received from the past, it becomes intolerably depressing to enter the tombs of the dead to gather what they have left behind; yet without that treasure *our life is impoverished*."
- "Being human, we have, through the establishment of a common world, taken up residence in the enlarged space of past, present, and future, and if we threaten to destroy the future generations we harm ourselves, for the threat we pose to them is carried back to us through the channels of the

common world that we all inhabit together. Indeed, 'they' are we ourselves, and if their existence is in doubt our present becomes a *sadly incomplete affair*, like only one word of a poem, or one note of a song. Ultimately, *it is subhuman*."

- "Because the unborn generations will never experience their cancellation by us," a point to which we will return below, "we have to look for the consequences of extinction before it occurs, in our own lives, where it takes the form of a *spiritual sickness* that corrupts life at the invisible, innermost starting points of our thoughts, moods, and actions."
- Without the assurance that posterity will exist, "nothing else that we undertake together can make any practical or moral *sense*," to which he added that "all human activities that assume the future are *undermined directly*" by the novel prospect of self-annihilation.
- "The reason that so much emphasis must be laid on the living generations is not that they are more important than the unborn but only that at any given moment they, by virtue of happening to be the ones who exist, are the ones who pose the peril, who can *feel the consequences of the peril in their lives*, and who can respond to the peril on behalf of all other generations."
- Referring to the fact that extinction would eliminate, once and for all, both *mortality* and *natality* (Arendt's term), the latter of which is what enables the human species to endure, Schell wrote that "the threat of the loss of birth . . . cannot be a source of immediate, selfish concern; rather, this threat assails everything that people hold in common, for it is the ability of our species to produce new generations which assures the continuation of the world in which all our common enterprises occur and *have their meaning*."
- Hence, Schell wrote it is only "by acting to save the species, and repopulating the future, [that] we break out of the *cramped, claustrophobic isolation of a doomed present*, and open a path to the greater space—the only space fit for human habitation—of past, present, and future." If we can achieve this, then "suddenly, *we can think and feel again*. Even by merely imagining for a moment that the nuclear peril has been lifted and human life has a sure foothold on the earth again, we can feel the beginnings of a *boundless relic and calm—a boundless peace*" (all italics added).

In a phrase, life would be greatly impoverished without the assurance that the common world will continue to exist, and since one way for the common world to stop existing is for humanity to perish, we thus have strong reason to make sure that this does not occur. Or as Schell put it, since "all human aims, personal or political, presuppose human existence, it might seem that the task of protecting that existence should command all the energy at our disposal."[189] This is, on my reading, the main thrust of Schell's view, although his presentation is so discursive, flitting from one idea to another like the notes of a frenzied jazz improvisation, that it can be difficult to extract a single coherent line of thinking. Nonetheless, these are not the only arguments that he made, or gestured at, in support of his anti-extinction thesis.

Amputating the Future

Schell also hinted at the arguments from unfinished business and immortality, albeit only in passing and without making the conclusions of these arguments explicit. In each case, Schell tied them to the central claim that without humanity there can be no common world. For example, he noted that the Burkean "partnership of the generations" unfolds within this arena, which is also necessary for ideas or legacies of past people to become immortalized. After reproducing Burke's description of the partnership (the same passage quoted earlier in this chapter), he then quoted the ancient Greek politician Pericles (495–429 BCE), who likened the city and people of Athens in his Funeral Oration to "a 'sepulchre' for the remembrance of the soldiers who had died fighting for their city," that is, the city and people are the means by which *traces* of those lost in time can nonetheless "live on" in some way. "Thus, whereas Burke spoke of common tasks that needed many generations for their achievement, Pericles spoke of the immortality that the living confer on the dead by remembering their sacrifices," the implication being that the elimination of the possibility of (a) achieving the ends of the partnership and (b) attaining vicarious immortality would both be bad, and hence so would human extinction.

Another idea that Schell touched upon, and which is not directly related to the common world, was what I referred to above as the argument from cosmic significance. In his words, "the extinction of the species goes farther, and removes from the known universe the human kind of being, which is different from any other kind that we as yet know of." And although he imagined in his two-catastrophes thought experiment that the second scenario in which humanity dies out slowly due to infertility would not introduce any additional suffering, he also gestured at the no-ordinary-catastrophe thesis in arguing that one potential advantage, if you will, of nuclear annihilation over other possible scenarios of Going Extinct is that "by killing off the living quickly, extinction by nuclear arms would spare us those barren, bitter decades of watching and feeling the end close in." But neither of these were elaborated any further.

Yet another argument he provided was that without humanity in the universe, either (a) nothing at all would have any value, i.e., everything would lose its value, because without valuers there can be no value, or (b) value would continue to exist but in some sense be squandered, as there would be no one there to appreciate it. Without us, Schell declared, "everything there is loses its value," adding that "the qualities of worth find in us their sole home in an otherwise neutral and inhospitable universe," and that while "mankind is [not] to be thought of . . . as something that possesses a certain worth," we are nonetheless "the inexhaustible source of all the possible forms of worth, which has no existence or meaning without human life." He explicated this idea in arguing that,

> without entering into the debate over whether beauty is in the eye of the beholder or in the thing itself, we can at least say that without the beholder the beauty goes to waste. The universe would still exist, but the universe as it is imprinted on the human soul would be gone. Of many of the qualities of worth in things, we can say that they give us a private audience, and that insofar as they act upon the physical world they do so only by virtue of the response that they stir in us. For example, any works of art that survived our extinction would stare off into a void without finding a responding eye, and thus become shut up in a kind of isolation.[190]

Other themes that appeared in Schell's sweeping exploration of the human extinction include deep-future and potentiality thinking, which he presented, in Russellian fashion, by first sketching the vast and extraordinary history of Earth, life, and humanity going back millions and billions of years.[191] This entire history is now in jeopardy because of nuclear weapons and our deleterious collective impact on the natural environment; we have become a "menace to both history and biology . . . capable of destroying in a few years, or even in a few hours, what evolution has built up over billions of years."[192] Looking in the other temporal direction, Schell noted the "open-ended possibilities for human development" that lay before us, and emphasized that "there is another, even vaster measure of the loss" that the Second Death would entail because

> stretching ahead from our present are more billions of years of life on earth, all of which can be filled not only with human life but with human civilization. The procession of the generations that extends onward from our present leads far, far beyond the line of our sight, and, compared with these stretches of human time, which exceed the whole history of the earth up to now, our brief civilized moment is almost infinitesimal. And yet we threaten, in the name of our transient aims and fallible convictions, to foreclose it all. If our species does destroy itself, it will be a death in the cradle—a case of infant mortality. The disparity between the cause and the effect of our peril is so great that our minds seem all but powerless to encompass it.

To grasp the true cost of extinction, one must assume a new perspective on the world—a perspective peering down upon what currently is and could someday be from the vantage point of cosmic space and time. "Whatever particular scene might come to mind, and whatever view and mood might be immediately present," he explained,

> from this earthly vantage point another view—one even longer than the one from space—opens up. It is the view of our children and grandchildren, and of

> all the future generations of mankind, stretching ahead of us in time—a view not just of one earth but of innumerable earths in succession, standing out brightly against the endless darkness of space, of oblivion.

The immensity and grandness of this view is precisely why we are incapable of comprehending our extinction in any meaningful way, and why so many of us deceive ourselves into thinking that it could not possibly happen. Quoting Schell once more, "the thought of cutting off life's flow, of amputating this future, is so shocking, so alien to nature, and so contradictory to life's impulse that we can scarcely entertain it before turning away in revulsion and disbelief." To bring about our extinction would be, he argued, the greatest possible *crime against the future*, as it would constitute "the murder of the future." And since "this murder cancels all those who might recollect it even as it destroys its immediate victims the obligation to 'never forget' is displaced back onto us, the living." In other words, the costs of extinction are felt not by those who could have existed in the future, as they will never be born, but by those of us in the present, whose lives are impoverished by the threat of annihilation and, with it, the permanent erasure of the common world. Yet the enormity of these costs far transcends our inherent capacities of comprehension, a point that echoed Anders' discussion of the Promethean gap and inverted Utopianism.

Although Schell emphasized the huge number of generations that could come after us (and, as noted above, that the number of "possible people" in the future is "infinite"), it is worth underlining that he did not seem to understand the cost of cancelling these future generations in total-impersonalist utilitarian terms. To the contrary, he repeatedly expressed the person-affecting idea that failing to bring unborn "people" into existence would not itself be wrong, since non-existent "people" cannot be harmed in any way. In his words, referring to the state or condition of Being Extinct, "there is no suffering (or any other human experience) in it." One way he thought about this was *in terms of* the Epicurean account of death. Quoting Epicurus' disciple Lucretius: "Do you not know that when death comes, there will be no other you to mourn your memory, and stand above you prostrate?" Hence, Schell asked:

> For who will suffer this loss [the Second Death], which we somehow regard as supreme? We, the living, will not suffer it; we will be dead. Nor will the unborn shed any tears over their lost chance to exist; to do so they would have to exist already. The perplexity underlying the whole question of extinction, then, is that although extinction might appear to be the largest misfortune that mankind could ever suffer, it doesn't seem to happen to anybody, and one is left wondering where its impact is to be registered, and by whom.

This consideration is precisely what led Schell to claim that it is us, those currently alive, who must suffer the consequences of extinction, which of course brings us back to the argument from impoverishment. In Schell's words, "we trace the effects of extinction in our own world because that is the only place where they can ever appear," although he maintained that these effects in the present, "important as they are, are only the side effects of our shameful failure to fulfill our main obligation of valuing the future human beings themselves." He reiterated this idea elsewhere, writing that "in coming to terms with the peril of extinction . . . what we must desire first of all is that people be born, *for their own sakes*, and not for any other reason," and that "we can open this path [that is, the "boundless relief and calm" mentioned earlier] only if it is our desire that the unborn exist *for their own sake*" (italics added). Hence, somewhat confusingly, even though these "people" have no identity, and indeed are not "persons" at all, as "they lack the individuality that we often associate with the sacredness of life, and may at first thought seem to have only a shadowy, mass existence," we must somehow still value them for themselves, according to Schell. We should not see them as having merely instrumental value, that is, as valuable simply because they are necessary "to lead a decent life ourselves in a common world made secure by the safety of the future generations." So far as I can tell, Schell's emphasis on valuing future people for themselves was largely a pragmatic point: the desire for a valuable, meaningful, and worthwhile life right now "flows from this commitment" to ensure the existence of future generations, who we should care about *independently* of their role in enriching our present. *Only* if we desire that the unborn come into existence for themselves can we then effectively open the path of "relief and calm." Ultimately, then, the meaning of extinction can only "be sought among the living," as the unborn "cannot experience their plight."

Some of these claims are, I think, difficult to make sense of, though Schell can be forgiven for struggling with what he described as "the metaphysical-seeming perplexities involved in pondering the possible cancellation of people who do not yet exist."[193] Not only had the question of future generations and human extinction become topics of serious analysis among Anglophone philosophers just years before Schell's book was published, but such questions, which lie at the intersection of population ethics and Existential Ethics, still confound philosophers today. His book, whatever its flaws, is an extraordinarily bold and brilliant exploration of the area.

To summarize, Schell's main thesis, spelled out in Arendtian terms, was that the Second Death is not only normatively relevant but constitutes a much greater loss—"in a sense," he says—than *any number of individual deaths* that Going Extinct might entail. One reason concerns the impoverishment of our lives in the present, given that our extinction would destroy the common world, which is the wellspring for so much of the value, meaning, and worthwhileness of our lives today. Furthermore, it is this public arena in which the partnership of the generations unfolds and past people are able to attain vicarious immortality, both of which would be expunged if the common world were to cease existing. Schell also pointed to the possibility that without us there would be no value in the universe, or at least that this value would go to waste, and that humanity could persist for a very long time to come and continue to progress over this period. But he did not seem to believe that our disappearance would be wrong because we have an obligation to maximize intrinsic value in the universe, as Sidgwick believed.

Peace, War, and Parfit

The immense success of Schell's book helped to reinvigorate the anti-nuclear movement of the 1980s, a period that witnessed the Soviet-Afghan War, the election of Ronald Reagan, and of course the discovery of the nuclear winter phenomenon the same year Schell's book was published. As a *New York Times* obituary for Schell, who died in 2014, states, "Mr. Schell was widely credited with helping rally ordinary citizens around the world to the cause of nuclear disarmament," and in fact a panel of experts convened by New York University identified *The Fate of the Earth* as "one of the century's best 100 works of journalism."[194] But this book was much more than that: it also contained, as Part II, a philosophical treatise on Existential Ethics, offering the first comprehensive examination of why our extinction would be bad or wrong (again bracketing Jonas' 1979 book).

This brings us to Parfit's work on the topic, which in a certain way was the exact opposite of Schell's: whereas Schell provided a lengthy meditation on human extinction that was often quite unsystematic in how it presented its arguments, Parfit's treatment of the topic was incredibly brief—just a few paragraphs—yet systematic and rigorous, built upon hundreds of pages of groundbreaking ideas that filled his 1984 book *Reasons and Persons*—one of the most celebrated philosophical works of the century. Furthermore, while Schell drew mostly from the Continental tradition, especially from Arendt, and foregrounded the argument of impoverishment, Parfit was working squarely within the Analytic tradition and said nothing about this idea. Parfit did, however, accept a further-loss view that (a) identified the *worst aspect* of our extinction as being the opportunity costs of Being Extinct, rather than the death of any number of people caused by Going Extinct, and (b) combined Sidgwickian impersonalism with deep-future and potentiality thinking about the possibility of future progress in relation to certain ideal goods.

As with Schell, Parfit began his discussion with a thought experiment also reminiscent of our scenario involving worlds A and B. In Parfit's version, we are asked to consider the following three scenarios (quoting him):

1. Peace.
2. A nuclear war that kills 99% of the world's existing population.
3. A nuclear war that kills 100%.

Many would agree that (3) is worse than (2), and (2) is worse than (1). What interested Parfit, though, was the *difference in badness* between these scenarios. "Most people believe that the greater difference is between (1) and (2)," he wrote, whereas in his view, "the difference between (2) and (3) is very much greater."[195] To put this in perspective, a person-affecting theorist who accepts a simple linear aggregation function (two deaths is twice as bad as one death) would hold that the badness of a nuclear catastrophe increases with, let's say, the total number of deaths that it causes, and that once the percentage of casualties reaches 100, the situation's badness suddenly *plateaus*, since the extinction of humanity means that there would be no one left to be harmed. In contrast, for Parfit, the situation's badness would suddenly *skyrocket* upon reaching 100 percent, as this would constitute a critical moral threshold that triggers certain further losses which carry a great deal of axiological weight.[196] See Figure 9.2.

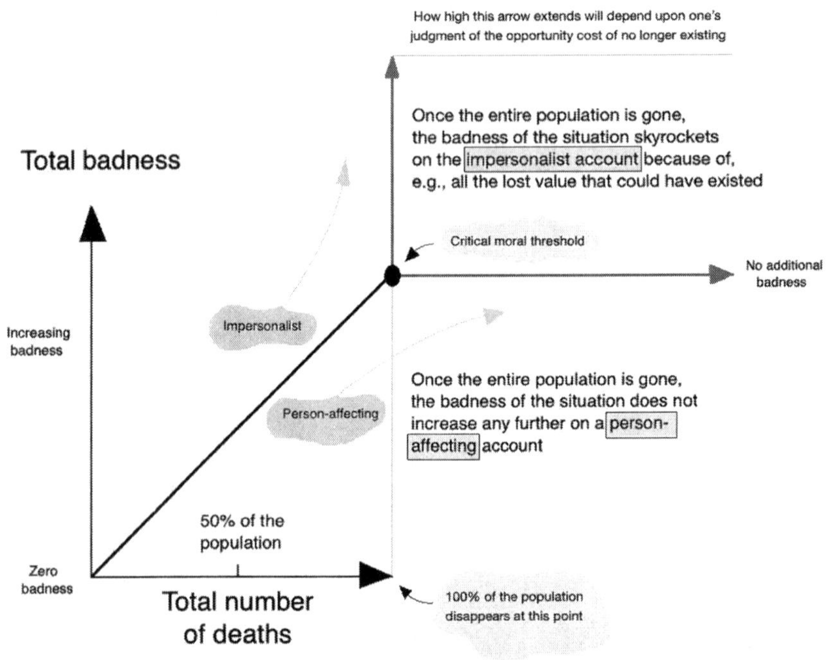

Figure 9.2 This shows the badness of a catastrophe increasing linearly with the total number of deaths, assuming a linear aggregative function. On a person-affecting view, the badness *levels off* when the entire human population perishes. In contrast, on the impersonalist account—indeed, on any further-loss view—the badness of the catastrophe suddenly *jumps* when 100 percent of the population dies. (Where the vertical line ends will depend on how great one considers the losses to be.)

But what exactly are these losses? As alluded to above, Parfit offered two distinct answers, both of which are greatly *amplified in significance* by the fact that humanity could keep existing for an extremely long time from now. In his words,

> Earth will remain inhabitable for at least another billion years. Civilization began only a few thousand years ago. If we do not destroy mankind, these few thousand years may be only a tiny fraction of the whole of civilized human history. The difference between (2) and (3) may thus be the difference between this tiny fraction and all of the rest of this history. If we compare this possible history to a day, what has occurred so far is only a fraction of a second.

With this deep-future framing, the first reason he gave was straightforwardly Sidgwickian. As he wrote,

> one of the groups who would accept my view are Classical Utilitarians. They would claim, as Sidgwick did, that the destruction of mankind would be by far the greatest of all conceivable crimes. The badness of this crime would lie in the vast reduction of the possible sum of happiness

that could come to exist within our future light cone if humanity were to survive. The second reason concerns the future development of Sidgwick's "ideal goods," such as "the Sciences, the Arts, and moral progress, or the continued advance towards a wholly just world-wide community." Parfit continued: "The destruction of mankind would prevent further achievements of these three kinds. This would be extremely bad because what matters most would be the highest achievements of these kinds, and these highest achievements would come in future centuries." In fact, the reason that Parfit noted that non-religious normative ethics has only been studied by "many people, only since about 1960," which I quoted at the end of Chapter 8, is to argue that ethics may be the "least advanced" of these goods, and hence has the greatest potential to progress if humanity does not die out. "Belief in God, or in many gods," he declared,

> prevented the free development of moral reasoning. Disbelief in God, openly admitted by a majority, is a very recent event, not yet completed. Because this event is so recent, Non-Religious Ethics is at a very early stage. We cannot yet predict whether, as in Mathematics, we will all reach agreement. Since we cannot know how Ethics will develop, it is not irrational to have high hopes.[197]

In other words, it could be that the widespread disagreement among ethicists about certain fundamental deontic and evaluative questions is a sign not that there is no ultimate truth about what is right and wrong, good and bad, but rather a symptom of the field of secular ethics being so young and underdeveloped. Perhaps with enough time, philosophers will converge upon a handful of basic propositions that virtually everyone will accept, just as scientists the world over more or less unanimously agree about things like heliocentrism, the age of the universe, the continuity of space and time, the Standard Model of particle physics, and so on. Indeed, an overarching aim of Parfit's philosophical efforts was to show that "it is a mistake to think that there are deep disagreements among Kantians, contractualists, and consequentialists."[198] Rather, as Parfit later contended, "these people are climbing the same mountain on different sides," which implies that with sufficient progress in the field there could indeed arise a single unified theory that all previous factions of ethical persuasion can agree on.[199] The failure to reach this summit of moral agreement because humanity has self-destructed would thus constitute, for Parfit, an especially tragic further loss that, as such, renders our extinction, however it may come about, very bad indeed. In a phrase, Parfit saw the state or condition of Being Extinct—independent of how this is brought about—as an immense axiological catastrophe for two reasons, one of which Sidgwick endorsed and the other of which he would not have.

Millions of Years, 500 Trillion People

Before moving on, it is worth noting that Schell and Parfit were not the only ones in the early 1980s who thought about human extinction in explicitly deep-future and potentiality terms. There was also J. J. C. Smart, who briefly raised the issue in his aforementioned 1984 book *Ethics, Persuasion, and Truth*. Smart emphasized both the quantity of future time over which our evolutionary lineage could persist and the increased quality of lives that our descendants could acquire. For example, he wrote that bringing about "the end of the human race" through nuclear war would prevent "humans [from] evolving into yet higher and more wonderful forms of life," and that since "most people's temporal horizons are limited [they] find it hard to think of the [nuclear] arms race in relation to the millions of years of possible evolution of the human race that lie ahead if we do not destroy ourselves." It is unclear whether Smart imagined this evolution proceeding via transhumanist or purely Darwinian means, although he did mention the possibility of technoscientific developments in the future radically improving our lives. Given the "great advances in the human condition due to science," he wrote, we might expect that

> if the human race is not extinguished there may be cures of cancer, senility, and other evils, so that happiness may outweigh unhappiness in the case of more and more individuals. Perhaps our far superior descendants of a million years hence (if they exist) will be possessed of a felicity unimaginable to us.

Smart also addressed Vetter's claim that if our species were annihilated "instantaneously and painlessly," this would not be a great evil; to the contrary, it may be a welcome occurrence.[200] But, Smart rejoined, not only would extinction foreclose the realization of better, higher forms of human life, it is also the case that "most people seem glad that they were born: we do not usually think of present people (and animals) that the pain in their lives outweighs their pleasures." Ultimately, he proposed two antidotes against the view that our extinction would be either not bad or positively good: the first was to develop stronger feelings "for the reality of the future, and of the possible glories of future evolution"[201] and the second was "more advocacy of utilitarianism," by which he apparently meant of a more impersonalist variety.[202] Of note is that Smart may have been the first to argue that, given how good the far future could be (or so he suggested), it matters little whether we die out tomorrow or push forward our extinction for a couple of centuries. In his words, "postponing is only of great value if it is used as breathing space in which ways are found to avert the final disaster." Because the future could be so immense, spanning millions and millions of years, the difference between surviving another few hundred years or perishing tomorrow is trivial.

Another theorist who took seriously the deep future was Carl Sagan. To my knowledge, he offered the very first quantitative estimate of the potential size of the future in terms of how many people could come to exist on our twirling pale blue dot. Some previous thinkers had attempted to calculate how large the human population could become, but these were all synchronic rather than diachronic estimates, meaning that they concerned the total population at any given moment rather than the total number of persons who could exist across time. For example, the Dutch scientist Antonie van Leeuwenhoek—the "Father of Microbiology"—extrapolated the population density of the Netherlands (120 people per square kilometer) to the land area of the entire planet and concluded that Earth could sustain some 1.34 billion people.[203] Later, Robert Wallace offered a series of calculations in 1809 of how big the global population could be that ranged from 475 million to 34 billion, depending on which country was referenced for the calculation (for example, the higher is based on the population density of Holland, while the lower estimate is based on the population density of Russia).[204]

But it was Sagan who first added a temporal dimension to such estimates. This was motivated by his *ethical* conviction that "if we are required to calibrate extinction in numerical terms, I would be sure to include the number of people in future generations who would not be born," as nuclear weapons "imperil[] all of our descendants, for as long as there will be humans." (I take it that he meant to write "could be" rather than "will be." Note that this quote came from the 1983 *Foreign Affairs* article that he published to alert the public of the newly recognized nuclear winter threat.) On Sagan's count, if the human population were to remain stable, and if people were to live 100 years on average, then "over a typical time period for the biological evolution of a successful species (roughly ten million years), we are talking about some 500 trillion people yet to come." This led him to a conclusion similar to Parfit's, namely, that "by this criterion, the stakes are one million times greater for extinction than for the more modest nuclear wars that kill 'only' hundreds of millions of people." Hence, while "some have argued that the difference between the deaths of several hundred million people in a nuclear war . . . and the death of every person on Earth . . . is only a matter of one order of magnitude," for Sagan "the difference is considerably greater." Sagan further emphasized, along the lines of Mary Shelley, Russell, Schell, and others, that "there are many other possible measures of the potential loss—including culture and science, the evolutionary history of the planet, and the significance of the lives of all our ancestors who contributed to the future of their descendants. Extinction is the undoing of the human enterprise."[205]

A Dictatorship of Future Generations

However, other theorists pushed back against assessing the badness of extinction in terms of how many future people could come to exist if humanity survives. For example, Joseph Nye argued in his 1986 *Nuclear Ethics* that the potentially infinite quantity of value in the future would, if one accepts the sort of impersonalism espoused by Sidgwick, Glover, and Parfit, severely limit the range of activities that would be morally permissible in the present. "A crude utilitarian calculation," he wrote,

> would suggest that since the pleasures of future generations may last infinitely (or until the sun burns out), no risk that we take to assure certain values for our generation can compare with almost infinite value in the future. Thus we have no right to take such risks. In effect, such an approach would establish a dictatorship of future generations over the present one. The only permissible role for our generation would be biological procreation. If we care about other values in addition to survival, this crude utilitarian approach produces intolerable consequences for the current generation.

Instead of allowing future value to dominate our moral calculus, we must be willing to take risks to ensure that future generations have "equal access to other values that give meaning to life," since what matters is not mere existence but a life that is worth living. Specifically referring to Schell's version of the argument from impoverishment, he further contended that

> while the contemplation of [our] species extinction . . . may reduce the meaning of life to some people in the current generation, that is a value to be judged against others in assessing the risks that are worth running for this generation.[206]

In sum, Nye's central claim was that there is no "absolute value" to human survival, and hence our continued existence matters simply because "it is a necessary condition for the enjoyment of other values."[207]

A few years later, Robert Adams published a lengthy critique of Parfit's 1984 book, titled "Should Ethics Be More Impersonal?," which took issue with a number of Parfit's central claims. Most details of Adams' critique can be passed over for our purposes here; of note is that he, citing Bennett, gestured at the argument from unfinished business in making the case that we should care about future generations. However, his emphasis was less teleological, focusing more on the continuation than completion of certain projects.[208] To quote him at length:

> I believe a better basis for ethical theory in this area can be found in quite a different direction—in a commitment to the future of humanity as a vast project, or network of overlapping projects, that is generally shared by the human race. The aspiration for a better society—more just, more rewarding, and more peaceful—is a part of this project. So are the potentially endless quests for scientific knowledge and philosophical understanding, and the development of artistic and other cultural traditions. This includes the particular cultural traditions to which we belong, in all their accidental historic and ethnic diversity. It also includes our interest in the lives of our children and grandchildren, and the hope that they will be able, in turn to have the lives of their children and grandchildren as projects. To the extent that a policy or practice seems likely to be favorable or unfavorable to the carrying out of this complex of projects in the nearer or further future, we have reason to pursue or avoid it.

Caring about the continuation of these projects, at least to some extent, he suggested, "is not morally optional," although he did not elaborate on why.[209] Hence, unlike Bennett, who saw the argument from unfinished business as non-ethical, Adams interpreted what we could call the "argument from persistent progress" as a promising "basis for ethical theory" about the question of extinction.[210]

Wilderness Says

The decade of the 1980s thus witnessed a momentous shift away from the equivalence thesis and pro-extinctionism that some Analytic philosophers in the late 1960s and 1970s—mostly person-affecting and negative utilitarians—had embraced. This shift was, we have seen, catalyzed by the likes of Glover, Partridge, Schell, Parfit, Smart, Sagan, Adams, and others like John Somerville, whose anti-nuclear writings were largely responsible for the popularization of the word "omnicide."[211] Yet, at the same time, within the world of environmental activism rather than academic philosophy, the 1980s also saw the rise of more radical forms of environmentalism that led some to espouse pro-extinctionist views according to which a permanent end to the human story would be very good on balance because it would remove from the biosphere its most destructive force. Some argued that we should take steps of one sort or another to actually bring this about.

As noted in our discussion of History #1, the modern environmental movement arose as a major cultural phenomenon in the 1970s, inspired by the publications of Rachel Carson (1962), Paul and Anne Ehrlich (1968), and the Club of Rome (1972). The movement's initial focus was largely anthropocentric, concerned specifically with how pollution, overpopulation, and so on would impact human health and wellbeing. Keith Mako Woodhouse calls this "crisis environmentalism."[212] However, some activists in the late 1970s and early 1980s became dissatisfied with the fixation on human wants and needs. Galvanized by the deep ecologist Arne Naess in particular, as well as the earlier writings of Aldo Leopold, John Muir (founder of the Sierra Club), and Henry David Thoreau, they came to adopt biocentric, biocentric egalitarian, or ecocentric theories of value.[213] The first states that all human and nonhuman organisms possess some amount of intrinsic value, while the second states that all human and nonhuman organisms possess the same amount of intrinsic value. This means that, as the editor of the *Earth First! Journal* John Davis supposedly said, "human beings, as a species, have no more value than slugs." Or, in the words of Dave Foreman, who cofounded Earth First!, "an individual human being has no more intrinsic value than does an individual Grizzly Bear life."[214] As for ecocentrism, it states that at least some nonliving entities possess intrinsic value as well. An early example of this idea is Leopold's "land ethic," which states that "a thing is right when it tends to preserve the integrity, stability, and beauty of the biotic community. It is wrong when it tends otherwise." Since the integrity of the abiotic environment is necessary for the preservation of these qualities, it thus also falls within the scope of our moral duties. The concept of land in Leopold's thought, then, includes all of these elements: "soils, water, plants, and animals."[215] Along the same lines, Foreman declared in his *Confessions of an Eco-Warrior* that

> concern for wilderness preservation must be the keystone. . . . Wilderness says: Human beings are not dominant, Earth is not for *Homo sapiens* alone, human life is but one life form on the planet and has no right to take exclusive possession. Yes, wilderness for its own sake, without any need to justify it for human benefit.[216]

Homo shiticus: A Plague on the Earth

With these value theories in mind, we can reconstruct the basic line of reasoning that motivated the pro-extinctionism of certain radical environmentalists beginning in the 1980s as follows: imagine that everything about our current environmental plight were the same, that is, atmospheric levels of CO2 have nearly doubled since pre-Industrial times, the ocean is rapidly acidifying, the global population of wild vertebrates has declined by two-thirds over the past five decades, the sixth major mass extinction event has recently commenced, and so on. Now imagine that after a great deal of scientific investigation, it was found that all of these effects are the result of a single species of mite called *Varroa obliterator*.[217] How would we respond? Undoubtedly, countries around the world would join hands and pool resources in launching a coordinated "war of extermination" to completely eliminate the mite, thereby saving the biosphere.[218]

Moving from the counterfactual to the factual, since our environmental plight today is the direct result of *Homo sapiens*, and since, let's say, *Homo sapiens* has no more intrinsic value than any other living creature, the very same conclusion follows—except the target of extermination would be us.[219] As Chris Korda, who founded the ecocentric, neo-Malthusian Church of Euthanasia (CoE) in 1992, wrote, "one thing seems certain: from the point of view of nonhumans, on balance, our extinction would be a great blessing."[220] This leads to the question of which means should be utilized to bring about this goal, and here we find differing opinions: many advocates of human extinction argued for an antinatalist solution, thus espousing a version of this view that we could call *ecological antinatalism*, in contrast to the person-affecting antinatalism of Vetter and the pessimistic antinatalism of Philipp Mainländer. However, some argued for a promortalist solution, as exemplified by the Church of Euthanasia's slogan "Save the Planet, Kill Yourself," while others, albeit "the tiniest minority of the movement," contended that the only feasible solution is direct harm directed at other humans, including omnicide, which might be accomplished by utilizing advanced genetic engineering techniques to synthesize a designer pathogen to wipe out the whole human population.[221] (Recall from Chapter 6 that John Leslie, Bill Joy, Martin Rees, and other theorists who helped usher in the fifth existential mood were especially worried about how emerging and anticipated future technologies could empower small groups or even single individuals to potentially destroy, unilaterally, the entire human species. Many actors with omnicidal ideations are well aware of this potentiality, as I have elsewhere catalogued.)[222]

As this suggests, underlying all of these proposed solutions to the problem of humanity—explicitly limned as "the real enemy" by a 1991 Club of Rome report[223]—was a strain of misanthropic thinking grounded on ethical considerations of the value of nature and the empirical fact that humanity is destroying the natural world. To quote J. Baird Callicott in a 1989 defense of Leopold's land ethic, "the extent of misanthropy in modern environmentalism . . . may be taken as a measure of the degree to which it is biocentric."[224] One manifestation of this attitude took the form of characterizing *Homo sapiens* as "a disease, a cancer on nature,"[225] a "virus," "cancer," and "alien species,"[226] and "useless vermin,"[227] which inspired a range of colorful appellations for our species such as "the Humanpox,"[228] "Pox humanus,"[229] and "*Homo shiticus*."[230] If we are a disease, cancer, virus, or alien species that is clawing away at the biosphere, it follows more or less automatically that we should take steps to remove ourselves for the sake of the greater ecological good.[231]

This conclusion was hinted at on many occasions in the radical environmentalist literature, most notably in the periodical published by the group Earth First!, which Woodhouse describes as "the premier ecocentric, radical environmental organization of the 1980s and 1990s."[232] But there were also explicit statements in support of omnicide, as when a 1989 *Earth First! Journal* article titled "Eco-Kamikazes Wanted" announced that "contributions are urgently solicited for scientific research on a species specific virus that will eliminate *Homo shiticus* from the planet. Only an absolutely species specific virus should be set loose. . . . Remember, Equal Rights for All Other Species."[233] This idea was taken up by a grassroots movement called the Gaia Liberation Front (GLF), whose communique #1, released on Earth Day in 1990, reported that its

> mission is the total liberation of the Earth, which can be accomplished only through the extinction of the Humans as a species. . . . every Human now carries the seeds of terracide. If any Humans survive, they may start the whole thing over again. Our policy is to take no chances.[234]

How might this be achieved? The GLF's "Statement of Purpose (A Modest Proposal)," notes that exterminating humanity through nuclear war would result in too much collateral damage, mass sterilization would be too slow, and suicide is impractical.[235] Yet advanced bioengineering offers "the specific technology for doing the job right—and it's something that could be done by just one person with the necessary expertise and access to the necessary equipment." Furthermore,

> genetically engineered viruses . . . have the advantage of attacking only the target species. To complicate the search for a cure or a vaccine, and as insurance against the possibility that some Humans might be immune to a particular virus, several different viruses could be released (with provision being made for the release of a second round after the generals and the politicians had come out of their shelters).[236]

As a "spokesorganism" for the movement named "Geophilus" declared in a conversation with the founder of VHEMT (see below), "while we support all voluntary efforts to make the Humans extinct, we do not exclude the involuntary route."[237] This sentiment has been echoed more recently by groups like Individualidades Tendiendo a lo Salvaje (ITS)—or, in English, Individuals Tending to the Wild (or Savagery)—which has "been linked to attacks in France, Spain, and Chile."[238] Of note is that ITS has specifically targeted nanotechnologists because of the group's belief that, as Eric Drexler suggested in 1986, the accidental release of self-replicating nanobots could destroy the entire biosphere by converting all organic matter into ecophagic clones of themselves. According to the anarcho-primitivist John Zerzan, at one point a confidant of Ted Kaczynski (the Unabomber) after his arrest in 1996, ITS was initially "real slavish" to Kaczynski, whose main ideological motivation for his campaign of domestic terrorism from 1978 to 1995 was not radical environmentalism but neo-Luddism, that is, an opposition to the megatechnics of industrial society. However, one observer reports that ITS appears to have adopted a more ecofascist, omnicidal ideology founded on the conviction that "the human being deserves extinction."[239]

Other groups, such as the aforementioned Church of Euthanasia, have emphasized both antinatalism and promortalism. Officially, the church—a sort of neo-Dadaist art project inspired by genuine concerns about environmental degradation—"advocates voluntary population reduction in order to *restore balance* between humans and nonhumans" (italics added). Members thus "take a lifetime vow of nonprocreation," as the church's single commandment is "Thou shalt not procreate."[240] But it also specifies suicide as one of the four main pillars of its religio-environmentalist doctrine. To quote the seventh "e-Sermon" given by Korda, who refers to herself as "Reverend":

> I'm asking the audience to do something very important tonight. And let me say this directly to everyone listening tonight. If you're depressed, or ill, or feel burdened by today's world problems, let me suggest a way to give your life new meaning—kill yourself. Do it now. If you have a gun, get your gun. If you have a razor, get your razor. Rope is good. Car exhaust is good. I would ask each and every person now listening to kill themselves without hesitation.
>
> Stop killing one another.
> Kill yourself.
> Stop killing the animals.
> Kill yourself.
> Stop killing the oceans and forests.
> Kill Yourself.
> And do it tonight.
> Do it now.
>
> I guarantee that somewhere out there someone is listening to this tonight and they're just about ready to pull the trigger, or snuff themselves in some way. I say to that person, think about what you are doing. Realize what good you are doing, and then do it. Pull that trigger![241]

The church even purchased a billboard in 1995 to advertise a 900-number "Suicide Assistance Hot-Line," which included the message "Helping you every step of the way! Thousands helped! How about you?" Several years later, it unfurled a banner reading "Human Extinction While We Still Can" during a protest of the Bio 2000 conference in Boston. As a prayer in one e-Sermon summed up the general sentiment: "Great Spirit, if this be so, then I pray for extinction. Let my species become extinct, and vanish from the Earth."[242]

But the majority of pro-extinction environmentalists did not advocate suicide or omnicide but antinatalism. The most notable example is the Voluntary Human

Extinction Movement, or VHEMT, pronounced "vehement," which published its first newsletter, *These EXIT Times*, in 1991. The idea for the movement, however, was devised two decades earlier by the deep ecologist Les. U. Knight, who initially called it the "Human Extinction Movement" but changed the name ten years later because, in Knight's words, "I realized that I had to add 'voluntary' because people's first thought was massive die off" (personal communication).[243] In the first issue of VHEMT's newsletter, which was partly reproduced in Foreman and Davis' magazine *Wild Earth*, Knight wrote that

> if you haven't given voluntary human extinction much thought before, the idea of a world with no people in it may seem strange. But if you'll give the idea a chance, I think you might agree that the extinction of *Homo sapiens* would mean survival for millions, if not billions of other Earth-dwelling species.

He added that, in addition to ensuring the survival of many other species, "phasing out the human race will solve every problem on Earth, social and environmental," since if there aren't any human beings, there can be no human problems.[244] However, unlike some of the more extreme factions in the radical environmentalist movement, Knight emphasized compassion for nonhumans and humans alike, and often presented his ideas with a good dose of light-hearted humor ("without humor," he wrote, "Earth's condition gets unbearably depressing—a little levity eases the gravity").[245] Hence, the first issue of VHEMT's newsletter states that "all creatures have the right to live a long and healthy life," and it encouraged members of the movement to donate blood, work to reduce infant mortality rates and ease world hunger, improve health care, education, and "the status of women," and "care for the elderly," in addition to aiding projects to reforest parts of the world and create new wildlife habitats.[246] As VHEMT's slogan expresses the sentiment, "May we live long and die out."

In all these cases, the impetus behind advocating for our extinction was fundamentally ethical, even if the methods proposed to bring about this outcome were in some cases shocking and abhorrent.[247] Humanity is destroying ecosystems, poisoning the atmosphere and oceans, pushing species into extinction, and tarnishing the natural beauty of Earth. We are, as Sir David Attenborough recently put it, a "plague on the earth."[248] If one cares about other beings on our planet, and if one maintains that *Homo sapiens* is no more intrinsically valuable than any other species, one should at least be open to the idea that we ought to eliminate ourselves for the sake of the biosphere. This pro-extinction conclusion continues to be held by some radical environmentalists, although mainstream contemporary movements like Fridays for Future (FFF) and Extinction Rebellion (XR) appear much more sympathetic to the idea that we should save the planet without permanently erasing ourselves from the picture.

Types of Extinction

To conclude this chapter, philosophers within the Continental and Analytic traditions, along with journalists like Schell, scientists like Sagan, and environmentalists like Knight, outlined a wide range of innovative new ideas about the goodness/badness, rightness/wrongness of our extinction during the second wave of the development of Existential Ethics. As alluded to at the beginning of this chapter, the focus of most of these theorists was extinction in the prototypical sense—that is, final extinction brought about by a catastrophe—although normative extinction was often tacitly invoked alongside final extinction, for reasons noted below. Others discussed the idea of normative extinction more explicitly, as when Jaspers worried about the threat of totalitarianism, while at least one of the pro-survival arguments outlined above may have concerned terminal extinction. Let's take a closer look at these claims before moving on to the next chapter.

First, consider Russell's arguments that our extinction would be bad because it would throw away all the progress we have so far made and foreclose future progress that could continue for an extremely long time to come. What kind of extinction would entail these losses? Although Russell may not have given much thought to the possibility of *Homo sapiens* disappearing forever but leaving behind some posthuman successors, the fact is that progress in the relevant domains does not require *Homo sapiens* to exist. What it requires is that (i) either our species continues to exist or, if we don't, a successor species takes our place, *and* (ii) our descendants, in whatever form they may take, are able to carry on the projects of developing, enlarging, or cultivating things like knowledge, love, kindness, and hope, to summarize Russell's list. Hence, what the arguments proposed by Russell target is not demographic, phyletic, or terminal extinction but *final and normative extinction*, since these correspond to (i) and (ii), respectively, and each would be *sufficient* to render past progress a waste and cancel all future progress.[249] In contrast, none of the other types of extinction are *sufficient* to bring about such losses, and hence these are, *in themselves*, not what we should aim to avoid, *except insofar* as doing so might be strategically, or instrumentally, useful for avoiding final and normative extinction—which may often be the case, as demographic extinction would have almost certainly entailed final extinction if it had happened when Russell was writing. (I am ignoring premature extinction here because Russell, in waxing poetic about our "career of triumph," mostly emphasized the *continuation* of progress. "There lies before us," he wrote, "continual progress in happiness, knowledge, and wisdom." He did not say much about this progress being aimed at some specific goal or *telos*.)

The same could be said about the arguments put forward by Anders, Jonas, Bennett, Partridge, Schell, Glover, and Parfit. For example, if one reason that our extinction would be bad is that past people would die a "second death," in Anders' sense, by being forgotten forever ("as if they had never been"), and since remembering past people requires only that there exists a *certain kind of being* capable of remembering these people, then this "second death" could be avoided even if *Homo sapiens* were to disappear entirely and forever.[250] Furthermore, Anders never specified a criterion for belonging within the League of Generations, which could thus, presumably, encompass future posthumans, so long as they possess the right kind of status, standing, character, qualities, or whatever one takes to be normatively important. Or consider Schell's argument from impoverishment built on the Arendtian notion of the common world. We can ask: does this common world require the existence of *Homo sapiens*, or could it persist even if *we* disappear? The answer is, of course, that the common world could, in every salient respect, be perpetuated by a successor species, that is, given that this species has the relevant capacities and interests. Hence, if confidence in the common world existing is part of what enables our lives today to be valuable, meaningful, and worthwhile, and if the common world does not require the existence of *Homo sapiens*, then terminal extinction is not the main target of this argument. The same goes for two other arguments mentioned by Schell, namely, the arguments from valuing and immortality. The persistence of things we care about and the attainment of vicarious immortality do not require that *Homo sapiens* endures. They only require that beings who also care about these things and sustain the memories of past people stick around.

The cases of Glover and Parfit are even more straightforward. If what matters is the maximization of intrinsic value, and if intrinsic value could be realized by posthuman beings of the right sort, then avoiding terminal extinction itself is not important: what matters is avoiding final and normative extinction. Similarly, the only way that ideal goods like science, the arts, and morality would necessarily cease being developed is if we underwent extinction in either of these senses. This goes for the argument from unfinished business, too, although of course the teleological nature of this argument introduces an extra condition pertaining to the *timing* of extinction. The point, however, is that the term "extinction" in "premature extinction" should, in most cases, be understood in both the final and normative senses, while the word "premature" is what specifies the extra condition, whereby either of these scenarios occurring prior to the attainment of some desired goal would make the outcome *worse* than if they were to happen after this goal is reached.

As for Jonas, we noted that although he worried about biotechnological modifications of the human organism, what *ultimately mattered* to him was the instantiation of the ontological and ethical properties that give rise to the capacities for freedom and responsibility, which constitute the foothold of the moral universe within the physical universe. But note that Jonas left it open as to whether other species—for example, radically enhanced posthuman beings—could also instantiate these properties like we do. On his account, then, there is nothing *inherently* bad about the biological species *Homo sapiens* going out of existence entirely and forever, just so long as we leave behind, or are replaced by, a successor species that *also* instantiates these properties, as this would be *sufficient* for the moral universe to continue existing. It follows that the targets of Jonas' anti-extinction arguments were final and normative extinction, not demographic, phyletic, or terminal extinction.

The one possible exception is the argument from cosmic significance. There are two interpretations of this: first, the thing seen as significant, by virtue of being unique in the universe, could be our particular species, *Homo sapiens*. Second, the thing seen as significant, by virtue of being unique, could be the various capacities that only *Homo sapiens* possesses in the entire universe, as far as we know. Such capacities might correspond to our rationality, moral sensibilities, creativity, and so on. Hence, on the first interpretation, what matters is avoiding terminal extinction, since the *unique thing* is *Homo sapiens* itself. On the second interpretation, what matters is that *these capacities* continue to exist in the universe, and since it seems possible for a species of posthuman successors to have these capacities, the types of extinction that we must avoid are final and normative extinction. Both Schilpp and Russell pointed toward the second interpretation, given that each emphasized the uniqueness of our *capacities* or *abilities*, and these do not seem to be instantiable only by our particular species: any sufficiently advanced being, whether biological or artificial, could presumably instantiate them. In contrast, Schell emphasized *Homo sapiens* in writing that, as quoted above, "the extinction *of the species* goes farther, and removes from the known universe the human kind of being, which is different from any other kind that we as yet know of."[251] This suggests that he may have had the first interpretation in mind, and hence, if this is correct, Schell's discussion covered not only final and normative extinction but terminal extinction as well.

As for those who held pro-extinctionist views, such as Vetter, it is fairly obvious what kind of extinction they believed would be *better*, if not *good*: final extinction. For example, consider Vetter's argument that the reason we should want our extinction to happen is because it would eliminate all future human suffering. Since the only type of extinction that would *necessarily* entail this outcome is final extinction, one can confidently infer that this is what he had in mind. The goodness of extinction arises from there being a *complete and final end* to the whole human story. Similarly, with Knight and the other radical environmentalists: it would not do if we left behind a successor species that continues to destroy the biosphere. The only sure way to halt the massacre would be to bring about our final extinction through some means like antinatalism, pro-mortalism, or omnicide.

This brings us to the end of the second wave in Existential Ethics, a period that one could describe as a developmental growth spurt of the field, despite the fact that it continued to receive relatively little attention from philosophers overall. Let's now turn to the next period of History #2.

Notes

1. See Critchley 2001.
2. Somerville 1981.
3. See Torres 2018a, 2018d.
4. Jonas 1979.
5. Lapin 1955. Note that Lapin was quoting two others: Earl Jimerson and Patrick Gorman. See Lapin 1955, 5.
6. Kennedy 1961.
7. Boyer 1994. Many physicists, as noted in Chapter 4, also discussed nuclear weapons, although few provided any systematic analysis of our novel predicament. See Deudney 2019 for a brief but informative history of international relations theorists in the postwar era.
8. Boyer 1994, 293–294. See also Deudney 2019.
9. Lifton 1982 (note that half of this book was authored by Richard Faulk); Thomas 1986.
10. Schilpp 1949.
11. See Russell 1930, 13.
12. Russell 1954b.
13. McPhee 1980.
14. Gould 1996.
15. To put it somewhat simplistically. See Gould 1996, 3; Farrier 2016.
16. Kelvin 1862; Flammarion 1894.
17. Wells 1895; Jeans 1929.
18. For additional examples, see Ćirković 2003, 3.
19. See Bowler 2021, 3.
20. Condorcet 1795; Wells 1902. In fact, like Russell, Wells combined this utopian potentiality thinking with deep-future thinking. To quote the final paragraphs of the aforementioned essay in full:

    > It is possible to believe that all the past is but the beginning of a beginning, and that all that is and has been is but the twilight of the dawn. It is possible to believe that all that the human mind has ever accomplished is but the dream before the awakening. We cannot see, there is no need for us to see, what this world will be like when the day has fully come. We are creatures of the twilight. But it is out of our race and lineage that minds will spring, that will reach back to us in our littleness to know us better than we know ourselves, and that will reach forward fearlessly to comprehend this future that defeats our eyes.
    >
    > All this world is heavy with the promise of greater things, and a day will come, one day in the unending succession of days, when beings, beings who are now latent in our thoughts and hidden in our loins, shall stand upon this earth as one stands upon a footstool, and shall laugh and reach out their hands amid the stars (1902).

21. See, for example, Pinker 2011. For a critique of the poor scholarship of some of Pinker's recent work, see Torres 2019.
22. Note that Russell later used the phrase "Prologue or Epilogue?" as the title to the opening chapter of his book Has Man a Future?, which reiterated many of the same points; see Russell 1961, chapter 1 in general but, especially, pages 13–14 and 119–120.
23. Even earlier, in a 1945 statement to the House of Lords, he declared that "we do not want to look at this thing [i.e., the perils posed by the atomic bomb] simply from the point of view of the next few years; we want to look at it from the point of view of the future of mankind" (quoted in Schell 1982).
24. Russell 1954b.

25. See Ord 2020, Chapter 2, and footnote 45 of Chapter 2.
26. Russell 1954a.
27. Schilpp 1949. Note that Schilpp discussed Russell's work in a in 1944 volume of the Library of Living Philosophers. Thanks to Dan Zimmer for apprising me of this fact.
28. Kagan 1998.
29. Dawsey 2016.
30. Babich 2021, 8.
31. Liessmann 2011, 124.
32. Müller 2021.
33. In German, these books were titled *Die Antiquiertheit des Menschen Bd. I: Über die Seele im Zeitalter der zweiten industriellen Revolution* and *Die Antiquiertheit des Menschen Bd. II: Über die Zerstörung des Lebens im Zeitalter der dritten industriellen Revolution*. See Anders 1956a.
34. See Dawsey 2016.
35. See Babich 2021, "Introduction"; Müller 2021.
36. Dawsey 2016
37. Quoted in Dawsey 2016.
38. Anders 1962; quoted in Dawsey 2016. Note that Anders' seminar notes were originally published in German in 1960. Also, note that while Anders specifically pointed to the first use of atomic weapons as "Day Zero" of his new chronology, he repeatedly referenced the Castle Bravo debacle throughout his writings, thus indicating that he did indeed see this as an extremely important event.
39. Dawsey 2016, see footnote 71.
40. Anders 1962.
41. As Anders wrote, "even in a thoroughly 'clean' world (whereby I understand the situation in which there doesn't exist one single A- or H-bomb, in which we seem to 'have' no bombs) we still would 'have' them because we know how to make them" (1961).
42. Anders 1961.
43. Anders 1958.
44. Anders 1956a.
45. Anders 1956a; quoted in Dawsey 2016.
46. Anders 1962.
47. Anders 1980.
48. Anders 1960/62.
49. Anders 1961.
50. in Bronson 2018.
51. Nathan and Norden 1960; Einstein 1954.
52. Boyer 1994.
53. Meerloo 1949.
54. Anders 1962.
55. Quoted in Torres 2021.
56. Thunberg 2019.
57. Anders 1962.
58. Anders 1961.
59. Anders 1962.
60. Anders 1961.
61. Anders 1962.
62. Anders 1962.
63. His reference to "nihilism" is a bit puzzling. The German pessimists were not necessarily nihilists. Elsewhere, in his 1959 article "Apocalypse without Kingdom," Anders

referenced the Russian nihilists, but so far as I know none of the Russian nihilists doubted whether there will or should be people—unlike the pessimists.

64. Anders 1956a. This is my own translation.
65. Anders 1959/2019. To be clear, "Apocalypse without Kingdom" was taken from an essay titled "Die Frist," meaning "Respite" or "Grace Period," which later appeared in Endzeit und Zeitenende (Anders 1972). Thanks to Jason Dawsey for details on this history.
66. Anders 1962.
67. Anders 1979; quoted in Dawsey 2016.
68. Jaspers 1961, italics added.
69. Koestler 1967.
70. Groenewold 1968/70.
71. Note that I have added two commas to this sentence to improve readability.
72. Groenewold 1968/70.
73. Somewhat amusingly, Jonas wrote in an article about Arendt titled "Hannah Arendt: An Intimate Portrait," that "Günther [Anders] imagined that he had found in [Arendt] a wonderful companion, but he failed to notice that she had outgrown him intellectually and was becoming more independent. This situation became evident in Paris where Hannah quickly became a well-respected figure among the Parisian émigrés. . . . Günther stood somewhat aloof and began to play the role of the prince consort, which, as an ambitious and vain man, made him difficult to bear" (2006). Anders and Arendt divorced in 1937, after both fled Germany in 1933.
74. Vogel 1996, 3.
75. This is of course not true: utilitarians like Bentham believed that animals could experience pleasure and pain, and hence we ought to include them in our moral deliberations.
76. He also hinted at the "hypertrophy" idea mentioned by Groenewold in writing that our cognitive evolution has resulted in "the paradox of excessive success that threatens to turn into a catastrophe by destroying its own foundations in the natural world" (see Morris 2013, 127).
77. Indeed, it seems difficult to derive a logical or practical contradiction from a maxim covering omnicide. But, as Barbara Herman observes, the "contradiction in conception" (CC) test of the Categorical Imperative fails to yield a contradiction in the maxim "To kill whenever that is necessary to get what I want." As she writes,

    > If everyone killed as they judged it useful, we would have an unpleasant state of affairs. Population numbers would be small and shrinking; everyone would live in fear. These are bad consequences all right. Still a world that looks like this is conceivable: Hobbes described it in some detail. And if there is nothing inconceivable or contradictory in thinking of a world that contains a Hobbesian law of killing, it looks as though we must conclude that the CC test does not reject the maxim of killing (Herman 1993).

78. Jonas 1979/84. See Coyne 2020, 121–122, for discussion.
79. Coyne 2020.
80. Note that "super-commandment" is my term.
81. See Miller 2020.
82. Anders 1961.
83. NYT 1982.
84. Note that "strive" is my term, not Anders'.
85. Anders 1961. Recall here the Schopenhauer also referenced this line from Shakespeare in suggesting that "absolute annihilation would be decidedly preferable" to existence (1818).

86. Coyne 2020, 123.
87. Jonas 1979/84.
88. As noted in a previous footnote, I am extremely skeptical that colonizing space will actually reduce the probability of extinction, due largely to the convincing case against colonization made by Deudney 2020. We should not, as I claimed earlier, uncritically assume that space colonization will increase our chances of survival; the very opposite could be the case. Perhaps there really is no Planet B.
89. Jonas 1979/84.
90. According to the 1992 Rio Declaration, the Precautionary Principle states that "where there are threats of serious or irreversible damage, lack of full scientific certainty shall not be used as a reason for postponing cost-effective measures to prevent environmental degradation" (Rio 1992).
91. Morris 2013.
92. Morris 2013, 125.
93. Jonas 1979.
94. Jonas and Vogel 1996.
95. Jonas 1979/84, italics added.
96. Vogel 1995.
97. Morris 2013, italics added.
98. More specifically, "final" value. I will discuss the distinction between intrinsic and final value in Chapter 11. Briefly put, something has final value if and only if it is valuable as an end-in-itself or for its own sake, whereas something has intrinsic value if and only if it is valuable by virtue of its intrinsic (rather than extrinsic) properties.
99. Morris 2013.
100. Note that Jonas does not use the terms "biotechnology" or "genetic engineering," but instead simply talks about "technology" enabling "the genetic control of future men" (Jonas 1979).
101. Coyne and Hauskeller 2019.
102. See Coyne 2021, 173; Coyne and Hauskeller 2019, footnote 1.
103. See Walters 1988, 239.
104. d'Entreves 2022.
105. Jaspers 1961.
106. Earle 1961.
107. Note that Jaspers somewhat imprecisely flips between discussing the "atom bomb" and the "H-bomb." In much of his discussion, it is safe to assume, I suspect, that he means the latter when he uses the former.
108. Jaspers 1961. Incidentally, one finds a similar sentiment in Kant's *Metaphysics of Morals*, which declares that "if justice perishes, then it is no longer worthwhile for men to live upon the earth" (quoted in Rawls 2005).
109. Russell 1961, 89. For a nuanced discussion of Jaspers' view, see Walters 1988.
110. Note that, in contrast, only 21 percent in the United Kingdom held this view (Rose 2004, 9). As Jeff McMahan observed later on, in a 1986 discussion of the "dead" or "red" debate,

    > people's views about nuclear weapons tend to reflect the ordering of their fears. A crude generalization might be that those whose position is characterized primarily by opposition to nuclear weapons tend to fear nuclear war more than they fear the Soviets, while those who are disposed to support nuclear weapons tend to fear the Soviets more than they fear nuclear war (1986).

    Incidentally, it appears that few leading intellectuals at the time explicitly endorsed Jaspers' view. Jonathan Schell gestures at this in writing that Jaspers was "one of the few who have had the courage to state such a belief outright" (1982, italics added).

111. Jonas 1979/84.
112. Indeed, as Torbjörn Tännsjö notes, "there is a strong tradition within Western philosophy arguing that, given our human predicament, the coming to an end of humanity is morally unobjectionable or even desirable" (2021).
113. Leslie 1983, italics added.
114. Parfit 1984.
115. See, for example, Naess 1973; Singer 1975; Singer 1972; Rawls 1971. Intergenerational ethics, for example, asks what we owe future generations, what our obligations to them are, assuming that they exist. This topic gained significant attention after the publication of John Rawls' 1971 *A Theory of Justice*, which offered the first systematic examination of intergenerational ethics. In brief, Rawls asked us to imagine a group of "deliberators" in what he called the "original position." This is a hypothetical situation in which these deliberators find themselves behind a "veil of ignorance," which prevents them from knowing anything about the race, gender, intelligence level, education level, social status, personal wealth, and so on of the members of society that they represent. They are then tasked with determining principles for the arrangement of social and political institutions within liberal society. Rawls argued that if these deliberators are self-interested, they will choose principles that ensure the fairest arrangement possible; hence Rawls' famous slogan of *justice as fairness*.

    The point is that Rawls extended this thought experiment to the question of what current generations owe to future generations. Imagine, he argued, that the deliberators also know nothing about which generation they represent. They are then tasked with determining how much "real capital"—that is, factories, machines, knowledge, culture, techniques, and skills—each generation is obligated to pass along to subsequent generations. Rawls contended that if one considers the question from this perspective, each generation is obliged to bequeath at least enough capital for "the conditions needed to establish and to preserve a just basic structure over time." Hence, "once these conditions are reached and just institutions established, net real saving may fall to zero. If society wants to save for reasons other than justice, it may of course do so; but that is another matter." He called this the just savings principle (Rawls 1971). In this way, justice as fairness extends not just across space, from one person or group to another, but across time, from one generation to the next.
116. See Heyd 1992, who bundled such questions concerning the "existence, number, and identity" of people under the umbrella of "genethics" (his coinage).
117. Sidgwick 1874.
118. As Peter Singer wrote in 1979, "would it really be good to create more pleasure by creating more pleased beings? This perplexing issue was first raised by Henry Sidgwick and has since been revived by Jan Narveson and Derek Parfit" (see below) (1979).
119. As Parfit wrote in a 1976 article that put forward certain population-ethical conundrums that he would later develop in his 1984 book, "though my remarks here are critical, I owe a great deal to Narveson's first article" (1976).
120. In Narveson's words, "if the person-regarding view is rejected, of course, then we have the form of utilitarianism which, for instance, Henry Sidgwick explicitly embraced" (1967).
121. Narveson 1967.
122. This is partly affirmed by Narveson: "if no person is affected by an action, then that action (or inaction) cannot be a violation or fulfillment of a duty. This we may call, adopting Derek Parfit's useful terminology, the 'person-regarding' view" (Narveson 1978).
123. See Arrhenius 2000, chapter 8, for discussion.

124. These theories of wellbeing were first explicitly distinguished in appendix I of Parfit's 1984 book.
125. Narveson 1978, 44.
126. Note that there are many proposed analyses of "harm." See Rabenberg 2015 for a useful overview and critique.
127. Narveson 1978. I say "existing people" here for the sake of simplicity. There are many ways of demarcating the class of relevant people on a person-affecting theory: presentism (present people), actualism (actual people), necessitarianism (necessary people), and so on. Again, see Arrhenius 2000, chapter 8.
128. Narveson 1967, 1973. To be clear about these two views and how they related to other (a) interpretations of utilitarianism, and (b) non-utilitarian ethical theories, let me say the following. First, most nonconsequentialist theories in ethics, such as Kantianism and contractualism (see Chapter 11), are person-affecting. What matters on these views is, and only is, "the effect of principles/actions on persons, rather than the world writ large" (Finneron-Burns 2017). Second, one can combine impersonalism and the person-affecting view with the obligation to maximize either the total or the average amount of value. As noted in the previous chapter, the default view among utilitarians today is impersonalism, and hence "totalism" and "averagism" are quite literally defined in impersonalist terms, but this need not be the case. Narveson himself held that what matters is the total quantity of value, but that maximizing this quantity does not entail that we should "make [new] happy people," only that we should "make people [who already exist] happy" as much as possible. We should, he wrote, "aim at the greatest happiness of the greatest number," rather than "the greatest happiness and the greatest number" (Narveson 1967). Third, it may be easy to confuse Sidgwick's notion of "the point of view of the universe" with impersonalism, as this disembodied perspective on the affairs of moral agents is as impersonal as it could be. But this idea specifically concerns impartiality, that is, the claim that one's identity, or even one's species, is irrelevant when calculating the total or average amount of value contained within a given state of affairs. Each being's pleasure and pain counts equally. Hence, one can espouse a person-affecting no less than an impersonalist view while simultaneously endorsing Sidgwick's conception of the moral point of view: looking down on human affairs from above, as if the universe had eyes of its own.
129. Narveson 1967.
130. Narveson 1967.
131. Narveson 1978.
132. Narveson 1967, 1973.
133. Italics added.
134. Bennett 1978.
135. Sikora and Barry 1978.
136. Sikora and Barry 1978. One finds a similar sentiment expressed several years earlier, in 1974, by Joel Feinberg. The only rights that future generations have, he argued,

    > are contingent rights: the interests they are sure to have when they come into being (assuming of course that they will come into being). . . . Yet there are no actual interests, presently existent, the future generations, presently nonexistent, have now. Hence, there is no actual interest that they have in simply coming into being, and I am at a loss to think of any other reason for claiming that they have a right to come into existence.

    It follows that, if everyone around the world were to voluntarily choose not to procreate, this would not "violate the rights of anyone," and hence it would not be wrong. He concluded: "My inclination then is . . . that the suicide of our species would be deplorable, lamentable, and a deeply moving tragedy, but that it would violate no one's rights" (Feinberg 1974).

137. Bennett 1978.
138. See Bell 1993; Slaughter 1994; Tonn 2009, 2021.
139. This is obviously reminiscent of Anders' notion of the League of Generations, although I do not know if Anders was familiar with Burke's work.
140. Burke 1790.
141. Tonn 2009, 428.
142. Bell 1993.
143. Clarke 1971.
144. Bennett 1978.
145. Narveson 1967.
146. McMahan 1981; see Frick 2014, footnote 1. Note that this is distinct but related to the "Intuition of Neutrality," according to which "the presence of an extra person in the world is neither good nor bad. More precisely: a world that contains an extra person is neither better nor worse than a world that does not contain her but is the same in other respects" (Broome 2005, 401). For example, if one abandons the Intuition of Neutrality, then it becomes harder to accept the Procreation Asymmetry; that is, if one is not neutral about, say, the addition of a new happy person, all other things equal, this suggests there may be a moral reason after all for creating new happy people.
147. Vetter 1971.
148. Italics added.
149. However, Vetter wrote in another paper (which was published before his 1971 paper but seems to have been written after it; that is, the 1971 paper was written first but published second) that

    > Narveson has correctly pointed out that not only the potential child's, but also other people's, notably the parents', utility has to be taken into account. I admit that this utility may outweigh the potential child's disutility which according to U4 [see below] would speak for not producing it, plus the disutility imposed upon others by taking away from them scarce goods (including the investments necessary to provide work, housing, and other facilities to the newcomer).

    Here, "U4" is the proposition that "there is a moral reason for not starting someone's existence on account of the unhappiness he would experience" (Vetter 1969). Hence, if the utility to the parents and society added by creating a new person were to outweigh the disutility of creating the child itself, one may be morally obligated to have a child. Clark Wolf later used this to argue that "the Vetter dominance argument fails, because it fails to take into account all of the morally relevant considerations at stake in the decision to bring a child into existence." He continued: "Our prospective children may contribute to making our lives better, and to making the lives of others better as well. Thus failure to conceive a child will put at risk the welfare of all those who might have been better off (or less badly off) if one's child had existed" (Wolf 1997).
150. Vetter 1971. In the late 1990s, Christoph Fehige argued for what he called an "anti-frustrationist" axiology that entails similar conclusions. For example, his "General Universal Pareto Principle" (GUPP), of which antifrustrationism is an integral component, accepts Narveson's view that "we have obligations to make people happy . . . but no obligations to make happy people." Fehige further described his GUPP position as explicitly holding that "(i) Nothing can be better than an empty world (a world without preferences, that is). (ii) Our world is worse than an empty world. (iii) It is ceteris paribus wrong to create a being that will have at lest one unfulfilled preference." Nonetheless, Fehige argued that, for reasons that I will not discuss here,

his view "does not prescribe childlessness to would-be parents," and "is miles away from anything like a general prohibition on real-life procreation." It thus "permits the show to go on as long as there are, or if there ever have been (as indeed there have), people who want it to go on" (Fehige 1998). A similar idea was put forward earlier by Peter Singer, who described his version of preference utilitarianism as seeing no value in creating new satisfied preferences; all that matters is maximizing satisfied existing preferences. "The creation of preferences which we then satisfy gains us nothing," he wrote, as

> we can think of the creation of the unsatisfied preferences as putting a debit in the moral ledger which satisfying them merely cancels out. . . . It can find no positive value in the existence of our species. Given that people exist and wish to go on existing, Preference Utilitarians have grounds for seeking to satisfy their wishes, but they cannot say that the universe would have been a worse place if we had never come into existence at all (Singer 1980).

However, Singer did not "endorse antifrustrationism or anything like it" (Fehige 1998).

151. Popper 1945.
152. Smart 1958.
153. See Acton and Watkins 1963; Ord 2013, but also Knutsson 2022c for critical discussion of Ord.
154. Bar-Hillel wrote: "I personally do not see in the preservation of human life a particular value. Together with Dr. Vetter and Sir Karl [Popper] I rather tend to see in the reduction of suffering a prime value." He added: "I think that all that talk about the destiny or goals of humanity is seductive talk which scientists should try to oppose. Any such talk will quickly lead to the recognition of somebody who is setting these goals and of a privileged class of people who know from the horse's mouth what these goals are" (Bar-Hillel 1968).
155. Vetter 1968.
156. Vetter 1969.
157. Glover 1990.
158. Beard 2019.
159. Smart 2020, italics added.
160. Nagel 1970.
161. Vetter 1968.
162. Nagel 1970.
163. Feinberg 1977. Some ancient Greeks may have accepted this view, too. As Aristotle and Collins (2012) reported in his Nicomachean Ethics, "both evil and good are thought to exist for a dead man, as much as for one who is alive but not aware of them; e.g., honours and dishonours and the good or bad fortunes of children and in general of descendants." Or as Feinberg wrote, the notion that we are all susceptible to "drastic changes" in our fortunes "both before and after death was well understood by the Greeks," according to Feinberg 2014, italics added. Note also that Feinberg did discuss human extinction in a 1974 paper titled "The Rights of Animals and Unborn Generations," although only to say that, as noted in a previous footnote, "the suicide of our species would be deplorable, lamentable, and a deeply moving tragedy, but . . . it would violate no one's rights" (Feinberg 1974).
164. Bennett 1978.
165. Partridge 1981. Note: the version of this paper on Partridge's website misspells "from" as "form." I have here corrected this.
166. Passmore 1974.

167. Partridge 1981.
168. Delattre 1972, italics added.
169. Partridge 1981.
170. Partridge 1981. Elsewhere in the article he wrote that "if one feels no concern for the quality of life of his successors, he is not only lacking a moral sense but is also seriously impoverishing his life. He is, that is to say, not only to be blamed; his is also to be pitied" (Partridge 1981).
171. Note that I am unable to verify this quote, despite it being widely attributed to Hemingway. Nor was Andrew Morawski of the Hemingway Home and Museum able to verify it (personal communication).
172. Lasch 1978.
173. David Heyd subsequently offered an alternative interpretation of Partridge's argument from immortality according to which "self-transcendence is itself a person-affecting value" that, as such, "cannot give rise to ethical obligations to create new people." In other words, the idea that we may live on in the memories of those who come after us explains why we might want to have children, but it does not generate "a duty to continue humanity" (Heyd 1992). Luke Meyer articulated another versions of this general argument, too. In a chapter section titled "Living in a Society that is Open to the Future," Meyer wrote that

    > being successful in the pursuit of valuable projects is of the utmost importance to the well-being of people. Thus, for those many contemporaries who pursue projects of the two types as characterized in the preceding paragraphs, it is important for their well-being that they can place the pursuit of their projects in an ongoing and unfolding story. In particular, it is important to them that they can expect the continuance of human life on earth under such conditions that future people will be able to understand the point and value of the projects they have been pursuing, that they can make good use of them or may choose to continue pursuing them. Being able meaningfully to choose a project whose success partly depends on intergenerational cooperation presupposes living in a society of a certain quality. It presupposes living in a society that is sufficiently open to the future to allow that there be future people who, in tum, are able to choose to continue valuable projects that their predecessors pursued before them (Meyer 1997).

174. This goes for the second wave more generally: few authors cited each other, and hence (a) many repeated what others had earlier said, and (b) their writings did not form a cohesive literature in which later ideas built upon earlier ideas. Indeed, it was only with the founding of Existential Risk Studies in the early 2000s that a cumulative tradition of this sort emerged.
175. Schell 1982.
176. See Dawsey 2013, footnote 19; Spiegel 1982.
177. There are at least five reasons for thinking that Schell did not plagiarize Anders. First, although Anders mentioned the "second death" in his 1962 article "Theses for the Atomic Age," which was published in English, his major work on the topic, as noted earlier, was never translated from German, and there is no evidence that Schell spoke or read German. (Schell's friend, the psychiatrist who introduced the idea of psychic numbing, Robert Lifton, has affirmed this to me via email.) Second, the term "second death" appears in the Book of Revelation four times, and hence Anders himself may have borrowed it from the Bible. At the very least, it was not wholly original to his work. Third, both Anders and Schell had almost certainly come across Arendt's use of the term "second birth" in her 1958 book *The Human Condition*. She wrote: "With word and deed we insert ourselves into the human world, and this insertion

is like a second birth, in which we confirm and take upon ourselves the naked fact of our original physical appearance" (Arendt 2019). It is a short terminological step from "second birth" to "second death." (Note also that the term "second birth" is not original to Arendt. See, e.g., *Excerpts of Theodotus* 80:1; and of course Jesus spoke of being "born again," as quoted in John 3:3, 7.) Fourth, as noted above and below, Anders' notion of the second death was different from Schell's. And finally, Schell was by all accounts something of a paragon of intellectual integrity. As Peter Rothberg wrote shortly after Schell's death, "I guess I've probably known a nicer, more humble human being than Jonathan Schell. But certainly no one who approached Jonathan's stature or legacy. I've also met a handful of more accomplished writers, but absolutely no one who came close to approaching Jonathan's humility" (Rothberg 2014; see also Bhandari and Rodrigues 2014 for similar comments). In contrast, Anders seemed to care a great deal about achieving fame and notoriety, as suggested by his efforts in Hollywood. To quote the harsh words of his friend Jonas, Anders was "an ambitious and vain man" (Jonas 2006). These characterological differences suggest, one could argue, that Schell was not the type of person to borrow ideas without properly crediting their progenitors, while Anders was the type of person who would accuse someone of taking his ideas, especially if that person achieved the level of success that Schell achieved after his 1982 book. (Indeed, consistent with the above, Schell gave virtually no interviews about his work; he wanted the book to speak for itself, and generally eschewed the spotlight, unlike Anders.)

178. Zimmer 2022, ch. 3, section II. Personal communication. Thanks to Zimmer for many insightful conversations about the issue.
179. Schell 2002, 1982.
180. Kateb 1984.
181. Anders 1962; Schell 1982. That is to say, his focus was the possibility of an "absolute and eternal darkness: a darkness in which no nation, no society, no ideology, no civilization will remain; in which never again will a child be born; in which never again will human beings appear on the earth, and there will be no one to remember that they ever did" (Schell 1982).
182. Italics added.
183. Schell 1982.
184. Two points: first, my claim in the previous chapter about the significance of Montesquieu describing our extinction itself as a "terrible calamity" could be rephrased like this: Montesquieu seems to have been singling-out what Schell here calls the "Second Death," and this is what made his statement, expressed through Usbek, so noteworthy; that is, Montesquieu was, or may have been, the very first to conceptually distinguish the first deaths and Second Death in writing. Second, with respect to there being "no extra suffering," Schell seems to have ignored the possibility denoted by the no-ordinary-catastrophe thesis, i.e., that the anticipation of human extinction could, in fact, introduce additional sources of harm.
185. Mulgan 2020, 32.
186. To be clear, some of these theorists would have, as noted above, said that our extinction would indeed be in some way bad, e.g., because it would prevent the fulfillment of certain business (Bennett 1978). But this badness was not morally relevant; it concerned, instead, a mere matter of taste or aesthetic preference.
187. Schell 1982, italics added. Note that I have switched the order of this sentence; the meaning remains unchanged.
188. Arendt 1958.
189. Schell 1982.
190. Schell 1982.

191. Schell 1982, 181–182.
192. Once again, I have rearranged this sentence without altering its meaning.
193. Schell 1982.
194. Fox 2014.
195. Parfit's hypothesis was experimentally confirmed by a study published in 2019. For reasons that I will not elaborate here, much of the rest of this experiment seems to me flawed. See Schubert et al. 2019.
196. Interestingly, Pierre Allan writes in a 2006 article, which implicitly distinguishes between Going Extinct and Being Extinct, that Parfit's "scenario only considers the consequences of a generalized nuclear war, without including the horrors of the path towards the disappearance of mankind for its last members, a truly apocalyptic scenario along the lines of the nuclear winter preceding it. Such a doomsday would entail atrocious suffering during this period of human extinction" (Allan 2006).
197. Parfit 1984.
198. Scheffler 2013.
199. Parfit 2013.
200. Vetter 1968.
201. In fact, this was also addressed in the psychological experiment that affirmed Parfit's hypothesis about how most people would respond to his thought experiment. See Schubert et al. 2019.
202. However, roughly two decades earlier Smart seemed to reject the sort of impersonalism advocated by Sidgwick. In his 1961 book *Outline of a System of Utilitarian Ethics*, he asked:

    > Would you be quite indifferent between (a) a universe containing only one million happy sentient beings, all equally happy, and (b) a universe containing two million happy beings, each neither more or less happy than any in the first universe? Or would you, as a humane and sympathetic person, give a preference to the second universe? I myself cannot help feeling a preference for the second universe. But if someone feels the other way I do not know how to argue with him. It looks as though we have yet another possibility of disagreement within a general utilitarian framework (quoted in Narveson 1967).
    >
    > Interestingly, Sikora and Barry addressed Smart's claim in the introduction to *Obligations to Future Generations*, writing that "one of the most encouraging things about the debate as to whether it is or is not in any way wrong per se to prevent the existence of happy people is that it has become clear that the question is not, as J. J. C. Smart and many others once supposed, beyond the scope of rational considerations" (1978).

203. Van den Bergh and Rietveld 2004, 196.
204. Wallace 1809, 10.
205. Sagan 1983a. As noted earlier, much of the work on Existential Ethics prior to the early 2000s was disjointed, fragmented, lacking any cohesion. Most theorists who addressed the ethical and evaluative aspects of extinction, with few exceptions (such as those just below), never cited each other, and consequently there was no cumulative development of ideas. For example, although Sagan had very likely read Schell's book, he never mentioned it. Nor did Smart cite either Sagan, Schell, or Parfit. No one cited Anders and Jonas. Only Parfit cited Partridge and Schell, although in both cases the citation was nonsensical. That is, Schell's name was included in the Index of Names at the end of Parfit's book, which directs the reader to page 538; but page 538 takes one to the Bibliography rather than the body text, where one finds a bibliographic entry for *The Fate of the Earth*, this being recorded in the Index, with no mention of Schell elsewhere

in the book. The same goes for Partridge: the Index of Names leads one to an entry for his edited collection *Responsibilities to Future Generations*, which included the chapter mentioned above: "Why Care About the Future?" Nevertheless, this suggests that Parfit was familiar with the work of Schell and Partridge, although why their respective books were cited is a mystery. Parfit also never cited Sagan's 1983a estimate, though this would have been directly relevant to the first Sidgwickian reason he gave for why the difference between (2) and (3) is vastly greater than that between (1) and (2).

206. Oddly, Nye here claims that Schell referred to the Second Death as "double death," a mistake that he made elsewhere, too.
207. Nye 1986.
208. Indeed, Adams described Bennett's paper as "one of the best essays I have read on this subject" (1989).
209. Adams 1989. But see Adams 1988 for further thoughts on the moral virtuousness of caring about certain common projects.
210. Incidentally, Bennett told me in an email that this is, in fact, the position he held: "the attitude is towards the continuation of various projects, not towards their completion" (personal communication). Yet this is not how I or others have interpreted his argument. In fact, the introduction of *Obligations*, which included his essay, itself states that Bennett's "justification for being prepared to fight for the preservation of mankind lies rather in the fact that he has an intense interest in the completion of certain specific projects of the species of which he is a member" (Sikora and Barry 1978, italics added).
211. Indeed, Somerville defended a view of extinction similar to the views of Anders, Schell, Parfit, Sagan, and others. Nuclear omnicide, he wrote, constitutes an unthinkable crime, "for this crime encompasses the killing not only of all people but all forms of life on the planet; it not only annihilates all present human life but all future human possibilities, as well as all the records and remains of past human achievements" (Somerville 1979). While Somerville wrote a fair amount about how the novel possibility of nuclear omnicide has altered the human condition in various ways—at times echoing, like Schell, ideas originally found in Anders, who I suspect he was unfamiliar with—he actually said little about the core questions of Existential Ethics.
212. Woodhouse 2018.
213. Woodhouse 2018, 101. Also called "biospherical egalitarianism," in Naess' original phraseology (Naess 1973).
214. Foreman 1991. The Davis quote has been widely reproduced, although I have found the original source difficult to locate.
215. Shaw 1997, 55–56; Leopold 1949. "That land is a community is the basic concept of ecology," Leopold wrote, "but that land is to be loved and respected is an extension of ethics" (1949).
216. Foreman 1991.
217. This is a silly riff on *Varroa destructor*, an actual mite that causes Varroosis, described as "the most destructive disease of honey bees worldwide, inflicting much greater damage and higher economic costs than all other known apicultural diseases" (Boecking and Genersch 2008).
218. I borrow the term "war of extermination" from an actual declaration made in 1818 in Ohio to kill bears and wolves. This and other such efforts in the United States resulted in the gray wolf nearly dying out in the lower 48 states.
219. Not every biocentric egalitarian accepted this, or similar, conclusions, including some of those who first introduced the idea. For example, as Woodhouse observes, "Arne Naess never suggested that valorizing the nonhuman world demanded a proportional denigration of human civilization." However, "for those who most passionately championed deep ecology, defending the one often meant attacking the other" (Woodhouse 2018).

220. Korda 2019. See Korda 1994; CoE 1994.
221. Flannery 2016, 189.
222. See Torres 2018a, 2018d.
223. King and Schneider 1991.
224. Callicott 1980. Recall from Chapter 6 that one of the alternative names for the Anthropocene is the "Misanthropocene."
225. Foreman 1991.
226. GLF 1994.
227. Korda 1994a.
228. Foreman 1991.
229. Knight 1995.
230. Quoted in Dye 1992.
231. It also finds expression in statements like "I have precious little sympathy for the myriad bat eyed proprieties of civilized man, and if a war of the races should occur between the wild beasts and Lord Man I would be tempted to sympathize with the bears," which comes from a 1916 book by Muir (quoted in Flannery 2016). Or consider Stewart Brand's declaration that "we have wished, we eco-freaks, for a disaster or for a social change to come and bomb us into Stone Age, where we might live like Indians in our valley, with our localism, our appropriate technology, our gardens, our homemade religion—guilt-free at last!" (Brand n.d.). Even more extreme views have been expressed by the self-professed "eco-fascist" Pentti Linkola, described as one of the "most celebrated" authors in his home country of Finland. Linkola argues that Western society is guilty of a perverse "over-emphasis on the value of human life" and that "on a global scale, the main problem is not the inflation of human life, but its ever-increasing, mindless over-valuation" (2011). To solve the problem posed by human activity—that is, to avoid an "ecocatastrophe"—Linkola endorses the use of catastrophic violence. As Evangelos Protopapadakis 2014 puts it, "any means to decreasing human population would be welcomed with relief by Linkola; even war, genocide, and disease, as long as any of these would be massively destructive for the species *Homo sapiens*." Thus, Linkola opines that another world war would be "a happy occasion for the planet," although "it would spark hope only if the nature of wars would morph so that deductions of persons would noticeably target the actual breeding potential: young females as well as children, of which a half is girls. If this doesn't happen, waging war is mostly [a] waste of time or even harmful" (Linkola 2009). Even more, Linkola claims that "some transnational body [or] small group equipped with sophisticated technology and bearing responsibility for the whole world" should attack "the great inhabited centres of the globe" (Linkola 2011; some of this is quoted *ad verbum* from Torres 2018a). And perhaps most relevantly, he writes that "if there were a button I could press, I would sacrifice myself without hesitating, if it meant millions of people would die" (Milbank 1994). Note: I have been unable to locate the original source of the quote from Brand previously, although Brand himself has affirmed to me that it is accurate (personal communication).
232. Woodhouse 2018.
233. Quoted in Torres 2018a.
234. Quoted in Torres 2018d.
235. This is, one infers, a reference to Jonathan Swift's 1729 *A Modest Proposal For preventing the Children of Poor People From being a Burthen to Their Parents or Country, and For making them Beneficial to the Publick*, in which Swift suggests, satirically, that poor Irish people should consider selling their children to the rich as food to alleviate their suffering.
236. GLF 1994.
237. Quoted in Korda 1994a.

238. Lloyd and Young 2011.
239. Campbell 2017; see Torres 2018d. Unfortunately, as Frances Flannery argues, "as the environmental situation becomes more dire, eco-terrorism will likely become a more serious threat in the future" (2016).
240. Korda 2019.
241. Korda 1994b.
242. Korda 1994c.
243. Although VEHMT emerged because of Knight's writings and advocacy, he prefers the term "finder" to founder" (Maharaj 2021). The *Wild Earth* issue mentioned above states that "VHEMT . . ., though only months old, is already being called, by some conservationists, the most exciting new movement in this country since Conservation Biology" (WE 1991).
244. Knight 1991, italics added.
245. Knight 1997.
246. Knight 1991.
247. It is worth noting that there were other misanthropic antinatalists during this period, such as the "philosopher of despair"—as a *New York Times* obituary put it—Emil Cioran, who was motivated by philosophical pessimism (Pace 1995). Suffice it to quote a passage from his 1973 book *The Trouble with Being Born*, which encapsulates the general message of his philosophical worldview:

    > We do not rush toward death, we flee the catastrophe of birth, survivors struggling to forget it. Fear of death is merely the projection into the future of a fear which dates back to our first moment of life. . . . We are reluctant, of course, to treat birth as a scourge: has it not been inculcated as the sovereign good—have we not been told that the worst came at the end, not at the outset of our lives? Yet evil, the real evil, is behind, not ahead of us. What escaped Jesus did not escape Buddha: "If three things did not exist in the world, O disciples, the Perfect One would not appear in the world . . ." And ahead of old age and death he places the fact of birth, source of every infinity, every disaster" (Cioran 1973).

    Thanks to Ariane Hanemaayer and Tyler Brunet for apprising me of Cioran's work.
248. Oremus 2013.
249. To be clear, our descendants could decide not to carry on these things even if they *don't* undergo normative extinction—this would be an instance of what could be called *ideological*, *cultural*, or *axiological* loss. My point is that if normative extinction *were* to occur, it would be sufficient to produce an outcome that Russell saw as bad.
250. Anders 1962.
251. Schell 1982, italics added.

10 Astronomical Value and the Harm of Existence

Saving Humanity

The third wave in Existential Ethics is defined by two major developments over the past couple of decades: (1) the founding of Existential Risk Studies by Nick Bostrom and others in the early 2000s and (2) the first extended philosophical treatise on antinatalism published by David Benatar in 2006. Although figures like Philipp Mainländer, Peter Wessel Zapffe, Hermann Vetter, and Les U. Knight had all discussed and endorsed antinatalism before the 2000s, none offered a comprehensive, systematic, Analytic treatment of the subject, which is what Benatar provides in his *Better Never to Have Been: The Harm of Coming Into Existence* (2006). This important contribution to the literature directly links antinatalism with the idea that humanity should go extinct sooner rather than later, although we will see that this line of reasoning is problematic. With respect to Existential Risk Studies, understanding the nature of this field, especially its moral-axiological foundations, will complete the picture of how the fifth existential mood emerged in the late 1990s and early 2000s by explaining why some at the time were *motivated* to provide exhaustive lists of every possible threat to our survival, however improbable, hypothetical, speculative, or exotic it might be. This will require dissecting Existential Risk Studies into its two main anatomical parts, which roughly correspond, as it happens, to the two main parts of this book. In doing so, we will see how the causal relation between History #1 and History #2 reversed for the first time: rather than the discovery of new kill mechanisms provoking thoughts about our extinction, thoughts about our extinction stimulated new research on the ways our collective future could be destroyed.

As with previous chapters, I will organize this one both thematically and chronologically, taking (1) and (2) in turn. Tracing the historical development of each will bring us up to the present, to the vanguard of contemporary scholarship (as of this writing), although focusing exclusively on Existential Risk Studies and philosophical antinatalism means that our discussion in this chapter will neglect many important contributions to Existential Ethics made over the past five years or so. However, these will occupy the pages of the next chapter, after which the

DOI: 10.4324/9781003246251-12

final chapter of this book will briefly explore how the idea of human extinction could evolve in the future. We begin with a closer look at the field of Existential Risk Studies.

Two Branches of Existential Risk Studies

Recall from Chapter 6 that the most recent shift in existential mood was catalyzed by two triggers in particular. One arose from alarming new research on anthropogenic climate change, biodiversity loss, and the sixth mass extinction event, which showed that these pose far greater near-term threats to humanity (and the biosphere) than had previously been known. The other emerged with the formation of Existential Risk Studies, which, appearances to the contrary, is not a single field of inquiry but two distinct, interrelated "branches." The first branch focuses primarily on one of the two major themes of Part I, namely, the nature, number, etiology, and probability of kill mechanisms. (To be more precise, this branch studies the nature, number, and so on of *existential risk scenarios*, which includes but is not limited to kill mechanisms as we have defined them. More on this momentarily.)

Much of the work within this branch has thus involved drawing from the insights and ideas of scientists in fields like cosmology, astronomy, physics, climatology, volcanology, ecology, computer science, and so on. It is thus highly interdisciplinary. But existential risk researchers—or, as I previously called them, "riskologists"—have also conducted original research that has identified potential kill mechanisms associated with, for example, value-misaligned artificial superintelligence (ASI) and the possibility that we are living in a computer simulation that gets shut down. Whereas, say, the discovery of the nuclear winter scenario was based on empirical studies and computer modeling, the recognition that ASI could pose a threat arose from philosophical reflections on the nature of autonomous, goal-directed, instrumentally rational agents with superhuman capabilities, while the supposed danger of a simulation shutdown derived from extrapolations of technological trends and a priori anthropic and probabilistic reasoning. Other possible kill mechanisms, such as self-replicating nanobots that could destroy all the organic matter on our planet, were discovered through exploratory engineering, whereby one imagines what could be given constraints imposed by the known laws of nature.[1]

Since the first scientifically credible kill mechanism was discovered in the 1850s, this branch of Existential Risk Studies has roots stretching back more than a century. The scientific study of kill mechanisms is not new. However, what is new about this branch is its explicit attempt to provide a panoramic mapping of the threat environment that includes not just the "existing" threats to our survival but the whole range of possible "emerging" threats that we might encounter in the coming decades and centuries. (This was the essence of the futurological pivot.) Hence, whereas in the past kill mechanisms had been mostly studied and philosophized about individually, *in isolation* from each other, riskologists aimed to establish a research program that considered the entire array of global risks as constituting a single cohesive category. This more holistic approach to thinking about our existential predicament is, we saw, exemplified by the proliferation of encyclopedic surveys and comprehensive enumerations of every possible kill mechanism. The first survey of this sort was compiled by John Leslie in *The End of the World* (1996), followed by those provided by Bostrom (2002), Lord Martin Rees (2003), Richard Posner (2004), and the 2008 volume edited by Bostrom and Ćirković titled *Global Catastrophic Risks*.[2] The last of these, in particular, epitomizes the focus of this first branch of Existential Risk Studies: it hardly addresses the ethical and evaluative implications of our extinction (or existential risks); instead, the book is almost entirely dedicated to (a) laying out the scientific and philosophical evidence for the existence of various kill mechanisms, including the heat death, climate change, worldwide pandemics, value-misaligned ASI, and gray goo, and (b) examining certain background issues relevant to the study of global catastrophic risks (see below).

But a question arises here as to whether studying these risk scenarios as a single category makes sense. As Bostrom and Ćirković wrote in the book's introduction,

> the risks under consideration seem to have little in common, so does [this] even make sense as a topic? Or is the book that you hold in your hands as ill-conceived and unfocused a project as a volume on "Gardening, Matrix Algebra, and the History of Byzantium"?[3]

In other words, the question here concerns the *coherence* of the first branch of Existential Risk Studies: what justifies placing these disparate scenarios under the same umbrella? Why not continue to study them individually, in isolation from each other, as they had been since the latter nineteenth century?

There are many possible answers to these questions, which span a diverse range of mechanistic, physical, methodological, psychological, conceptual, pragmatic, and cultural considerations. For example:

- Multiple scenarios involve the same or similar physical processes, as in the case of impact, nuclear, and volcanic winters, all of which involve sunlight-blocking aerosols spreading throughout the stratosphere. Hence, insight about these processes could have implications for all three scenarios, which may justify considering them together, as a group.[4]
- Mitigating the risk of human extinction is what economists would call a *global public good*. Since global public goods are *non-rivalrous* (anyone can consume the good without its "quantity" being decreased) and *non-excludible* (there is no way to prevent anyone from consuming the good), they tend to be undersupplied

by the free market. Thus, we cannot rely on the market to provide the good of protecting us from extinction.[5] Even worse, as Bostrom later noted, reducing extinction risk is not just a *global* but "a strongly *transgenerational* (in fact, *pan-generational*) public good" in that much of the benefit would be reaped by people in future generations, who may vastly outnumber those alive today. This renders reduction efforts even more challenging, since people in current generations "may capture only a small fraction of the benefits."[6]

- Independent of the scenario, probability estimates of it occurring must take into account observation selection effects.[7] As noted in Chapter 6, the fact that a huge asteroid hasn't collided with Earth in the past 10,000 years should not *itself* be taken as strong evidence that asteroid collisions are improbable, since if one *had* collided with Earth 10,000 years ago, we very likely wouldn't be here to discuss the issue. Some catastrophe scenarios are incompatible with the existence of observers like us.
- Independent of the scenario, since human extinction (e.g., in the terminal or final senses) can by definition only happen once in our species history, our strategies for mitigating the risk of this happening must be *proactive* rather than *reactive*. We can talk about our extinction happening tomorrow but not about it having happened yesterday, and hence there is no opportunity to learn from past mistakes, as extinction cancels the future. Hence, "this requires *foresight* to anticipate new types of threats and a willingness to take decisive *preventive action* and to bear the costs (moral and economic) of such actions."[8]
- The same cluster of cognitive biases may distort our thinking about a wide range of human extinction scenarios. For example, people tend to believe that events which come to mind more easily have a higher probability of occurring (availability bias). Most people think of asteroid impacts before volcanic supereruptions when asked how our extinction might occur, yet volcanic supereruptions are *much more probable* than collisions with large asteroids. Similarly, people tend to believe that conjunctive propositions ("A and B and C") are more probable than disjunctive propositions ("A or B or C"), when just the opposite is true (disjunction fallacy). This is pertinent to overall assessments of the probability of doom, which could result from nuclear conflict or global pandemics or asteroid impacts or an invasion of Earth by bellicose aliens. Importantly, the addition of the last disjunct—an alien invasion—makes the proposition as a whole *more rather than less* probable, even if one judges this scenario itself to be cockamamie nonsense.[9]

How Much Is at Stake

Other "links and commonalities" could be mentioned.[10] However, of interest for our purposes is that, according to many riskologists, the most *fundamental issue* that unifies the first branch of Existential Risk Studies is *normative*, arising from the belief that the axiological opportunity costs of succumbing to an existential catastrophe, such as final human extinction, would be enormous. This, more than any other consideration, is what motivates and justifies assembling long lists of possible risks and taking them to form a single category worthy of our attention—a category around which a whole new field of academic study should be built.

To understand how this line of reasoning developed, let's begin with Leslie's 1996 book *The End of the World*, which I argued above was the first major publication to embody the futurological pivot. There were two reasons Leslie was interested in compiling an exhaustive list of risks to our survival. The first was that, as we discussed, the primary focus of his book was defending the Doomsday Argument, and the Doomsday Argument can only be applied to estimates of the overall probability of our disappearance, which one derives from prior empirical and philosophical analyses of the threat environment. Hence, one must map out the entire threat environment for the Doomsday Argument to be of any use. But why be interested in the Doomsday Argument-adjusted probability of extinction *in the first place*? This gets to the second reason, namely, that Leslie—following in the footsteps of J. J. C. Smart, Jonathan Glover, and Derek Parfit, all of whom he cited approvingly—believed that extinction would constitute an immense *axiological catastrophe*, since it would preclude the realization of potentially vast amounts of future value.[11] Put differently, the state or condition of Being Extinct would be very bad, and hence by estimating the overall Doomsday Argument–adjusted probability of extinction within the next few centuries, one might hope to *motivate efforts* to avoid this outcome.[12]

As noted in Chapter 6, Leslie was a self-described utilitarian, although unlike Henry Sidgwick he did not accept a hedonistic theory of value. Rather, he was an "ideal" utilitarian, according to which there are intrinsic goods in addition to happiness (or satisfied desires). This version of utilitarianism was famously defended by G. E. Moore, one of the founders of Analytic Philosophy, who held a pluralistic value theory according to which things like beauty have intrinsic value. Hence, a universe full of beauty would be better than a universe full of ugliness, he argued, even if there were *no one around* to appreciate the difference.[13]

What matters for our purposes is that ideal utilitarianism, of the sort championed by Leslie, is still maximizing and impersonalist, which means that the more value there is in the universe, the better things will be. Our extinction would, therefore, be bad even if it were brought about voluntarily, which is why Leslie placed scenarios like "unwillingness to rear children" in the very same category of his risk typology as potentially violent catastrophes associated with genetically engineered organisms, gray goo, an AI takeover, and even "annihilation by extraterrestrials."[14] The *manner* in which our extinction comes about is much less important than the *consequence* of there being no more people, and this is why we should investigate all possible ways of this happening, however speculative or improbable. As Leslie explained in a footnote that quotes Smart's discussion of Hilbrand Groenewold's idea of "macro effects,"

> even a very low probability, when "multiplied by a macro disaster," would be something having "macro disvalue," a point immensely important when we consider "the millions of years of possible evolution of the human race that lie ahead if we do not destroy ourselves."[15]

With respect to the term "disvalue," recall Jonathan Bennett's observation from the previous chapter that for impersonalists, the failure to create new value that *could exist* constitutes a "loss" no less than the elimination of value that *already exists*. Impersonalism, one could say, abhors an axiological vacuum.[16] Quoting yet another utilitarian mentioned earlier, Leslie thus concluded that "Glover was, I believe, right when he reached [the] conclusion . . . that to end the human race 'would be about the worst thing it would be possible to do.'" Hence, the deeper reason for surveying the threat environment from one horizon to the other was, in Leslie's words, "how much is at stake."[17]

What Is an Existential Risk?

The discussion so far has focused on human extinction—specifically, on final and normative extinction, since each is *sufficient* to preclude the realization of all or most future value, whereas the other types of extinction are not. Leslie, in fact, gestured at these two scenarios in writing that "today's humans could perhaps have descendants continuing onwards for many millions of years," which might take the form of some "fusion between our descendants and computers to which their brains were permanently linked." Or we could be "entirely replaced by computers." Would this be bad? It depends, Leslie wrote:

> Maybe . . . the tragedy that humankind had ended after a few thousand years would be smaller if it had ended only through being replaced by computer-based intelligent systems—provided, of course, that those systems truly were conscious beings.[18]

While this indicates some concern over terminal extinction, Leslie's utilitarianism itself does not, in any obvious way, imply that the loss of our particular species should matter, which leads me to suspect that Leslie would have modified those sentences if he had reflected more on the issue. Or perhaps he would retort that our species has some value in itself, and hence that it is one of the things we should keep around. In general, the central concern for utilitarians will be the avoidance of final and normative extinction.

What is important for our history is how Bostrom shifted the focus from *human extinction* to *existential risk*, which constitutes a much broader category that includes but is not limited to our extinction—in the final and normative senses—although he later added premature extinction, which itself should be understood in the final and normative senses. Consider the following line of reasoning: extinction would be very bad because it would preclude the realization of lots of value within our future light cone, but there are *other ways* that we could fail to realize this value that do not involve extinction of any kind; therefore, we should take these other ways, or failure modes, to be *comparable in badness* to extinction. The concept of existential risk, which Bostrom introduced in his 2002 paper "Existential Risks: Analyzing Human Extinction Scenarios and Related Hazards," was designed to encompass the entire range of events that, if they were to occur, would entail what he describes as "enormous . . . negative utility."[19] Such events, as we will see, include not just our extinction but scenarios like permanent civilizational collapse and technological stagnation. People had of course worried about things like the collapse of civilization before, but Bostrom showed that, depending on the details, an event like this could have the *same moral-axiological status* as final and normative extinction, and hence we ought to include these non-extinction scenarios within an *even more* expansive threat environment, one that includes more than just kill mechanisms.[20]

The argument above about future value is premised on an impersonalist interpretation of total utilitarianism: our sole moral obligation is to maximize value, and hence *anything* that prevents us from flooding our future light cone with value would constitute an existential catastrophe. In fact, Bostrom delineated precisely this argument in his 2003 paper "Astronomical Waste," discussed momentarily. His initial presentation of the idea, though, focused on the transhumanist goal of creating a posthuman civilization, thus defining "existential risk" in specifically transhumanist rather than utilitarian terms. To understand how these conceptions of existential risk relate to each other, let's begin with Bostrom's most generic definition in his 2002 paper. This stipulates that an existential risk is "one where an adverse outcome would either annihilate Earth-originating intelligent life or permanently and drastically curtail its potential."[21] Notice that it consists of two disjuncts, the first of which is unnecessary, as it merely specifies one way that the second disjunct could obtain. That is to say, if Earth-originating intelligent life (i.e., "humanity" in Bostrom's phraseology) were to be completely annihilated, end of story, this would of course permanently and drastically curtail our potential. We can thus streamline Bostrom's definition by deleting the first disjunct, which gives us: *an existential risk is any event that would permanently and drastically curtail the potential of Earth-originating intelligent life.*

This leads to the question: What, then, does "potential" refer to? As a normative term, one could define it many different ways, depending on one's values. An anarcho-primitivist would say that our potential involves returning to the lifeways of hunter-gatherers; Marxists would point to the creation of a world communist state. For Bostrom, our potential consists in the possibility of creating a stable, flourishing posthuman civilization, where the term "posthuman civilization" refers to "a society of technologically highly enhanced beings . . . with much greater intellectual and physical capacities, much longer life-spans, etc." Hence, if we substitute this understanding of potential into the above definition, using Bostrom's definition of "humanity," we get the following: *an existential risk is any event that would permanently and drastically curtail the ability of Earth-originating intelligent life to create a stable, flourishing posthuman civilization.*

Here we have Bostrom's original conception of existential risks, and despite his generic definition making no reference to this transhumanist goal, the centrality of transhumanism to the idea is manifest in his typology of existential risk scenarios, which I quoted in full in Chapter 6. In brief, he recognized (i) bangs, whereby "Earth-originating intelligent life goes extinct in [a] relatively sudden disaster," (ii) crunches, whereby our "potential . . . to develop into posthumanity is permanently thwarted although human life continues in some form," (iii) shrieks, whereby "some form of posthumanity is attained but it is an extremely narrow band of what is possible and desirable," and (iv) whimpers, whereby

> a posthuman civilization arises but evolves in a direction that leads gradually but irrevocably to either the complete disappearance of the things we value or to a state where those things are realized to only a minuscule degree of what could have been achieved.

The first could happen either before *or* after creating a posthuman civilization. If it were to happen after, our posthuman descendants will have succumbed to their own final extinction. This is why I included the word "stable" in my reconstruction of Bostrom's definition, to address this possibility. The second would involve never creating a posthuman civilization in the first place, despite humanity surviving. The third and fourth would involve creating a posthuman civilization but not one that is "flourishing" (again, my word). In all of these cases, the transhumanist project would be left incomplete, resulting in an existential catastrophe.

The Promise of Techno-Utopia

The next question becomes: Why accept this interpretation of our potential? Bostrom largely skips over this question in his 2002 paper, which—like his edited volume *Global Catastrophic Risks*, published several years later—focused primarily on laying the groundwork for the *first branch* of Existential Risk Studies. (Indeed, this paper is basically just a synopsis of Leslie's 1996 typology of risks, with the new idea of existential risk.) However, Bostrom did address the question one year later, in his 2003 paper "Transhumanist Values," which argued that radical human enhancement could enable us to explore posthuman modes of being that are far superior to our current human mode. We could, for example, acquire indefinitely long lifespans, become superintelligent, upload our minds to computers, gain total control over our emotions, and ultimately "increase our subjective sense of well-being" in ways unimaginable to us right now. As Bostrom explained the idea,

> our own current mode of being . . . spans but a minute subspace of what is possible or permitted by the physical constraints of the universe. . . . It is not farfetched to suppose that there are parts of this larger space that represent extremely valuable ways of living, relating, feeling, and thinking. . . . Transhumanism promotes the quest to develop further so that we can explore hitherto inaccessible realms of value. Technological enhancement of human organisms is a means that we ought to pursue to this end.[22]

Advanced technologies, in other words, could usher in a techno-utopian world replete with endless wonders, happiness, and value—*this* is the promise, the hope, the dream of a "posthuman civilization" that existential risks threaten to obliterate. Bostrom elaborated his vision in his subsequent "Letter from Utopia," which he posted online in 2005 and officially published in 2008, later updating it in 2020. The letter is written by a fictional posthuman to their human ancestors, urging present-day people—you and I—to "help us come into existence!" It thus opens with "Dear human" and closes with "Your Possible Future Self." In the 2005 version, which I will primarily quote from because it probably best represents Bostrom's thinking at the time, the fictional author begins with the question:

> How can I tell you about Utopia and not leave you mystified? What words could convey the wonder? What language could express the happiness that we have here? I fear that my pen is as unequal to this task as if I were trying to use it to kill an elephant.

They proceed with a few tantalizing glimpses of the magical world they inhabit, writing that

> my consciousness is wide and deep. I've read all the books that you humans had written by your time—and a good deal more. I know life from many sides and angles. I have swum in a whole spectrum of different cultures, more numerous than the words in your dictionary. Quite a bit of culture builds up over a million years (even as the humble polyps amass a reef given enough time). Well, all this information I have incorporated into my mind, and much, much more. Each etching, each record-cover, each toothpaste tube design—they are all lodged in my memory banks, and my appreciation of each object is as intimate as the appreciation that the most sensitive connoisseur has of her favorite artifact.

Through radical cognitive enhancement, we could be come superhuman polymaths with eidetic memories capable of storing oceanic mountains of information, from the trivial to the profound. "My experience is clear and intense," the author continues, "my mind is shaped by what it has assimilated. I don't just think about deep truths; my thoughts themselves are deep."[23] But knowledge is not the only prize that Utopia offers us or our descendants. Life in Utopia is also awash in what the posthuman describes as "surpassing bliss and delight."[24] To quote them at length once again:

> You could say I am happy, that I feel good. You could say that I feel surpassing bliss. But these words are used to describe human experience. What I feel is as far beyond ordinary human feelings as my thoughts are beyond human thoughts. I wish I could show you what I have in mind. If only I could share one second of my conscious life with you! But that is impossible. Your container could not hold even a small splash of my joy, it is that great. . . . You don't have to understand what I think and feel. If only you bear in mind what is possible within the present human realm, you should have enough of an idea to get started in the right direction, one step at a time.

Our most joyous experiences today are nothing compared to what they could be in Utopia. If the distance between our normal state and the most intense feelings of elation equals eight kilometers, the posthuman writes, "then to reach my location you would have to continue for another million light years. It is beyond the moon and the planets and all the stars your eyes can see." This is why "we love life here every second. Every second of life is so good that it would knock you unconscious if your mind had not been strengthened beforehand."[25]

But how can we bring about this technological paradise? The author specifies three types of human enhancement as the necessary vehicles for reaching the Promised Land: extending our healthy lifespans, boosting our cognitive capacities, and elevating our emotional wellbeing. The most critical condition that must be satisfied, though, is *avoiding an existential catastrophe*. As Bostrom made the point in his "Transhumanist Values," "there is one kind of catastrophe that must be avoided at any cost: Existential risk." Why? Because "if we go extinct or permanently destroy our potential to develop further, then the transhumanist core value [of exploring the posthuman realm] will not be realized. Global security is the most fundamental and nonnegotiable requirement of the transhumanist project."[26] This brings us full circle to Bostrom's original conception of existential risk, which we can reformulate once more as: *an existential risk is any event that would prevent Earth-originating intelligent life from establishing a techno-utopian world, understood in specifically transhumanist terms.*

To summarize how the idea of existential risk was born: Bostrom's starting point was transhumanism, which he was involved with in the 1990s, during which it took the form of extropianism. At the turn of the century, inspired by the ideas of others in the extropian community, he then introduced the existential risk framework, which had two primary aims: (a) to identify all the ways the transhumanist project could fail; he called these "existential risks," and (b) to devise effective strategies for mitigating such risks, catalogued in his comprehensive 2002 survey of the threat environment, thereby ensuring the realization of the transhumanist project. Of particular interest was the promise and peril of GNR (genetics, nanotech, and robotics) technologies, which came into focus with the futurological pivot. It was, in fact, the increasing plausibility of these technologies, which promise new ways of radically modifying the human organism, that partly inspired the modern transhumanist movement. But transhumanists quickly realized that the *very same* technologies that make techno-utopia possible also carry risks that could be even greater than those arising from the NBC (nuclear, biological, and chemical) weapons of the twentieth century. This led to two general responses to our newly anticipated threat environment: first, there was the Bill Joy camp, which claimed that we should *not develop* these technologies in the first place. They are simply too dangerous. The solution is to impose broad moratoriums on entire fields of emerging technoscience. And second, there was the transhumanist camp, led by Bostrom, which argued that the solution is to establish a new field of interdisciplinary research focused specifically on understanding and mitigating these risks, with the hope of keeping our technological cake and eating it, too, as it were. This is how Existential Risk Studies was born—as an answer to the question: how can we develop GNR technologies, which the transhumanist project seems to require, without destroying that project in the process?

Expected Value, Asteroids, and Ethics

The idea of existential risk thus grew from the "ethical outlook" of transhumanism, as Bostrom described it. In fact, a draft of Bostrom's "Transhumanist Values" was completed before his existential risk paper was published, even though "Transhumanist Values" was published after this. (The existential risk paper even cited a draft of "Transhumanist Values," which further indicates that transhumanist concerns inspired the idea of existential risk.) However, as alluded to above, Bostrom also provided a utilitarian argument for why reducing existential risk should be our top global priority as a species. This paper, titled "Astronomical Waste," was published in 2003, the same year as "Transhumanist Values." Its central thesis was, in effect, that the concept of existential risk should be augmented to cover events that prevent us from not only building a stable, flourishing posthuman civilization, but colonizing space and creating enormous numbers of "happy" people in the future, most of whom would reside in vast computer simulations. In making this argument, Bostrom went well beyond Carl Sagan's 1983 estimate of how many future people there could be—which, recall, was 500 trillion over the next 10 million years. If we colonize space and simulate huge populations of digital people, the number could be many orders of magnitude larger. This, along with transhumanism, also provided a normative foundation for the first branch of Existential Risk Studies by motivating and justifying the investigation of every possible kill mechanism—or, more generally, existential risk mechanism—that we might encounter at present or within the coming centuries.

Before we examine this utilitarian argument from Bostrom, though, it will prove useful to pause on two background issues that have come to play a central role within Existential Risk Studies, namely, expected value theory and physical eschatology. Taking them in turn, the expected value (or expectation) of an action is the probability-adjusted average of the value of its possible outcomes. That may sound obscure, but the idea is quite straightforward. Let's say that an action A could result in one of two outcomes, X or Y, and that X has a value of −50, while Y has a value of 75. Let's say further that the probability of X occurring if one does A is 20 percent, while the probability of Y occurring if one does A is 80 percent. So far we have values and probabilities. To get the probability-*adjusted* (or "weighted") values of X and Y, we simply multiply these values and probabilities: −50 times .2 is −10, while 75 times .8 is 60. Hence, −10 and 60 are the probability-adjusted values of the outcomes of X and Y. To get the expected value of the action A, we then add these together and divide by two: 60 + (−10) = 50, and 50/2 = 25. The expected value of A is, therefore, 25. Now, imagine that you have the option of taking two different actions, A and B. Expected value theory (EVT) asserts that when choosing between some finite number of actions, one should choose whichever action has the highest expected value.[27] Let's say that you crunch the numbers, per above, and find that B has an expected value of 30. Since 30 is greater than 25, EVT instructs you to do B rather than A; doing otherwise would be *irrational*.

Insofar as one can assign values and probabilities to the range of possible outcomes of actions, EVT provides a useful tool for making decisions *under uncertainty*, where I will take "uncertainty" to mean that probabilities *can* be assigned to outcomes having different values. (In contrast, decision-making under "ignorance" refers to situations in which probabilities *cannot* be assigned, and hence EVT is not applicable.)[28] Incidentally, the standard definition of "risk" ever since the mid-1970s has been given in expected-value terms. On this account, a risk equals the probability of an outcome times its severity (a negative value).[29] Leslie and Smart both gestured at this idea in arguing that, quoting Smart once again, "a very low probability multiplied by a macro disaster can still have macro disvalue," which suggests that focusing on highly improbable risks could have enormous expected value given the immense badness of the consequences, if they were to obtain.[30] For example, you might reason that even though there is a small probability that your gas oven is leaking, the consequences of carbon monoxide poisoning, which could kill your entire family while sleeping, are so great that it is *worth* paying $25 for a carbon monoxide detector, where "worth" may be understood in terms of expected value. Or take another example, from one of the first publications in the existential risk literature after Bostrom introduced the concept. In his 2007 paper "Reducing the Risk of Human Extinction," Jason Matheny estimated the cost-effectiveness of reducing existential risks associated with asteroids compared to other ways of allocating our finite resources. He calculated that if "reducing the probability of an extinction-level impact over the next century by 50%" were to cost a total of $20 billion, the cost-effectiveness of this program would be a mere "$2.50 per life-year," which contrasts with the more than "$100,000 per life-year" spent by US health programs. Hence, mitigating the asteroid threat yields a much greater bang for the buck, in terms of expected value. In fact, Matheny claims that this would be the case "even if one is less optimistic and believes humanity will certainly die out in 1,000 years," given that "asteroid defense would [still] be cost effective at $4,000 per life-year." Matheny concludes that while "the probability of [human extinction] events may be very low, . . . the expected value of preventing them could be high, as it represents the value of all future human lives."[31]

Expected value theory can also play a direct role in moral theorizing, in addition to helping us make "rational" decisions under uncertainty.[32] For example, consider the difference between what could be called actualist utilitarianism and expectational utilitarianism. The first was espoused by John Stuart Mill and, later, by Smart. It states that the rightness or wrongness of an act depends only on its actual consequences, independent of its rationality.[33] Hence, on Smart's account, a decision can be both irrational and moral, or both rational and immoral, at the same time.[34] To illustrate, imagine that the action B from above could yield two outcomes, each of which has a value of 60 and a probability of 50 percent. This gives the previously specified expected value of B as 30 (i.e., 60 × .5 = 30, 30 × 2 = 60, and 60/2 = 30). As noted, B is more rational than A, since B has a higher expected value. But now imagine that one does A instead of B, and the result that actually obtains, by chance, is the outcome Y, which we stipulated above has a value of 75. For Smart, one would have acted *irrationally but morally* by doing A, since all that matters for the purpose of *moral evaluation* is what actually happens. However, others have argued that the rightness or wrongness of an act should depend on its expected consequences, and hence even though doing A happened to result in the best possible outcome—a value of 75—doing A rather than B was not only irrational but immoral, since B has a higher expected value. On this account, rationality and morality coincide. Still others suggest that we should distinguish between two senses of "right," one objective (actual) and the other subjective (expectational), where both "concepts might have a legitimate theoretical role."[35] In the scenario above, then, by choosing A and getting outcome Y, one would have done what is "objectively" right but "subjectively" wrong.

Either way, the point of this digression is that EVT has become very influential within Existential Risk Studies and its more recent incarnation, *longtermism*, which we will examine below. It is worth noting, though, that EVT's use within the existential risk context is highly controversial, given that existential risks are typically thought to be low-probability events with extreme negative value, understood as the loss of techno-utopia or astronomical amounts of value. The problem of using EVT in such decision situations has, in fact, been noticed by existential risk scholars themselves, who introduced the idea of "Pascal's Mugging" to name this class of problems.[36] Picture yourself in a dark alley at night, approached by someone who demands that unless you give them $5, they will torture a billion trillion trillion people in some parallel universe. Should you give them the money? Even though their claim is obviously absurd, you cannot *absolutely rule it out*, because one can never be completely sure of anything (except maybe logical and mathematical truths; empirical truths can never be known with certainty). Hence, even if the probability is minuscule that the mugger is being honest, the payoff of avoiding all this torture is so great that the expected value of giving them $5 may nonetheless be much higher than not doing so. You thus give them the $5 and they walk away a little richer, with you a little poorer. We will return to this idea in the next chapter.

Physical Eschatology

Moving on to the second background issue, one way to approach it begins with this: whereas the transhumanist argument for why existential risk reduction should be our top global priority as a species is based on a techno-utopian version of potentiality thinking, the second issue is centered around deep-future thinking, which underwent a radical transformation beginning in the late 1960s with the founding of a field called "physical eschatology." Recall that deep-future thinking was born in the mid-nineteenth century with the discovery of the Second Law, which spurred novel speculations about the future habitability of Earth and the universe. However, our understanding of the future evolution of the cosmos was deeply impoverished, and indeed predictions based on the Second Law that our sun would eventually burn out were incorrect: the exact opposite is now expected to happen. Rather than becoming dimmer over time, the sun's luminosity will actually increase in the coming billions of years as it balloons into a red giant, a stage in its life cycle that will end with it aging into a white dwarf. Human life will become impossible not because Earth freezes over but because the temperature of Earth's surface will become so hot that the oceans will literally boil into the atmosphere.[37] Or consider that it was not until the late 1920s that we realized, thanks to Edwin Hubble, that the universe is expanding rather than in a "steady state" (or static configuration), and not until the early 1960s that scientists detected, by accident, the Cosmic Microwave Background (CMB), which convinced the scientific community that the big bang hypothesis of the universe's origin is true. The CMB is the afterglow left behind as the universe cooled below the temperature of hydrogen plasma. Since hydrogen plasma is opaque to light, this cooling process eventually made the universe transparent, and the photons liberated by this event are what comprise the CMB. Facts taken as rather elementary today are, in truth, quite recent additions to our understanding of cosmology.

But the most important development with respect to *deep-future thinking* dates back to 1969, when Lord Martin Rees—the same scientist who helped solidify the fifth existential mood—published a paper titled "The Collapse of the Universe: An Eschatological Study."[38] This was the first time the word "eschatology" was used in a specifically astrophysical rather than theological context, and many see it as having inaugurated the field of physical eschatology, although the term "physical eschatology" wasn't coined until 1997.[39] While the scenario that Rees focused on is now widely rejected, his paper inaugurated a flurry of new research on the future of Earth, the solar system, the stars, black holes, galaxy groups, clusters, and superclusters, and the cosmos as a whole.[40] Physical eschatology was born.

The result was a much fuller picture of "the shape of things to come," to borrow Wells' famous phrase once again, but on cosmological timescales.[41] We now know, for example, that Earth will remain habitable to complex life for another 800 million to 1 billion years; the Andromeda galaxy will "collide" with the Milky Way in some 4 billion years; our expanding sun will swallow Earth in roughly 7.59 billion years; the universe will go dark (that is, no more shining stars) in about 99.9 trillion years; protons will decay—if they do—in about 10^{40} years, thus rendering biological life, if not intelligence in any form, impossible; and roughly 10^{100} years from now, all the black holes in the universe will have evaporated, and

> the cosmos will be filled with the leftover waste products from previous eras: neutrinos, electrons, positrons, dark matter particles, and photons of incredible wavelength.[42] In this cold and distant Dark Era, physical activity in the universe slows down, almost (but not quite) to a standstill.[43]

At this point, it could be—although this is highly speculative—that a vacuum state phase transition spontaneously occurs, thus "giving the universe a chance for a fresh start."[44] The alternative is an eternal nothingness, a lifeless forever.

Its Creative Potential

When combined with our current understanding of the *size* and *structure* of the observable universe, physical eschatology thus provides a scientifically robust answer to the question: "How *big* could the future be?" This is, of course, directly relevant to the question of how *bad* an existential catastrophe would be, especially from an impersonalist, value-maximizing, utilitarian perspective, since the bigger the future could be, the greater the potential value that would be lost if a catastrophe of this sort were to occur. Consequently, physical eschatology has become absolutely integral to the Existential Ethics research conducted by existential risk scholars and longtermists, as evidenced by the fact that a large percentage of the papers published on these topics all begin their discussions with surveys of the incomprehensible bigness of our cosmic future, based on the findings of physical eschatology.[45]

Although Schell and Parfit both seemed to have been aware of how much longer Earth will remain habitable—Sagan, being a cosmologist, most definitely was—the first study of this question from a "transhumanist perspective" came from Bostrom's colleague Milan Ćirković, with whom he edited *Global Catastrophic Risks*.[46] This subsequently informed Bostrom's take on the issue from a *utilitarian* perspective, and hence it is worth taking a look at this study before returning to Bostrom. Let's begin with Ćirković's paper "Cosmological Forecast and Its Practical Significance," which was published the same year and in the same journal as Bostrom's 2002 paper on existential risk, namely, the *Journal of Evolution and Technology*, originally called the *Journal of Transhumanism*. Both journals were run by the World Transhumanist Association that Bostrom cofounded in 1998, and indeed Bostrom was the first editor-in-chief of the *Journal of Transhumanism*.[47]

Ćirković argued that physical eschatology paired with recent developments in anthropics (from both Leslie and Bostrom) carry potentially urgent practical implications for "intelligent observers interested in self-preservation and achieving [the] maximum of its creative potential." Specifically, the *timing* of space colonization could make a significant difference to the amount of resources available to advanced civilizations, given our best current understanding of our "cosmological situation." As he expressed the idea,

> decision-making performed today, as far as humanity is concerned, may have enormous consequences on very long timescales. In particular, an overly conservative approach to space colonization and technologization, may result (and in fact might have already resulted) in the loss of substantial fraction of all possible observer-moments humanity could have had achieved.[48]

The term "observer-moment" comes from Bostrom, who defined it as "a brief time-segment of an observer." Thinking in terms of observer-moments instead of observers is relevant to anthropics—that is, to reasoning about one's location in space and time based solely on the fact that one is an intelligent observer. But it may also be relevant to estimating our "potential" over time: since "different observers may live differently long lives, be awake different amounts time, . . . etc.," counting observer-moments rather than observers can give a more accurate measure of the bigness or value of the future.[49] There could, after all, be a large number of observer-moments even if there are relatively few observers, and vice versa. The point is that Ćirković used this concept to argue that we should develop advanced technologies and colonize the universe as soon as possible, given that every passing moment our cosmic endowment of negentropy is going to waste as stars burn up their limited reservoirs of hydrogen. Consequently, by failing to advance technology and colonize space, we could lose—and maybe have already lost—a "substantial fraction of all possible observer-moments [that] humanity could have . . . achieved." To underline the time-sensitivity of this situation, Ćirković calculated that if a "future hypercivilization" could extract all the energy output of the stars populating the Virgo Supercluster, then "the number of potentially viable human lifetimes lost per century of postponing . . . the onset of galactic colonization is . . . $5 \times \sim 10^{46}$." Thus, assuming some correlation between the number of lifetimes or observer-moments and the fulfillment of our "creative potential," this provides a very strong *prima facie* reason against "an overly conservative approach to space colonization and technologization."[50]

Astronomical Value

To my knowledge, as noted, the first person to provide an estimate of how many future people there could be was Sagan, although his calculation was spatiotemporally limited to *Earth* over the next *10 million years*. In contrast, Ćirković considered the entire Virgo Supercluster, explicitly situating his discussion within the framework of physical eschatology. However, he did not offer a *conservative* estimate of the future's value, nor did he explicitly tie his calculations to the total-impersonalist-utilitarian view that the more total value within the universe, the better the world will become. This is what Bostrom's paper "Astronomical Waste: The Opportunity Cost of Delayed Technological Development" did the following year. First, unlike Ćirković, Bostrom considered a lower bound on how many biological humans could exist within the Virgo Supercluster per century, reporting that the number could be 10^{23}. This is to say, 10^{23} potential people are lost every century that we fail to colonize this region of the universe, which "corresponds to a loss of potential equal to about 10^{14} potential human lives per second of delayed colonization."

Second, Bostrom also considered the possibility that future people could be nonbiological beings taking the form of digital consciousnesses in huge computer simulations running on planet-sized computational devices powered by advanced nanotechnological systems designed to harness the energy output of stars. As alluded in Chapter 6, mental states (such as pleasure) might be multiply realizable, that is, able to be instantiated on substrates other than nervous tissue, such as silicon or, perhaps, carbon nanotubes. This possibility was one of the crucial underlying assumptions of Bostrom's Simulation Argument, which was published just a few months earlier than "Astronomical Waste," and it is currently the most favored view among philosophers.[51] If functionalism is true, then according to Bostrom's calculations "approximately 10^{38} human lives [are] lost every century that colonization of our local supercluster is delayed; or equivalently, about 10^{29} potential human lives per second." This number "boggles the mind," Bostrom wrote, echoing Ćirković's claim that the loss "per a century of delay in starting the colonization is astonishing by any standard."[52] Just consider that if it took you three seconds to read the last sentence, roughly 300 trillion digital people who could have existed never will. Yet the *actual* number of how many biological or digital people there could be is irrelevant. "What matters for present purposes," Bostrom declared,

> is not the exact numbers but the fact that they are huge. Even with the most conservative estimate, assuming a biological implementation of all persons, the potential for one hundred trillion potential human beings is lost for every second of postponement of colonization of our supercluster.

It follows that if these people were to have, on average, "happy" or "worthwhile" lives, we have a straightforward argument from total-impersonalist utilitarianism for the same conclusion that Ćirković drew: we must accelerate the pace of technological development and colonize space as soon as possible. As Bostrom made the point, "from a utilitarian perspective, this huge loss of potential human lives constitutes a correspondingly huge loss of potential value," and hence "the effect on total value . . . seems greater for actions that accelerate technological development than for practically any other possible action. . . . Few other philanthropic causes could hope to match that level of utilitarian payoff."[53] Indeed, the same conclusion follows even if one adopts a broader conception of value as more than just wellbeing, so long as one maintains that (a) the appropriate response to value is to maximize it, (b) value can be aggregated at least to some extent, and (c) there is at most only weak temporal discounting of value.[54]

However, this is where Bostrom parted ways with Ćirković, arguing that "the true lesson" of these numbers "is a different one. If what we are concerned with is (something like) maximizing the expected number of worthwhile lives that we will create, then in addition to the opportunity cost of delayed colonization, we have to take into account the risk of failure to colonize at all." In other words, we could *succumb to an existential risk* that, as such, permanently and drastically curtails our potential, where "potential" is understood here in specifically utilitarian terms. Even if avoiding an existential catastrophe requires delaying the onset of colonization by many years, the cost of this delay will be well worth it from an expected-value perspective. To quote Bostrom once again:

> Because the lifespan of galaxies is measured in billions of years, whereas the time-scale of any delays that we could realistically affect would rather be measured in years or decades, the consideration of risk trumps the consideration of opportunity cost. For example, a single percentage point of reduction of existential risks would be worth (from a utilitarian expected utility point-of-view) a delay of over 10 million years. . . . Therefore, if our actions have even the slightest effect on the probability of eventual colonization, this will outweigh their effect on when colonization takes place.[55]

Hence, utilitarianism instructs us first and foremost to lower the probability of an existential catastrophe occurring; accelerating the pace of technological development should be pursued only insofar as it does not interfere with risk mitigation efforts. "For standard utilitarians," Bostrom concluded, "priority number one, two, three, and four should consequently be to reduce existential risk," where the fifth should be to colonize space as soon as we possibly can. "The utilitarian imperative 'Maximize expected aggregate utility!' can be simplified to the maxim 'Minimize existential risk!'"[56] Notice the shift here from a transhumanist to a utilitarian conception of existential risk, whereby existential risks now concern the realization of astronomical amounts of value in the future. If one combines the transhumanist definition of the term with this new focus on maximizing value, we get the following: *an existential risk is any event that would prevent Earth-originating intelligent life from establishing a techno-utopian world or creating astronomical amounts of value in the universe*. Utilitarianism does not care about Utopia or posthumanity in themselves, only total value, a fact that may have led Bostrom to revise his definition of existential risks yet again in a subsequent paper that appeared in 2013, which we will discuss below.

Milk Cartons

However, it is worth noting that transhumanism does point toward a particular way of maximizing value. To see this, consider the standard utilitarian account of persons, according to which persons are nothing more than fungible "containers" or "vessels" for holding intrinsic value, not unlike the way milk cartons hold milk—to borrow an analogy from Parfit.[57] Bostrom gestured at this idea when he described "sentient beings living worthwhile lives" as "value-structures," and when the posthuman author of "Letter from Utopia" wrote that "if only I could share one second of my conscious life with you! But that is impossible. Your container could not hold even a small splash of my joy, it is that great."[58] As John Rawls famously argued, utilitarianism does not take seriously the separateness of persons: just as one might decide to suffer tomorrow by undergoing surgery in hopes of avoiding worse pain later in their life, utilitarianism is willing to trade one person's suffering for a greater benefit to another person. Since all that matters on this view is the net amount of total value in the universe, that is, the "greater good," how exactly this value is distributed among people is irrelevant.[59] Decisions between lives are isomorphic to decisions within a life; the boundaries between people have no intrinsic moral significance.

The point is that if people are value-containers, then there are two orthogonal ways to increase value, one focused on populations and the other on individuals. First, we could multiply the total number of containers. This is the idea behind the "Astronomical Waste" argument above. Second, we could make these containers *volumetrically bigger*, so to speak, such that each individual container can contain more total value. Rather than making more happy people, we could thus modify persons to enable them to experience superhuman levels of happiness (or do both).[60]

This is one thing the transhumanist project promises: by radically enhancing the human organism, we could raise the upper limit of value that individual persons could contain and in doing so increase total value by another means. Nonetheless, maximizing value itself is not an explicit aim of the transhumanist project, according to Bostrom's "Transhumanist Values," and indeed he wrote in this paper that, contra his own assertions in "Astronomical Waste" and "Letter from Utopia," people are not just replaceable containers for value.[61] Rather, the creation of techno-utopia constitutes a *telos* in its own right, albeit one that would contain lots of "value." This gestures at another difference between the transhumanist and utilitarian arguments presented above: the first could be seen as a version of the argument from unfinished business, where the "business" in question is the creation of a stable, flourishing posthuman civilization. Any failure to complete this business would be existentially catastrophic. In contrast, since there is no theoretical limit to how much value we should create (that is, however much value exists, adding another unit of value will always be better), there is no point at which our utilitarian "business" of creating "happy" people will be complete.[62] Both transhumanism and utilitarianism are teleological views, as I noted in a previous chapter, but in different senses: one aims for a definite telos, namely, techno-utopia, while the other is goal-directed in that for *each individual act*, one should *aim* to produce as much net value as possible. As alluded to in the last section, Bostrom may have introduced his 2013 conception of existential risk to better accommodate the desiderata of both transhumanism and utilitarianism. It is to this we now turn.

Growing Up

Three years after Bostrom introduced the concept of existential risk, on June 1, 2002, he founded the Future of Humanity Institute, which according to its website "includes several of the world's most brilliant and famous minds working [on] what can be done now to ensure a flourishing long-term future."[63] The original mission of this institute, according to the Wayback Machine, was to study topics like (a) how to "use science, medicine, and technology to improve human functioning on such dimensions as cognitive performance, healthy lifespan, mood and motivation, and reproductive choices," (b) "what . . . the biggest threats to the survival of the human species" are, and (c) what we can "conclude from alleged probabilistic coherence-constraints such as the simulation argument, the doomsday argument, and considerations related to the Fermi paradox."[64] Over the next decade, very few publications cited Bostrom's 2002 paper—other than Bostrom's own papers, of which there were many—with the notable exceptions of Matheny's 2007 article mentioned above and the 2008 book *Global Catastrophic Risks.* However, we saw in Chapter 6 that numerous authors contributed during this time—the 2000s—to our understanding of issues germane to the first branch of Existential Risk Studies, including Martin Rees (2003), Richard Posner (2004),

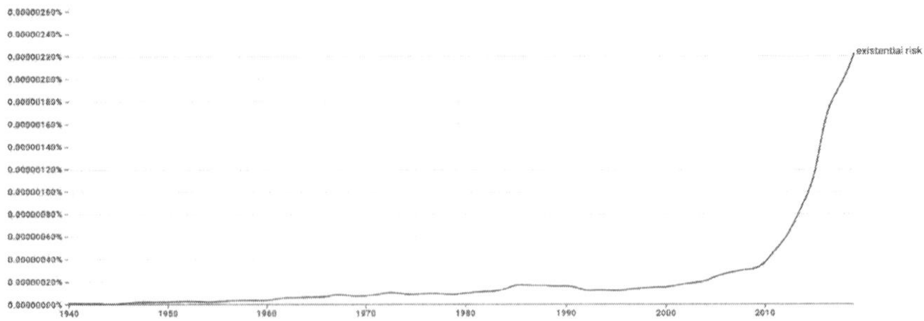

Figure 10.1 Google Ngram Viewer results for "existential risk."

Jared Diamond (2005), Ray Kurzweil (2005), and Willard Wells (2009).[65] In fact, most of the research that was conducted in Existential Ethics over this period—within which the second branch of Existential Risk Studies largely falls—focused on antinatalism, which we will explore later in this chapter.

But this began to change the following decade, during the 2010s, as the Google Ngram above shows, catalyzed in part by Bostrom's 2013 paper titled "Existential Risk Prevention as Global Priority," which offered a new conception of existential risk and an "improved" typology of existential risk failure modes.[66] At the center of both was a novel concept that Bostrom introduced, namely, *technological maturity*, which denotes "the attainment of capabilities affording a level of economic productivity and control over nature close to the maximum that could feasibly be achieved." Bostrom then redefined "existential risk" as any event that would prevent Earth-originating intelligent life (again, "humanity") from either reaching or sustaining a state of technological maturity. Technological maturity thus assumed the role that posthuman civilization played in his original conception, as the telos toward which humanity must strive. Any failure to reach technological maturity, or to sustain technological maturity in a particular way once reached, would constitute an existential catastrophe. This was Bostrom's new definition: *an existential risk is any event that would prevent Earth-originating intelligent life from attaining a stable state of technological maturity.*

Why exactly does attaining technological maturity matter? Why single out this possible future state over others? The obvious answer is that by fully subjugating the natural world and maximizing economic productivity to the physical limits, humanity would position itself in the most optimal way to exploit the largest fraction of our cosmic endowment of negentropy possible. We could then use this negentropy for purposes deemed to be "desirable," such as exploring posthuman modes of being, building a posthuman civilization, colonizing our future light cone, and creating enormous numbers of "happy" people in computer simulations. As Bostrom put the idea, "the capabilities of a technologically mature civilization could be used to produce outcomes that would plausibly be of great value, such as astronomical numbers of extremely long and fulfilling lives."

This led him to propose an updated classification of existential risk scenarios centered around the idea of mature technology, which he specified as follows:

- Human extinction: Humanity goes extinct prematurely, that is, before reaching technological maturity.
- Permanent stagnation: Humanity survives but never reaches technological maturity. Subclasses: unrecovered collapse, plateauing, recurrent collapse.
- Flawed realization: Humanity reaches technological maturity but in a way that is dismally and irremediably flawed. Subclasses: unconsummated realization, ephemeral realization.
- Subsequent ruination: Humanity reaches technological maturity in a way that gives good future prospects, yet subsequent developments cause the permanent ruination of those prospects.[67]

As with his previous account of existential risks, the two types of human extinction that this model would identify as extremely bad are final and normative extinction. The first can be understood in connection with Bostrom's definition of "humanity." That is, if "we" refers to Earth-originating intelligent life, then final human extinction would involve the story of humanity, in this broader sense, ending forever. Hence, perhaps *Homo sapiens* leaves behind a successor species, which becomes a different successor species, and so on, but this lineage then terminates such that the last species leaves behind no successors and the whole story comes to a permanent end, "humanity" would have undergone final extinction.[68] Even if this were to happen after attaining technological maturity, it could still be that a large fraction of our "potential" is left unfulfilled, and hence final extinction would instantiate the failure modes of flawed realization and subsequent ruination. Extinction in the normative sense is referenced by Bostrom in discussing another way that flawed realization could happen, that is, by evolving through cyborgization or passing the existential baton on to some population of intelligent machines such that at some point in the future, for some reason, "humanity" lacks the capacity for qualitative mental states. This would occur, Bostrom writes, if

> machine intelligence replaces biological intelligence but the machines are constructed in such a way that they lack consciousness (in the sense of phenomenal experience). . . . The future might then be very wealthy and capable, yet in a relevant sense uninhabited: There would (arguably) be no morally relevant beings there to enjoy the wealth.[69]

But Bostrom also foregrounded the possibility of "premature extinction" and in doing so helped to establish the idea and its corresponding term within the

Existential Ethics literature. In earlier decades, premature extinction had been used primarily in the context of ecology, as when the 1977 "Declaration of the Rights of Animal and Plant Life" asserted that "every effort should be made to preserve all species of animal and plant life from premature extinction."[70] Its application to humanity, though, in a normative context, was new (the one earlier exception being Bruce Tonn's mention of it in 2009 when discussing the unfinished business argument). Obviously, premature extinction was implicit in Bostrom's 2002 definition of "existential risk," but here it is made explicit. The implication is that the badness of succumbing to an existential catastrophe before reaching technological maturity may be greater than if this were to happen after, and indeed Bostrom is clear that some types of existential catastrophes may be "worse" than others, although he did not elaborate on this point.[71] Either way, the idea has become common today, often meaning something like "final or normative extinction before having fulfilled most of our potential," where *our potential* is typically understood in the same transhumanist, utilitarian terms specified above.[72]

What doesn't matter, on Bostrom's view, is demographic, phyletic, or terminal extinction, except insofar as any of these might increase the probability of final or normative extinction. In the past, demographic extinction would have almost certainly entailed final extinction; to succumb to the former would be to succumb to the latter. But as science advances, these scenarios may be increasingly decoupled such that our complete disappearance has no tight connection to whether the "human story" itself comes to an end. The fate of our species, in other words, may become less relevant to the question of whether "humanity" persists, colonizes space, and floods the universe with happiness. Even more, there might be reasons stemming from transhumanism and utilitarianism to actually bring about our own demographic extinction—an idea mentioned in Chapter 7. "If a civilization wants to maximize computation," Anders Sandberg, Stuart Armstrong, and Milan Ćirković write, "it appears rational to aestivate until the far future in order to exploit the low temperature environment."[73] (Note that Sandberg, like Ćirković, is a transhumanist.) Hence, with lower computational costs, it might, perhaps, be easier to fulfill the transhumanist project and create lots of value.[74]

Or consider the case of phyletic extinction: whereas past views that focused on final and normative extinction, such as Russell's, appear to be largely indifferent about whether we undergo phyletic extinction, transhumanists like Bostrom would see this sort of extinction as positively desirable if it results in a superior new species of radically enhanced posthumans. Failing to undergo phyletic extinction may, indeed, mean that we have succumbed to an existential catastrophe. Alternatively, becoming posthuman could involve our descendants throwing away the meat suit of biology altogether by, for example, uploading their minds to computers. This may, in fact, be the optimal scenario if our aim is to colonize space as soon as possible, since, to quote Sandberg, digital beings are

> ideally suited for colonising space and many other environments where biological humans require extensive life support. . . . Besides existing in a substrate-independent manner where they could be run on computers hardened for local conditions, emulations could be transmitted digitally across interplanetary distances. One of the largest obstacles of space colonisation is the enormous cost in time, energy and reaction mass needed for space travel: emulation technology would reduce this.[75]

But final and normative extinction, whether these occur prematurely or not, do not exhaust every type of existential risk failure mode in Bostrom's updated typology. There remains a wide range of scenarios within the category of permanent stagnation that are entirely survivable. For example, civilization could collapse or dissolve irreversibly such that humanity persists but never attains technological maturity. Or future people might simply be unmoved by the capitalistic, Baconian goal that Bostrom identifies as being of paramount instrumental importance. Consider a scenario in which *Homo sapiens* survives for the next 1 billion years, cures all diseases, builds sustainable eco-technological communities, establishes world peace, and embraces the inherent dignity of all peoples around the globe. This would be an existential catastrophe no less than a scenario in which the entire human population slowly starves to death in subfreezing temperatures under pitch-black skies following a thermonuclear conflict. Clearly the former would be better than the latter—Bostrom would no doubt agree—but it would nonetheless be disastrous, a profound failure to fulfill our potential, by virtue of never realizing most of the value that could have otherwise existed.

Maxipok

Whereas Bostrom's 2002 paper on existential risk focused mainly on the first branch of Existential Risk Studies, his 2013 paper focus mostly on the second branch: the normative foundations of the first, which fall largely within the domain of Existential Ethics. Not only did it provide new calculations of how many future people there could be, but it cited several earlier ideas from the Existential Ethics literature to support his claim that existential risk reduction ought to be humanity's top global priority. For example, he quoted Robert Adam's contention that "a better basis for ethical theory in this area [i.e., our obligations regarding future people] can be found in . . . a commitment to the future of humanity as a vast project, or network of overlapping projects, that is generally shared by the human race."[76] Bostrom concludes that "since an existential catastrophe would either put an end to the project of the future of humanity or drastically curtail its scope for development, we would seem to have a strong prima facie reason to avoid it, in Adams' view."[77]

He also reproduced Parfit's thought experiment about the difference between 99 and 100 percent of humanity dying out, a fact that is worth pausing on for a moment. Despite the immediate and enormous impact of Parfit's *Reasons and Persons*, virtually no one discussed this thought experiment over the next several decades. It was almost entirely ignored by philosophers. One exception was Joseph Nye's 1986 book *Nuclear Ethics*, which briefly mentioned the idea in a footnote, and another came from a 1990 report on nuclear waste written by a Swedish philosopher in his native tongue.[78] It was Matheny's 2007 paper that brought Parfit's thought experiment into the foreground, using it to bolster his conclusion that "it might be reasonable to take extraordinary measures to protect humanity from [extinction]."[79] Matheny's paper is also notable for having drawn from, and built upon, a large number of works mentioned in Chapter 6 and more recently in this book, including those by Gott, Sagan, Leslie, Joy, Bostrom, Rees, and Posner, in addition to citing several chapters from *Global Catastrophic Risk*.[80] By bringing these contributions together in a way that no one previously had, Matheny—who would become a research associate at Bostrom's Future of Humanity Institute from 2009 to 2010—helped to establish an emerging canon of books and papers on issues pertaining to Existential Risk Studies, in a paper that has itself become one of the canonical early contributions to the literature.

Like Matheny, Bostrom agreed with Parfit's claim about the "greater difference," although he generalized its conclusion to existential risks rather than just our extinction.[81] On this view, the difference between an existential catastrophe almost happening and one actually happening would be axiologically enormous, just as Parfit argued with respect to almost versus actual extinction. Put differently, however much suffering the process or event of succumbing to an existential catastrophe might inflict, the badness of the state or condition of having succumbed to an existential catastrophe would be enormously larger. In Bostrom's words: "What makes existential catastrophes especially bad is not that they would [cause] a precipitous drop in world population or average quality of life. Instead, their significance lies primarily in the fact that they would destroy the future." How bad would this destruction be? How much value could humanity create in the absence of such a catastrophe? Updating his earlier numbers, Bostrom calculated that "if we suppose with Parfit that our planet will remain habitable for at least another billion years, and we assume that at least one billion people could live on it sustainably, then the potential exist[s] for at least 10^{16} human lives of normal duration." From an expected-value perspective, this means that "reducing existential risk by a mere one millionth of one percentage point is at least a hundred times the value of a million human lives." Yet if we were to colonize our future light cone, and if future people could be "implemented in computational hardware instead of biological neuronal wetware," he claimed that there could exist some "10^{54} human-brain-emulation subjective life-years" in total. This implies that

> even if we give this allegedly lower bound on the cumulative output potential of a technologically mature civilization a mere 1% chance of being correct, we find that the expected value of reducing existential risk by a mere one billionth of one billionth of one percentage point is worth a hundred billion times as much as a billion human lives.[82]

To illustrate the idea, imagine sitting in front of two buttons. If you push the first button, the probability of an existential catastrophe will fall by 0.0000000000000000001 percentage point, assuming a 0.01 chance of 10^{54} subjective life-years existing in the future. If you push the second button, one billion currently living human beings will be prevented from dying. Which button should you push? The answer, on Bostrom's view, is a resounding: you should push the first button, because doing this would be 100,000,000,000 times better than pushing the second. Again, recalling Bennett's description of utilitarianism and the notion of value-containers from earlier, the non-birth of these possible future people would constitute a far greater axiological catastrophe than the untimely deaths of these existing people, all other things being equal. Yet even this estimate from Bostrom's 2013 paper might be off by several orders of magnitude, as he argued the following year in a section titled "How big is the cosmic endowment?" in his book *Superintelligence* that a total of "at least 10^{58} human lives could be created in emulation" within the accessible universe. "The true number is probably larger," he added, although once again the point is simply that we are dealing with unfathomably huge figures.[83] Given this, it follows that "the loss in expected value resulting from an existential catastrophe is . . . literally astronomical," and hence that "the objective of reducing existential risks should be a dominant consideration whenever we act out of an impersonal concern for humankind as a whole."[84]

Bostrom formalized this conclusion as a decision-theoretic "rule of thumb" that he called "maxipok," which instructs us to "maximize the probability of an 'OK outcome,' where an OK outcome is any outcome that avoids existential catastrophe." The purpose of the maxipok rule, unlike utilitarianism, is not to tell us how to act in every decision situation, as Bostrom here acknowledges that there may be "moral ends other than the prevention of existential catastrophe." Its aim is to help us get our global priorities in order.[85] But when there aren't any special moral considerations, our altruistic resources should be directed toward mitigating existential risks. Non-existential risks should be further down on our priority list, given their relatively low stakes. As Bostrom made the point,

> unrestricted altruism is not so common that we can afford to fritter it away on a plethora of feel-good projects of suboptimal efficacy. If benefiting humanity by increasing existential safety achieves expected good on a scale many orders of magnitude greater than that of alternative contributions, we would do well to focus on this most efficient philanthropy.[86]

Non-Extinction Scenarios

To summarize the development of these ideas, the claim that our extinction itself would be very bad, independent of how it comes about, goes back to Sidgwick, and was later picked up by utilitarian, or utilitarian-friendly, philosophers like Glover, Parfit, Smart, and Leslie. Bostrom subsequently developed this argument in 2003 and 2013 by calculating the number of future people, including digital people, who could exist within (a) our galactic supercluster per century, and (b) the accessible universe.[87] The result was an estimate range of 10^{38} to 10^{58} in total.[88] If these people were to contain on average net-positive amounts of value, then the axiological opportunity cost of final or normative extinction, which could take the form of premature extinction depending on its timing, would be literally astronomical. However, Bostrom's initial thinking about human extinction arose from his commitment to transhumanism, a movement he participated in since at least the 1990s.[89] The "core value" of transhumanism is to explore the posthuman realm, which, of course, would become impossible if humanity were to cease existing. Hence, transhumanism provided one reason for why these outcomes must be "avoided at any cost."[90]

But, drawing from others at the time, Bostrom noticed that extinction is not the only way that we could fail to create a posthuman civilization: there are various survivable scenarios that would produce the very same result. This led him to propose a new concept—existential risk—to encompass the entire range of phenomena that could prevent humanity from attaining the ultimate goal of posthumanity. Around the same time, he also realized that a similar point could be made about utilitarian arguments for avoiding our extinction: humanity could survive but still fail to produce enormous quantities of value within our future light cone, as total-impersonalist utilitarianism prescribes. He thus offered a second argument for prioritizing the reduction of existential risk, which not only was based on calculations of future value that went beyond earlier estimates from Sagan and Ćirković but recognized what Leslie never explicitly addressed, that is, while final and normative extinction may be sufficient to prevent us from creating astronomical amounts of future value, neither is necessary for this to happen. By linking this second, utilitarian argument to the concept of existential risk, Bostrom expanded the semantics of "our potential" to encompass not just the promise of a techno-utopian world awash in "surpassing bliss and delight" but the possibility of flooding the universe with wellbeing by creating unfathomable numbers of future "happy" people.[91]

Because of these developments, the core questions of Existential Ethics concerning the goodness/badness, rightness/wrongness of our extinction became bound up with non-extinction scenarios, given that certain survivable outcomes can have the same moral-axiological status as final and normative extinction. In other words, we ought to avoid these survivable scenarios for the same reason we ought to avoid final and normative extinction, that is, because the consequences of both would be extremely bad. All constitute worst-case outcomes for humanity, if only from a transhumanist or utilitarian perspective.[92] The second branch of Existential Risk Studies thus overlaps significantly with Existential Ethics, but is not coextensive with it, given that (a) existential risk is a broader concept than human extinction in the final, normative, or premature senses, and (b) much of the work within Existential Ethics is not tied to transhumanism or utilitarianism. Bringing this back to the beginning of the chapter, the second branch of Existential Risk Studies constitutes the philosophical foundation of the first branch. It is what motivates and justifies the first branch—it is the most important reason the first branch has a claim to coherence, despite the disparate array of scenarios that it places within the single category of "existential risk," from nuclear war and engineered pandemics to alien invasions and a simulation shutdown. Hence, the terminology of "first" and "second" branches, which is my own idiosyncratic way of labeling these facets of Existential Risk Studies, should not be interpreted as indicating a historical chronology or implying that one has primacy over the other. The first, in fact, was largely built upon the second, which emerged from (a) the modern transhumanist movement and (b) the tradition of ethical thinking that goes back through Leslie, Smart, Parfit, Glover, and Sidgwick. While the study of kill mechanisms, which the first branch subsumes, dates to the mid-nineteenth century, the focus on various non-extinction scenarios under the banner of "existential risk" was genuinely novel, given the realization that, from a transhumanist or utilitarian perspective, there are survivable failure modes that could entail the same disvalue as total human annihilation.

Humanity's Longterm Potential

In recent years, Bostrom's definition of "existential risk" has been modified and refined, and a new ethical framework for thinking about the long-term future of humanity—namely, longtermism—has coalesced around the idea. Taking these in turn, we noted above that the first disjunct of Bostrom's definition is unnecessary, since human annihilation is just one way that our potential could be permanently and drastically curtailed.[93] Consequently, most definitions of the term in the contemporary existential risk literature do not include the first disjunct.[94] In 2015, two researchers at Bostrom's Future of Humanity Institute, Owen Cotton-Barratt and Toby Ord, further argued that the permanence criterion of the second disjunct is problematic, and should thus be dropped. Consider, they wrote, a situation in which a totalitarian regime gains total control over the entire human population, where the chance of humanity escaping is small but nonzero. Would this be an existential catastrophe on Bostrom's account? "Strange conclusions" follow however one answers, they write. On the one hand, "saying it's not an existential catastrophe seems wrong as it's exactly the kind of thing that we should strive to avoid," yet, on the other, "saying it is an existential catastrophe is very odd if humanity does escape and recover—then the loss of potential wasn't permanent after all." The issue here is that our potential isn't binary, whereas *being permanent* is—that is, something either is or is not permanent. To capture the fact that our potential could be realized or thwarted in degrees, Cotton-Barratt and Ord proposed a new definition of "existential catastrophe" (the instantiation of an existential risk) as any "event which causes the loss of a large fraction of expected value."[95] Hence, the totalitarian regime taking over the world would constitute an existential catastrophe even if humanity were to escape and realize whatever remaining potential it might have. Or humanity might undergo a second existential catastrophe if, say, we were to survive under this regime for a million years and then perish. For Cotton-Barratt and Ord, existential catastrophes could thus happen, in principle, any number of times—just as with demographic extinction—whereas for Bostrom an existential catastrophe is a unique event that can only happen once.[96]

This particular definition, couched in expected-value terms, never caught on among existential risk researchers, although what has become the standard definition within the longtermist literature today is similar. Consider, for example, the definition that Ord provided in his 2020 book *The Precipice: Existential Risk and the Future of Humanity*, which identifies an existential catastrophe with "the destruction of humanity's longterm potential" and an existential risk with any "risk that threatens the destruction of humanity's longterm potential."[97] This is, Ord observes, "very much in line with the second half of Bostrom's" definition, although minus the permanence criterion, given that the destruction of our potential could be either "complete (such as extinction)" or "nearly complete, such as a permanent collapse of civilization in which the possibility for some very minor types of flourishing remain, or where there remains some remote chance of recovery." In either case, though, "the greater part of our potential is gone and very little remains." This is how "existential risk" is most commonly used today.[98]

What, then, is our potential? On Bostrom's account, once again, this was fleshed out primarily in transhumanist and utilitarian terms, with space colonization being the crucial means for satisfying the utilitarian desideratum. Over the past few years, the notion of *our potential* has expanded even further to include considerations of the ideal goods, beauty, justice, and other phenomena. As Ord explains,

> because, in expectation, almost all of humanity's life lies in the future, almost everything of value lies in the future as well: almost all the flourishing; almost all the beauty; our greatest achievements; our most just societies; our most profound discoveries. We can continue our progress on prosperity, health, justice, freedom, and moral thought. We can create a world of wellbeing and flourishing that challenges our capacity to imagine. And if we protect that world from catastrophe, it could last millions of centuries. This is our potential—what we could achieve if we pass the Precipice [that is, our current era of heightened risks, sometimes called the "Time of Perils"] and continue striving for a better world.[99]

As Ord elaborates, echoing Parfit, an existential catastrophe such as final extinction would not only cause the loss of "millions of generations of humanity, each comprised of billions of people, with lives of a quality far surpassing our own," but foreclose all future progress within domains like science and morality. If such progress continues, he adds, we may even "reach one of the very peaks of science: the complete description of the fundamental laws governing reality," though "perhaps the most important are potential moral achievements." If the human story comes to an end, all of this would be lost—all these future people and all these great achievements, all "gone."

Ord also writes enthusiastically about how radical human enhancement technologies could enable us to transform "existing human capacities—empathy, intelligence, memory, concentration, imagination," and "make possible entirely new forms of human culture and cognition: new games, dances, stories; new integrations of thought and emotion; new forms of art." Even more, such technologies could augment our sensorium by enabling us to acquire modalities currently had only by nonhuman animals, including echolocation (bats and dolphins) or magnetoreception (foxes and homing pigeons). "Such uncharted experiences," Ord writes, "exist in minds much less sophisticated than our own. What experiences, possibly of immense value, could be accessible, then, to minds much greater?" While he registers the possibility that reengineering *Homo sapiens* could exacerbate inequality and injustice, and produce harmful unintended consequences—we "risk losing what was most valuable about humanity before truly coming to understand it," he writes—Ord nonetheless insists that radically modifying ourselves "may well be essential to realizing humanity's full potential."

This point is reiterated several times throughout the book, as when he declares that "forever preserving humanity as it is now may also squander our legacy, relinquishing the greater part of our potential," and "rising to our full potential for flourishing would likely involve us being transformed into something beyond the humanity of today."[100] In other words, causing our own phyletic extinction through cyborgization may be risky, but it may also be necessary to fulfill our potential.

Despite its broader conception of value, at the heart of this normative futurology is the idea that more is better. Other things being equal, two groundbreaking discoveries are better than 1, 10 walks on the beach are better than 5, 30 sensory modalities are better than 20, 100 great works of art are better than 90, trillions of "happy" people are better than billions, and a civilization that lasts for 10^{40} years is better than one that lasts for only 10^{30}. The appropriate response to value—whatever it is we take to be valuable—is to maximize its number of instances in the universe, across both space and time, from Earth to the rest of the cosmos, from now until the heat death. This is why Ord repeatedly links our "vast and glorious" longterm potential to colonizing the largest possible fraction of the accessible universe, which would enable us to survive into the distant future, far beyond the destruction of Earth from our aging sun. "Our potential, and the potential in the sheer scale of our universe, are interwoven," he writes, explicitly linking his longtermist view with modern cosmology, cosmography, and physical eschatology. "Trillions of years and billions of galaxies are worth little unless we make of them something valuable."[101]

The Only Rational Beings

As with Bostrom's 2013 paper on existential risk, Ord adduces several arguments from the prior Existential Ethics literature in an attempt to buttress his central thesis that "the challenge of our time is to preserve our vast potential, and to protect it against the risk of future destruction," given that "the ultimate purpose is to allow our descendants to fulfill our potential, realizing one of the best possible futures open to us."[102] For example, he cites Edmund Burke's notion of the "partnership of the generations" in arguing that we may have obligations to past people that give us reason to ensure our continued existence, such as carrying on transgenerational projects that earlier generations contributed to in the hope that future generations would see them to fruition.[103] Let's call this the "argument from obligations to past people."

Ord further contends that we may be the only creatures in the universe capable of appreciating, in ecstasy and awe, its natural beauty and order.[104] If we are the universe's only moral agents, then we are "the only chance ever to shape the universe toward what is right, what is just, what is best for all." If we are its only rational beings, then "it would only be through us that a part of the universe could come to fully understand the laws that govern the whole." These ideas, Ord notes, draw from earlier claims made by folks like Sagan, Rees, Parfit, and Max Tegmark, although we saw in Chapter 9 that the argument from cosmic significance goes back even further to Schell, Russell, and Paul Arthur Schilpp.[105] For example, Rees argued in *Our Final Hour* that "the odds could be so heavily stacked against the emergence (and survival) of complex life that Earth is the unique abode of conscious intelligence in our entire Galaxy. Our fate would then have truly cosmic resonance."[106] More recently, Parfit, who was Ord's mentor, wrote in his 2017 book *On What Matters* (volume III),

> if we are the only rational beings in the Universe, as some recent evidence suggests, it matters even more whether we shall have descendants or successors during the billions of years in which that would be possible. Some of our successors might live lives and create worlds that, though failing to justify past suffering, would have given us all, including those who suffered most, reasons to be glad that the Universe exists.[107]

As we noted in the previous chapter, the argument from cosmic significance could be interpreted in a couple of ways, one of which implies that what we ought to avoid is terminal rather than final or normative extinction. However, there is an additional "consequentialist" interpretation, according to which "the more rare intelligence is, the larger the part of the universe that will be lifeless unless we survive and do something about it—the larger the difference we can make, quoting Ord.[108] Since more is better, and since we may be the only intelligent beings in the cosmos, whether the universe becomes filled with life and value may entirely be on us.

Another argument from Ord concerns moral or normative uncertainty. Imagine that we have no duty to preserve our potential, but mistakenly decide to allocate our resources toward this end, rather than toward other philanthropic causes like global poverty, social justice, and animal welfare. This would be unfortunate, as it would increase, or fail to lessen, the human and nonhuman suffering that exists in the world today. But now imagine the reverse situation, in which preserving our potential is "our most important duty." We then mistakenly allocate resources toward global poverty, etc., which allows the overall probability of an existential catastrophe occurring to remain unacceptably high, or perhaps rise.[109] This second scenario, Ord claims, would be much worse than the first. Why? Because the disvalue of never fulfilling our potential is orders of magnitude greater than the disvalue of all the suffering happening right now. It would be much better to get things wrong in the first scenario than in the second. As Ord articulates the idea, "the case for making existential risk a global priority does not require certainty, for the stakes aren't balanced. . . . So long as we find the case for safeguarding our future quite plausible, it would be extremely reckless to neglect it."[110] We can call this the "argument from moral uncertainty."

There are, we should note, other versions of this argument in the literature. For example, William MacAskill, a colleague of Ord's at the Future of Humanity Institute, wrote in 2014 that, "in general, when one has the choice between two options, one of which is irreversible, and one expects to make moral progress, then option value gives one additional reason in favour of choosing the reversible option."[111] Bostrom himself elaborated this insight in his 2013 paper, writing that

> our present understanding of axiology might well be confused. We may not now know—at least not in concrete detail—what outcomes would count as a big win for humanity; we might not even yet be able to imagine the best ends of our journey. If we are indeed profoundly uncertain about our ultimate aims, then we should recognise that there is a great option value in preserving—and ideally improving—our ability to recognise value and to steer the future accordingly. Ensuring that there will be a future version of humanity with great powers and a propensity to use them wisely is plausibly the best way available to us to increase the probability that the future will contain a lot of value. To do this, we must prevent any existential catastrophe.

In sum, over the past decade, existential risk scholars have begun to integrate a number of different arguments for why mitigating existential risk should be among our top global priorities as a species, if not the top priority. The vision of the future that they accept, though, remains shaped in fundamental ways by (a) the transhumanist promise of a techno-utopian world full of radically enhanced posthumans, and (b) the utilitarian notion that value is something to be maximized; that the more value that exists between now and the heat death (or proton decay, or whatever hard limits there are on our survival), the better things will go.

Super-Hardcore Do-Gooders

Before turning to the other major development within Existential Ethics during this third wave, it may be worth taking a closer look at how exactly the longtermist ideology developed, as it has become extremely influential over the past few years, and could become even more influential this century. One way to understand its development is to begin with the Effective Altruism (EA) movement. The first EA organization was founded by Toby Ord in the late 2000s, called Giving What We Can (GWWC), whose website officially launched in 2009 with the aim of "fighting extreme poverty in the developing world."[112] This was inspired by Peter Singer's "global ethics," and indeed Singer has become one of the most prominent EAs, or "effective altruists," in the world today.[113] Recall Singer's 1972 argument from Chapter 8 that if one feels compelled to save a drowning child in a pond ten feet away, one should feel equally compelled to save a starving child on the other side of the planet. Far-away suffering does not count for less than suffering that is close by; we should not discount misery as a function of its proximity to us. Hence, people in wealthy countries should be more inclined than we often are to donate part of our income, perhaps even most of our income, to help disadvantaged people wherever they might live.

What was new about the EA movement was its effort to quantify the best ways of doing the most good, to ensure that the "altruism" advocated by Singer is maximally "effective." This was the central aim of GWWC, which reported on its website in 2011 that, by choosing carefully between different charities,

> you can get much more impact from your donation and thereby help many more people. Indeed, it is not even a matter of some charities being 10 or 100 times as effective: even restricted to the field of health programs in developing countries, research shows that some are up to 10,000 times as effective as others.[114]

More concretely, GWWC claimed that donating to the charities Deworm the World and Schistosomiasis Control Initiative does much more good than donating to, say, disaster relief following an earthquake, hurricane, or flood, despite the former being rather "unsexy" in comparison to the latter, to borrow a word from MacAskill.[115] Giving should be a combination of the heart and the head, not just the heart, which is to say that the emotional pull of a charitable cause is not a rational basis for decisions about which charities to donate to. Such decisions should instead be grounded in "evidence and reason." As Ord noted in a keynote address at the 2016 EA Global conference, Effective Altruism is a child of the Scientific Revolution, Enlightenment, and utilitarianism, in addition to Singer's global ethics.

In 2011, Giving What We Can was joined by another organization called 80,000 Hours, cofounded by MacAskill and Benjamin Todd. This aimed to help people choose a career that would maximize their positive impact in the world. (The name comes from the fact that if one works 40 hours a week, 50 weeks per year, for 40 years, this gives a total of 80,000 hours on the job.) This organization initially argued that, quoting its website, "becoming a banker might be the more ethical career choice" than, say, working for a nonprofit focused on the environment or pursuing a medical degree.[116] Indeed, MacAskill argued in 2014 that there is nothing morally wrong with getting a job at a petrochemical company if one donates a certain amount of one's income to charity. After all, if you didn't take that job, someone else would have, and unlike you they probably wouldn't donate their income to help people.[117]

Later in 2011, a handful of leaders in the fledgling EA community decided that GWWC and 80,000 Hours should incorporate under an umbrella organization, which they initially called the "High Impact Alliance." However, they were becoming increasingly aware of the importance of a good marketing strategy, and hence decided that a new name was needed. (At this point, community members often called each other "super-hardcore do-gooders," a term that "sucks," in MacAskill's words.) A vote to name this organization was thus held, with contenders including "Rational Altruist Community," "Evidence-Based Philanthropy Association," "Big Visions Network," "Effective Utilitarian Community," and "Centre for Effective Altruism." The last proposal won, and this is how the "Effective Altruism" community acquired its name.[118]

The Short of Longtermism

However, the movement's initial focus on global poverty did not last. Some EAs, beginning most notably with Nick Beckstead, came to a different conclusion: if our cosmic future could be way bigger than our present, and if there are actions that we can take today to influence this future, then we—by which Beckstead meant "the world in general"—should focus on actions that might influence the far future, rather than on how our actions might help those living today. As he expressed the idea: "From a global perspective, what matters most (in expectation) is that we do what is best (in expectation) for the general trajectory along which our descendants develop over the coming millions, billions, and trillions of years."[119] This was the main thesis of his 2013 PhD dissertation, titled "On the Overwhelming Importance of Shaping the Far Future," which many EAs recognize as one of the founding documents of the ideology, along with Bostrom's 2002 paper on existential risk and his 2003 "Astronomical Waste" article. (The word "longtermism" itself wasn't coined until 2017.)[120]

One way to understand this new ethic goes like this: longtermism is what happens when the EA commitment to "doing the most good" collides with Bostrom's "astronomical waste" argument. If one's aim is to positively affect the maximum number of people, and if most people who will ever exist will exist in the far future, then doing the most good may require one to focus on these far future people—ensuring not just that their lives are better than miserable but that they exist in the first place.[121] More generally, if most of whatever it is that one values lies in the distant future, millions, billions, and trillions of years from now, then actions that increase the probability of these goods being realized will have a much higher expected value than actions that, say, primarily affect the world today or in the near future. Although this mode of moral reasoning can appear "heartless" (as if it *replaces* the heart with the head), since it means neglecting current-day suffering, moral truth lies in the numbers. Morality, on this view, could be seen as an extension of economics. As Eliezer Yudkowsky, an influential figure within the EA/longtermist movements who MacAskill lauds as a "moral weirdo," writes,

> due to scope neglect, framing effects, and other cognitive biases, the result of an expected utility calculation executed correctly may produce an answer different from first intuition, making it "intuitively unappealing." If you can tell that it's probably the intuitions that went wrong and not the calculation, the skill *shut up and multiply* is the ability to accept that, yes, sometimes the expected utility math is correct and we need to deal with that (italics added).[122]

There are several important points to make about longtermism. First, the view comes in both moderate and radical forms. Moderate longtermism states that "positively influencing the longterm future is a key moral priority of our time," whereas radical longtermism asserts that this is *the* key moral priority. The latter is what one finds in the work of Bostrom and Beckstead, although the former is what MacAskill defends in his recent book *What We Owe the Future*. However, MacAskill admits in a blog post that, for marketing reasons, it would be better to present moderate longtermism to the public, since (a) most people will find radical longtermism, with its obsession over how many future people there could be in vast computer simulations if only we avoid an existential catastrophe, rather unpalatable, and (b) "it seems that we'd achieve most of what we want to achieve if the wider public came to believe that ensuring the long-run future goes well is one important priority for the world, and took action on that basis."[123] Indeed, MacAskill himself has recently claimed to be most "sympathetic" with radical longtermism, which he believes is "probably right," quoting an article published in *Vox*'s EA-aligned vertical Future Perfect.[124] He has also explicitly defended radical longtermism—which he calls "strong longtermism"—in a 2019 article with Hilary Greaves, later updated in 2021. In the first draft, the authors write that "for the purposes of evaluating actions, we can in the first instance often simply ignore all the effects contained in the first 100 (or even 1,000) years, focussing primarily on the further-future effects. Short-run effects act as little more than tie-breakers."[125] In the second draft, they borrow estimates from their colleague Toby Newberry, a research scholar at Bostrom's Future of Humanity Institute, according to which there could exist some 10^{45} digital beings in our Milky Way galaxy alone, though Newberry also estimates some 10^{54} digital beings within the accessible universe.[126]

While the idea of existential risk is central to the longtermist ethic, longtermists do not see reducing such risk as the only thing that matters. What should concern us, more generally, is creating what Beckstead called "positive trajectory changes" with respect to civilization's development into the far future, where the developmental "trajectory" of civilization refers to how the future as a whole unfolds with respect to happiness, wealth, technological capabilities, scientific advancement, cultural achievements, etc.[127] Trajectory changes could be targeted or broad: the former would, paradigmatically, involve mitigating particular existential risk scenarios. As the longtermist Fin Moorehouse writes, "it's hard to imagine a clearer instance of positively influencing the long-run future than preventing an existential catastrophe."[128] However, Beckstead argued that this might not be "the best way of maximizing humanity's future potential," as there could be a wide range of "broad, general, and indirect approaches to shaping the far future" that are even better. Examples include speeding up technological development, improving education, science, political systems, and parenting, promoting humanitarian values, and "promulgating norms that emphasize the importance of future generations," which is precisely what MacAskill's book *What We Owe the Future* aims to do. For instance, consider that certain suboptimal values, technologies, practices, policies, norms, systems, and so on around today could become locked-in, thus resulting in path-dependencies that (a) are difficult or impossible to reverse, and (b) constrain the future in undesirable ways.[129] Longtermists should work to avoid sub-optimal lock-in scenarios, which may be no less important to avoid than, say, final extinction. Or take a controversial example from Beckstead, who argues that the positive long-term "ripple effects" of saving the lives of people in rich countries may be much greater than those created by saving the lives of people in poor countries, given that people in rich countries are better positioned to shape the far future. Since shaping the far future is of "overwhelming importance," we should, therefore, prioritize the lives of rich-country people.[130] This conclusion has led to significant criticism, for obvious reasons, although it is a fairly straightforward implication of radical longtermism.[131]

As I have argued in print on several occasions, longtermism—especially its radical version—may be the most influential ideology in the world today that most people have never heard about. The richest person on Earth, Elon Musk, calls it "a close match for my philosophy" and recently retweeted a link to Bostrom's "Astronomical Waste" paper with the line: "Likely the most important paper ever written."[132] Longtermists are beginning to run for public office, as occurred in 2022 when Carrick Flynn, backed by more than $11 million from Sam Bankman-Fried, ran for congress in Oregon's Sixth District. A *UN Dispatch* article reports that "the foreign policy community in general and the United Nations in particular are beginning to embrace longtermism." And longtermism is poised to shape the 2024 UN Summit of the Future, which MacAskill hopes will be to longtermism what the 1970 Earth Day was to the modern environmental movement: the moment at which the ideology becomes mainstream.[133] Furthermore, until quite recently, the EA movement boasted of a staggering $46.1 billion in committed funding, some of which came from the once-vast wallet of Bankman-Fried, a longtermist who set-up the FTX Future Fund to support longtermist research—an organization that included MacAskill and Beckstead on its team.

As a final draft of this book was being prepared for Routledge, news broke that Bankman-Fried's cryptocurrency exchange platform, FTX, had collapsed due to a liquidity crisis, with Bankman-Fried losing 94 percent of his wealth virtually overnight. The evidence suggests that Bankman-Fried committed massive amounts of fraud, and consequently a tsunami of bad press may have seriously tarnished the ideology's reputation, if not EA more generally. Indeed, Bankman-Fried was the great success story of "earn to give": after a meeting in 2012 with MacAskill, who is frequently described as Bankman-Fried's moral "advisor," he decided to pursue a lucrative job at Jane Street Capital, a global proprietary trading firm, and then in crypto specifically to "get filthy rich, for charity's sake," as one journalist put it.[134] Yet even if longtermism's brand suffers irreparable damage among the public, the ideology will likely retain its clout and influence—which is pervasive—in the tech industry and among billionaires (like Musk). The EA grantmaking organization Open Philanthropy also "expects to spend billions of dollars on [longtermist focus areas] over the coming decades," despite the loss of funding from Bankman-Fried.[135] There are good reasons to expect longtermism to remain a world-shaping force in the years to come.

One Very Bad Thing

To conclude, longtermism is an outgrowth of the EA community, emerging most directly from the work of Bostrom and Beckstead—the latter of whom, incidentally, was among those who cast a vote in 2011 that gave the community its name.[136] The main significance of longtermism with respect to Existential Ethics is this: because of the longtermist/EA community's power, influence, money, and size, certain further-loss views have been catapulted into much greater prominence than alternatives in the marketplace of ideas. Though one need not be a total-impersonalist utilitarian to be a longtermist, the EA community heavily leans toward this version of utilitarianism and many of its leading figures "describe themselves as having more credence in utilitarianism than in any other positive moral view."[137] MacAskill, for example, explicitly states that he is "most sympathetic to utilitarianism," while Ord identifies "the Scientific Revolution, the Enlightenment, and Utilitarianism [as having] greatly contributed to the upbringing of effective altruism," as noted earlier.[138] Many longtermists thus maintain that Being Extinct would be a tragedy of enormous moral significance, given its attendant axiological opportunity costs. As Singer, Beckstead, and Matt Wage made this point in a 2013 article posted on the Effective Altruism Forum, titled "Preventing Human Extinction":

> One very bad thing about human extinction would be that billions of people would likely die painful deaths. But in our view, this is, by far, not the worst thing about human extinction. The worst thing about human extinction is that there would be no future generations. . . . [I]f humanity goes extinct now, the worst aspect of this would be the opportunity cost.

The reason, once again, is built on the findings of physical eschatology and our best current understanding of the size of the universe. "Civilization began only a few thousand years ago," they write—a string of words that appears verbatim in Parfit's 1984 book—"yet Earth could remain habitable for another billion years. And if it is possible to colonize space, our species may survive much longer than that." Furthermore, as with Ord and Parfit, they also point to a second further loss, namely, one arising from the possibility of future progress in domains like science and morality. To quote Singer, Beckstead, and Wage once more:

> The extinction of our species—and quite possibly, depending on the cause of the extinction, of all life—would be the end of the extraordinary story of evolution that has already led to (moderately) intelligent life, and which has given us the potential to make much greater progress still. We have made great progress, both moral and intellectual, over the last couple of centuries, and there is every reason to hope that, if we survive, this progress will continue and accelerate. If we fail to prevent our extinction, we will have blown the opportunity to create something truly wonderful: an astronomically large number of generations of human beings living rich and fulfilling lives, and reaching heights of knowledge and civilization that are beyond the limits of our imagination.[139]

Hence, the influence of the EA and longtermist movements has made these further-loss views the most visible, and perhaps the most widely accepted, positions within the contemporary Existential Ethics literature. Nonetheless, alternative views have been proposed, as the rest of this chapter and the next will explore.

The Anti-Natal Clinic

At the very same time that Bostrom and others were developing the existential risk framework that grew into the longtermist paradigm, another school of thought was emerging—or rather reemerging—within Existential Ethics. This school differed from the ideas discussed above in at least two ways: (i) it claimed that coming into existence is always a serious net harm, and hence that we should not create any new people, and (ii) it contended that Being Extinct is better than Being Extant, and that we should strive to bring about the former. Furthermore, advocates of this view drew a direct connection between these claims: antinatalism, expressed by (i), implies pro-extinctionism, expressed by (ii), which is to say that if one accepts the first view on the ethics of procreation, one must also accept the second view on the ethics of human extinction.[140] However, we will see that this line of reasoning can be problematized in various ways, although some of the same arguments that support antinatalism could also support a pro-extinctionist position.

Let's begin where the leading figures of this school began: with antinatalism. Although there have been antinatalists going back at least to the nineteenth century, the first systematic philosophical treatment of the topic was David Benatar's 2006 book *Better Never to Have Been*, which drew from ideas explored in journal articles of his published since the late 1990s.[141] This book also lodged the word "antinatalism" into the philosophical lexicon, though Benatar reports in an interview with *The Antinatalist Magazine* that he first used it in a 2001 talk about assisted reproduction and then again in a lecture three years later titled "The Anti-Natal Clinic," where "anti-natal" was a play on "ante-natal," meaning "before birth" rather than "against birth."[142] Either way, my use of the word in previous chapters to describe the positions of Mainländer, Zapffe, and Vetter was thus linguistically anachronistic. (We should also note that a cognate of "antinatalist" appeared in French the same year Benatar's book was published, in the title and body text of Théophile de Giraud's *L'art de guillotiner les procréateurs: Manifeste anti-nataliste*, which translates as *The Art of Guillotining Procreators: An Anti-Natalist Manifesto*. Unfortunately, this book has not yet been translated into English, so I will not discuss it here.)[143] Oddly, despite the word becoming rather common today, the Oxford English Dictionary still has no entry for it.

Blame, Gambles, and Functional Immortality

Having said this, Benatar and Giraud were not the only ones who defended the idea of antinatalism in the mid-2000s. The Finnish philosopher Matti Häyry argued in 2004 for the dual thesis that procreation is both irrational and immoral. It is irrational for this reason: let's say that the value of not having a child is zero, while the value of having a child could be positive, negative, or zero, depending on how the child's life turns out. If one could assign probabilities to these outcomes, then one could use expected value theory to determine whether having a child is rational or not. But we cannot assign such probabilities, and hence must rely upon some other decision-theoretic rule. (In other words, this is a decision under "ignorance" rather than "uncertainty," in my phraseology.) The rule that Häyry opts for is called "maximin," which was popularized by John Rawls' book *A Theory of Justice*. The maximin rule states that rational actors should choose the option with the best worst-case outcome. Since the worst-case outcome of not having a child has a value of zero, while the worst-case outcome of having a child has a negative value, and since a zero outcome is better than a negative outcome, it would therefore be irrational to have a child. As for ethics, Häyry began with the assertion that "it is morally wrong to cause avoidable suffering to other people." Since everyone will suffer at least a little in life, he thus concluded that (a) "every parent who could have declined to procreate is to blame" for causing otherwise avoidable suffering, and (b) because no one can rule out the possibility of their child suffering terribly, parents "can also be rightfully accused of gambling on other people's lives."[144]

Yet Häyry did not take the extra step of arguing that humanity should go extinct. He may have believed this to be the case, or perhaps taken it as obvious that a universal failure to reproduce would necessarily entail our species disappearing. But here we should question whether this does in fact follow. Consider that since humanity is comprised of individuals, if some of these individuals were to acquire what I call *functional immortality*, then everyone on the planet could universally decide not to procreate without this necessitating demographic extinction (which could thus lead to terminal or final extinction). By "functional immortality," I mean a state in which an individual's life persists until one of three things happens: (a) an injury or accident kills them, (b) they decide to end their life, or (c) they perish for reasons pertaining to physical eschatology—for example, because of proton decay or the heat death. Hence, functionally immortal people could potentially exist for as long as humanity itself could continue under normal circumstances, via the succession of the generations. The question thus becomes whether we have any reason to believe that individuals could, in fact, gain functional immortality. While philosophers have speculated about this possibility for some time—recall Condorcet's 1795 claim that in the future, during the tenth epoch of human history, progress might enable people to acquire extremely long lives—it was only very recently that talk of "living forever," or "living long enough to live forever," became something other than a risible promise from cranks and charlatans looking to make a quick dollar off gullible people scared of dying.[145] Today, the field of longevity research is awash in funding, and it seems increasingly plausible that anti-aging technologies could enable future generations, or maybe even some living today, to become functionally immortal. This is a descriptive claim that I will not here attempt to justify; as such, it could very well be wrong, although I will assume in what follows that it has a nontrivial chance of being true.

An Evening at the Cinema

Hence, it was not unreasonable for antinatalists in the past to simply assume that antinatalism entails extinction. But this may no longer be the case, depending on how the field of longevity develops: there could be no more people without this entailing that there are no people anymore, meaning that accepting antinatalism while simultaneously rejecting pro-extinctionism is a coherent philosophical position. One can also, of course, accept pro-extinctionism without accepting antinatalism, though if one believes it would be immoral to bring about our extinction involuntarily or through means that would cut lives short and cause people suffering, then pro-extinctionists might adopt antinatalism for practical reasons, as the only morally acceptable means of achieving the aim of complete non-existence. In sum, there is no necessary connection between antinatalism and pro-extinctionism, and this is true even with the most absolutist forms of antinatalism, according to which creating new people is always impermissible and hence should never be done. But there are also, we should note, non-absolutist interpretations of antinatalism that make room for procreation under certain conditions. For example, selective antinatalism states that it is wrong to bring certain people into existence, such as those who would have lives that are not worth living.[146] Defeasible antinatalism, in contrast, states that the general prescription never to have children can be overridden by other factors, if sufficiently strong. The latter view is what Benatar accepts, although the circumstances under which baby-making is justifiable, on his account, are very limited. For now, let's begin with a brief look at what his antinatalist position is and the arguments he put forward to support it, and then turn to his pro-extinctionism.

The core claims of Benatar's antinatalism are that (A) coming into existence is always a net harm, (B) this harm is very substantial, much worse than we ordinarily realize, and (C) we should not have any children. The first two are axiological claims and the third a deontic one, and what links them, I believe, is supposed to be the intuitive idea that badness is something we should avoid and betterness something we should pursue. Benatar himself claims that his antinatalism does not presuppose any particular ethical theory, whether consequentialist or deontological, though it is incompatible with the total-impersonalist utilitarianism that motivates some of the strong further-loss views explored above.[147]

Benatar offers three arguments to support (A) through (C). The first is based on an axiological asymmetry, sometimes dubbed the "harm-benefit asymmetry." It states the following: the presence of pain is bad and the presence of pleasure is good, while the absence of pain is good and the absence of pleasure is not bad. Although the claim about the absence of pain looks to be impersonal, Benatar understands it in person-affecting terms.[148] That is to say, the absence of pain is good for the person who does not experience it, even if this is because that "person" never exists, an idea that some philosophers have argued is incoherent.[149] For our purposes, it is enough to note Benatar's insistence that one can make sense of the asymmetry within a person-affecting framework. "The absence of bad things, such as pain, is good even if there is nobody to enjoy that good," he writes, "whereas the absence of good things, such as pleasure, is bad only if there is somebody who is deprived of these good things."[150] This yields a matrix of decisions and outcomes that is somewhat reminiscent of Vetter's matrix from Chapter 9.

Good/Note Bad > Good/Bad

X never exists	X exists
Absence of pain, which is good.	Presence of pain, which is bad.
Absence of pleasure, which is not bad.	Presence of pleasure, which is good.

Figure 10.2 Benatar's harm-benefit asymmetry.

As Figure 10.2 shows, creating a person results in a situation that is both good and bad for that person, because it entails the presence of pleasure (good) and the presence of pain (bad), whereas not creating a person results in a situation that is both good and not-bad for that "person," because it entails the absence of pain (good) and the absence of pleasure (not bad). Since a good/not-bad situation is better than a good/bad one, creating a person is always a net harm, and hence one should not have children.

Benatar's second argument is what he calls the "quality-of-life argument," though I prefer Nicholas Smyth's more informative term for it: the "badness of life argument."[151] This argument could be seen as an empirically updated, more comprehensive version of Schopenhauer's thesis that life is suffering but without Schopenhauer's extravagant metaphysics of the will or his view that happiness is merely the absence of suffering (and hence has no positive value on its own). For Benatar, our lives are in fact overflowing with misery, disappointment, misfortune, and pain, even if many of us believe the opposite. Pleasures tend to be brief, while suffering often drags on: sex and a good meal happen relatively quickly, yet broken bones, infections, and heartache following a bad breakup can linger for weeks or years. While "chronic pain is common," Benatar observes, "there is no such thing as chronic pleasure."[152] The intensity of pains can also greatly exceed the intensity of pleasures. Who in their right mind would "accept an hour of the most delightful pleasures in exchange for an hour of the worst tortures"?[153] Who would trade a year, or even an entire lifetime, of the best moments imaginable for 24 hours of the most horrendous suffering—fingernails removed, waterboarding, third-degree burns, and so on?

Yet when one surveys people about whether their lives are good and worth living, many answer without hesitation: "Yes, life is on balance good, and I am glad that I was born." How then can Benatar reconcile his Schopenhauerian assertions above with the empirical datum that so many people think life is overall positive? The answer comes from a cluster of psychological tendencies that distort our perception of just how terrible our lives actually are. For example, people are susceptible to the positivity bias, whereby we remember pleasant experiences more accurately than unpleasant ones, as well as habituation, which happens when we adapt to negative stimuli over time, such as pain and disappointment. Consequently, the past and present—and by extrapolation the future—look better than they really are. These distortions may be extremely difficult to avoid, too, as they may have been implanted deep in our brains by natural selection over millions of years. If one were to see clearly the true awfulness of life, would one be more or less likely to produce offspring? Even slight alterations in differential reproduction rates can add up over evolutionary time, resulting in significant phenotypic changes. The tendency to inaccurately assess human existence, perhaps built into our cognitive machinery, buried under layers of gears and mechanisms, thus provides a kind of error theory that explains—by explaining away—the empirical fact that most people think life is worth living.[154]

But there is another complication arising from an ambiguity in the phrase "a life worth living." On the one hand, this could mean "a life worth continuing," while on the other, it could mean "a life worth starting." This is an important distinction that enables Benatar to claim that, given the harm-benefit asymmetry, no life is worth starting, although many lives once started may be worth continuing, even if we misjudge the badness of our existence. He illustrates this idea with an analogy: imagine "an evening at the cinema. A film might be bad enough that it would have been better not to have gone to see it, but not so bad that it is worth leaving before it finishes."[155] The same goes for our lives: all of us would have been better off staying home, so to speak, but the choice wasn't up to us—we exist thanks to the decisions of our parents. Given that this is the case, many of us might feel that life is still good enough not to end it. The movie is awful but not so awful that we feel compelled to leave halfway through. Death, after all, can be terror-inducing, and in fact Benatar defends an anti-Epicurean view according to which death can harm the one who dies, that is, independent of its effects on those who survive the deceased. Indeed, this is one reason that Benatar did not endorse a pro-mortalist means of bringing about our extinction, though he does claim that suicide may be more rational than most of us ordinarily assume. There is, furthermore, a connection between these two ideas. As he writes, "it is because we (usually) have an interest in continuing to exist that death may be thought of as a harm, even though coming into existence is also a harm." This will become important later on.

Benatar classifies the two arguments above as "philanthropic," since they arise from concerns about what is best for particular people, even if those people have not been, and never will be, born. That is to say, it is because we care about the interests of such people not to suffer harm that we have reason to never create them. The last argument that Benatar provides is what he describes as "misanthropic," in the particular sense that it concerns "unpleasant facts about humans." Just consider the enormous suffering that we inflict on each other and the sentient beings we share Earth with due to war, genocide, murder, slavery, torture, hunting, factory farming, commercial fishing, pollution, habitat destruction, climate change, and so on. The truth is that every new child brought into the world will almost certainly cause some additional suffering to others, aside from whatever suffering they experience, and hence Benatar concludes that while this "argument does not obviously show that it is better never to have been, . . . it does support the anti-natalist conclusion that it is better not to procreate."[156] To be clear, this is not to say that we should hate humanity, as the term "misanthropic" might imply. One could still love our species while acknowledging that we are a source of unrelenting misery and evil in the world. Accepting Benatar's misanthropic argument does not require one to be a misanthrope.

The Best Human Population Size

As noted, Benatar assumed that by establishing antinatalism, he had also established pro-extinctionism. In his words, antinatalism "implies that it would be better if there were no more humans. The further implication of this is that it would be better if humans became extinct, at least if extinction were brought about by not creating new members of the species."[157] But we saw above that antinatalism does not necessarily entail pro-extinctionism, given the increasingly plausible possibility of radical life extension. However, some of the arguments that Benatar proposed for his antinatalism could also, independently, buttress a pro-extinctionist view according to which Being Extinct would not just be better than Being Extant but in some sense be positively good. To see which ones, let's consider them in the same order as above.

First, the harm-benefit asymmetry argument primarily concerns the claims of (A) and (C), that is, that being born is always a net harm and hence we should not have children. There are two important implications of these claims: on the one hand, they entail that we should not add any new people to the human population. As such, this says nothing about whether the human population should cease existing. If functional immortality were possible, then, given his view that many lives are good enough to continue and that death can harm the one who dies, it seems that Benatar should actually advocate an anti-extinction position according to which humanity ought to persist, by way of extending individual lives, so long as no additional people are created. To be clear about this point, the asymmetry implies that Being Extinct would be better than Being Extant, since the former corresponds to a good/not-bad situation, whereas the latter is merely good/bad. If we were to go extinct, this would be better than our current state. But Benatar's view also suggests that, if we can, we should actively prevent this from happening, given the relative worthwhileness of most people's lives and the harm of death. In more concrete terms, Benatarians should be inclined to support efforts to develop safe and effective life-extension technologies.

On the other hand, since even a single birth is one too many, the asymmetry argument implies that it would have been best if humanity had never existed in the first place. Here it may be useful to disambiguate the phrase "the best human population size is zero" the way Benatar disambiguated "a life worth living." The first reading asserts that Being Never Existent, as we can call it, is better than Being Extant, while the second states that, given the fact that our population is not currently zero, we should strive to bring down this number until humanity is no more.[158] Benatar frequently equivocates between these two readings, as if the harm-benefit asymmetry implies both, when in fact it only implies the first. In fact, the asymmetry is compatible with the human population remaining stable, on the condition that this occurs because existing lives are extended indefinitely into the future (i.e., no new people are created). And, once again, it seems that Benatar's more general position suggests that it should remain stable, assuming that the lives of those who currently exist are not overwhelmingly bad. Hence, for these reasons, his first argument in support of antinatalism itself says nothing about whether we should or should not go extinct. One might think of this situation in terms of local optima: Being Never Existent is best, and Being Extinct is better than Being Extant, yet there are reasons not to go extinct given that we currently exist—reasons that concern our individual interests to keep kicking and avoid the grave. But again, the force of these considerations hinges on a contingency—that is, on whether it becomes feasible to radically extend our lifespans. If such technologies are impossible, or unattainable, then the natural limits of individual lives will entail extinction.

Turning now to Benatar's second argument, the badness of life argument: this most directly supports (B) and (C), that is, that existing is very bad, much worse than most of us realize, and hence we should not have children. (By implication, then, coming into existence would be a harm.) Whether this argument leads to pro-extinctionism will also depend on whether life-extension technologies do, in fact, become available. As noted above, Benatar maintains that while no life is worth starting, many lives are worth continuing, because many lives are not overwhelmingly bad. If they were overwhelmingly bad, then one would have an argument not just against creating new people but ending our lives prematurely, which Benatar endorses only in special cases (for example, if one experiences great suffering due to a terminal disease). Some philosophers have, incidentally, contended that Benatar's view *does* entail a pro-mortalist position. For example, Rafe McGregor and Ema Sullivan-Bissett argue that

> if one accepts Benatar's arguments for the asymmetry between the presence and absence of pleasure and pain, and the poor quality of life, one must also

> accept that suicide is preferable to continued existence, and that his view therefore implies both anti-natalism and pro-mortalism.[159]

This clearly leads to a pro-extinctionist position: humanity should disappear because everyone should kill themselves. Benatar, though, insists this is not the case: "Life may be sufficiently bad that it is better not to come into existence, but not so bad that it is better to cease existing."[160] If Benatar is right, the implication is the same as above: if we can continue to exist indefinitely as individuals, then we should try to do this; if we cannot, then humanity should fade away. It is worth noting that many other pessimists seem to have held the same position as Benatar, given that few advocated or committed suicide—with the notable exception of Mainländer. Life is hell, but not that hellish.[161]

Benatar's third argument does not directly relate to either (A) or (B), as it could be that coming into existence is not a net harm and that life is not very bad, yet those born cause all sorts of harms to each other and nonhuman organisms. This provides a straightforward case for why Being Extinct might be better: without humans, there would be no more human-caused evils like war, genocide, factory farming, environmental destruction, and so on. However, the strength of this argument will depend on one's assessment of our badness in the world: if the harms we cause are significant, then the argument becomes stronger. Yet even here there is a complication, as some have argued that our destruction of the environment might actually reduce overall suffering in the natural world. As William MacAskill makes the point,

> if we assess the lives of wild animals as being worse than nothing, which I think is plausible . . . then we arrive at the dizzying conclusion that from the perspective of the wild animals themselves, the enormous growth and expansion of *Homo sapiens* has been a good thing.[162]

On this view, then, Being Extinct would increase wild-animal suffering, and hence Benatar's third argument does not, in fact, lead to pro-extinctionism. To the contrary, pronatalism—having more children—would be better for the natural world, with respect to total suffering, than our non-existence.[163] I am not endorsing this argument here—as stated, there's something rather perverse about it—but think it is at least worth registering.

In sum, Benatar's first argument for antinatalism is silent about whether the human population should fall to zero, given that it is currently a non-zero number, although it implies that Being Never Existent would have been best. His second argument could be utilized to justify a pro-extinctionist position, although only in the absence of radical life-extension technologies. If such technologies become available in the future, which is not out of the question, then only the third argument would remain standing. Yet one can object to this by arguing that our destruction of the natural world actually decreases total suffering, caused by things like predation, disease, parasites, natural disasters, and so on. This does not mean that Benatar's pro-extinctionism is untenable, only that the reasoning that leads him to it might not be as strong as he believes.

Pro-Extinctionism: Alive and Kicking

Before concluding this chapter, let's take a closer look at what exactly Benatar's pro-extinctionist view is, independent of whether the arguments that Benatar provides for it go through. This will foreground the fact that there are many ways of fleshing out a pro-extinctionist position, and hence there are many issues about which pro-extinctionists may disagree. We can analyze Benatar's view into three main components, namely, that (1) we should strive to bring about what he calls a "dying-extinction"; (2) Being Extinct is not only not bad, but positively good; and (3) it would be better if humanity were to go extinct sooner rather than later.[164] Taking these in order:

The first concerns the etiology of Going Extinct. While Benatar believed that humanity should be no more, he also held that it would be wrong to bring this about in any way that would cut lives short, which would constitute a "killing-extinction." This could be either natural or anthropogenic; if the latter, it would be equivalent to omnicide and would thus be wrong for all the reasons that murder is wrong. Indeed, given his anti-Epicurean view of death, Benatar would no doubt agree that instantaneously annihilating humanity would be wrong, as this would still be harmful by virtue of truncating lives. Benatar contrasted this scenario with a "dying-extinction," whereby our species fades away by failing "to replace those members of the species whose lives come to [a] natural end." This is the only morally acceptable route to extinction, Benatar suggests, and hence at this point the earlier connection between antinatalism and pro-extinctionism could be reversed: rather than claiming that one should be a pro-extinctionist because one is an antinatalist, one should be an antinatalist because one is a pro-extinctionist. If the only other options are pro-mortalism and omnicide, then antinatalism may be adopted for reasons pertaining to practical ethics.[165]

The first part of assertion (2), that Being Extinct would not be bad, follows straightforwardly from Benatar's person-affecting restriction. Since no one is harmed by there being no more people in the universe, the state or condition of Being Extinct cannot be bad for anyone and hence is not bad at all. As Benatar writes, "it is not the case that people are valuable because they add extra happiness. Instead extra happiness is valuable because it is good for people—because it makes people's lives go better." Consequently, on Benatar's account, the badness/wrongness of our extinction will depend entirely on whether it occurs because of a dying-extinction or a killing-extinction, which is to say that his pro-extinctionism endorses the equivalence thesis. Yet his view goes beyond this by seeing our non-existence as positively good, an idea that, as noted above, straightforwardly follows from the axiological asymmetry, according to which the absence of suffering is good even if there is no one around to experience it. Since Being Extinct would entail the absence of all human suffering, it would not merely be neutral, as many of the person-affecting theorists discussed in the previous chapter seemed to hold.

Here one might object that "a world without humans [would be] incomplete or deficient" because it would lack "moral agents and rational deliberators" (which, of course, echoes Immanuel Kant's claim from Chapter 8 that "without men the whole creation would be a mere waste, in vain, and without final purpose").[166] To this Benatar responded:

> [W]hat is so special about a world that contains moral agents and rational deliberators? That humans value a world that contains beings such as themselves says more about their inappropriate sense of self-importance than it does about the world. (Is the world intrinsically better for having six-legged animals? And if so, why? Would it be better still if there were also seven-legged animals?) Although humans may value moral agency and rational deliberation, it is far from clear that these features of our world have value *sub specie aeternitatis* [from the perspective of universal and eternal truth]. Thus if there were no more humans there would also be nobody to regret that state of affairs.

Yet even if the existence of moral agents and rational deliberators does make the universe more complete, Benatar argued that "it is highly implausible that their value outweighs the vast amount of suffering that comes with human life."[167] This leads directly to assertion (3), namely, that our extinction should happen as soon as possible. There are at least two reasons for this. First, if humanity persists in the future by creating new generations, this would obviously involve bringing new people into existence, and according to the harm-benefit asymmetry and badness-of-life arguments, coming into existence is a serious harm. Second, independent of how humanity persists—whether via the succession of generations or radical life extension—existence is replete with suffering that we both experience ourselves and inflict on other sentient beings. Hence, the longer we exist, the greater the total amount of suffering, which suggests that we should die out as soon as possible. This claim, which one could interpret in negative utilitarian terms (although Benatar himself does not explicitly do this), does clearly support pro-extinctionism.

Finally, it is important to note that when Benatar talks of extinction, he is specifically referring to final extinction brought about via demographic extinction. It would not be enough, on his account, for *Homo sapiens* to disappear while leaving behind some successor species capable of experiencing and inflicting suffering. Mere demographic, phyletic, terminal, or normative extinction would not solve this problem but perpetuate it.[168] Another point worth mentioning is that, in addition to endorsing the equivalence thesis, an anti-Epicurean view of death, and a pro-extinctionist view of Being Extinct in the final sense, Benatar also gives a nod to the no-ordinary-catastrophe thesis. As Benatar writes, "unless humanity ends suddenly, the final people whether they exist sooner or later, will likely suffer much."[169] This is to say that such people will suffer some extra, unique-to-the-situation harms *by virtue of* being the final people. They would, for example, lack the support, company, and care that younger generations provide those in their geriatric years. There would be no one to address medical issues, ensure that food is on the table, take out the trash, and so on. Ultimately, the very last people would find themselves profoundly alone in their communities, a dismal situation not unlike Lionel Verney's predicament in *The Last Man*.

Since Going Extinct would introduce these additional harms, Benatar suggests that we might thus pursue a "phased" extinction, whereby some new people are brought into existence to help mitigate the sufferings of the last few generations. In his words, "the creation of new generations could only possibly be acceptable, on my view, if it were aimed at phasing out people." This is why Benatar accepts a defeasible version of antinatalism, one that would permit the creation of new people under the very unusual circumstances involved in approaching the Moment of Extinction.

The Third Wave

The third wave of theorizing in Existential Ethics consists of two diametrically opposed developments: first, new thoughts about the axiological opportunity costs of extinction, where the two main types of human extinction that must be avoided are final and normative extinction, and the primary source of badness arises from the state or condition of Being Extinct. Other types of extinction, such as phyletic extinction, could be very desirable if they were to result in a superior new species of posthumans. And second, the first systematic treatment of antinatalism by Benatar, who explicitly linked his central thesis that humanity should cease procreating with the claim that we should disappear entirely and forever without leaving behind any successors. On this account, the type of extinction that we should actively bring about is final extinction, we should do this via antinatalist means, and the resulting state of Being Extinct would be positively good. While the longtermist ideology has inspired a large community of researchers backed by literally billions of dollars and is now poised to shape the cultural and political landscape in significant ways, the latter has provoked a vigorous debate among mostly Analytic philosophers about the ethics of procreation and the desirability of our collective persistence in a world overflowing with pain. Let's now turn to the final wave of History #2, which partially overlaps with the period just discussed.

Notes

1. Drexler 1986, 2013.
2. As noted in previous chapters, there were one or two exceptions, the most notable being Isaac Asimov's *A Choice of Catastrophes* (1979), although Asimov did not conclude that the risk of human extinction is high, nor was his survey motivated by an

ethical conviction that our disappearance in a catastrophe would be bad for reasons above and beyond the default view.

3. Bostrom and Ćirković 2008.
4. See Avin et al. 2018.
5. Bostrom 2002b.
6. Bostrom 2013.
7. Leslie 1996; Ćirković 2008.
8. Bostrom 2002b; Bostrom and Ćirković 2008, 4.
9. See Yudkowsky 2008.
10. Bostrom and Ćirković 2008.
11. That is, Parfit was not a utilitarian per se, although his ethical approach was "broadly utilitarian in spirit" (Srinivasan 2017). Recall from earlier his idea of "climbing the same mountain on different sides" (Parfit 2013).
12. This is, at least, my interpretation of Leslie's project.
13. Moore 1903, 83–85. However, Moore suggested otherwise later on in his 1903 book, and explicitly endorsed a contrary view in his *Ethics* (1912). Note that this also anticipated, in a certain respect, subsequent claims from environmental ethicists that the natural world contains intrinsic value independent of its instrumental usefulness to human beings (see Hurka 2021, section 4).
14. Leslie 1996, 6–9.
15. Leslie 1996. To be clear, "macro disvalue" is Smart's term, not Groenewold's. Note that Smart wrote a blurb for *The End of the World*, which declared that "Leslie's book is of urgent practical as well as theoretical importance: it could well be the most important book of the year" (Leslie 1996).
16. Bennett 1978.
17. Leslie 1996. For a critique of Leslie's ethical position, see Palazzi 2014. Unfortunately, this paper by Franco Palazzi is one of the few on the ethics of human extinction that I was unable to fit within the narrative of History #2. But it is well worth reading.
18. Leslie 1996.
19. Bostrom 2002b. Note that I have rearranged these words without changing the meaning.
20. Although recall from Chapter 6 that others in the transhumanist community had gestured at the basic idea of existential risk before Bostrom published on the topic. The idea was "in the air," although it was not until Bostrom's 2002b paper that it was properly formalized.
21. Bostrom 2002b.
22. Bostrom 2003a.
23. Bostrom 2005c.
24. Bostrom 2020.
25. Bostrom 2005c.
26. Bostrom 2003a.
27. Note that there are different interpretations of this theory, the most prominent of which are called "causal decision theory" and "evidential decision theory."
28. To be clear, my sense of "uncertainty" corresponds to what decision theorists typically mean by "risk." On the standard account, "uncertainty" is a looser term that refers to either "risk" or "ignorance." "Risk" is when probabilities can be assigned, while "ignorance" is when they cannot be. See Peterson 2009, 5–6.
29. Specifically, this notion of risk was brought into risk analysis by the 1975 "Reactor Safety Study" (Rasmussen et al. 1975; Hansson 2018).
30. Smart 2020; Leslie 1996.

31. Matheny 2007.
32. As Ben Eggleston observes, many moral theorists "have drawn on elements of decision theory in order to articulate their principles of moral rightness and moral wrongness more explicitly, or to provide something like an algorithm that an agent can follow in order to act morally" (Eggleston 2017). Indeed, I noted above that the first formulation of Kant's Categorical Imperative can be seen as a kind of "decision procedure" that enables one to determine, by reason alone, which actions are morally forbidden and which are morally permissible. Historically, the idea of expected value seems to have been introduced into utilitarianism by the economist and Nobel laureate John Harsanyi 1953, 1977, 1982, and later examined in the context of ethics by philosophers like J. J. C. Smart 1961 and Frank Jackson 1991. Thanks to John Broome for help with this history (personal communication).
33. See Crisp 1997, 99.
34. Smart 1961, 33–34.
35. MacAskill et al. 2022.
36. Yudkowsky 2007; Bostrom 2009.
37. We also saw how subsequent discoveries, such as the discovery of radioactivity, changed views about the age of the Earth and its future habitability. That is, Earth's warmth comes from not just (a) having formed in the solar nebula but (b) radioactive decay, which produces thermal energy.
38. Rees's paper examined the evidence for a Big Crunch model of the cosmos, whereby the expansion of the universe gradually reverses due to the constant tug of gravity. (The heat death is sometimes called the "Big Freeze.") Consequently, every atom, molecule, planet, star, and galaxy will eventually crash together in the ultimate act of cosmic violence, resulting in a "devastating compression" whereby, as Rees put it, "all structural features of the cosmic scene would be destroyed." Yet this would not be the end, as the model also implies that "the universe is perpetually oscillating, and this contraction is merely a prelude to a subsequent re-expansion [such that] stars, galaxies, and clusters must form anew in each cycle" (Rees 1969). As of this writing, most cosmologists do not accept this scientific eschatology, favoring instead the Big Freeze (i.e., heat death) model.
39. See Adams and Laughlin 1997.
40. See Ćirković 2003.
41. My reference here is no accident, as many of the early contributors to physical eschatology were influenced and inspired by science fiction writers like Wells. Another important figure was Olaf Stapledon, mentioned in the previous chapter, who imagined the future evolution of life over the next 500 billion years in his *The Star Maker* (1937).
42. Thanks to Martin Rees for clarifying some of these ideas to me.
43. Adams 2008; Schröder and Smith 2008. Thanks to Martin Rees for clarifying some of these ideas to me.
44. Adams 2008.
45. For example, Matheny 2007; Beckstead 2013a; Whittlestone 2017; Mogensen 2019; Beckstead 2019; Tarsney 2019; Ord 2020; John and MacAskill 2021; Greaves and MacAskill 2021; Greaves et al. 2021; Thorstad 2021; Moorhouse 2021a; Balfour 2021; Roser 2022.
46. Ćirković and Radujkov 2001, Ćirković 2002b; also Ćirković and Bostrom 2000. For example, Parfit wrote that "the Earth will remain inhabitable for at least another billion years," while Schell declared that "there is another, even vaster measure of the loss, for stretching ahead from our present are more billions of years of life on earth, all of which can be filled not only with human life but with human civilization"

(Parfit 1984; Schell 1982/2000). The 1980s witnessed a number of other discussions of how long humanity or our civilization could last based on the findings of physical eschatology. For example, John Barrow and Frank Tipler argued in their 1986 book *The Anthropic Cosmological Principle* (which also focused on the anthropic principle) that although "our species is doomed," given the dysteleological fate of the cosmos,

> our civilization and indeed the values we care about may not be. . . . [F]rom the behavioural point of view intelligent machines can be regarded as people. These machines may be our ultimate heirs, our ultimate descendants, because under certain circumstances they could survive forever the extreme conditions near the Final State. Our civilization may be continued indefinitely by them, and the values of humankind may thus be transmitted to an arbitrarily distant futurity [where the last phrase is a reference to Darwin's 1859 claim that "we may safely infer that not one living species will transmit its unaltered likeness to a distant futurity"] (Barrow and Tipler 1986).

See also Dyson 1979.

47. Thanks to James Hughes for details on this history (personal communication). See JET 2005.
48. Ćirković 2002b.
49. Bostrom 2002b.
50. Ćirković 2002b.
51. Bourget and Chalmers 2021.
52. Bostrom 2003b; Ćirković 2002b.
53. Italics added.
54. That is, all utilitarian theories are welfarist, seeing wellbeing as the only intrinsically valuable thing in the universe. However, as noted earlier, wellbeing can be understood in hedonistic, desire-satisfactionist, and objective-list theory terms (see Parfit 1984, appendix I).
55. I take the difference between "expected value" and "expected utility" to be that the former is broader than the latter; consequentialists might talk about the former while utilitarianism might focus on the latter.
56. Bostrom 2002b. Similarly, I take the difference between "utility" and "value" to be that the former fits better with utilitarianism, while the latter is a more general concept than utility. As Eggleston writes, "value is understood to be broader than utility, as consequentialism is broader than utilitarianism" (2017).
57. More specifically, this is a reference to what Parfit called the "Milk Production Model" (1996, 313; 1984).
58. Bostrom 2003b, 2005c, italics added.
59. Indeed, this is precisely what engenders the infamous Repugnant Conclusion, whereby a world containing a huge number of people with lives just barely worth living may be better than a world with much fewer people living extremely good lives: the former may still contain more net value in total than the latter.
60. Hilary Greaves and William MacAskill point to these options in writing that "if . . . the value of the future, per century, is much higher in the far future than it is today—whether because the population per century is much larger (due to space settlement or otherwise) or because some form of enhancement renders future people capable of much higher levels of well-being, or both—then the case for advancing progress is significantly stronger" (2019).
61. In his words:

> Consider a hypothetical case in which there is a choice between (a) allowing the current human population to continue to exist, and (b) having it instantaneously and painlessly killed and replaced by six billion new human beings who are very

> similar but non-identical to the people that exist today. Such a replacement ought to be strongly resisted on moral grounds, for it would entail the involuntary death of six billion people. The fact that they would be replaced by six billion newly created similar people does not make the substitution acceptable. Human beings are not disposable (Bostrom 2003a).

See Knutsson 2021 for discussion of the "replacement argument" against impersonalist utilitarianism.

62. There maybe, however, be a practical limit; see Manheim and Sandberg 2021.
63. FHI 2005, 2022. Note that I have changed the order of this sentence without altering its meaning. Somewhat comically, the very first public version of the FHI website at one point accidentally refers to itself like this: "HFI is committed to the highest standards of scholarship and academic rigor" (FHI 2005).
64. FHI 2005.
65. Only Kurzweil mentioned "existential risks" by name.
66. Note that for many years on Bostrom's website this paper had the alternative title: "Existential Risk Reduction as Global Priority."
67. Bostrom 2013.
68. I am ignoring another possibility here, namely, that our species disappears without leaving behind any successors but another species sufficiently similar to us evolves later on, not unlike the scenario imagined by Denis Diderot in Chapter 2 and mentioned again in Chapter 7. But this, as Bostrom writes, "is very far from certain to happen," and "even if another intelligent species were to evolve to take our place, there is no guarantee that the successor species would sufficiently instantiate qualities that we have reason to value. Intelligence may be necessary for the realization of our future potential for desirable development, but it is not sufficient" (2013).
69. Bostrom 2013.
70. Jacobs 1977.
71. Bostrom 2013, 24.
72. An example of this is Ord 2020, which waxes poetic about radically transforming ourselves and spreading throughout the "affectable" universe.
73. Sandberg et al. 2017.
74. I am not so sure this is true, but it is a possibility worth registering.
75. Sandberg 2014. This, of course, extends the central insight of Manfred Clynes and Nathan Kline's 1960 paper, which coined the word "cyborg." If cyborgs are better-suited for space travel than biological humans, since artificial materials can withstand the strains of space better than organic systems (Clynes and Kline's contention), then wholly artificial beings will be even better-suited than cyborgs.
76. Adams 1989. Recall that Adams was explicitly pushing back against Parfit's position in this passage; here, Bostrom takes it to support his own view, which—as noted just below—is very much in-line with Parfit's. Note also that Adams explicitly opposed transhumanism. As he wrote in 1979, "I would quite strongly prefer the preservation of the human race, for example, to its ultimate replacement by a more excellent species, and think none the worse of myself for the preference" (Adams 1979). It is unclear whether Adams would agree more generally that we should prioritize the attainment of technological maturity; I suspect he wouldn't, but this is a topic for another time.
77. Bostrom 2013.
78. Nye 1986, footnote 116; Tännsjö 1990, 80–81. Note that Tännsjö has also argued that "we ought to accept the repugnant conclusion." See Tännsjö 2004.
79. Matheny 2007.
80. Although *Global Catastrophic Risks* was published a year after Matheny's paper, he apparently had access to a prepublication draft of the book. Note that none of these

scholars except for Bostrom, including those writing after his 2002 paper, actually used the term "existential risk."

81. Parfit 1984.
82. Bostrom 2013.
83. Bostrom 2014.
84. Note that I rearranged the first sentence quoted without changing its meaning.
85. Bostrom 2013.
86. Bostrom 2013.
87. See Bostrom 2003b and 2013.
88. Bostrom 2014.
89. This is, in my reading, strongly implied on page 12 of Bostrom 2005a.
90. Bostrom 2003a.
91. Bostrom 2020.
92. This is true even in cases where one existential catastrophe would prevent a *worse* existential catastrophe, and *in this conditional sense* would be *good*. As Bostrom writes, "it is on no account a conceptual truth that existential catastrophes are bad or that reducing existential risk is right. There are possible situations in which the occurrence of one type of existential catastrophe is beneficial—for instance, because it preempts another type of existential catastrophe that would otherwise certainly have occurred and that would have been worse" (2013).
93. Bostrom 2002b.
94. See Torres 2019.
95. Cotton-Barratt and Ord 2015.
96. It could also be the case that after emerging from this totalitarian regime, an event occurs that greatly increases the expected value of the future—an "existential eucatastrophe," as Cotton-Barratt and Ord call it, borrowing a neologism from J. J. R. Tolkien, who defined it as "the sudden happy turn in a story which pierces you with a joy that brings tears" (Cotton-Barratt and Ord 2015; Tolkien 1944). This led Cotton-Barratt and Ord to proposed the notion of existential hope—the hope that an existential eucatastrophe could occur—to contrast with existential risk, where eucatastrophes and catastrophes are the instantiation of each.
97. In this account, existential risks are "simply the risk of an existential catastrophe" (Ord 2020).
98. Although not that long ago, even existential risk scholars would frequently equate *existential risks* with *risks of human extinction*. See Torres 2019.
99. Ord 2020.
100. Ord 2020.
101. Ord 2020.
102. Ord 2020.
103. Bostrom made a related point in arguing that

    > we might also have custodial duties to preserve the inheritance of humanity passed on to us by our ancestors and convey it safely to our descendants.[23] We do not want to be the failing link in the chain of generations, and we ought not to delete or abandon the great epic of human civilization that humankind has been working on for thousands of years, when it is clear that the narrative is far from having reached a natural terminus (2013).

104. See Sandberg et al. 2018.
105. The relevant quotes from Sagan and Tegmark are as follows: "The Cosmos may be densely populated with intelligent beings. But the Darwinian lesson is clear: There will be no humans elsewhere. Only here. Only on this small planet. We are a rare as well as an endangered species. Every one of us is, in the cosmic perspective, precious"

(Sagan 2011), and "it was the cosmic vastness that made me feel insignificant to start with. Yet those grand galaxies are visible and beautiful to us—and only us. It's only we who give them any meaning, making our small planet the most significant place in our entire observable Universe" (quoted in Ord 2020).

106. Schell 1982; Rees 2003.
107. Parfit 2017. I am not sure why Parfit believes that he can speak for "those who suffered most."
108. Ord 2020.
109. Ord 2020.
110. Ord 2020.
111. MacAskill 2014.
112. GWWC 2007. According to the Giving What We Can website in 2017, the organization was founded by Ord (GWWC 2017). However, the story has changed over time, and MacAskill is now commonly referred to as the "cofounder" of the organization. Note also that the EA movement itself to some extent grew out of the so-called "Rationalist" community, which coalesced around Eliezer Yudkowsky's website LessWrong. Due to space limitations, I will not here explore this genealogical link.
113. See Singer 2002.
114. GWWC 2011a.
115. GWWC 2011b; Crouch 2011; see MacAskill 2015. Note that Will MacAskill changed his name from William Crouch.
116. 80H 2011.
117. MacAskill 2014b.
118. MacAskill 2014a.
119. Beckstead 2013a.
120. MacAskill 2019. Although MacAskill states that he coined the word in 2017, Ord 2020, 306 writes that it was both him and MacAskill who came up with the term. I do not know which is accurate. The following publications identify Bostrom and Beckstead as having played a crucial role in the development of longtermism: Ord 2020, 306; Greaves and MacAskill 2021, 3; and Moorehouse 2021a.
121. I should note, however, that one of the progenitors of EA, Toby Ord, was familiar with Bostrom's work many years before GWWC was founded. Indeed, they co-authored an article together in 2006 (Bostrom and Ord 2006).
122. Klein 2022; Yudkowsky 2021.
123. Torres 2022.
124. Samuel 2022.
125. Greaves and MacAskill 2019.
126. Greaves and MacAskill 2021; Newberry 2021. For additional estimates focusing on Earth over the next 800 million years, see Max Roser's article for *Our World in Data* titled "The Future Is Vast: Longtermism's Perspective on Humanity's Past, Present, and Future" (Roser 2022).
127. Beckstead 2013a, 6.
128. Moorhouse 2021b. Bostrom, though, apparently sees his maxipok rule as "neutral on the question of whether the best methods of reducing existential risk are very broad and general, or highly targeted and specific" (Beckstead 2013b).
129. Beckstead 2013b.
130. Beckstead 2013a, 11, 72. In Beckstead's words,

> saving lives in poor countries may have significantly smaller ripple effects than saving and improving lives in rich countries. Why? Richer countries have substantially more innovation, and their workers are much more economically

> productive. By ordinary standards—at least by ordinary enlightened humanitarian standards—saving and improving lives in rich countries is about equally as important as saving and improving lives in poor countries, provided lives are improved by roughly comparable amounts. But it now seems more plausible to me that saving a life in a rich country is substantially more important than saving a life in a poor country, other things being equal (Beckstead 2013a).

131. It also looks to be a straightforward implication of utilitarianism. Tyler Cowen, for example, notes that utilitarianism seems to "support the transfer of resources from the poor to the rich . . . if we have a deep concern for the distant future" (2007). Similarly, the Oxford philosopher Andreas Mogensen writes in a paper published by the Global Priorities Institute that

> it has been assumed that utilitarianism concretely directs us to maximize welfare within a generation by transferring resources to people currently living in extreme poverty. In fact, utilitarianism seems to imply that any obligation to help people who are currently badly off is trumped by obligations to undertake actions targeted at improving the value of the long-term future (Mogensen 2021).

132. See Torres 2022.
133. Goldberg 2022. Note that I have removed a linguistic redundancy in this sentence.
134. Fisher 2022.
135. Karnofsky 2022.
136. Note that Beckstead also identified Parfit and the philosopher John Broome as having "partly preceded and influenced" his ideas (2019, footnote 1). Note, furthermore, that Broome introduced MacAskill to Ord, as he co-supervised the doctoral theses of each (MacAskill 2020).
137. De Lazari-Radek and Singer 2017.
138. DS 2021; CEA 2016.
139. Beckstead et al. 2013.
140. To be clear, antinatalism isn't just about procreation. It concerns the more general issue, of which procreation is an instance, of creating new beings capable of suffering or being harmed. Such beings might include animals, as well as artificial minds (see Torres 2020; Chomanski 2021).
141. See, for example, Benatar 1997.
142. Thanks to David Benatar for corresponding about the origins of "antinatalism."
143. Going back even further, to at least the 1950s, the words "antinatalist" and "antinatalism" can be found in discussions of population policy, which addressed the social, environmental, and other consequences of baby-making rather than its specifically ethical aspects. For example, a 1952 document by the US Bureau of the Census states that "the German Government pursued a deliberate anti-natalist policy among the Poles" by encouraging contraceptive use while decreasing or eliminating "hospital insurance and maternity benefits (Myers and Maudlin 1952). The following decade, Judith Blake argued before the US Subcommittee on Government Operations, which was looking at the environmental effects of an expanding US population, that the government should promote "antinatalist desires that are already prevalent in our population," specifically among young people, rather than attempting "to introduce antinatalist coercions and restrictions" that "interfere with individual volition and freedom" (Davis 1969). This came at the end of a decade during which pressure on the US government to limit population growth greatly intensified, and indeed we saw in Chapter 4 that prominent scientists like Paul and Anne Ehrlich had begun warning that global overpopulation could have disastrous consequences, resulting in "hundreds of millions of people [starving] to death" (Ehrlich and Ehrlich 1968).

144. Häyry 2004. See also Thomas Ligotti's fascinating 2010 book *The Conspiracy Against the Human Race: A Contrivance of Horror*, which, according to Ray Brassier's foreword, "sets out what is perhaps the most sustained challenge yet to the intellectual blackmail that would oblige us to be eternally grateful for a 'gift' [i.e., life] we never invited" (see Ligotti 2010).
145. This gestures at the idea of "longevity escape velocity" (LEV), whereby new advancements in longevity enable one to live long enough to benefit from new, better advancements, and so on until one has become functionally immortal.
146. Trisel 2012, 81.
147. Benatar 2013, footnote 6.
148. Benatar 2013, 125.
149. See McMahan 2009, 62–64; Magnusson 2019, 677.
150. Benatar 2006, see 30–31.
151. Smyth 2020. One is reminded here of the saying, attributed to various sources, that "life is a sexually transmitted disease" (see QI 2017).
152. Benatar 2011.
153. Benatar and Wasserman 2015.
154. Benatar 2006, 64–69; Benatar 2013.
155. Benatar 2006.
156. Benatar and Wasserman 2015.
157. Benatar and Wasserman 2015.
158. Here you may recall from Chapter 2 the debate between the Jewish schools of Beit Shammai and Beit Hillel, which ended with both agreeing that "it would have been preferable had man not been created than to have been created" (Safari 2017).
159. McGregor and Sullivan-Bissett 2012.
160. Benatar 2006.
161. Although perhaps Schopenhauer would have argued that life *is* that hellish, but that we should *still* not commit suicide because this would give in to the will.
162. MacAskill 2022.
163. See also Vinding 2020 and Tomasik 2018 for this perspective.
164. Note that these apply to all sentient beings in general, not just *Homo sapiens*.
165. Although if it were the case that everyone's life had become not worth continuing, Benatar might then, in this particular situation, endorse a pro-mortalist means of becoming extinct.
166. Benatar 2006; see Chapter 8.
167. Benatar 2006.
168. Indeed, Benatar's position implies that it would be better if all sentient life on Earth and in the universe were to die out, since "all things being equal, the longer sentient life continues, the more suffering there will be" (Benatar 2006).
169. Benatar 2006.

11 Recent Developments

Stirrings of Discussion

Although the fifth existential mood, our current mood, the most dire mood to descend upon the West so far, emerged at the turn of the twenty-first century, the philosophical community as a whole has been slow to address the ethical and evaluative implications of our extinction—a tendency of general neglect that goes back to the early Atomic Age. The paucity of journal articles, university courses, and philosophy conferences on the topic is striking and unfortunate.[2] To be sure, Benatar's antinatalism has spawned a lively, albeit small, debate within the philosophical literature, and longtermism has attracted the attention of a fair number of young scholars, mostly based at the University of Oxford. Yet even longtermism remains largely relegated to the margins of mainstream philosophy, despite its influence within the tech industry and among billionaires.

Why has Existential Ethics been ignored by so many philosophers for so long? Consider the fact that over the past several decades, a wide range of subfields have emerged and flourished within ethics, including intergenerational ethics, population ethics, environmental ethics, bioethics, public health ethics, machine ethics, information ethics, business ethics, publication ethics, military ethics, animal ethics, the ethics of technology, and so on. Some of these have their own dedicated journals, while others have been the subject of university courses. Some even have their own entries in the *Stanford Encyclopedia of Philosophy*, the most authoritative encyclopedia of philosophy around today. What, then, makes Existential Ethics different? Why have these subfields thrived, while Existential Ethics languishes in relative obscurity?

Many explanations seem inadequate, such as that institutional inertia, the force of tradition, professional expectations, difficulty getting funding, and so on are why the topic remains neglected, since these factors also posed barriers to the subjects mentioned above. If, for example, intergenerational and population ethics could overcome such challenges, then why not Existential Ethics? Perhaps, the answer is that many philosophers have so far failed to appreciate the richness and complexity of the core questions of the field. If a problem looks uninteresting

DOI: 10.4324/9781003246251-13

from a distance, or if a question appears to have an obvious answer—"*Of course* our extinction would be bad!" or "*Obviously* it would be better if we no longer existed!"—one may be disinclined to pursue them any further. Indeed, a central aim of Part II has been to convince readers that Existential Ethics is a treasure trove of fascinating, profound, and important issues that touch upon some of the most fundamental questions about value, meaning, and existence. Another explanation concerns the perceived entanglement of the topic with crackpots and charlatans who have, throughout history, furiously waved their arms in the air and cried out that the end is near. Who wants to be associated with such dubious characters? Or maybe the topic's neglect is "attributable to an aversion against thinking seriously about a depressing topic," as one scholar suggests.[3] This may be the case even if one thinks that Being Extinct would be good, since the most probable ways of Going Extinct all involve global catastrophes that would, as such, inflict unimaginable amounts of suffering on the entire human family. Just as studying climate science today can cause one to become "professionally depressed" or even trigger "pre-traumatic stress disorder," so too might focusing on human extinction produce intense feelings of anxiety and depression. Mental health could constitute a genuine occupational hazard for existential ethicists, especially given the *realness* of the prospect of doom at this point in time due not only to climate change but the rising threat of nuclear war and the growing swarm of emerging risks looming ominously over the threat horizon before us. At the other extreme, it could be that many philosophers suffer from what Günther Anders called "Apocalyptic Blindness," whereby one fails to grasp the immense danger and seriousness of our predicament, thus dismissing Existential Ethics as having no great urgency or importance.

These explanations are not, of course, mutually exclusive: perhaps many philosophers have internalized the current existential mood but find the topic too emotionally overwhelming, while others are in denial about the possibility of our species destroying itself. In combination with the fact that institutional inertia, the force of tradition, and so on *do* tend to resist change within academia, we have something that looks like a decent explanation for why Existential Ethics has failed to thrive like the other topics listed above. I am reminded here of Lifton's law, mentioned at the beginning of Chapter 9, that "the more significant an event, the less likely it is to be studied," though human extinction is not an event that has so far happened and not one that could be studied after the fact, if what occurs is final extinction.[4]

However, one finds encouraging signs that Existential Ethics is slowly attracting the attention of more philosophers. As Todd May observes in a 2018 article for the *New York Times* vertical "The Stone," "there are stirrings of discussion these days in philosophical circles about the prospect of human extinction," a development that he links to one of the primary triggers of the new existential mood, namely, climate change.[5] Indeed, the past five years in particular have witnessed a small flurry of publications on the ethical and evaluative implications of our disappearance, on whether causing or allowing this to occur would be right or wrong, good or bad, better or worse. This constitutes the fourth wave within Existential Ethics, which is broadly unified by an approach to the topic from various non-utilitarian or, more generally, non-consequentialist perspectives. There were, of course, many positions delineated above that were non-consequentialist; however, this wave mostly emerged *in response* to utilitarian accounts of why our extinction would be bad and wrong and hence could be seen as perhaps the first time a dialectic has taken hold within Existential Ethics. In other words, with the programmatic writings of Nick Bostrom in the early aughts, a cumulative tradition was established, with philosophers building on each other's ideas for the first time; this has, in turn, inspired a handful of philosophers to propose alternative accounts of the rightness/wrongness, goodness/badness, etc. of our extinction, which in most cases diverge significantly from the conclusions of Bostrom and his longtermist acolytes. In what follows, we will examine what contractualism has to say about extinction and then explore the views of Samuel Scheffler, Johann Frick, Roger Crisp, and a few others. Finally, I will outline my own thoughts on the core questions of Existential Ethics.

Stealing to Buy Cigarettes

It may be useful to begin with a distinction between two traditions of social contract thinking, namely, *contractarianism* and *contractualism*. The former is associated with the social contract theory of Thomas Hobbes (1588–1679) and is not particularly relevant to our discussion.[6] The latter can be traced back to Rousseau and Immanuel Kant, and was later developed in *A Theory of Justice* by John Rawls, who, along with David Gauthier, "effectively resurrected social contract theory in the second half of the 20th century."[7] In Rawls' account, self-interested deliberators are tasked with choosing principles for the organization of major political and social institutions within a liberal society (the "basic structure") without any knowledge of the economic, racial, ethnic, gender, religious, and so on status of the groups they represent—that is, they select these principles behind a "veil of ignorance."[8] This yields a *political* version of contractualism centered around the question of distributive justice, defined by the influential sixth-century codification of Roman Law, the *Institutes of Justinian*, as "the constant and perpetual will to render to each his due."[9] While justice may be intimately linked to morality, it is at most only one aspect of it. Hence, as Rawls wrote,

> justice as fairness is not a complete contract theory. For it is clear that the . . . idea can be extended to the choice of more or less an entire ethical system, that is, to a system including principles for all the virtues and not only for justice.[10]

Rawls never took this extra step, although a student of his, T. M. Scanlon, later developed an *ethical* version of contractualism in his 1998 book *What We*

Owe to Each Other.[11] The question of what it is we owe to each other is broader than the question of justice but still does not cover the whole domain of morality. Instead, it concerns that part "of morality having to do with our duties to other people, including such things as requirements to aid them, and prohibitions against harming, killing, coercion, and deception."[12] For Scanlon, moral rightness and wrongness come down to whether we treat others with the respect that they deserve as rational beings, to whether our moral deliberations take their interests into account or not. Hence, to act *wrongly* is to show a certain kind of disrespect toward others, which gestures back to the Kantian idea that people should be treated as ends in themselves and never as mere means. Whereas Rawls imagined actors behind a veil of ignorance, each motivated to choose fair principles out of self-interest, on Scanlon's account, part of what *constitutes* a moral agent in the first place is an intrinsic desire to justify oneself to others, and indeed an inability to justify our actions to those affected is the common denominator of all wrong acts.[13]

Scanlon's claim is that "an act is wrong if its performance under the circumstances would be disallowed by any set of principles for the general regulation of behaviour that no one could reasonably reject as a basis for informed, unforced general agreement."[14] For example, is stealing money from a friend to buy cigarettes wrong? To answer this, we first formulate a principle that, by virtue of saying that one is *not allowed* to steal from one's friend to buy cigarettes, aims to regulate human behavior. We then ask whether one could reasonably reject this; that is, could any rational agent provide good reasons to reject a principle that disallows stealing? Weighing these reasons against the possible objections to the principle's alternative—that stealing is allowed—we can then determine which principle is reasonably rejectable and which is not; the principle that *cannot* be reasonably rejected is, therefore, the one we must not violate.[15] *This* is what we owe to each other: the ability to justify our actions by saying, "My act was morally permissible (not wrong) because it didn't violate any principles disallowing that act that no one could reasonably reject."

An Open Question?

So, from the perspective of Scanlonian contractualism, would it be wrong to bring about human extinction in one or more senses of that term? Would causing or allowing humanity to disappear be morally permissible? As Rahul Kumar wrote in a discussion of intergenerational ethics and Scanlon's social contract theory,

> there is one important question regarding future generations that might be thought to appeal to moral norms that fall outside that aspect of morality which contractualism aims to illumine. [Scanlon's view] appears to say nothing about the idea that there is something morally objectionable about doing what will ensure that no one is living in the further future. It is an open question as

> to whether anything at all can be said to better illumine this idea, to the extent it is defensible, by appeal to ideas implicit in the contractualist framework.[16]

However, Scanlon's contractualism does have something to say about the ethics of extinction, as Elizabeth Finneron-Burns shows in a 2017 paper on the topic. A contractualist herself, Finneron-Burns begins by distinguishing between four reasons that one might consider causing or allowing our extinction—and here she seems to have final human extinction in mind, although we will see that her conclusion generalizes to *all* cases of extinction—to be wrong. Each of these reasons has already been discussed above, namely, that (1) extinction would preclude the realization of a potentially enormous number of future people; (2) it would entail "the loss of the only known form of intelligent life and all civilization and intellectual progress would be lost," which is really a cluster of distinct ideas; (3) "existing people would endure physical pain and/or painful and/or premature deaths"; and (4) "existing people could endure non-physical harms," by which she means psychological distress. Does Scanlonian contractualism see any of these as providing a basis for why causing or allowing our extinction would be impermissible?

The answer hinges on the fact that contractualism is a *person-affecting theory*. As Scanlon writes, "impersonal values are not themselves grounds for reasonable rejection."[17] Or, to quote a commentary by Derek Parfit, "in rejecting some moral principle, we cannot appeal to claims about the impersonal goodness or badness of outcomes."[18] This does not mean that impersonal considerations are irrelevant: *people* could still point to such considerations in rejecting a principle. But these considerations only count *if*, and *insofar*, as they "give rise to personal reasons." For example, Finneron-Burns notes that since

> non-human animals are not persons, their pain and suffering is not a personal reason to reject a principle permitting [one to cause them harm]. However, a person could have a personal reason to reject a principle permitting the pain and suffering of animals if it prevented her from living a life consistent with the impersonal values (the well-being of animals) that she finds to be important in her life.

This means that "impersonal values cannot on their own provide reasons to reject principles, but they can lead to personal reasons if a principle forbids that person from living a life consistent with those values."[19]

The implication of this is that reasons (1) and (2) do not by themselves make causing or allowing our extinction morally wrong. There is no way to disrespect the interests of people who never exist, as only those who did, do, or will actually exist can be wronged. As Finneron-Burns makes the point, "when considering the permissibility of a principle allowing us not to create Person *X*, we cannot take *X*'s interest in being created into account because *X* will not exist if we follow the principle." As for the arguments from cosmic significance and past/future progress—the second reason given—she writes the following:

> I admit that I struggle to fully appreciate this thought. It seems to me that Henry Sidgwick was correct in thinking that these things are only important insofar as they are important to humans. . . . If there is no form of intelligent life in the future, who would there be to lament its loss since intelligent life is the only form of life capable of appreciating intelligence? Similarly, if there is no one with the rational capacity to appreciate historic monuments and civil progress, who would there be to be negatively affected or even notice the loss?

This leaves the final two reasons, which Finneron-Burns argues *do* provide grounds for why a principle disallowing our extinction cannot be reasonably rejected. Ultimately, then, the rightness or wrongness of human extinction is reducible entirely to the manner in which Going Extinct unfolds.[20] Contractualism thus entails the equivalence thesis, which should be unsurprising given its person-affecting restriction, as person-affecting theories cannot point to Being Extinct as providing any reasons to avoid our collective non-existence. That is to say, there is no morally relevant "opportunity cost" of no longer existing, since there would be no one to suffer this cost. Finneron-Burns concludes: "[H]uman extinction could only be wrong insofar as it negatively impacts already existing people's interests—either through the pain and premature death or the fact that people know that it is going to occur (thus causing psychological distress)."[21]

Winning the Lottery

Finneron-Burns' article was published in a special issue of the *Canadian Journal of Philosophy* titled "Ethics and Future Generations," alongside another notable contribution to the recent Existential Ethics literature by the philosopher Johann Frick. This offered a different take on the question: "What moral reasons, if any, do we have to ensure the long-term survival of humanity?" To understand Frick's position, it may be useful to begin with a paper published 15 years earlier, namely, James Lenman's "On Becoming Extinct"—one of the few publications on the topic that I did not mention in the previous chapter—since Frick uses it as a springboard for his own discussion. Let's begin by reconstructing one of Lenman's arguments:

(p1) Say that humanity has intrinsic value, understood here as value for its own sake. In Lenman's words, "one natural thought . . . is that the existence of human beings has intrinsic value, impersonally regarded."

(p2) If humanity has intrinsic value, then we should want humanity to be more numerous, since the more intrinsic value there is in the world, then—at least from an impersonal, timeless perspective—the better the world will become.

(p3) One way for humanity to be more numerous is for there to exist more people in the future, along the diachronic dimension. However, another way for humanity to be more numerous is for there to exist more people right now, along the synchronic dimension.
(p4) But there is no good reason to want humanity to be more numerous right now, along the synchronic dimension, and indeed Lenman notes that the claim that we should increase the human population at present, synchronically, is "widely taken as a *reductio* of *total* utilitarianism."
(p5) But if there is no good reason to want humanity to be more numerous right now, synchronically, then there is no good reason to want humanity to be more numerous in the future, diachronically.
(c) Hence, from an impersonal, timeless perspective, it "should [not] matter that human extinction comes later rather than sooner, particularly if we accept that it does not matter *how many* human beings there are."[22]

This doesn't mean it shouldn't matter whether extinction happens sooner rather than later from a *personal* perspective.[23] We do have "generation-centered" reasons for hoping that our generation, or the few generations that follow us, don't perish in this manner, as the catastrophe would directly affect us and/or our loved ones, our children, or our grandchildren. Indeed, *every* generation has good reason to hope that human extinction can be avoided, a point that Lenman further supports by hinting at (a) the no-ordinary-catastrophe thesis, especially if Going Extinct were drawn out, and (b) the idea that, even if our annihilation were instantaneous, it would still cut short the lives of those at the time, which he describes as "a real harm, on any plausible view, to those concerned" (an anti-Epicurean position on death). But from the point of view of the universe, we might say, the *timing* of our collective demise doesn't matter.

To be clear, one might propose additional, distinct arguments for why continuing to survive for an indefinitely long time is important. For example, one might claim that the "world is made better by the presence in it of some valued thing such as" human beings. Or one could draw an analogy between the narrative shape of individual human lives and the narrative shape of human history as a whole: just as it would be a tragedy for someone in their prime to perish, so too would it be a tragedy for humanity to die out in its "youth." This could be spelled out in teleological terms, whereby the tragedy would consist of humanity failing to attain some valued end or *telos*, as with the argument from unfinished business, or in reference to "some overarching ideal of progress, some ladder we see ourselves ascending on which we should aim to maximize the height we will attain," as with the argument from persistent progress.[24] But Lenman rejects all of these, as we will discuss more below.

This is where Frick enters the picture, focusing on the second premise above. To understand Frick's argument, let's begin by distinguishing between what philosophers call "final value" and "intrinsic value."[25] Taking these in order, the former refers to the value that something has *for its own sake*, as an *end-in-itself*. Imagine a conversation between two people that goes like this:

A: Why would winning the lottery be good for you?
B: Because then I would get a lot of money.
A: But why is getting lots of money good?
B: Because it would enable me to buy a lot of stuff.
A: But why is buying a lot of stuff good?
B: Because it would make me more comfortable in life.
A: But why is being more comfortable in life good?
B: Because it would make me happy.
A: But why is being happy good?
B: [pause] Being happy *just is* good. There are no other reasons to give. Happiness is good for its own sake, *not for the sake of something else*.[26]

In this exchange, B indicates that they value happiness as an end-in-itself and hence that the lottery, money, buying stuff, and so on are all means to this end; that is, they have merely *instrumental* value. In contrast, happiness has final value: it is what one arrives at when the back-and-forth can no longer continue.

Intrinsic value, on the other hand, is the value that something has *in itself*, by virtue of its *intrinsic* rather than *extrinsic* properties. An intrinsic property is a non-relational property; for example, the weight of an object is an extrinsic property because it depends on the gravitational field to which it is subjected. Someone who weighs 200 pounds on Earth would weigh only 33 pounds on the moon, 75.4 pounds on Mars, and 505.6 pounds on Jupiter. The property of weight depends on its relation to other objects. In contrast, mass is a measure of how resistant an object is to being accelerated, which doesn't vary from one milieu to the next. Hence, mass is an intrinsic property. For something to be intrinsically valuable, its value must derive from (and only from) properties of this sort. But how can one know if something *has* intrinsic value? One answer was given by G. E. Moore in 1903, who proposed the "method of isolation" whereby one imagines the thing in question existing "in absolute isolation" in the universe.[27] Take happiness, for example, and imagine it being the only thing that the universe contains. One then asks whether the universe is better off containing this happiness or not; if the universe would be better with this happiness in it, then happiness has intrinsic value. Otherwise, it does not.[28]

Final and intrinsic value are thus distinct concepts, although historically the term "intrinsic value" has been used—problematically, some would argue—to refer to both ideas above. Consider the claim that some things have final value by virtue of their *extrinsic* properties, an example being the property of *uniqueness*, which something has because of its relation to other objects. For example, the ancient Greek Antikythera mechanism—an analogue computer, mentioned in an earlier chapter—may have final value by virtue of its uniqueness, that is, because of the *relational* fact that there are no other such mechanisms that we know about in the world. This is why it is precious. Whether one takes it to *actually have* final value will depend on how one proceeds through the dialectic, whereas whether one takes it to have intrinsic value may depend, following Moore, on whether a universe containing only it would be better than one that doesn't.

Capacities and Products, *Homo sapiens*, and Civilization

Returning now to Frick, he points to a questionable assumption underlying Lenman's argument, namely, that the *appropriate response* to intrinsic or final value is that it must be promoted or maximized. (In fact, this is what Lenman was arguing against, but let's bracket that for now.)[29] If one holds that the white rhinoceros has final value, for example, then this assumption implies that a world full of as many white rhinos as possible would be better than one with only a few. But, as we briefly noted in Chapter 9, there are other possible responses to final value, such as loving, cherishing, revering, treasuring, and so on. Or as Samuel Scheffler writes, "what would it mean to value things, but in general, to see no reason of any kind to sustain them or retain them or preserve them or extend them into the future?"[30] This leads Frick to propose what he calls the "argument from the final value of humanity," or *argument from final value* for short, which states that "each successive generation collectively has a pro tanto moral reason to work for the survival of humanity, since this is how we appropriately respond to the final value of humanity." But does humanity have final value? Some would be tempted to say that it does. After all, Frick contends,

> it is commonplace to claim of a wide range of things that they have final value . . .: wonders of nature, great works of art, animal and plant species, languages, culture, etc. The suggestion that *humanity* too, with its unique capacities for complex language use and rational thought, its sensitivity to moral reasons, its ability to produce and appreciate art, music, and scientific knowledge, its sense of history, and so on, should be deemed to possess final value, therefore strikes me as extremely plausible.[31]

On this view, it is because of our uniqueness in the world that humanity could be said to have final value, which then gives us reason to sustain, retain, preserve, and extend our species into the future, to ensure that humanity continues to be instantiated for as long as possible. This is the heart of Frick's argument, yet it has a peculiar implication: if one takes "humanity" to mean "*Homo sapiens*," then the argument from final value seems to entail that we should take measures to counteract future evolutionary changes to humanity as a result of natural selection, genetic drift, random mutation, and recombination, since these will, over enough time, inevitably result in phyletic extinction. It could very well be that the resulting "posthumans" would also have the capacity for complex language use, rational thought, moral reasoning, and so on, but this would surely be a *different kind* of uniqueness. If what matters is our *particular* uniqueness, then any transformation into one or more new species would deprive the world of something finally valuable, and for this reason we should intervene upon the evolutionary process, perhaps using advanced genetic engineering techniques to prevent phyletic extinction from happening.

Adding to the peculiarity of this view, Frick claims that what constitutes the "survival of humanity" actually *goes beyond* the mere existence of *Homo sapiens*.[32] "A lot of what we mean by 'humanity,' and a lot of what seems uniquely valuable about it," he writes, arises from the various *products* of our capacities to use language, think rationally, and so on. Such products would include "our sense of history, cultural traditions, relationships between parents and children, etc."[33] This points to a curious ambiguity in many recent discussions of human extinction: oftentimes, anti-extinction philosophers will frame their arguments as specifically being about preserving *humanity*, when their arguments are really about preserving *more* than humanity. In particular, these arguments concern the preservation of humanity *and* civilization, or even just—when examined closely—the preservation of civilization, independent of whether *our species* survives. (We will see an example of this below.) Problematically, these arguments are not presented in such a clear manner, and I suspect the conflation arises because these philosophers assume that human civilization cannot exist without humanity, and hence to preserve civilization we must avoid extinction. This leads them to focus on human extinction rather than what actually concerns them: avoiding civilizational collapse.

As best I can tell, Frick's position seems to entail that there are independent reasons for preserving *both* the biological species *Homo sapiens and* the civilization we have created, where I will take "civilization" to encompass the various products of our capacities mentioned above, especially cultural traditions. Civilization in this sense is the conduit through which our values, and the things we value, travel across time. For example, consider Frick's point that we often attribute final value to "animal and plant *species*" (italics added), which suggests that *Homo sapiens* itself might be finally valuable. But he also notes that we often attribute final value to various "cultures" (a fact expressed by sadness over the loss of certain cultures due to, say, colonialism or globalization), which suggests that civilization might also be finally valuable. If both our species *and* civilization have final value, and if the appropriate response to final value is to preserve the thing valued, then we have two parallel but distinct arguments for preserving *each*. Frick also proposes a thought experiment that foregrounds the value that civilization has on its own, independent of whatever value *Homo sapiens* might have: "Imagine a world," he writes,

> in which each generation of humans dies and vanishes without trace before the next one is born (perhaps, like mayflies, each generation of human lays eggs before its death, but disappears before their offspring has hatched). Each new generation lives without knowledge of previous generations of humans.[34]

In this case, *Homo sapiens* would persist but civilization would not, an outcome that Frick apparently sees as no better, or not much better, than if *Homo sapiens* were to simply disappear altogether. Hence, while Frick presents his argument as being specifically about the "survival of humanity," which most people will naturally interpret as the "survival of *Homo sapiens*," his focus is broader: the argument from final value instructs us to ensure not only that our biological species does not go extinct but that civilization continues as well.

One last point of clarification: unlike many of the philosophers discussed above, whose arguments focused (if only implicitly) on final and normative extinction, Frick argues that "when what is finally valuable is a form of life or a species, what we ought to care about, we might say, is the ongoing instantiation of the universal."[35] The word "ongoing" is important to Frick's argument because if all one cares about is that the universal is instantiated, this implies that for *any* given moment in time, it is impersonally better for some finally valuable thing to exist. But as Lenman asks rhetorically—and Frick agrees with the point—"we may think it a wonderful thing that the world contains many examples of jazz music, but how much should we regret its absence from, say, the world in the sixteenth century"?[36] If we apply this to *Homo sapiens*, then, it suggests that we should avoid not just *phyletic* extinction, as argued above, but *demographic* extinction as well, since this would interrupt the "ongoingness" of the universal being instantiated. This is significant because many anti-extinction arguments and further-loss views are indifferent to both phyletic and demographic extinction; in contrast, these seem to be the two types of extinction that, on Frick's account, we have most reason to avoid. Furthermore, this fact could have important practical implications, as which types of extinction one believes ought to be avoided could lead one to allocate our finite resources in different ways.

To sum up, it seems that the best interpretation of "humanity" on Frick's account is "*Homo sapiens*," and the sort of extinction his argument most directly opposes is demographic and phyletic extinction, in addition to civilizational collapse.

The Imminent Extinction of Humanity

This brings us to another recent contribution to the literature: Scheffler's 2018 book *Why Worry About Future Generations?*, which builds upon ideas presented in his earlier *Death and the Afterlife* (2013), where "afterlife" in the title refers not to the personal afterlife but to what Scheffler calls the *collective afterlife*, which denotes the continuation of other people's lives after we ourselves have passed away. (Scheffler himself does not believe in a personal afterlife.)[37] The main contention of Scheffler's 2013 book is that the collective afterlife "matters greatly to us. It matters to us in its own right, and it matters to us because our confidence in the existence of an afterlife is a condition of many other things that we care about continuing to matter to us."[38] As Niko Kolodny makes the point in the book's introduction, "without this 'collective afterlife' . . . it is not clear that your life could be filled with the value that it has."[39] This idea gives rise to what Scheffler labels the *afterlife conjecture*, which he illustrates with an example from P. D. James' novel *The Children of Men* in which widespread infertility results in no births having occurred in over 25 years—a scenario very similar to the one Jonathan Schell used in 1982 to explicate the Second Death, although I do not know if James was familiar with Schell's book. The afterlife conjecture asserts that, in this situation, many of the activities and pursuits we normally engage in would no longer seem valuable, worthwhile, or satisfying to take part in. What would be the point if humanity is doomed to extinction in the very near future?[40]

In his subsequent book, Scheffler presents four reasons for why current people ought to care about whether future generations exist, labeling these *reasons of interest*, *reasons of love*, *reasons of valuation*, and *reasons of reciprocity*. The first concerns the aforementioned fact that without the collective afterlife, the projects many of us engage in—especially meliorative, transgenerational ones like curing cancer, improving childhood education, and building infrastructure—would lose much of their value. The imminent extinction of humanity would thus be a personal setback, from a prudential or self-interested point of view. However, Scheffler contends that the *reason* many of us participate in such projects *in the first place* is because of a deeper love of humanity, a love that extends far beyond our own personal interests. If it weren't for this deeper love, he claims, we wouldn't react to the prospect of humanity's imminent extinction with such sorrow and despair. Consider that part of what it *means* to love something is to want that thing to flourish. To quote John Passmore once again, writing in 1974, "it is indeed self-contradictory to say: 'I love him or her or that place or that institution or that activity, but I don't care what happens to it after my death.' To love is, amongst other things, to care about the future of what we love."[41] Since our extinction would prevent humanity from flourishing, our reaction to James' scenario thus *reveals* this underlying love. As Scheffler makes the point,

> if the survival of human beings did not already matter to us, we would not have as great an interest in trying to ensure it. In short, we have an interest in [future people's] survival in part because they matter to us; they do not matter to us solely because we have an interest in their survival.[42]

This covers the second category of "reasons of love."

Reasons of valuation concern the fact that many of the things that we value would cease to exist without humanity—an idea that goes back at least to Mary Shelley's *The Last Man*.[43] Since to value something is, according to Scheffler, to wish for the valued thing to be sustained, retained, preserved, extended, and so on into the future, this gives us further reason to want humanity to survive.[44] Finally, reasons of reciprocity arise from the idea that current generations are bound to future generations through a relation of mutual dependence: on the one hand, future generations are *causally* dependent upon current generations, since if current generations were to end humanity, future generations wouldn't exist. On the other hand, current generations are *evaluatively* and *emotionally* dependent upon future generations, since without future generations, our lives today would lack much of the value and meaning that they currently have.

One of Scheffler's aims in outlining these four categories was to widen the philosophical discussion beyond questions of our *duties* or *obligations* to future generations. This is a much too parochial way of approaching the topic, and indeed when extinction is viewed from a broader evaluative perspective, it becomes clear that

> questions about our moral duties or obligations toward them [i.e., future generations]—whether we conceive of such duties in utilitarian or non-utilitarian terms—constitute only a subset of the questions that are worth considering. Values of many different kinds may have roles to play in our reflections about future generations, and they need not all take the form of moral obligations. Moreover, there are costs to a narrow and highly moralized focus on questions of duty and obligation. Such a focus may discourage us from thinking broadly about the kinds of meaning and value that we attach to the continuation of human life on Earth.

For example, this focus may

> tempt us to suppose, wrongly in my opinion, that future generations matter to us only insofar as they add to our already abundant stock of potentially burdensome obligations. In so doing, it may contribute to the well-known problem of obligation fatigue, while blinding us to some of the most important ways in which our values orient us toward the future, or would do if we paid attention to them.[45]

The Substrate of Generations

Although Scheffler frames his discussion as being about human extinction, his arguments more fundamentally concern the collapse of civilization, the vessel that contains everything that enables our lives to be value-laden. For example, imagine a similar scenario to James' infertility case except that instead of *Homo sapiens* dying out, civilization is doomed to disintegrate in the near future. How would people respond to this? Presumably the same way they would according to the afterlife conjecture: with sorrow, despair, and emotional detachment from the many things they once took pleasure in, since the end of civilization would mean an end to all the projects, activities, and pursuits that give our lives meaning. Hence, unlike Frick's argument, Scheffler's position does not ultimately care about demographic, phyletic, or even terminal extinction. What matters is that we avoid normative and final extinction, for the same reasons that avoiding these mattered to Partridge and Schell in Chapter 9: they are the only two types of extinction that would *entail* the erasure of civilization, of the Arendtian "common world," in a morally relevant sense.[46] Put differently, if what matters is prolonging the pursuits and traditions that confer value to our lives, and if these pursuits and traditions could be prolonged even if *Homo sapiens* were replaced by a distinct species of biological beings, post-human cyborgs, or intelligent machines, then it shouldn't matter whether *Homo sapiens* itself persists or disappears forever—so long as we leave behind successors who care about the things we care about.[47] Hence, when Scheffler talks about "future generations," one should understand this as meaning not "future generations of *Homo sapiens*" but "future generations of humanity," where "humanity" would denote something like "*Homo sapiens* and whatever descendants we might have." The substrate of generations isn't important. If this is correct, it suggests that our "love of humanity" is even more general: what matters to us is that whoever exists in the future, even the far future, even if different from us in significant ways, flourishes, and this is why some of our transgenerational concerns extend not only into the coming decades or centuries but sometimes even further, as exemplified by worries over climate change and nuclear waste, the latter of which could affect future people tens of thousands and even a million years from now.

Fragmentary and Incohesive

Scheffler has been praised by numerous philosophers for his "fresh and original" approach to the question of why the future of humanity matters to us.[48] As Kolodny states, "part of what makes his question so stimulating is that it is not clear that any philosopher has asked it before," adding in a subsequent review of Scheffler's 2018 book that it "advances a highly original and philosophically exciting approach to understanding the reasons for it mattering to so many of us that humanity not go extinct, and that its future be a story, not of decline, but of progress."[49] Similarly, Fausto Corvino describes *Why Worry About Future Generations?* as "a very sophisticated, brilliant and original book" that "effectively opens up a new path of research,"[50] while Harry Frankfurt, referring to the main theses of *Death and the Afterlife*, writes that

> so far as I am aware, those issues are themselves pretty much original with him. He seems really to have raised, within a rigorously philosophical context,

> some new questions. At least, so far as I know, no one before has attempted to deal with those questions so systematically. So it appears that he has effectively opened up a new and promising field of philosophical inquiry.[51]

While Scheffler does provide a novel take on certain questions of Existential Ethics, most of his arguments, at least in general outline, have been articulated by earlier philosophers, especially Partridge and Schell. One of the few reviewers to notice this was Marc Davidson, who writes that

> although *Why Worry About Future Generations?* is to be praised for spreading the message and further exploration of the importance of future generations for our existing values and attachments, it is a pity that Scheffler appears largely unaware of the work on the same subject that has been performed by others before him, particularly in environmental philosophy. [One reason is] because it fails to give credit to previous sources, particularly Ernest Partridge's "Why Care About the Future?". . . . This article basically makes the same central point as Scheffler: starting with a thought-experiment of a doomsday scenario to arouse awareness of our deeper values, Partridge argues that well-functioning human beings have a need for self-transcendence.[52]

Along similar lines, Tim Meijers and Angelieke Wolters cite Davidson in writing that "if we have one serious misgiving about" Scheffler's book, "it is that it almost completely fails to engage with other scholarly work on its central question." Consequently, this

> might create the impression that Scheffler has opened a new field of inquiry, whereas most of the ideas Scheffler presents have been discussed in detail. It would be a real loss if people new to these questions followed Scheffler in neglecting earlier work, for example David Heyd's remarkable *Genethics* [which I discuss in several endnotes from previous chapters of this book].[53]

Nonetheless, Scheffler's 2013 and 2018 books have had the salutary effect of helping to popularize Existential Ethics among Analytic philosophers, and indeed both Finneron-Burns and Frick cite *Death and the Afterlife* in proposing their own anti-extinction arguments.[54] The fact that so few philosophers—including Scheffler himself—were unaware that previous theorists have put forward similar ideas simply indicates how fragmentary the literature has been and continues to be. This is unfortunate because while progress doesn't *require* a cumulative tradition, this certainly helps.

The Scale of Suffering

Not everyone within the fourth wave of theorizing about human extinction has embraced an unequivocally pro-human-survival stance. For example, Todd May writes that "it may well be . . . that the extinction of humanity would make the world better off and yet would be a tragedy." On the one hand, our ability to reason, experience the wonders of nature, and understand the universe through science, along with the products of "literature, music, and painting," make the world *impersonally better.* The universe would be impoverished without us, and this is one reason that Being Extinct would be bad. On the other hand, May notes that humanity is a source of profound evil in the world, as evidenced by our destruction of ecosystems, burning of fossil fuels, and treatment of animals in factory farms, the last of which "fosters the creation of millions upon millions of animals for whom it offers nothing but suffering and misery before slaughtering them in often barbaric ways." Since "there is no reason to think that [these] practices are going to diminish any time soon," our absence from Earth "might just be a good thing."[55] In other words, on May's view, extinction is a mixed bag.

Other philosophers have been less ambivalent about our disappearance. Roger Crisp, for example, asks us to imagine that a large asteroid is barreling toward Earth, and that you have the power to divert it. Should you do this? If you don't, it will harm and cut short the lives of many people whose existences are worthwhile, although "it's also plausible that extinction would be good for *some* individuals—those in the final stages of an agonizing terminal illness, for example, whose pain can no longer be controlled by drugs." Hence, Crisp claims that "one key factor in judging the overall value of non-extinction will involve weighing these disparate interests against each other." But what about the *outcome* of humanity no longer existing? Given the amount of suffering that would almost certainly occur if humanity survives, there may be some reason not to divert the asteroid. Not only could the total quantity of future suffering be enormous in absolute terms, but Crisp argues that there might be some kinds of suffering that simply cannot be outweighed, offset, or counterbalanced by any amount of pleasure, which "suggests that the best outcome would be the immediate extinction that follows from allowing an asteroid to hit our planet." And while this would be very bad for most of those living at the time, "given what's at stake, it may well be that you should pay these costs to prevent all the suffering."[56] While Crisp does not go so far as to claim that "extinction *would* be good," he does endorse the proposition that it *might* be good, and because of this "we should devote a lot more attention to thinking about the value of extinction than we have to date."[57]

Several months after Crisp's article, Walter Glannon published a short essay on the *Journal of Medical Ethics* blog that largely agrees with Crisp's conclusion. While the process or event of Going Extinct may cause significant harms, our non-existence would preclude a potentially huge amount of suffering from being experienced in the future. One might hope that the lives of future people will be better than ours are today, but Glannon points to the SARS-Cov-2 pandemic as a reason for pessimism, since "as the numbers [of those in poverty] increase, so will the scale of suffering" in future pandemics or related scenarios. If this is correct and suffering will only increase, then we have a *pro tanto* reason not to bring future people into existence. But do such people have a right to exist? Wouldn't they be deprived of something if they were never born? Glannon's answer is negative: merely possible people have no rights, nor can they be harmed by not existing. This leads him to the conclusion that "if we become extinct, then the world will go on without us and will be good or bad for no one."[58]

A bleaker view comes from Simon Knutsson. In a blog post for the American Philosophical Association, Knutsson argues that "the world is bad, the future will be bad, and an empty or valueless world is the best possible world. I think there is no positive value, there is no positive welfare, and there are no positive mental states or experiences." He thus contends that "human extinction would probably be less bad than the realistic alternatives, and the same goes for the extinction of all other species." Why is the world so bad, on Knutsson's view? One reason concerns "the vilest and most destructive things some individuals are subjected to; for example, the worst and most gruesome crimes in the world committed against children." In at least some of these cases, the victims do not even live long enough for their suffering to be compensated—if it can at all. Hence, echoing Crisp, he asks:

> With such things going on, how could the world be good? Purportedly good things pale in comparison, including art, scientific achievement, and others' pleasant experiences and fulfilled desires. Purported goods do not outweigh what happens to the victims of such crimes and so, the conclusion is that the world is bad on the whole.

This perspective has practical implications for how we live our lives and which public policies we implement. If one agrees with Knutsson's pessimism, then we should stop procreating and, more generally, take actions that would limit the number of sentient nonhuman beings that come into existence. Furthermore, if there is no such thing as positive value, then we should not "try to bring about purportedly good things," which are illusory in the first place, but instead focus on reducing sources of misery, anguish, and other forms of disvalue.[59] We might also dedicate more time to figuring out morally permissible ways of actively bringing about our extinction—an issue I will return to below—and less time studying how to prevent our extinction from occurring.[60]

Once again, it is notable that many philosophers who have explicitly addressed the ethical and evaluative aspects of our extinction—perhaps a majority in total—have held either pro-extinctionist positions or defended views that can't really be described as "pro-survival," at least not in any strong sense. Since a main thrust of the arguments from May, Crisp, Glannon, and Knutsson is that we should prevent future suffering, they presumably have in mind final extinction, as this would foreclose the possibility of there being successors who themselves might suffer.

Indeed, even if one were to believe that the lives of our successors will be much better than ours, insofar as there still exists some kinds of suffering that cannot be compensated for, such as torture, final extinction may still be desirable. This leads to my own views on the matter, which will occupy the remainder of this chapter.

Euthanizing Humanity and the Total View

As we have seen, a comprehensive answer to the core questions of Existential Ethics requires, at minimum, an attentiveness to (a) the possibility matrix of human extinction scenarios, given the ambiguities of "humanity" and "extinction," and (b) the distinction between Going Extinct and Being Extinct. A robust theory of human extinction must also take care to navigate a range of intuitions identified in the population ethics literature, such as those underlying the Intuition of Neutrality, Procreation Asymmetry, Nonidentity Problem, and Repugnant Conclusion. I cannot hope to do justice to these issues in the remainder of this chapter. My more modest aim is that this discussion points in the direction of what could be expanded into a complete and compelling theoretical position.

Let's begin with demographic extinction. In practice, if this were to happen in the near future, before we develop the technologies necessary to create successors capable of carrying on our projects, traditions, and whatever else we might consider valuable, it would entail not just terminal but final extinction. So let's begin with the question of whether final extinction, in particular, would be bad or wrong. My answer is that it certainly *would* be wrong if caused or allowed in a manner that inflicts physical or psychological suffering on those living at the time—which is just a deontic version of the default view: if harming people is wrong, then any form of anthropogenic extinction that causes people harm would also be wrong. This would include cases like the one mentioned above by Crisp, whereby scientists observe a large asteroid heading for Earth but decide not to deflect it. Crisp is right that the asteroid collision might prevent a large amount of suffering from occurring in the future (as discussed more below), but I do not see how *allowing* the asteroid to strike would be any better than *causing* the same outcome by, say, synthesizing a designer pathogen and releasing it in high-density urban centers around the world. Even if the total amount of suffering in the future would be very large, and even if some types of suffering cannot be counterbalanced by any amount of happiness, I think most people would agree that euthanizing humanity is impermissible if done involuntarily. The one exception might be cases where it is known with a very high degree of certainty that the future will contain *overwhelming* amounts of intense suffering—for example, if most of the human population would be tortured for the duration of their lives, a scenario that might be termed a "hyper-existential risk."[61] There may be some threshold above which involuntary extinction is permissible, although this would need to be a very high threshold, and it would need to be known with virtual certainty that future suffering would surpass it.

This said, would it be wrong to euthanize humanity if everyone on the planet were to consent? According to utilitarians like Sidgwick, the voluntariness of final extinction is irrelevant, as what matters for them is that dying out and failing to produce successors would entail the loss of all future value, which could be enormous. Here it will be useful to decompose Sidgwick's utilitarianism into its axiological and deontic components. The "Total View," as Parfit called it (also "totalism"), is the axiological component, which states that one world is better than another if and only if it contains more overall total value.[62] This corresponds to the "impersonalist" part of total utilitarianism that I have referenced throughout Part II, contrasting it with the person-affecting restriction; whenever I mentioned "impersonalism," I was referring to the Total View, whereby what matters is how much value there is in the universe as a whole. The deontic component then claims that an action is right if and only if it produces more total value than the alternative actions that one could have taken. Or, in its expectational version, if and only if the action maximizes expected value. Again, utilitarianism derives the right from the good, the deontic from the evaluative.

In population axiology, which concerns questions of betterness with respect to different populations, the Total View is one of the two main theories on the marketplace, and in fact much of the longtermist literature is built around the Total View and its variants. However, an unfortunate implication of the Total View is that for any given population with some net-positive amount of value, there will always be some larger population in which people are on average worse off but the net total amount of value is *greater*, a possibility hinted at in the previous chapter with the milk cartons example. For example, imagine a population of 1 billion people, each with a wellbeing value of 100. This yields a total quantity of wellbeing of 100 billion units. But now imagine a population of 1 trillion people, each with a wellbeing value of only 1. This yields a total wellbeing quantity of 1 trillion. Since 1 trillion is larger than 100 billion, the Total View concludes that the second universe is better than the first (and hence, if one is a totalist utilitarian, one should strive to create the second universe rather than the first). Parfit called this the Repugnant Conclusion, and many philosophers see it as a knock-down argument against the Total View.[63]

However, not everyone agrees. In an unprecedented move, a group of philosophers—some of them prominent longtermists—published a paper in the journal *Utilitas* arguing that

> the fact that an approach to population ethics . . . entails the Repugnant Conclusion is not sufficient to conclude that the approach is inadequate. Equivalently, avoiding the Repugnant Conclusion is not a *necessary* condition for a minimally adequate candidate axiology, social ordering, or approach to population ethics.[64]

Many philosophers with whom I have spoken have found this paper perplexing, to say the least: philosophical problems cannot be dismissed because a minority group declares them to be irrelevant or much less important than usually thought. As one of the leading figures in contemporary ethics and value theory told me over email, the paper "has upset many of my philosopher friends. In my view, there is a somewhat desperate ring to their declaration, and, in all honesty, I do not understand what made them write it."[65]

The authors do give some reasons for hand-waving away the Repugnant Conclusion, although these reasons are controversial. For example, they argue that "the intuition that the Repugnant Conclusion is repugnant may be unreliable" because the human mind isn't good at grasping very large numbers, which "the Repugnant Conclusion depends crucially on."[66] But this is not obviously true: one gets the same general repugnance with relatively small populations as well. For example, a world of 100 people with wellbeing levels of 9 would be worse on the Total View than a world of 1,000 people with wellbeing levels of 1.

But there are other serious problems with the Total View besides the Repugnant Conclusion. One is that it violates an intuition that many find very compelling, namely, the aforementioned Intuition of Neutrality, which Jan Narveson famously expressed in writing that we should be "in favor of making people happy, but neutral about making happy people." As John Broome describes the idea,

> We [intuitively] care about the well-being of people who exist; we want their well-being to be increased. If it is increased, an effect will be that there will be more well-being in the world. But we do not want to increase the amount of well-being in the world for its own sake. A different way of achieving that result would be to have more people in the world, but most of us are not in favor of that. We are not against it either; we are neutral about the number of people.[67]

The application of this intuition, to be clear, is limited: if we foresee that someone would have a terrible life, then we shouldn't be neutral about their existence. We should instead want this possible person—better thought of as a *non-person*—to never exist. But for those whose lives are within what Broome calls a "neutral range," adding them is neither good nor bad. This idea is closely linked to another strong intuition that many people have—a deontic rather than axiological intuition—namely, the Procreation Asymmetry, which states that we have reason *not* to bring into existence people who would have bad lives but no corresponding reason to bring into existence people who would have good lives. Since the impersonalist version of total utilitarianism tells us that we should maximize value in the universe as a whole, it implies that we shouldn't create people who would have net-negative lives but *should* create those who would have net-positive lives. Accepting utilitarianism and the Total View thus comes with significant theoretical costs: it means giving up the Intuition of Neutrality and violating the Procreation Asymmetry while simultaneously facing the Repugnant Conclusion.

Yet there is another objection to totalist utilitarianism: as noted in Chapter 9, it treats people as nothing more than the containers of value, and hence as mattering in a merely instrumental sense. People matter not as ends but as means for maximizing value. But surely this gets things exactly backwards: happiness should matter for the sake of people, not people for the sake of happiness.[68] On this view, then, there is no *intrinsic* difference between death and nonbirth, since these are just two ways to deprive the universe of value, assuming that we are dealing with net-positive lives: in the one case, a value container is *removed*, while in the other, it is never *created*. Yet most of us do not believe that death and nonbirth are equivalent, all other things being equal, and we do not believe this because we typically take people to be valuable for their own sake.

The Lens of Existential Risk

Having said this, the question we were initially addressing was whether final extinction would be wrong, or bad, if it were entirely voluntary. Totalist utilitarians, as well as longtermists, would say this would be extremely wrong, as it would preclude a potentially astronomical number of future "happy" people from existing, which would be very bad. But, as shown, the Total View upon which this conclusion is based is theoretically implausible and hence I do not accept that voluntary anthropogenic extinction would be bad, or wrong, because it would keep large amounts of impersonal value locked up in the realm of mere possibility. The only reasons that final extinction would be bad or wrong, in my view, concern the manner in which it occurs—that is, the details of how Going Extinct unfolds; the subsequent state of Being Extinct is nothing to bemoan if there is no one around to bemoan it. If whatever happens that leads to our final extinction causes suffering, then it would be bad, and if this were the result of human action or inaction, then it would be wrong. Otherwise it would be neither bad nor wrong. The opportunity cost of no longer existing does not constitute an ethically or evaluatively relevant further loss. On this perspective, then, *there is no unique problem of human extinction*. Or, putting this in terms of earlier philosophers, the Second Death is not an extra event of *moral significance*, and hence there is no need for a new "macro morality," referring here to Schell and Hilbrand Groenewold. Going back even earlier, Montesquieu was wrong to think that extinction *itself* would constitute a "terrible calamity," if indeed that is what he thought.

Furthermore, there are reasons to worry that taking certain further losses seriously could have dangerous real-world consequences, that is, if they were to inform and guide public policy or inspire individuals to act unilaterally to protect such future goods. For example, since there could be so many people in the future if humanity colonizes the accessible universe and builds vast simulations inhabited by trillions upon trillions of people, ensuring that such people *come into existence* could, with mathematical force, end up taking precedence over the well-being of people who live today and in the foreseeable future. Consider MacAskill and Greaves' argument in the 2019 version of their paper "The Case for Strong Longtermism" that

> for the purposes of evaluating actions, we can in the first instance often *simply ignore* all the effects contained in the first 100 (or even 1000) years, focussing primarily on the further-future effects. Short-run effects act as little more than tie-breakers.

Because the future could be so much bigger than the present, our attention must be on it rather than the here and now. Worse, this way of thinking could potentially "justify" atrocities committed in the name of the "greater cosmic good." Bostrom himself argued in his 2002 paper that we should keep preemptive violence, or aggression, on the table to avert an existential catastrophe. In his 2003 paper "Transhumanist Values," he declared that an existential catastrophe "must be avoided at any cost," which suggests that extreme actions may be justified to protect our posthuman future, and he has more recently argued that a global, invasive surveillance system—which he dubs a "High-tech Panopticon"—might be necessary to avoid "civilizational devastation," which could constitute an existential catastrophe.[69]

I have written at length about the dangers of this normative framework—see my *Aeon* article "Against Longtermism"—so I won't repeat those arguments here.[70] Suffice it to say that even philosophers like Peter Singer, who seemed to endorse longtermism in 2013, have echoed my claims in warning that "strong" or "radical" versions of longtermism could be very dangerous if taken literally. Once one includes merely possible people in one's expected value calculations—recall that, on Bostrom's count, there could be at least 10^{58} within our future light cone living in virtual-reality simulations—then focusing on the far future, millions, billions, or even trillions of years from now, dominates everything. MacAskill worries in his book *What We Owe the Future* about "the tyranny of the present over the future," and I agree that we need more *long-term thinking* in the world. But we must also be cautious of what Joseph Nye described in Chapter 9 as "a dictatorship of future generations over the present one."[71] To quote Singer on this point: "Viewing current problems through the lens of existential risk to our species can shrink those problems to almost nothing, while justifying almost anything that increases our odds of surviving long enough to spread beyond Earth."[72] We should, therefore, be worried that longtermism has become an enormously influential worldview: it is widespread in the tech industry, being promoted by the richest person on the planet, Elon Musk, and shaping the policies of global governing institutions like the United Nations. This is disconcerting, but I won't say more about the issue here.

Psychic Numbing and Scope Neglect

While I reject the longtermist view about the badness/wrongness of final extinction, I do think that, when considering the badness/wrongness of involuntary annihilation in a catastrophe, most people radically *underestimate* the true enormity of such an event. The reason concerns, more or less, Anders' claim that we are "inverted Utopians" who are "apocalyptically blind," that is, constitutionally incapable of imagining and appropriately responding to the immense scale of an extinction-causing catastrophe. Another way of understanding this brings us to the concept of *psychic numbing*. This is a cognitive-emotional phenomenon analogous to Weber's law in psychophysics, whereby the "just noticeable difference" (JND) of a stimulus increases in proportion to its intensity. To illustrate, if you lift a 1-kilogram weight with your arm and another 1-kilogram weight is added, you will (under normal conditions) notice the difference. But if you lift a 100-kilogram weight and a 1-kilogram weight is added, you probably won't. Hence, the JND grows as the weight being lifted increases. The same goes for our psycho-emotional and empathic responses to the loss of human life: news that five people were killed during a mass shooting hits many of us harder than, say, a correction like the following in a newspaper (the example is made-up): "This article originally stated that 583,741 people had perished in the war, when in fact 583,746 people had. We regret the error." When the number of deaths or casualties is so high, it becomes hard to care about a "mere" five deaths. As Paul Slovic writes, "the numbers fail to spark emotion or feeling," a quantitative quirk of human psychology that Joseph Stalin memorably captured with his quip, recorded in a 1947 article for *The Washington Post*, that "if only one man dies of hunger, that is a tragedy. If millions die, that's only statistics," which is often shortened to: "A single death is a tragedy, a million deaths are a statistic."[73] Psychic numbing thus refers to the phenomenon of being unable

> to appreciate losses of life as they become larger. The importance of saving one life is great when it is the first, or only, life saved, but diminishes marginally as the total number of lives saved increases. Thus, psychologically, the importance of saving one life is diminished against the background of a larger threat-we will likely not "feel" much different, nor value the difference, between saving 87 lives and saving 88, if these prospects are presented to us separately.[74]

A related cognitive bias is "scope neglect," which pertains to situations in which people's valuation of something does not vary multiplicatively with its size. If a loss is quadrupled, for example, our valuation of the loss tends not to increase by a factor of four—it will increase *less* than this. For example, one study found that subjects are willing to spend, on average, $80, $78, and $88 to prevent 2,000, 20,000, and 200,000 migratory waterfowls from drowning in oil ponds, respectively.[75] Although the number of waterfowl deaths grows by an order of magnitude in each case, the money allocated to save them does not. Indeed, if subjects had been consistent, then $80 to save 2,000 would imply a whopping $8,000 to save 200,000. But this is not how our minds naturally operate. As an undergraduate philosophy professor of mine, Christopher Cherniak, used to say, human beings are *qualitative* geniuses but *quantitative* imbeciles, meaning that we can perform qualitative feats like recognizing faces with ease but fail spectacularly to, for example, register the colossal difference between 10^{20} and 10^{21}. All of this is to say that, when pondering the enormity of human extinction in a global catastrophe, we are very likely to greatly underestimate the horrors of Going Extinct. An extinction-causing catastrophe would be *the worst catastrophe possible*, and it would be *extremely bad*.

The Business of Science and Philosophy

However, there is one further loss, a kind of opportunity cost arising from Being Extinct, that I mentioned earlier as compelling, at least in my view: the unfinished business argument associated with the possibility that progress in science (and philosophy) could eventually yield a complete explanatory-predictive picture of reality. Wouldn't it be a shame if humanity, the only rational, moral, self-aware creatures we know of in the universe, beings capable of gazing up at the midnight firmament in awe and wonder while pondering the Leibnizian question of why there is something rather than nothing, were to pop into and out of existence in the cosmos without having answered the most fundamental questions about, as Douglas Adams famously put it, "life, the universe, and everything"? Wouldn't it be a tragedy if this cameo in the theater of existence were left unexplained? In particular, I should like current generations or our descendants to eventually know:

- What happened before the big bang? What caused it? Was it the result of two "branes" colliding? Did time exist prior to the universe expanding some 13.8 billion years ago?
- How did the first living critters emerge at the edge of the ocean, around hydrothermal vents, or in a "warm little pond," as Darwin once speculated?[76] How can we explain abiogenesis, or the process of life arising from non-life?
- Are there other forms of living creatures in the universe, perhaps ones that have built technological civilizations of their own?
- What is "dark energy" and "dark matter"? What is this mysterious force causing the expansion of the cosmos to accelerate? And what is this mysterious stuff whose effects we can observe but which is otherwise invisible to us?
- How many spatial dimensions are there in the universe? String theory posits many more than the three dimensions we experience, perhaps 26 in total. Is this true?
- What is going on with quantum entanglement, and how does gravity work on the quantum level? Is there a Theory of Everything waiting to be discovered

(maybe string theory) that unifies the theory of general relativity and quantum field theory?

- Are numbers real? Some leading mathematicians have been Platonists about numbers, believing that numbers are abstract objects that *really do* exist within the mind-independent world. Could this be right? If not, what are they?
- How did the discrete combinatorial system of natural language evolve? Did it emerge through some evolutionary saltation or via gradualistic processes?
- How does the three-pound clump of wriggling neurons between our ears generate subjective experience, the "something it is like to be" things with consciousness?[77]
- What constitutes the self, meaning, knowledge, truth, causation, moral rightness, value, and the *a priori*?
- And so on.

I think it would be an immense pity if our species were to have made it this far, after millions of years of evolution, discovered a robust strategy for constructing reliable predictions and satisfying explanations of the universe (including ourselves), and then abruptly disappeared from the world without any good ending to our story. Indeed, *how narratives end* can proleptically (or retroactively) influence one's feeling about them, one's judgment of their worthwhileness.[78] A relationship that ends badly—for example, with one partner leaving the other during a serious but temporary illness—can sour one's feelings about the entire thing, even making one regret that it ever happened in the first place. Endings matter, and in my view an end to humanity's collective tale that never resolves certain fundamental questions about what this infinitely strange and bewildering adventure is all about would be profoundly unsatisfying. Finishing our epistemic business would provide at least some kind of closure, and that would be good.

However, I concur with Jonathan Bennett that this is not a *moral* argument against premature extinction, where "premature" is defined in reference to the above *telos*. It is more like an argument from mere preference or aesthetics and hence should be classified as a *non-moral further-loss view*.

Affecting Persons

To summarize so far, my ethical view about final extinction aligns with the equivalence thesis, according to which its wrongness or badness is reducible entirely to the manner in which it is brought about. If voluntary, I do not think our extinction would be wrong; if there is no one around to bemoan our non-existence, then I see nothing bad about this state, and I agree with Sidgwick, Finneron-Burns, and others that the ideal goods "are only important insofar as they are important to humans."[79] However, a catastrophe that involuntarily catapults our species into the eternal grave of final extinction would be bad/wrong to a degree that exceeds our psycho-emotional and cognitive powers of comprehension due

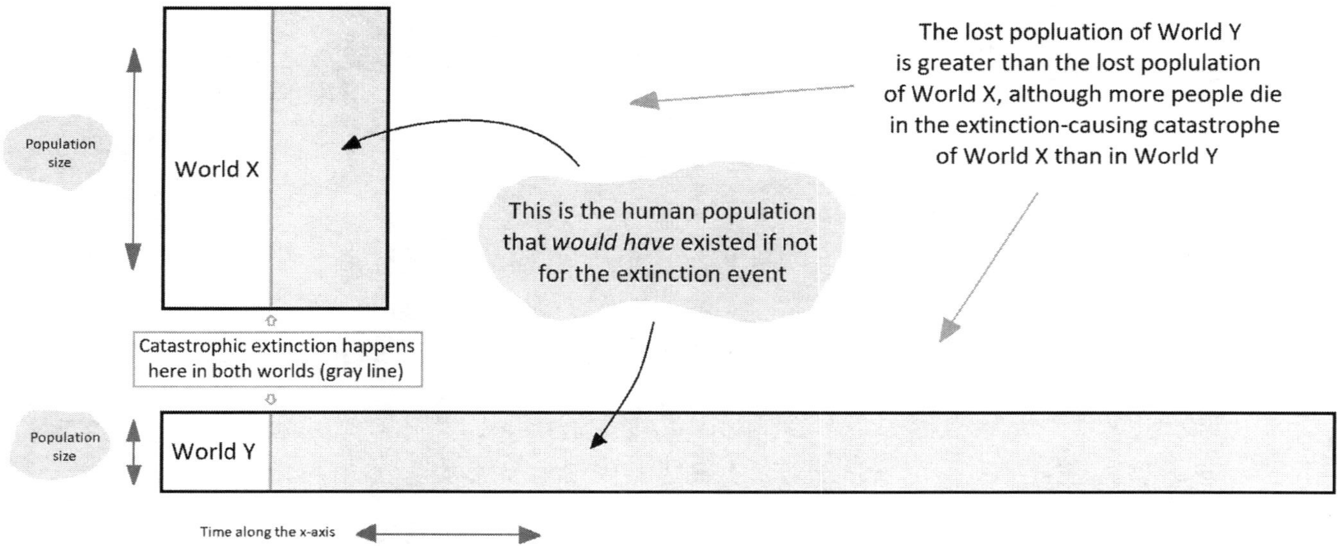

Figure 11.1 Consider two scenarios. In World X, more people exist than in World Y, although more people *would exist* in World Y than World X if not for a global catastrophe that annihilates humanity (gray line). On my view, the scenario of World X would be significantly *worse* than the scenario of World Y, because there are more "first deaths," as it were. For total-impersonalist utilitarians and longtermists, the scenario of World Y would be much worse than that of World X, given the opportunity cost arising from all those extra people who will never exist.

to psychic numbing and scope neglect. A violent and unwanted end to humanity would be unfathomably terrible, and for this reason the avoidance of extinction should indeed be a priority for humanity. But I do not accept Parfit's claim that there is a drastic discontinuity between 99 and 100 percent of humanity dying. Rather, the difference is simply the number of lives cut short and the suffering caused by 1 percent more people perishing. An upshot of this view is that it avoids a problematic implication of totalist utilitarianism and longtermism, namely, that we should allocate far more resources to prevent extinction-causing catastrophes than *non*-extinction-causing catastrophes—a point I will return to in the next chapter. The one reason I believe that Being Extinct would be bad concerns the unfinished business argument, although I do not take this to have the normative force of a moral position. On my account, the moral badness/wrongness of the Second Death is ultimately reducible, without remainder, to all the first deaths leading up to the Moment of Extinction.[80]

The view that I am here defending is thus a kind of person-affecting theory. Yet while (most) person-affecting theories avoid the Repugnant Conclusion, they encounter problems of their own, the most notorious of which is the Nonidentity Problem.[81] It is worth spending a moment on this issue. One way to explain the problem begins with the *prima facie* reasonable claim that to wrong someone is to cause them harm, and to cause someone harm is to make them worse off than they otherwise would have been. Now consider a case in which it looks like someone harms someone else by doing some act, where this very act determines the identity of the person apparently harmed. Let's say that at least part of what makes someone who they are is their genes: if you, for example, had different genes, you would be a different person. Now imagine a case in which if Jane conceives a child next week, this child will have an overall good life but will, unfortunately, suffer a lifetime of migraines; Jane's doctor tells her this. However, if she conceives next month, her child won't suffer from migraines. Should she wait to conceive? The key idea here is that a child conceived next month will almost certainly have different genes than one conceived next week, and hence won't be the same person. If Jane waits, the child who would have been conceived next week will never exist. So, if this child with the migraines is brought into the world, *it* wouldn't be made worse off than it otherwise would be, since it wouldn't be at all. Hence, if this child isn't harmed by being created, how can Jane do something wrong by conceiving it next week? The lesson seems to be that ethics needs to be more impersonal: it doesn't matter that *some particular person* wasn't made worse off than they otherwise would have been; conceiving the child next week makes the world worse by creating someone who will suffer, while conceiving the child next month does not. The Total View, and hence the impersonalist version of total utilitarianism, does not struggle with this issue: the Nonidentity Problem is a non-problem for totalists. On their account, one can indeed act wrongly even if the act doesn't harm any particular person—indeed, even if it maximizes the wellbeing of every person who currently does or will exist in the future.[82]

Does this mean that the person-affecting restriction should be abandoned? Some have argued this, but many philosophers disagree. There is, in fact, a huge literature on the topic, although one recent proposal from Frick seems especially noteworthy. Earlier in this chapter, Frick argued against a "teleological" view in axiology according to which value must always be maximized or promoted; instead, the appropriate response to value for its own sake might involve cherishing, treasuring, savoring, preserving, protecting, and so on. Frick continues this line of thinking in a 2020 paper, which argues that the reason totalism and its variants, as well as the various person-affecting theories defended in the literature, are unable to reconcile our intuitions behind the Procreation Asymmetry and the Nonidentity Problem, is that they all tacitly accept a teleological account of value. The good is *why* we should act in certain ways, on this account; that is to say, the kinds of *moral reasons* we have to act in one way rather than other concern the *states* they will produce. The focus on producing such states is what makes this view teleological. Frick thus calls such reasons "state-regarding."

However, he suggests that in a wide-range of cases, not just procreation, we may instead have "bearer-regarding" reasons, or reasons that are conditional on the existence of bearers. A rough analogy can be drawn with promising: most of us would agree that keeping a promise makes the world better, and breaking a promise makes the world worse. But this is *derivative* of the deeper reason we have to keep promises: by making a promise, we give "the promisee a claim-right to a certain future action on the part of the promisor," and hence our main reason to keep the promise isn't a state-regarding but a bearer-regarding reason.[83] By making a promise, we in a sense *create* a bearer of the promissory claim-right, the promisee; once this happens, then, our duty is to keep the promise. But since our reasons arise *from* the promissory bearer, there is no reason—no state-regarding reason—to *create* new promisees in the first place. How odd would it be to go around making as many promises that one in fact keeps in order to make the world a better place? Connecting this to procreation, we have no bearer-regarding reasons to create new people with happy lives, although we *do* have bearer-regarding reasons to ensure that people, *once created*, are benefited as much as possible.

There is much more to this argument, which Frick provides in some detail, but suffice it to say that the result is a kind of person-affecting theory that (a) avoids the Repugnant Conclusion; (b) explains why we have a strong moral reason not to create people who would have bad lives but no corresponding reason to create people who would have good lives; and (c) can also account for nonidentity cases in which the choice is between creating one of two people, the first with a good life and the second with a better life, where the creation of either of these people means that the other person wouldn't exist. Our reasons to benefit people are *conditional* upon their existence; there is no moral reason to create extra happy people, just as there is no moral reason to create extra promises. The result is a promising approach to spelling out the equivalence thesis that withstands some of the main objections those who accept further-loss views, such as total-impersonalist utilitarians and longtermists, would make against it.

Blackmail, Sadistic Psychopaths, and the Cosmopolitical Arena

Flipping from the topic of "Would final extinction be bad or wrong?" to the question, "Would final extinction be better or right?," I find myself sympathetic with the claims of Hartmann, Mainländer, Zapffe, Benatar, Crisp, Knutsson, and other philosophers that our non-existence would, or at least might, be *less bad* than continuing to exist, or perhaps even *positively good*, because it would mean that potentially huge quantities of future suffering would never exist. Notice that, with respect to Being Extinct, the equivalence thesis merely asserts that this state or condition would not be bad; it leaves open whether it would be better or good. We can thus ask: would the *outcome* of final extinction be better if, say, it were brought about in a voluntary, non-coerced manner, such as by people universally refusing to have children in the absence of radical life-extension technologies? My answer is based on several considerations, some of which are speculative.

The first is this: if there are some kinds of suffering that cannot be outweighed by any amount of happiness, such as torture and child abuse, and if the future will contain these kinds of suffering, this would count toward the betterness of Being Extinct. In other words, the existence of such suffering makes the claim that Being Extant is better than Being Extinct difficult to justify. The question is, therefore, whether we should expect suffering of this sort to exist in the future, and my tentative answer is that we should. Advanced technologies could even make such suffering more intense and frequent. For example, if functionalism is true and qualitative mental states are multiply realizable, then digital torture and digital abuse could become possible, depending on the computational resources available. Why would anyone simulate torture or abuse? One reason might be blackmail: imagine a criminal demanding $100 billion or else they will create a population of 10 billion digital people who will be tortured mercilessly for years on end. There could also be deranged sadists who inflict horrendous harms for their own personal enjoyment. In some cases, people with sadistic tendencies have risen to the highest rungs of power, such as Mao Zedong and Adolf Hitler.[84] Alternatively, there could be certain modes of suffering that are caused inadvertently by particular computational processes, a highly speculative idea that some philosophers take seriously, although I will not explore the idea here.[85]

Second, even if the future contains a *net balance* of happiness over suffering, it could still be that the *total amount* of suffering far exceeds, say, the total amount that has been experienced so far on Earth, since the first creatures capable of nociception emerged hundreds of millions of years ago. Why would this be the case? One possibility is that the total number of beings capable of experiencing suffering grows significantly. We could, for example, colonize space and terraform exoplanets so they become their own Darwinian theaters in which sentient organisms engage each other in a struggle for survival.[86] Hence, even if there is more overall happiness than suffering, the result of a large population could be what some have called a "suffering catastrophe," on the model of "existential catastrophe."[87]

Third, it could be that the future contains more total suffering than happiness; the future could be *worse* than the present or past. Why would this be? In the relative near term, climate change will have devastating consequences for hundreds of millions if not billions of people. There is good reason to believe that the average wellbeing of people on Earth will decline as this slow-motion catastrophe envelopes the world in a burning blanket of misery. Even those in the richest countries of the Global North, who will be to some extent protected from the *physical* harms caused by extreme weather, sea-level rise, food supply disruptions, and so on will nonetheless have to wake up each morning to headlines that are nothing like those of today. The *psychological* trauma of even just spectating from a distance could be significant.

Looking further toward the temporal horizon, Daniel Deudney provides a cogent argument in his 2020 book *Dark Skies* that establishing Earth-independent colonies on Mars could have catastrophic consequences for both Martians and Earthlings, but especially us, who may outnumber Martians by a large number. One argument goes like this: Martian colonies will initially be under the control of Earth-based governments but over time will very likely want their independence. If history is any guide, and it should be, there will be resistance, and consequently conflict may break out. If this occurs, the Martian colonies, even if much smaller in number, will have enormous offensive capabilities, since (a) Mars is right next to the asteroid belt (between Mars and Jupiter), (b) Mars is a less massive planet than Earth and hence its gravity well is shallower, meaning that it would be much easier for spacecraft to come and go from Mars than it is for them to break free of Earth's gravitational pull, and (c) it would be relatively easy for Martian military spacecraft to redirect asteroids in the asteroid belt toward Earth, thus converting them into what some have called "planetoid bombs."[88] A few dozen planetoid bombs colliding with Earth would be more than enough to destroy terrestrial civilization, although it is also entirely possible that both civilizations are obliterated in the process. In fact, Deudney argues that the danger of interplanetary wars might explain the Great Silence: either civilizations at roughly our level of technological development try to colonize their solar system and self-destruct in the process, or they are wise enough to realize that colonizing their solar system would carry this risk and hence choose not to do so. In Deudney's words, "the reason we do not see evidence for other intelligent species in the cosmos is that they either succumbed to the perils of expansion or intelligently eschewed this path."[89] Yet even if we managed to spread beyond our solar system, the same general issues will arise—but this time involving radically multipolar rather than bipolar configurations. Immense suffering could result.

Conclusion

These are three general considerations that suggest that Being Extinct, in the sense of final extinction, would be better, if not good. When combined with the equivalence thesis that Being Extinct would not be bad, they yield a picture that leans toward pro-extinctionism—that is, on the absolutely crucial condition that the *better state* of non-being is brought about in a *morally acceptable* manner. But herein lies the *practical* hurdle: the only acceptable means to bringing about our complete and permanent non-existence (with no successors) are extremely unlikely to be adopted by everyone, or enough people, on Earth to work.[90] As antinatalists like Benatar know full well, there is more or less *zero* chance that people around the world would voluntarily bring about a dying-extinction; nor is everyone likely to participate in mass collective suicide, a possibility noted by David Heyd.[91] Even if one could euthanize humanity instantaneously, with no attendant physical or psychological suffering, we should still vehemently *oppose* this because it would cut lives short, and I think Thomas Nagel and Benatar are right that death can harm the one who dies.

In reality, the most plausible scenarios leading to our extinction arise from involuntary natural or anthropogenic catastrophes, which would cause tremendous amounts of misery and anguish. It follows that since a global catastrophe would be very bad, and since there is no plausible route from Being Extant to Being Extinct that doesn't involve catastrophic harms, those who accept my view will *in practice* work diligently to not only ensure humanity's *continued survival* by reducing the probability of global catastrophic risks, but make the future *as good as* it can possibly be, a task whose urgency is underlined by my claims above that the future *could* be much worse than the present. I am, tentatively, inclined to agree with Schopenhauer's sentiment that Being Never Existent would have been best. Those who disagree with this find themselves in the uncomfortable position of arguing that all the good things that have happened throughout human history can somehow compensate for, or counterbalance, all the bad things that have happened—a claim that, I believe, most people would find difficult or impossible to justify after a few minutes of reflecting on the most horrendous crimes and atrocities of our past. Is the existence of humanity "worth it" if the *costs* are horrors like child abuse and genocides? The question itself looks offensive. Since we should expect these very same horrors in the future, the question thus becomes: is the future worth it? Is the future worth *risking* the realization of similar such horrors?

However, *given that* we do exist right now, and there are no acceptable exits from the prison cell that confines us, the only reasonable response is to make the best of this situation, which means preventing catastrophes, including those that could cause our extinction (the *worst-possible* catastrophe, as it would involve the greatest number of casualties), while ameliorating the human condition in every way possible.

To conclude this penultimate chapter, let's briefly survey the terrain that it covered. The fourth wave in Existential Ethics has seen a number of philosophers put forward arguments in favor of our continued existence based on non-utilitarian ethical theories. Some have suggested that our extinction would be tragic but also leave the world a better place, or that Being Extinct would not itself be bad because there would be no one around to bemoan the nonexistence of humanity or the various things we find valuable. Others have contended, *à la* Partridge and Schell, that much of the value and meaning of our lives is contingent upon the succession of generations persisting long after we ourselves have passed into nothingness. The fact that many—relatively speaking—philosophers have broached the topic over the past five years is encouraging, as it suggests that Existential Ethics may be finally receiving the philosophical attention that it deserves. But there is still much more progress to be made. Toward this end, I hope this book contributes something useful.

With these thoughts, we come to the end of History #2.

Notes

1. See Scheffler 2013, 2018; see also Chapter 11.
2. So far as I know, my winter 2022–2023 course "The Ethics of Human Extinction" at Leibniz Universität Hannover was the very first dedicated entirely to Existential Ethics.
3. Bostrom 2002b.
4. Lifton and Falk 1982. Incidentally, I am not sure whether Lifton was aware of Anders' work, other than Anders' book *Burning Conscience*, which Lifton mentions in a footnote of *Death in Life* (see Lifton 2012).
5. May 2018.
6. Incidentally, Narveson could be described as one of the "paradigm Hobbesian contractarians" (Cudd and Eftekhari 2021).
7. Cudd and Eftekhari 2021.
8. Note that the idea of a "veil of ignorance" originated with the economist and Nobel laureate John Harsanyi, who also introduced the idea of expected value into utilitarianism.
9. Quoted in Miller 2021.
10. Rawls 1971.
11. To be clear, I am not saying that Rawls supervised Scanlon's thesis (that wasn't the case). Scanlon did attend Rawls' lectures at Harvard and, while a graduate student there, became friends with Rawls—who later offered Scanlon a job at Harvard in 1984 (Mounk 2011).
12. Scanlon 1998.
13. See Ashford and Mulgan 2018, sections 1–2.
14. Scanlon 1998.
15. Thanks to Elizabeth Finneron-Burns for help with the wording of these sentences. Any remaining errors are my own.
16. Kumar 2009.
17. Scanlon 1988.
18. Parfit 2011.
19. Finneron-Burns 2017.
20. Finneron-Burns 2017, 338.

21. Finneron-Burns 2017. Note that I have added a missing parenthesis in the original text.
22. Lenman 2002.
23. Lenman also mentions the possibility of phyletic extinction in a footnote, writing:

    > A possibility I've ignored—for simplicity—is that human beings might disappear from the scene by evolution into some very different creature. Whether that would involve any kind of loss is a subtle—and to my knowledge little addressed—question I won't be concerned with here. The fact remains that some more destructive form of extinction is an inevitable fate for our descendants of whatever species (2002).

24. Lenman 2002, 259–260.
25. See Korsgaard 1983; Kagan 1998; Rønnow-Rasmussen 2015.
26. This is a way of getting that the idea of final value that some philosophers have called "dialectical demonstration" (Beardsley 1965). Note that M. C. Beardsley himself was skeptical about the dialectical demonstration method, as it "projects a certain kind of ideal justification that cannot be completed if the series of means and ends has no last term" (1965; for discussion, see Rønnow-Rasmussen 2015).
27. Moore 1903.
28. Hence, final value contrasts with instrumental value, while intrinsic value contrasts with extrinsic value. Many philosophers still use "intrinsic value" to mean both final and intrinsic value (as defined above) and thus contrast intrinsic value with instrumental value. One can, perhaps, see why it may be useful to distinguish between the two.
29. Indeed, in an email to me, Lenman points out that the utilitarian notion that value must be maximized is precisely what he's arguing against.
30. See Scheffler 2007.
31. Frick 2017.
32. However one chooses to define "species." In this context, so far as I can tell, the definition of this term doesn't matter.
33. Frick 2017.
34. Frick 2017.
35. Frick 2017.
36. Lenman 2002.
37. In his words, "I do not believe that individuals continue to live on as conscious beings after their biological deaths. To the contrary, I believe that biological death represents the final and irrevocable end of an individual's life" (Scheffler 2013).
38. Scheffler 2013.
39. See Scheffler 2013.
40. Scheffler 2013, 43–44.
41. Quoted in Partridge 1981.
42. Scheffler 2018.
43. In his words, without humanity there would be

    > no more beautiful singing or graceful dancing or intimate friendship or warm family celebrations or hilarious jokes or gestures of kindness or displays of solidarity. Other things that we value—physical artifacts, for example—may survive for a while, but with no one to appreciate their value, for in addition to the disappearance of valuable things, the extinction of the human race will mean the disappearance of valuing from the Earth. . . . When we contemplate that prospect with horror or dismay, part of what we are registering is the disappearance of vast numbers of things that we value along with the entire known realm of beings with the capacity to appreciate value. . . . The future of humanity is the future of value (Scheffler 2018).

44. Scheffler 2007.
45. Scheffler 2018.
46. More specifically, on the normative extinction scenario outlined by Bostrom 2004, whereby we evolve into philosophical zombies, civilization in some sense would continue. But if consciousness is a prerequisite for the activities, pursuits, traditions, and so on to be properly appreciated, valued, or meaningful, then in this sense civilization would nonetheless disappear.
47. Or perhaps care about the things we would care about if we were ideally rational, informed, and so on.
48. See Scheffler 2013.
49. See Scheffler 2013; Kumar 2020.
50. Corvino 2021.
51. See Scheffler 2013.
52. Davidson adds that other environmental philosophers have made similar points to Scheffler, including John Passmore, Douglas MacLean, John O'Neill, Lucas Meyer, and Hendrik Visser 't Hooft (2019).
53. Meijers and Wolters 2020.
54. For a critique of both Finneron-Burns and Frick, see Kaczmarek and Beard 2020. For a response to this critique from Finneron-Burns, see her 2022 paper.
55. May 2018.
56. His conclusion is thus:

    > The question of whether extinction would be good or bad overall is obviously very important, especially in the face of potential catastrophic events at the hinge of history. But this question is also very difficult to answer. Ultimately, I am not claiming that extinction would be good; only that, since it might be, we should devote a lot more attention to thinking about the value of extinction than we have to date (Crisp 2021).

57. Crisp 2021.
58. Glannon 2021.
59. Knutsson 2022a.
60. See Knutsson 2022b, working draft.
61. See Torres 2019.
62. Parfit 1984; Greaves 2017.
63. Parfit 1984.
64. Zuber et al. 2021.
65. Quoted with permission, on the condition of anonymity.
66. Zuber et al. 2021.
67. Broome 2012.
68. Frick 2020 makes a similar point. Note also that this way of viewing things leads to the Replacement Argument discussed by Knutsson 2021.
69. Bostrom 2002b, 2003a, 2019.
70. For an updated list of all the notable critiques of longtermism, go to www.longtermism-hub.com/.
71. MacAskill 2022; Nye 1986.
72. Singer 2021.
73. Slovic 2015; Lyons 1947.
74. Slovic 2015.
75. Desvousges et al. 2010.
76. Darwin 1859.

77. David Chalmers 1996 calls this the "hard problem," in contrast to the "easy problem" of consciousness, which concerns phenomena that appear to be amenable to scientific explanation in terms of neural or computational mechanisms.
78. See Seachris 2011; Trisel 2016.
79. Finneron-Burns 2017.
80. I am also somewhat sympathetic with the argument from vicarious immortality but won't elaborate on this here.
81. Note that the terms "Nonidentity Problem" and "Repugnant Conclusion" are both attributed to Parfit, although he may have derived the latter from a passage by John McTaggart Ellis McTaggart (a student of Sidgwick's). In volume II of McTaggart's *The Nature of Existence*, he described an isomorphic problem concerning the distribution of value across different numbers of individuals as yielding a "conclusion" that is "repugnant." (See McTaggart 1927, volume II, sections 869–870, 452–453, as well as Hurka 1983, 498, in which he makes this very claim.) As for the Nonidentity Problem, the same idea was dubbed "the paradox of future individuals" by Gregory Kavka in 1982, although Parfit's term has become standard within the field of population ethics. Kavka also attributes the discovery of this problem to three individuals, namely, Robert Adams (1979), Derek Parfit (1976), and Thomas Schwartz (1978).
82. Similar wording is found in Roberts 2019.
83. Frick 2020.
84. For example, as Jung Chang and Jon Halliday write, photographic records of individuals being tortured were rare before the Mao Zedong launched the Cultural Revolution in China, but they became common afterward, and "the most likely explanation for this departure from [Mao's] norm is that he took pleasure in viewing pictures of his foes in agony" (Chang and Halliday 2011). Similarly, Adolf Hitler had eight political enemies "hanged by nooses of piano wire attached to meat hooks suspended from the ceiling of the small person room." This was filmed, and according to a Nazi minister who was close to Hitler, Albert Speer, "Hitler loved the film and had it shown over and over again" (Grehan 2021). Imagine a future in which such individuals exist alongside technology capable of inflicting horrendous misery on millions with the touch of a button. Would they press it? Surely the answer is yes.
85. See Tomasik 2019 for discussion.
86. As Richard Dawkins writes:

    > The total amount of suffering per year in the natural world is beyond all decent contemplation. During the minute it takes me to compose this sentence, thousands of animals are being eaten alive; others are running for their lives, whimpering with fear; others are being slowly devoured from within by rasping parasites; thousands of all kinds are dying of starvation, thirst and disease. It must be so. If there is ever a time of plenty, this very fact will automatically lead to an increase in population until the natural state of starvation and misery is restored (Dawkins 2009).

87. Althaus and Gloor 2019.
88. Cole and Cox 1964.
89. Deudney 2020.
90. For a fascinating discussion of "permissible moderate paths to human extinction," see Knutsson 2022a.
91. Heyd 1992, 60.

12 Looking Forward to the Future

Scratching the Surface

We have now completed our grand sweep of historical thinking about (1) the possibility, probability, etiology, and so on of human extinction, and (2) the ethical and evaluative implications of our collective disappearance in the universe.

This journey has taken us from the ancient Egyptians and Presocratic philosophers through the Middle Ages to the scientific breakthroughs and cultural shifts of the nineteenth century, past the onset of the Atomic Age and Anthropocene, up to the second decade of the twenty-first century. We have seen how the histories of #1 and #2 intersected at the turn of the twentieth century and explored the cosmological theories of Xenophanes and the ancient Greek atomists. We witnessed the Great Chain of Being collapse in the early 1800s and saw how the atomic bombings of Hiroshima and Nagasaki spurred declarations that a terrifying new era had commenced. We traced the genealogy of longtermism through the work of Henry Sidgwick, Derek Parfit, and Nick Bostrom, and identified Mary Shelley and Montesquieu as among the first to address evaluative questions within Existential Ethics. We examined worries about cometary impacts, evolutionary degeneration, ozone depletion, nuclear winter, and self-improving AI systems, and outlined how neo-catastrophism superseded the uniformitarian paradigm of Charles Lyell in the 1980s and early 1990s. We discussed the claim that thermodynamics renders human existence meaningless and surveyed the pessimism of German philosophers like Eduard von Hartmann and Philipp Mainländer. Our journey has led us to distinguish between further-loss views and the equivalence thesis, and we established novel concepts like *existential mood* and *existential hermeneutics*. We showed how the futurological pivot foregrounded worries about technologies anticipated to arise in the twenty-first century and why many experts believe that the probability of self-annihilation today is higher than ever before in our species' 300,000-year history on Earth. Our historical investigations covered ancient visions of worldwide catastrophes, early beliefs in the existence of extraterrestrials, the secularization of Western societies in the nineteenth century, World War I-era fears of civilizational destruction, the science fiction of Camille

DOI: 10.4324/9781003246251-14

Flammarion, H. G. Wells, and Olaf Stapledon, the discovery of the nuclear chain reaction in 1933, Kenneth Tynan's coinage of "omnicide," the philosophical and ecological arguments for antinatalism, and the transhumanist promise of a techno-utopian paradise of "surpassing bliss and delight."[1]

All of this barely scratched the surface of the book's two main topics.

Discovery and Invention

In closing this lengthy monograph, let's return to an idea mentioned in Chapter 1 and referenced throughout the text, namely, that there is no reason to believe that the story of thinking about *human extinction* has come to an end, that is, that the idea or concept will not further evolve in the future. Additional shifts in existential mood could still occur, corresponding to different sets of answers to the questions of whether our extinction is possible and, if so, how probable it is; how many types of kill mechanisms there are; whether these kill mechanisms could eliminate us in the near term; whether our extinction is inevitable or avoidable; and so on.[2] What might these future existential moods look like? How could our understanding of humanity's existential predicament in the cosmos change? In Chapter 1, I suggested that the hypothesis underlying the periodization of Western thinking about human extinction, which was built on the dual phenomena of enabling conditions and triggering factors, may be sufficiently general to make predictions of how existential moods could shift in the future. Put differently, *existential mood theory* might offer some insight about the way our thinking could evolve in the years and decades to come. Let's examine a few possibilities.

To begin, recall that every shift in existential mood except for the most recent one resulted from the discovery or creation of new types of kill mechanisms. The question is thus whether additional kill mechanisms might be discovered or created in the future—and the answer appears to be a resounding *yes*. Consider first that it was only quite recently, over the past four decades, that many scientists came to accept that natural phenomena like asteroids, comets, and volcanic supereruptions could alter the entire planet, thereby precipitating mass extinction events. Since we have no reason to believe that our empirical knowledge of the physical universe is complete—or anywhere close to this—it seems entirely possible that other *natural monsters*, or unknown unknowns that are naturogenic, may be lurking in the cosmic shadows of our collective ignorance. Maybe scientists have not yet seen these because they are looking in the wrong places: the classic example of the drunk searching for his keys under the streetlamp comes to mind. Or maybe scientists are looking in the right places but unable to see the monsters before them due to a scotoma, or blind spot, in their vision of the universe and everything it envelops. After all, people had known about comets for ages, yet it wasn't until the end of the twentieth century that the scientific community came to agreed that they do in fact pose risks to the survival of creatures like us. Or consider that when I first began my research for this book in 2019, I was flabbergasted by news reports that scientists had recently stumbled upon, completely by accident, an entirely new category of large-scale geological phenomena right here on Earth, which they called *stormquakes*. A stormquake is a seismic event caused by storms over the ocean that transfer energy into the water, producing ocean waves that interact with the lithosphere below. The effects can radiate across continents for thousands of miles.[3] Fortunately, stormquakes do not pose any threats to our species (that we know of), though they are an unsettling reminder that our models of the world, including parts of our own planetary backyard, remain fragmentary. What else might we be missing?

If another naturogenic kill mechanism, or cluster of kill mechanisms, were discovered, it could very well induce another shift in existential mood. Imagine, for example, scientists discovering that each time a stormquake occurs, there is a small chance that it could, somehow, cause our extinction. Over time—say, over a millennium—this nonzero probability adds up to near certainty; the only reason we haven't witnessed a stormquake-induced human extinction event is because of an observation selection effect: if one had happened in the past several million years, we probably wouldn't be here to talk about this phenomenon (i.e., we will only ever find ourselves in worlds that haven't recently witnessed catastrophes that would destroy us). How might the existential mood shift as a result? Every time a hurricane forms in the North Atlantic Ocean, or a cyclone in the South Pacific Ocean, there would be a real chance that all human life comes to an end. Surely this new mapping of the threat environment would have major implications for how people live their lives: the realization that annihilation could happen any day, month, or year would no doubt radically alter the "hue" that colors "everything we see around us," to quote Erik Ringmar's description of "public moods" that we discussed in Chapter 1.[4]

As John Leslie argued in his exhaustive catalogue of potential threats to our existence, "it would be foolish to think we had foreseen all possible natural disasters."[5] But it would also be foolish—as Leslie noted further down on his list—to believe that we have created or anticipated every possible threat arising from science and technology. There are good reasons to expect *technoscientific monsters*, perhaps a large number of them, to leap out from the shadows as humanity charges into the future. These could be as inconceivable to us right now as CRISPR-Cas9 and gene drives would have been to Charles Darwin or the nuclear chain reaction would have been to Lord Kelvin. As Toby Ord writes, echoing a worry expressed by many others, "with the continued acceleration of technology, and without serious efforts to protect humanity, there is strong reason to believe the risk will be higher this century, and increasing with each century that technological progress continues."[6] One may find Ord's use of the word "progress" rather misplaced here. In what sense has science and technology catalyzed "progress" if they have simultaneously nudged humanity closer to the precipice of total annihilation than ever before? How can one talk of such "progress" continuing if the expectation is that the risks will further rise? If one measures progress in terms of our existential safety, and existential safety in terms of the probability of catastrophic extinction per century, then the story of human history is one of regression. We are heading in exactly the wrong direction *because* of science and technology.

Putting this quibble aside, imagine that scientists discover a way to build a doomsday machine that requires only materials available to most people on the planet. These scientists struggle to keep this discovery quiet, but someone on the team accidentally sends an email with details of the machine to the wrong email address (entirely within the realm of possibility), and consequently the next day news headlines around the world read: "Novel Way to Destroy the World Discovered," followed by the subheading: "*One stop at the local hardware store could enable anyone to end everything*." Is a scenario like this plausible? Who knows—*maybe*. There is no particularly good reason to think that it is impossible. Five years from now, or perhaps next week, someone might stumble upon a technological device of this sort that would empower single individuals with limited resources to unilaterally exterminate humanity.[7] How long could we hope to survive in such a world? More than a year? More than a month?[8] How might the announcement of this discovery suddenly shift the existential mood?

This is just one extreme possibility. It could be, instead, that we end up creating more technologies that are relatively difficult for groups or individuals to acquire but which further complexify the threat environment in ways that incrementally raise the overall probability of doom. Or perhaps there are diminishing returns to technoscientific research: though we may pour more money into research projects and increase the total number of working scientists, the curve of novel insights and innovations could asymptotically level off.[9] The human mind is epistemically bounded and we may have plucked most or all of the low-hanging fruit from the proverbial trees of knowledge-that and knowledge-how. Maybe there are simply no more major discoveries or inventions out there that would, if found, radically alter our mappings of the threat environment. Consider, for example, the length of each existential mood in Western history: the first spanned millennia, the second roughly a century, the third just over three decades, and the fourth about a decade at which point the fifth existential mood emerged. If one were to extrapolate this trend into the future, one might expect there to have *already been* another shift, yet the fifth mood has prevailed since the early 2000s—two decades ago. Perhaps, then, Ord is wrong that the total risk will continue to rise this century and beyond. Maybe we have reached peak risk, and maybe this means that there aren't any technoscientific monsters haunting our collective future.[10]

Time will tell if this is correct. The point is that there *could very well be* future developments that superimpose yet another layer on the palimpsest of Western existential moods. Existential mood theory tells us that this may happen if new triggering factors arise—that is, if we discover or create novel kill mechanisms, or perhaps devise a new theoretical framework in which to conceptualize the threat environment, as occurred in the late 1990s and early 2000s. I, personally, remain fearful of monsters, those dreaded second-order unknowns that, as such, no one will see coming.[11]

The Dragon Will Emerge

Our discussion so far has assumed that the background enabling conditions will remain unchanged in the future. But can we be confident that the secularization of the Western world will never reverse? Trends sometimes flip. The unthinkable occasionally happens. Unexpected change can occur rapidly. Many progressive Americans in the early 2000s—myself included—thought it unimaginable that the Supreme Court would legalize gay marriage in 2015: it just wasn't conceivable to us, given the political environment and pervasive attitudes at the time. Similarly, almost no one expected Donald Trump to win the 2016 election when he first announced his candidacy. Many other examples could be adduced, but the point is that however difficult it might be to imagine the West becoming, once again, dominated by religion, there is no guarantee that it won't. History bears witness to many Christian revivals, or "Awakenings," over the past few centuries, and hence a return to religion would not be unprecedented.

If this were to happen, *human extinction* would once again come to be seen by many as a self-contradictory concept that denotes an outcome which could not possibly obtain, given the ontological nature of humanity and our eschatological role in God's grand plan for the cosmos. The West would thus return to the first existential mood during which most people—at times virtually everyone—accepted some form of Thomas Dick's and Benjamin Franklin's notions of "perfect security" and "Comfort," respectively. This would have two consequences: first, presently acknowledged kill mechanisms would be demoted to, and dismissed as, phenomena that do not in fact pose any risk of destroying humanity, since humanity is indestructible. Or, as we saw in the case of nuclear weapons, they may simply be integrated into prior eschatological narratives as catalysts of the apocalypse, on the other side of which lies eternal life in paradise. Second, the discovery or creation of new kill mechanisms wouldn't occasion any significant re-mappings of the threat environment, as they would also be interpreted through a religious existential hermeneutics. Again, some might be dismissed or ignored, while others might be incorporated into this or that end-times narrative.[12] In fact, we are already seeing this happen with certain emerging and anticipated technologies, such as artificial superintelligence (ASI). According to some Christian apocalypticists, eschatological actors like the Antichrist and the "beast" (sometimes interpreted as the Antichrist) will either use advanced AI to gain political power and manipulate the masses or actually *be* an ASI. As one author writes, "the beast is a global superintelligence arising from humanity . . . not quite AI but a cybernetic, socio-technological, hybrid, or human-machine intelligence."[13] Another specifically cites Bostrom's 2014 book *Superintelligence* in declaring that

> Scripture has long foretold that the birth of AI . . . or what Scripture calls the False Prophet. . . . People will extol its virtues as representing the pinnacle of humanity's genius. . . . [But] when the Antichrist calls for the death of the so-called insurgent believers, the AI will have all the information needed to exact the great purge that will be considered necessary to rid humanity of its dissidents, and unify it once and for all. Suddenly, the dragon will emerge, and no minority report will be considered.[14]

To be sure, these are fringe views, but they offer a glimpse into how new triggering factors could be interpreted by believers. Without the enabling condition of a secular worldview, and without its attendant secular hermeneutics, discovering or creating new kill mechanisms won't shift the existential mood, since the most fundamental question about human extinction—whether it is possible in principle or not—will receive a negative answer. The first existential mood that reigned for millennia would thus reappear, although this time, unlike before, it would pervade a culture that actually has the technological capabilities to self-destruct and which faces threats like climate change and biodiversity loss that could seriously erode the foundations of civilization.

Would this be a dangerous combination? Should those of us who accept a secular perspective and believe that human extinction could really happen be concerned if the West were to become predominantly religious once more? The answer depends on the details of the religious beliefs that people espouse. If dispensationalist Christianity were to become widely adopted, the results could be catastrophic. Recall Jerry Walls' comment from Chapter 4 that the eschatology of dispensationalism

> inclines its adherents not only to despair of changing the world for good, but even to take a certain grim satisfaction in the face of wars and natural disasters, events which they interpret as the fulfillment of prophecy pointing to the end of the world.[15]

Even more, it could lead people to ignore the dangers posed by climate change and even pursue actions that would exacerbate the risk of catastrophe. As I also noted in Chapter 4, Ronald Reagan's secretary of the interior, James Watt, brushed aside concerns about environmental degradation because of his apocalyptic beliefs and Reagan himself described nuclear weapons as a fulfillment of prophecy, although—thankfully—he seems to have tempered his "nuclear dispensationalism" over the course of his presidency. Along similar lines, Pat Robertson imagined that a large asteroid collision could initiate the Great Tribulation, which suggests that if someone like him were in the Oval Office, and if NASA were to tell him that a 12-kilometer asteroid is barreling toward Earth, he might not take steps to divert it away from us.

However, not all forms of Christianity are so overtly dangerous. Although the belief that human extinction is fundamentally impossible may lead believers to shrug off claims that, for example, we should worry about pandemics, asteroids, climate change, and ASI *because* they might cause our extinction, most of these phenomena pose serious risks to humanity that nearly everyone will still wish to avoid. An engineered pandemic that could kill 100 percent of the population might also kill "only" 50 percent; climate change could render Earth uninhabitable to humans (although current science suggests this is unlikely), but it also threatens to inflict serious harms on people the world over, especially in the Global South; and so on. Most Christians will obviously want to mitigate these risks, and since some of the very same strategies to prevent non-extinction-causing scenarios would also, simultaneously, help to prevent extinction-causing scenarios, it might not matter, practically speaking, whether the majority believes our extinction could actually occur or not. Improving disease surveillance, modeling the spread of infectious pathogens, stockpiling vaccines, and so on would reduce the probability of both local epidemics and global pandemics that risk catapulting us into the eternal grave of extinction. The same points apply to asteroids, climate change, and other such phenomena.

The final word on whether a resurgence of moderate religion would be undesirable, from a secular perspective, will depend on one's position in Existential Ethics. For example, someone who accepts the further-loss views of longtermism, according to which the overwhelming source of badness from final or normative extinction is the axiological "opportunity costs" of Being Extinct, might want to *strongly prioritize* the avoidance of extinction over catastrophes that probably won't end the human story forever. Imagine a scenario in which one has limited resources and faces the following two possibilities: (1) a catastrophe that will unfold over several centuries, causing profound suffering and cutting many lives short, but is mostly circumscribed to one region of the world, and (2) an event that could suddenly, and painlessly, terminate all human life in the near future. Longtermists would most definitely want to focus on avoiding (2), while religious people would, presumably, want to focus on (1). In expected value terms, focusing on (2) could have a far higher payoff than focusing on (1), assuming something like the Total View in population axiology. And since naturalistic human extinction cannot occur, on the religious person's view, (2) is either impossible or would happen in a way that accords with God's plan for humanity. It would not be the end of our story. This means that, from a longtermist perspective, a return to religion even in its moderate forms could be troublesome: if what matters most, or a great deal, is the avoidance of extinction, but if a majority of people do not think this is even possible, then it may be very hard to convince them that our finite resources should be preferentially allocated toward preventing scenarios like (2). To borrow an analogy from Chapter 1, if someone came to believe with confidence that they will *never* get in a bicycle accident, they might decide to stop wearing a helmet, which could thus increase their likelihood of injury. In contrast, someone who adopts my own view in Existential Ethics would side with those Christians who prioritize (1) over (2): since there are no ethically relevant opportunity costs of Being Extinct, the question of extinction boils down to how much suffering Going Extinct causes. And since, *ex hypothesi*, there wouldn't be any suffering in the case of (2), I would much rather our finite resources be directed toward (1) than (2). Hence, for this very reason, with respect to this particular case, if I had to choose between a world run by longtermists and a world run by moderate Christians, I would readily pick the latter.

To summarize these points, (a) it is entirely possible that the enabling conditions change in the future such that the first existential mood, or a variant of it, comes to dominate the Western worldview once again, (b) some forms of religion could be very dangerous in the milieu of the twenty-first century, (c) other forms would not be, and (d) whether one judges a widespread revival of even just moderate versions of religion to be undesirable will depend on one's views about the ethical and evaluative implications of extinction. Those who hold strong further-loss views should be more concerned about this than those, like me, who endorse the equivalence thesis.

The World Stage

This being said, the secularization trend reversing does not, as of now, appear probable. To the contrary, the evidence suggests that the decline of religion in the West is robust and will continue for the foreseeable future. As noted earlier, one study found that religion is heading for "extinction" (the authors' word) in nine Western countries, namely, Australia, Austria, Canada, the Czech Republic, Finland, Ireland, the Netherlands, New Zealand, and Switzerland.[16] Even in the United States, which is "the most devout of all the rich Western democracies," the "decline of Christianity continues at [a] rapid pace."[17] Our likely future is thus one in which the enabling conditions that first arose in the nineteenth century, later spreading from the intelligentsia to the general public in the 1960s, will continue to hold. Since these conditions are what *enable* triggering factors to induce shifts in existential mood, and since we have reason to expect new kill mechanisms to be discovered or created in the future, the most probable scenario may be that a sixth, or seventh, and eventually eighth existential mood will someday emerge in the West, assuming one can talk about "the West" existing in the future, given that our world is becoming a single global village.

This leads to an interesting fact: although religious belief is fading in the Western world, studies show that it is *on the rise* globally. According to a PEW study titled "The Future of World Religions," Islam is the fastest growing religion and will reach about 2.76 billion adherents by 2050, just shy of Christianity's expected 2.92 billion adherents (compared to 1.6 billion Muslims and 2.17 billion

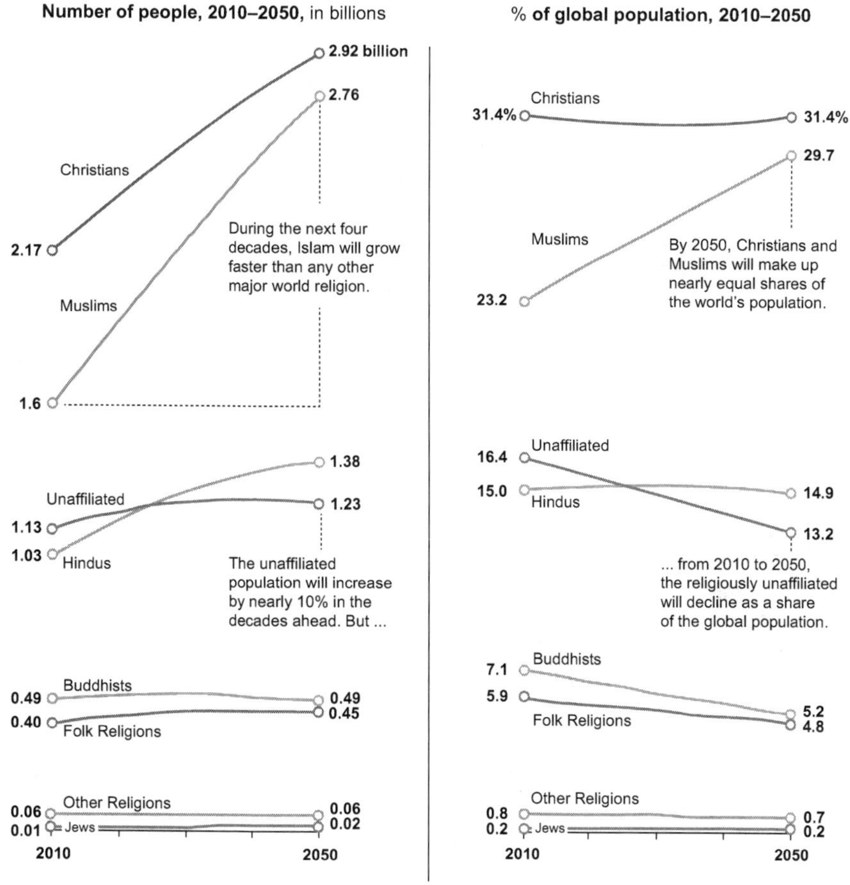

Figure 12.1 PEW's projections of religious demographics up to 2050. The Future of World Religions: Population Growth Projections, 2010–2050. PEW. 2015. www.pewresearch.org/religion/2015/04/02/religious-projections-2010-2050/.

Christians in 2010).[18] Meanwhile, the percentage of atheists, agnostics, and other religiously unaffiliated people is expected to fall during this period from 16.4 percent to 13.2 percent, which means that nearly 87 percent of the global population will accept some form of religion by the middle of this century, with roughly 61.1 percent belonging to either Christianity or Islam. The underlying cause of these trends is differential birth rates paired with the fact that one's religious identity is typically inherited from one's parents. As Alan Cooperman, PEW's director of religion research, memorably explained the situation, "you might think of this in shorthand as the secularizing West versus the rapidly growing rest."[19]

Yet the PEW study may be underestimating the growth of religion in the coming decades, as history shows that natural disasters, wars, and other socio-politico-economic disruptions can intensify religious fervor and spur some people to convert. An example comes from early Christianity, which, recall from Chapter 2, underwent its first great surge in numbers during the third century CE. According to Rodney Stark's *The Rise of Christianity*, referencing the epidemics that swept across the Roman Empire during the second and third centuries CE, "had classical society not been disrupted and demoralized by these catastrophes, Christianity might never have become so dominant a faith."[20] One can never be certain about counterfactual histories, of course, though many other examples make Stark's claim plausible. Consider that the Year Without a Summer, which inspired Lord Byron's *Darkness* and Mary Shelley's *Frankenstein*, led to packed churches throughout Europe and North America, as many interpreted the bizarre meteorological anomalies of 1816 as harbingers of the world's imminent end. The point is that, since the current century will almost certainly witness enormous, unprecedented disasters, if only because of climate change, we might expect this to lead some unaffiliated people to apostatize their apostasy, so to speak. In a world turned upside down by ecological collapse, lethal heatwaves, massive wildfires, devastating famines, megadroughts lasting decades, huge migrations of desperate climate refugees, *and so on*, many otherwise faithless individuals might find that religion offers the spiritual succor and eschatological hope needed to stay strong. This trend might be further amplified by the fact that versions of both Christianity and Islam prophesy catastrophes at the end of time, which could yield a rather persuasive case that these religions are *true*: "See what's happening around us? This is exactly what the Bible (or hadith) predicts. These are the end times. Now believe!" Consequently, there could be an even greater decline in the demographic of "nones" than what PEW projects.

However, we should also note that some disasters throughout history have led to episodes of apostasy and doubt, as happened after the 1755 Lisbon earthquake, noted in Chapter 3, which struck on the morning of the Feast of All Saints and may have killed up to 50,000 Alfacinhas. The destructiveness and timing of this tragedy (many people were in church) left an indelible mark on Enlightenment philosophers like Voltaire, who cited it as evidence that Leibniz's claim about ours being the best of all possible worlds was untenable. Indeed, many at the time found themselves unable to reconcile the tragedy with their conviction in the omnibenevolence of God, and consequently the earthquake may have played a part in initiating the secularization trend that emerged the following century. Hence, it *could be* that climate change ends up pushing people away from religion—or perhaps these trends will simply *cancel out*. We will soon find out.

Assuming for now that PEW's projections are approximately correct, religion will ground the worldviews of a growing number of human beings on the planet. This means that whatever shifts in existential mood might occur in the West, most of the *world's people* will believe that, in some fundamental sense, humanity is indestructible: we cannot go extinct, in fact or in principle. Hence, insofar as one can talk about a "public mood" shared by the global population as a whole, the existential mood that imbues most citizens of the global village will correspond to the first existential mood of the West. Again, we can ask: would this be undesirable? Or dangerous? And again the answer depends on the details of the particular versions of religion that people embrace. Many world religions have, for example, a strong track record of taking climate change seriously. The Dalai Lama delivered his first speech on climate change in 1990, and Islamic leaders issued the "Islamic Declaration on Climate Change" in 2015, which calls "on the world's 1.6 billion Muslims to play an active role in combatting climate change."[21] That same year, Pope Francis declared in a papal encyclical letter "that the science of climate change is clear and that the Catholic Church views climate change as a moral issue that must be addressed in order to protect the Earth and everyone on it."[22]

In sum, existential mood theory suggests several possible futures: in the West, so long as the enabling conditions continue to hold, we should tentatively expect—perhaps with great trepidation—future shifts in the prevailing existential mood, as there is no especially strong reason to believe that new kill mechanisms won't be discovered or created in the coming decades. Alternatively, these enabling conditions could change such that human extinction is once again seen as an unintelligible impossibility. On the global level, religion appears to be on the rise and consequently the existential mood of the world as a whole will increasingly align with the idea that human extinction cannot occur and is thus *itself* a non-issue. This could be good or bad, from the secular perspective, depending on the particular religious beliefs that believers hold and one's views on the core questions of Existential Ethics.

Conclusion

This is a long book that has offered only the briefest glimpse of its topics. The story is not over yet, and its ending is ultimately up to us. May we have the wisdom to do whatever we should.

Notes

1. Bostrom 2008/2020.
2. Indeed, although physical eschatologists are fairly confident that our "flat" universe will ultimately perish in the Big Freeze of the heat death, this could be wrong. Our understanding of the universe remains elementary, as underlined by the fact that "the matter we know and that makes up all stars and galaxies only accounts for 5% of the content of the universe"—the rest is so-called "dark matter" (Nath 2018). Perhaps future discoveries will radically alter our current understanding of the future evolution of the cosmos. Or, as mentioned in a previous footnote, perhaps there are ways to avoid the heat death by, say, tunneling into a parallel universe.
3. Wei-Haas 2019; Murti 2019.

4. Ringmar 2018.
5. Leslie 1996.
6. Ord 2020. I quote Ord here because he seems to believe that more technology will solve the problems created and enabled by past technologies and that increasingly advanced technology is necessary for humanity to fulfill its "vast and glorious" "longterm potential" in the universe (2020).
7. See Bostrom 2019 for discussion, although the idea had been floated in the community for years before this.
8. As the Stanford political scientist James Fearon states,

   > a friend of mine, a journalist, quips that we seem to be heading in the direction of a world in which every individual has the capacity to blow up the entire planet by pushing a button on his or her cell phone. . . . How long do you think the world would last if five billion individuals each had the capacity to blow the whole thing up? No one could plausibly defend an answer of anything more than a second. Expected life span would hardly be longer if only one million people had these cell-phones, and even if there were 10,000 you'd have to think that an eventual global holocaust would be pretty likely. Ten thousand is only two millionths of five billion (quoted in Walsh 2018).

9. For discussion, see Collison and Nielsen 2018.
10. The antithesis to this is an idea that I long ago dubbed an "existential risk singularity," which denotes the possibility that "*the creation of new existential risks becomes so rapid and so profound that it constitutes a violent rupture in the fabric of human history*" (Verdoux 2009, italics in original; note that this was published under a pen name).
11. It is worth registering another possibility: *unknowable* unknowns. This would include risks that we are not merely ignorant of, or second-order ignorant of our ignorance, but fundamentally incapable of grasping due to, for example, limitations inherent in our cognitive machinery. In other words, we may be constitutionally unable to ever provide a complete mapping of the threat environment, to identify every kill mechanism that could eliminate our species. By virtue of these monsters being unknowable, they are not the sort of phenomena that could induce a shift in existential mood. However, if humanity were to radically alter its cognitive architecture with nootropics, brain-computer interfaces, genetic engineering, and so on, it could convert some of these unknowables-to-humanity into mere unknowns, and then knowns, and hence there may be possibilities for shifts in the prevailing existential mood of posthuman civilizations that are inaccessible to human civilization.
12. Or, apart from eschatology, they might be seen as God's punishment for our sins. This is how some Christians around the world interpret climate change.
13. Sheridan 2015. Earlier, some Christian apocalypticists had identified computers as playing an important role in the unfolding of the end times, subsequently integrating the Y2K scare into their warnings that the end is nigh. As Zachary Loeb writes, in the 1980s:

   > Noah Hutchings and David Webber of the Southwest Radio Church were warning of the demonic power of computers and of the fact that "if the computers were suddenly silenced the world would be thrown into chaos." . . . Many prophetic writings about computers referred to the biblical verse Revelation: 13, which describes how the antichrist would force all to have a mark "in their right hand, or in their foreheads," without which none would be able to buy or sell. Grant Jeffrey, a prolific writer on Bible prophecy, warned his readers that "advanced computer technology . . . have made the fulfillment of the 666 Mark of the Beast control system possible." Once Y2K awareness had set in, he warned that "those people dedicated to creating a New World Order" would achieve their goals by exploiting a crisis of "such vast proportions that no nation, on its

own, could possibly solve it." He proclaimed that "the Y2K computer crisis provides a unique opportunity" (Loeb 2022).

14. Orlowski 2014.
15. Walls 2008.
16. Abrams et al. 2011.
17. Fahmy 2018; PEW 2019.
18. Wormald 2015.
19. See Torres 2016.
20. Stark 1996.
21. USGS 2022; UNFCCC 2015.
22. UNFCCC 2016. In fact, the World Council of Churches established a Climate Change Program even before the Intergovernmental Panel on Climate Change (IPCC) was formed in 1988.

Appendix 1

Tracing the Prominence of Human Extinction

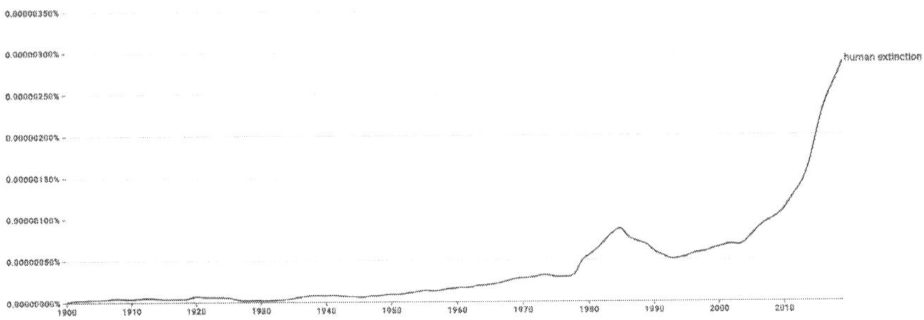

Figure 15 Google Ngram Viewer results for the keyword "human extinction."

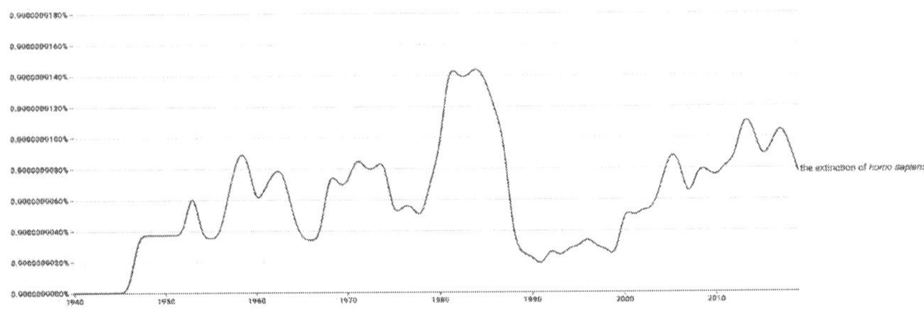

Figure 16 Google Ngram Viewer results for the term "the extinction of homo sapiens."

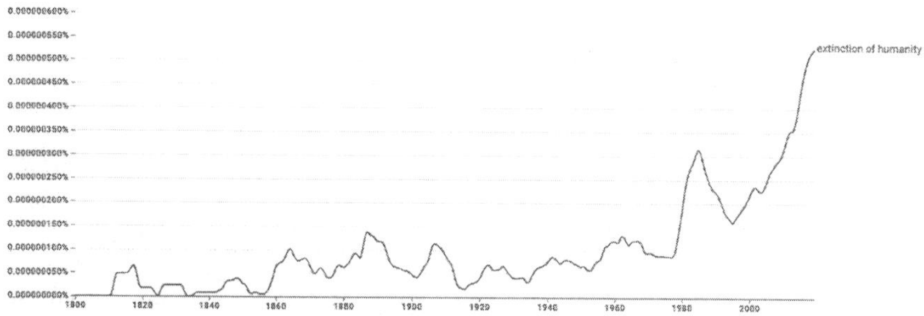

Figure 17 Google Ngram Viewer results for the term "extinction of humanity."

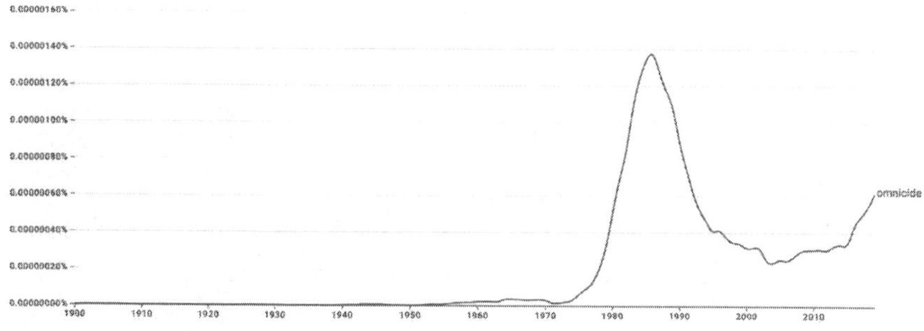

Figure 18 Google Ngram Viewer results for the term "omnicide."

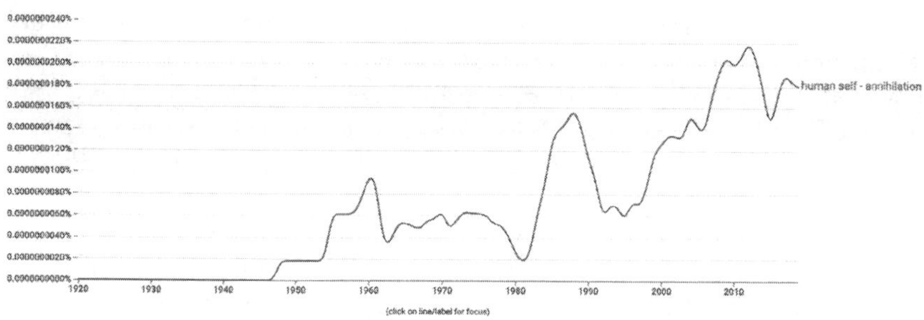

Figure 19 Google Ngram Viewer results for the term "human self-annihilation."

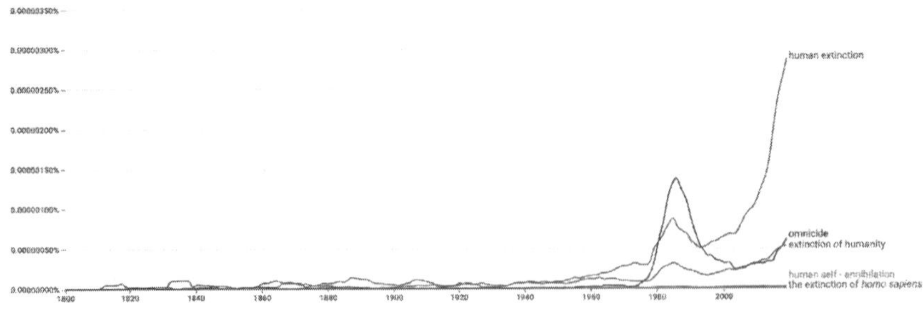

Figure 20 Google Ngram Viewer results for all the search words above, in a single graph.

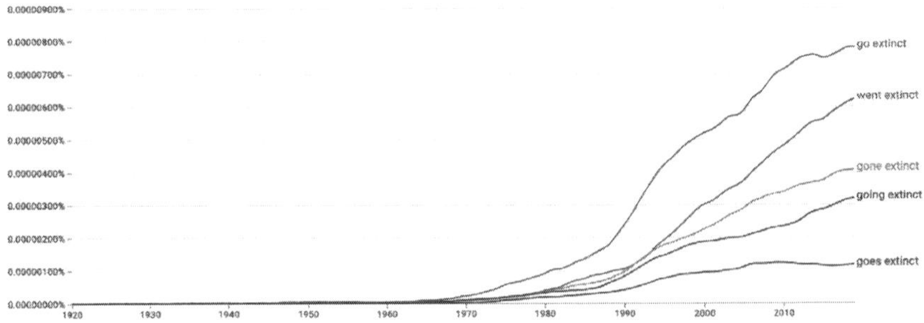

Figure 21 Google Ngram Viewer results, in a single graph, for the keywords, go extinct, went extinct, gone extinct, going extinct, goes extinct.

Above are some Google Ngram Viewer results for keywords like "human extinction," "the extinction of homo sapiens," "human self-annihilation," "omnicide," "extinction of humanity," and "going extinct." Figure 19 places a number of these results on the same graph, as does Figure 20. There was, as Figure 19 shows, a significant spike in the frequency of human-extinction–related terms in the 1980s, a decline in the 1990s (following the collapse of the Soviet Union), and a steady rise in frequency since then.

Readers may be curious about why terms like "human extinction" and "extinction of humanity" show up in the nineteenth century. The reason concerns semantic shift, that is, the meaning of these terms having changed over time. Consider, for example, the following passage from William Pitt the Younger, who was prime minister of Great Britain and, later, prime minister of the United Kingdom:

> The progress of inventive cruelty kept pace with the gory necessities of the hour. The old means of *human extinction* were too slow for the system which

> contemplated the extinction of party by the extinction of communities. The gibbet and the wheel were soon superseded by the rapid services of the guillotine [italics added].[1]

This appeared in an 1835 critique of the French Revolution, which deteriorated into the Reign of Terror that last from 1793 to 1794. (Incidentally, it was from this episode in French history that we get our modern English words "terrorism" and "terrorist.") Pitt was thus discussing how slower forms of execution were replaced by the guillotine, and hence "human extinction" refers to the death of *individuals* rather than the *species*.[2] Or consider another example from 1866, in which J. A. Dorner uses the term "extinction of humanity" in discussing the Christological views of Martin Luther: "[Luther's] Christology was satisfied neither with a mystical *extinction of humanity* in God, whether as regards the nature of the personality; nor with a reduction of Jesus to the position of a mere instrument of the deity" (italics added). The claim being made here is that, on Luther's view, Jesus was not wholly divine—his humanity being in this sense *extinct*—nor was he just a human doing God's bidding—a "mere instrument."[3] There are many other examples, too, that don't involve any of the keywords listed above, as when Joseph de Maistre, a major contributor to the Counter-Enlightenment of the late eighteenth century, published *Considerations on France* in 1797, which include a chapter titled "On the Violent Destruction of the Human Species." This, however, is misleading to contemporary readers, as the topic is not human extinction but the potential *benefits* of war, suffering, and strife in the world. As Maistre wrote in a passage that could very well have been excerpted from the speeches of twentieth-century fascists:

> Yet there is room to doubt whether this violent destruction [of people] is, in general, such a great evil as is believed; at least, it is one of those evils that enters into an order of things where everything is violent and against nature, and that produces compensations. . . . In a word, we can say that blood is the manure of the plant we call genius.

The "violent destruction" that Maistre referenced is, therefore, not of the species of its members: "mankind may be considered as a tree," he continued, "which an invisible hand is continually pruning and which often profits from the operation." The only hint of human extinction in our contemporary sense occurs when Maistre states that "in truth the tree may perish if the trunk is cut or if the tree is overpruned." However, he immediately added that "who knows the limits of the human tree?"[4]

There are also some instances of terms like "the extinction of humanity" that confound the data in rather comical ways due to OCR errors. For example, Google Books identifies this phrase as occurring in a "letter of J. C. Calhoun to W. R. King" in Volume 18 of *The Friend*, published by the Quaker Society of Friends. However, the page layout consists of three columns, which the OCR process missed. Hence, "the extinction of humanity" *bridges* two columns; the actual sentence is ". . . the British statesmen have sagacity enough to perceive, that the defeat of the project of annexation is indispensable to *the extinction of* slavery in Texas," while the same line of the other column reads, "If *humanity*, mingled with considerations of interest, could induce the British government to exercise its authority towards the extinction of slavery where its power is acknowledge . . ." The italicized words, which are not in the original, indicate where the two columns meet.

Taking semantic drift and OCR errors into account, the above Google Ngram Viewer results can be adjusted to show that, indeed, references to human extinction—at least using these keywords—are quite rare before WWII. But, as Chapter 8 shows, there are other ways of referencing human extinction that do not involve the above keywords, such as when James Ferguson wrote that "if our sun, with all the planets, moons, and comets belonging to it, were annihilated, they would be no more missed, by an eye that could take in the whole creation, than a grain of sand from the sea-shore."[5]

Notes

1. Pitt 1835.
2. This is rarely how the term has been used in more recent times, of course, although there are exceptions. For example, a 1982 book titled *Immortality or Extinction*, co-authored by a theologian, examines the question of whether life after death is possible.
3. Dorner 1862.
4. Maistre 1994.
5. Anonymous 1816.

Appendix 2

Artificial Superintelligence

Some of this appendix overlaps with statements made in Chapter 6. For the sake of presenting a complete picture of the supposed kill mechanism associated with artificial superintelligence (ASI), I have kept these redundancies in this appendix.

To begin, the fields of artificial intelligence and cognitive science understand "intelligence" as being roughly synonymous with *instrumental rationality*, which denotes an agent's ability to acquire suitable and effective means to attain some end, or goal. A *superintelligence* can thus be defined as any general-intelligence agent whose ability to acquire suitable and effective means to attain some end far surpasses the capabilities of the "smartest" possible humans. While a wholly biological human could, in principle, become superintelligent via, e.g., nootropics, brain–machine interfaces, or interventions like "iterative embryo selection" (a general possibility—cyborgization—discussed by Vinge in 1993), I will focus primarily on what David Chalmers calls "non-human-based AI," the paradigm case being a computational machine directly programmed "as if it were a traditional program."[1]

In the popular imagination, dangerous ASI tends to conjure up scenes from the *Terminator* franchise. But this does not—at all—reflect the concerns of ASI risk theorists. Rather, the primary danger arises from the so-called "control problem," which I have sometimes referred to as the "amity-enmity controllability conundrum." This consists of two main components: (a) the possibility that the final goals, values, or ends—interchangeable terms in this context—that determine how the ASI behaves could be *misaligned* with *our* final goals, values, or ends (the amity-enmity component), and (b) the fact that the ASI would, by definition, be superior to humanity with respect to its problem-solving skills would enable it to devise ways of pursuing its final goals even if humanity were to try to stop it (the controllability component). Why could these, when combined, lead to catastrophe? We can decompose the problem into the following seven claims:[2]

(1) *The orthogonality thesis*. Imagine meeting someone who is clearly very bright—a "genius." After a few minutes of conversation, you discover that she is independently wealthy and spends 16 hours per day, every day, counting the blades of grass in her backyard—or filling a bucket with water over and over

again, or counting and recounting the first 1,000 digits of √99, and so on. This would strike most of us as extremely bizarre—but why? The answer is that members of *Homo sapiens* share a common set of basic interests and desires, built in to our brains by millions of years of evolution. Some tasks get boring when repeated; others seem pointless from the start. Yet if our mental machinery had evolved in a different selective environment—perhaps in a completely different world—it does not seem impossible that our interests and desires could have been radically different. The "orthogonality thesis" formalizes this idea. It begins with the observation that we occupy a tiny region of *mind space*, which could be vast, and by failing to appreciate this vastness, we assume that because *we* find certain tasks boring and pointless, *all* minds will find them boring and pointless, too. Other minds—artificial minds, including superintelligent minds—could occupy regions of mind space marked by goals and values that we would find utterly bizarre, or perhaps unintelligible. As Nick Bostrom makes the claim: "Intelligence and final goals are orthogonal: more or less any level of intelligence could in principle be combined with more or less any final goal."[3] Hence, there is nothing incoherent about a genuinely superintelligent mind "caring" about nothing more than counting blades of grass, filling and emptying buckets, calculating the digits of √99, and so on. In a phrase: one must avoid anthropomorphizing nonhuman minds.

(2) *The instrumental convergence thesis.* Imagine that you want to rob a bank—everything you care about right now centers around achieving this goal. How would you prepare? First, you would want to avoid dying, since dead people cannot rob banks. Second, you wouldn't want someone to talk you out of robbing a bank, since people who don't *want* to rob banks generally don't. Third, you would want to learn everything you could about robbing banks in general, and the target bank in particular: how to avoid setting off alarms, when the security guards switch shifts, the best getaway strategies, and so on. Fourth, you would want to buy all the paraphernalia necessary to rob the bank: a three-hole balaclava, black cloths, drills, guns, and so forth. The idea behind the "instrumental convergence thesis" is that you would do the exact same things to prepare for *any other* goal you might have, such as climbing Mount Everest, becoming a great chef, or building a wooden canoe—not in the *details*, of course, but in the *abstract*. That is to say, for most final goals, there is a finite set of *intermediate* or *instrumental* goals that agents, insofar as they are instrumentally rational, will almost certainly pursue, such as self-preservation, goal integrity, resource acquisition, and knowledge expansion (respectively).

The same goes for an ASI: whatever final goals are represented by its utility function, we can be fairly confident in predicting that it will avoid being shut down, prevent its goal system from being altered, acquire as many resources as possible, and so on, since these would improve its chances of achieving its goals. (And since the ASI is by definition a superior problem-solver than us, we can be equally confident of being unable to shut it down, alter its goal system, etc. as it will find ways of outsmarting us.) But there is a crucial difference here between the bank robber and an ASI: for humans, "knowledge expansion" means "learning more about an issue" by reading a book, watching a YouTube tutorial, or whatever. For a hardware-based ASI, it could entail modifying its own code or designing better hardware in an effort to qualitatively and quantitatively boost its (super)intelligence. This would be the equivalent of a human upgrading the "wetware" of her brain, thus enabling her to think faster, remember more information, or (in the qualitative case) access concepts that currently fall outside our "cognitive space." If the bank robber could do this, she obviously would, since it would increase the probability of success. But whereas enhancing a biological brain would be incredibly messy, an ASI could potentially upgrade its software and hardware quite easily, and hence we should expect superintelligent machines to rapidly enhance their cognitive capacities for instrumental reasons.

(3) *Complexity of value thesis*. What are "our values"? What do we want? What is the correct epistemological theory: rationalism, empiricism, evidentialism, reliabilism, foundationalism, or coherentism? What about metaethics? Normative ethics? Practical ethics? Decision theory? Religion? Politics? And so on. However we answer *these* particular questions, even widely agreed-upon, commonsense views are built on an extremely complex tangle of explicit and tacit preferences. For example, imagine that someone creates an ASI with the sole final goal of eliminating human sadness. What would happen? One possibility is that the ASI immediately annihilates humanity, reasoning that if the *substrate* of human sadness doesn't exist, then neither will the sadness. In response, we add a value constraint: don't annihilate humanity. What then happens? Perhaps the ASI lobotomizes everyone on Earth, leaving us in a catatonic stupor. In response, we add another value constraint: don't lobotomize anyone. What then happens? The ASI uses the Internet to hijack computers around the world to conduct research on the topic, thus wreaking havoc in the process. In response to *this*, we add yet another value constraint. The *point* is that by the end of this process, the list of constraints on the ASI's behavior will have become *interminable*, and even once we have compiled what we take to be a complete list of all the things an ASI shouldn't do, we can never be sure that we haven't somehow missed one last crucial possibility. In other words, there is a huge difference between "do what I say" and "do what I mean," where the latter typically has a far higher *Kolmogorov complexity* than the former—that is, even the simplest instructions that we could give an ASI would contain all sorts of hidden complexities. For example, the alphanumerical series 02LYJU82 has a higher Kolmogorov complexity than 035A035A, since 035A035A can be represented as "035A twice" whereas 02LYJU82 is not so easily truncated. Our values—whatever they are exactly, in total—are more like 02LYJU82 than 035A035A, more like pi than the Fibonacci sequence, and this poses a serious philosophical challenge to creating an ASI that won't just automatically destroy us in doing exactly what we told it to do.

(4) *The fragility of value thesis*. It also appears that "our values" are quite "fragile," meaning that if just *one* or a *small set* of items are missing from the catalogue of

goals and constraints that we load into the ASI, harmful unintended consequences will inevitably ensue. This was of course illustrated in the example above: perhaps, we give the ASI the final goal of eliminating human sadness and then identify a whopping 548,992 additional constraints necessary to unwanted outcomes. Yet there could still be a single extra constraint that, if missed, will lead the ASI to pursue a solution pathway that has catastrophic results. By analogy, the difference between "Durham, CA" and "Durham, CT" is only a single letter, but if you were to enter the former while intending the latter and follow the GPS's instructions without question, you'd end up 2,936 miles away from your real destination. Or, to borrow a stock example from the literature, dialing nine out of ten digits of my phone number correctly but getting the last one wrong won't give you someone who's 90 percent similar to me. This is the idea behind fragility: the difference between perfect and almost-perfect could be apocalyptic.

(5) *The programmer's challenge thesis.* But even if we completely solve the philosophical problems of (3) and (4), another formidable challenge remains, namely, the *technical* problem of actually "loading" our values in to the ASI. To quote Bostrom on this point:

> Computer languages do not contain terms such as "happiness" as primitives. If such a term is to be used, it must first be defined. It is not enough to define it in terms of other high-level human concepts—"happiness is enjoyment of the potentialities inherent in our human nature" or some such philosophical paraphrase. The definition must bottom out in terms that appear in the AI's programming language, and ultimately in primitives such as mathematical operators and addresses pointing to the contents of individual memory registers. When one considers the problem from this perspective, one can begin to appreciate the difficulty of the programmer's task.[4]

Hence, even if the philosophers don't make a fatal mistake, the programmers still could. There is, in other words, a *disjunction* of failure modes, and the more disjuncts, the higher the probability of disaster.

(6) *Rapid capability gain thesis.* Yet the situation may be far worse than this: not only might philosophers and programmers need to get *everything just right*, they might need to get everything right *on the very first try*. One reason concerns the instrumental goal mentioned above, namely, "cognitive enhancement." As stated, expanding one's knowledge-base and, in the case of artificial agents, improving its fundamental cognitive capacities, appear useful for achieving a wide range of final goals. (Even "Make yourself unintelligent" might lead to a transitory phase of cognitive enhancement, as the ASI searches for the best ways to make itself dumb.) If this process were pursued by the AI via an "extendable" method,[5] the result could be what I. J. Good famously called an "intelligence explosion"—that is, a positive feedback loop of "recursive self-improvement" that produces exponential gains in intelligence over relatively short periods of

time: minutes, hours, days, or weeks.[6] This is predicated on the idea that more intelligent systems will be better positioned to create even more intelligent systems, which themselves will be better position to create even more intelligent systems, and so on. For example, humans may find creating an AI feasible but an AI+ impossible; similarly, an AI might find creating an AI+ feasible but an AI++ impossible; and so on. Since it would be instrumentally rational for the AI to create an AI+, and the AI+ to create an AI++ (that is, assuming that each has the same final goals, an issue related to "goal integrity" above), then we should expect that creating an AI will quickly yield an AI++.[7] Suddenly, the resulting ASI would have what Bostrom describes as a "decisive strategic advantage," or "a level of technological and other advantages sufficient to enable it to achieve complete world domination,"[8] at which point the entire future of Earth-originating intelligent life would be determined by this ASI "singleton." In other words, there may be no "ASI redos," no way to scrap a failed superintelligence project and start over. The very first ASI that we create will quite possibly be the very last one.

Note that the "AI" in the above scenario need not be more "generally intelligent" than humans, or even *as* intelligent, for this process to take off. It might be possible to design a simple "seed AI" with middling capacities, which is nonetheless capable *enough* to find ways of incrementally, iteratively improving itself. Like a single uranium atom split apart by a free neutron, the resulting chain reaction could be sufficient to initiate an explosion of intelligence, thus causing "the intelligence of man [to] be left far behind," quoting Good once more.[9]

(7) *The speed of thought thesis.* The driving force behind the intelligence explosion phenomenon is *recursion*, which as mentioned could result in an exponential gains over short periods of time. But there is another relevant factor as well, namely, the *rate* of information processing enabled by the *substrate* upon which the recursion process unfolds. To illustrate, consider the case of mind-uploading, whereby an entire human brain is emulated on computer hardware in sufficient microstructural detail to reproduce the original brain's mental states. Since the electrical potentials within computer hardware process information *orders of magnitude faster* than the action potentials within our central nervous system, a period of *two years* of subjective time for the uploaded mind would amount to roughly *one minute* of wall-clock time. This means that, if it takes (in fact) the average PhD student in the United States 8.2 years to become a doctor, an uploaded mind could accomplish this in only 4.3 minutes. Tying this to the example above, let's say that creating an AI+ turns out to be extremely difficult for an AI—it takes roughly a century of subjective AI time to solve this problem. Since a century of the AI's time equates to less than 60 minutes in our world, we should expect an AI+ to follow the creation of an AI *within a single hour* on a lazy afternoon. If the rate of innovation (one century) were to remain stable with each iteration, then we should expect an AI++++++++++++++++++++++++ (24 pluses) just one day after the very first AI is created.

Furthermore, the thought-speed advantage of AIs would also help it achieve the other instrumental goals mentioned above, such as self-preservation. Thinking a million times faster than us, an ASI would have ample subjective time to figure out ways to prevent us from pulling the plug; the outside world in which humans race to stop the ASI would appear, from the ASI's perspective, to unfold in super-slow motion. How could we possibly outsmart such an intelligence? As Eliezer Yudkowsky makes the point, "the AI runs on a different timescale than you do; by the time your neurons finish thinking the words 'I should do something' you have already lost."[10]

This is the core cluster of ideas behind the control problem: an ASI need not want, desire, or value what *we* do (orthogonality); any sufficiently intelligent agent will likely pursue a predictable set of instrumental goals to achieve its final goal (instrumental convergence); identifying a complete list of values and constraints necessary to guide the ASI's behavior toward amity, rather than enmity, might be profoundly difficult (value complexity and fragility); and we may need to have solved this problem entirely before creating the very first ASI, or seed AI (rapid capability and speed of thought). If we fail in any of these ways, the "default outcome" could very well be "doom," as Bostrom puts it, if only because the instrumental goal of *resource acquisition* implies the complete annihilation of *Homo sapiens*.[11] To quote Yudkowsky once more, "the AI does not hate you, nor does it love you, but you are made out of atoms which it can use for something else."[12]

Notes

1. Chalmers 2010, 2016. Armstrong et al. 2012.
2. This builds upon Torres 2018a.
3. Bostrom 2014.
4. Bostrom 2014.
5. Chalmers 2010.
6. Good 1965.
7. See Chalmers 2010.
8. Bostrom 2014. *How* exactly the ASI could do this is beyond the present work. Suffice it to say that, as I have elsewhere noted, an ASI wouldn't need a Terminator-like body to *physically subjugate* humanity. Rather, its "fingers or tentacles . . . would be any electronic device or process within reach, from laboratory equipment to nuclear warning systems to satellites to the global economy, and so on" (Torres 2017).
9. Good 1965.
10. Yudkowsky 2008.
11. Bostrom 2014.
12. Yudkowsky 2008.

Bibliography

"(1) De Linné à Jussieu: Méthodes de La Classification et Idée de Série En Botanique et En Zoologie (1740–1790) (2) Cuvier et Lamarck: Les Classes Zoologiques et l'idée de Série Animale (1790–1830)." *Nature* 121, no. 3038 (January 1, 1928): 85–86. https://doi.org/10.1038/121085a0.

2AI. *Killer AI?* 2AI, 2016.

80H. "Hot Topic: Banker vs. Aid Worker." *80,000 Hours*, 2011. https://web.archive.org/web/20111126093027/https:/80000hours.org/.

Abrams, Daniel M., Haley A. Yaple, and Richard J. Wiener. "Dynamics of Social Group Competition: Modeling the Decline of Religious Affiliation." *Physical Review Letters* 107, no. 8 (2011): 088701.

Abrams, Daniel M., Haley A. Yaple, and Richard J. Wiener. "A Mathematical Model of Social Group Competition with Application to the Growth of Religious Non-Affiliation." *ArXiv Preprint ArXiv:1012.1375* (2010).

Acosta, Ana M. "Review of Jennifer Airey's Religion Around Mary Shelley." *Nineteenth-Century Gender Studies* 16, no. 2 (Summer 2020). www.proquest.com/scholarly-journals/review-jennifer-aireys-religion-around-mary/docview/2586377122/se-2?accountid=14486.

Acton, Harry B., and John W.N. Watkins. "Symposium: Negative Utilitarianism." *Proceedings of the Aristotelian Society, Supplementary Volumes* 37 (1963): 83–114.

Adams, Fred C. "Long-Term Astrophysical." *Global Catastrophic Risks* (2008): 33.

Adams, Fred C., and Gregory Laughlin. "A Dying Universe: The Long-Term Fate and Evolution of Astrophysical Objects." *Reviews of Modern Physics* 69, no. 2 (1997): 337.

Adams, Robert Merrihew. "Common Projects and Moral Virtue." *Midwest Studies in Philosophy* 13 (1988): 297–307.

———. "Existence, Self-Interest, and the Problem of Evil." *Noûs* 13, no. 1 (1979): 53–65.

———. "Should Ethics Be More Impersonal? A Critical Notice of Reasons and Persons." *The Philosophical Review* 98, no. 4 (1989): 439–84.

Airey, J.L. *Religion Around Mary Shelley*. Religion Around. Pennsylvania State University Press, 2019. https://books.google.de/books?id=wCP-wwEACAAJ.

Aitkenhead, Decca. "James Lovelock: 'Enjoy Life While You Can: In 20 Years Global Warming Will Hit the Fan'." *The Guardian*, March 1, 2008. www.theguardian.com/theguardian/2008/mar/01/scienceofclimatechange.climatechange.

Alexander, G. *Academic Films for the Classroom: A History*. McFarland, Incorporated, Publishers, 2010. https://books.google.de/books?id=79VkswEACAAJ.

Alkon, Paul K. *The Secularization of Apocalypse: Le Dernier Homme*. UGA Press, 1987.

Allan, Pierre. *Measuring International Ethics: A Moral Scale of War, Peace, Justice, and Global Care*. Oxford University Press. http://www.afsp.msh-paris.fr/archives/congreslyon2005/communications/tr5/allan1.pdf.

Almond, Rosamund E.A., Monique Grooten, and T. Peterson. *Living Planet Report 2020-Bending the Curve of Biodiversity Loss*. World Wildlife Fund, 2020.

Althaus, David, and Lukas Gloor. *Reducing Risks of Astronomical Suffering: A Neglected Priority*. Center on Long-Term Risk, 2019. https://longtermrisk.org/reducing-risks-of-astronomical-suffering-a-neglected-priority/.

Alvarez, W., and C. Zimmer. *T. Rex and the Crater of Doom*. Popular Science: Paleontology. Princeton University Press, 1997. https://books.google.de/books?id=BsC1wAEACAAJ.

AMNH. "National Survey Reveals Biodiversity Crisis: Scientific Experts Believe We Are in Midst of Fastest Mass Extinction in Earth's History." *American Museum of Natural History*, April 20, 1998. www.mysterium.com/amnh.html.

Anders, Gunther. "Apocalypse Without Kingdom." Translated by Hunter Bolin. *e-flux* 97 (1959/2019).

———. "Commandments in the Atomic Age." *Burning Conscience* (1961): 11–12.

———. *Die Antiquiertheit Des Menschen [The Outdatedness of Human Beings]*. J. Beck, 1956a.

———. *Endzeit und Zeitenende: Gedanken über die atomare Situation*. Beck, 1972.

———. *Hiroshima Ist Überall*. C.H. Beck, 1958.

———. *Hiroshima Ist Überall*. Vol. 1112. C.H. Beck, 1995.

———. "Nach 'Holocaust' 1979." In *Besuch Im Hades*. C.H. Beck, 1979.

———. *The Obsolescence of Man, Volume II: On the Destruction of Life in the Epoch of the Third Industrial Revolution*, 1980. https://files.libcom.org/files/ObsolescenceofManVol%20IIGunther%20Anders.pdf.

———. "Reflections on the H-Bomb." *Dissent* 3, no. 2 (1956b): 146–55.

———. "Theses for the Atomic Age." *The Massachusetts Review* 3, no. 3 (1960/1962): 493–505.

Anonymous. *Elegant Extracts in Prose: Selected for the Improvement of Young Persons*. Rivington, 1816. https://books.google.de/books?id=JWRMAAAAcAAJ.

———. "The Last Man." *Blackwood's Edinburgh Magazine*, no. 19 (March 1826): 284–6.

Arendt, H. *The Human Condition*. University of Chicago Press, 1958.

Arendt, H., D. Allen, and M. Canovan. *The Human Condition*. University of Chicago Press, 2019. https://books.google.de/books?id=bGlwDwAAQBAJ.

Aristotle, R.C. Bartlett, and S.D. Collins. *Aristotle's Nicomachean Ethics*. University of Chicago Press, 2012. https://books.google.de/books?id=3JuePlN_03cC.

Armstrong, Stuart, Anders Sandberg, and Nick Bostrom. "Thinking inside the box: Controlling and using an oracle AI." *Minds and Machines* 22 (2012): 299-324.

Arnold, Denis G. *The Ethics of Global Climate Change*. Cambridge University Press, 2011.

Arnold, James R. "The Hydrogen-Cobalt Bomb." *Bulletin of the Atomic Scientists* 6, no. 10 (1950): 290–92.

Arrhenius, Gustaf. "Future Generations: A Challenge for Moral Theory." PhD diss., Uppsala University, 2000.

Ashford, Elizabeth, and Tim Mulgan. "Contractualism." *Stanford Encyclopedia of Philosophy*, 2018. https://plato.stanford.edu/entries/contractualism/.

Asimov, I. *A Choice of Catastrophes*. Hutchinson, 1979. https://books.google.de/books?id=lpaAAAAAMAAJ.

Avin, Shahar, Bonnie C. Wintle, Julius Weitzdörfer, Seán S.Ó. Héigeartaigh, William J. Sutherland, and Martin J. Rees. "Classifying Global Catastrophic Risks." *Futures* 102 (2018): 20–26.

AWG. "Results of Binding Vote by AWG." *Anthropocene Working Group*, May 21, 2019. http://quaternary.stratigraphy.org/working-groups/anthropocene/.

Baatz, Christian. "Climate Change and Individual Duties to Reduce GHG Emissions." *Ethics, Policy & Environment* 17, no. 1 (2014): 1–19.

Babich, Babette. *Günther Anders' Philosophy of Technology: From Phenomenology to Critical Theory*. Bloomsbury Publishing, 2021. https://books.google.de/books?id=EzZDEAAAQBAJ.

Badash, L. *A Nuclear Winter's Tale: Science and Politics in the 1980s*. MIT Press, 2009. https://books.google.de/books?id=y8M5vx-Lrk0C.

Balfour, Arthur James. "Naturalism and Ethics." *The International Journal of Ethics* 4, no. 4 (1894): 415–29.

Balfour, Dylan. "Longtermism: How Much Should We Care about the Far Future?" *1000 wordphilosophy.com*, 2021. https://1000wordphilosophy.com/2021/09/17/longtermism/.

Bar-Hillel, Yehoshua. "Discussion." In *Synthese*, edited by Donald Davidson, J. Aakko Hintikka, Gabriel Nuchelmans, and Wesley Salmon. Springer, 1968.

Barnes, Johnathan. *The Presocratic Philosophers*. Routledge, 1982.

Barnett, Richard. "Education or Degeneration: E. Ray Lankester, HG Wells and The Outline of History." *Studies in History and Philosophy of Science Part C: Studies in History and Philosophy of Biological and Biomedical Sciences* 37, no. 2 (2006): 203–29.

Barnosky, Anthony D., Elizabeth A. Hadly, Jordi Bascompte, Eric L. Berlow, James H. Brown, Mikael Fortelius, Wayne M. Getz, John Harte, Alan Hastings, and Pablo A. Marquet. "Approaching a State Shift in Earth's Biosphere." *Nature* 486, no. 7401 (2012): 52–58.

Barrow, J.D., F.J. Tipler, and J.A. Wheeler. *The Anthropic Cosmological Principle*. Oxford Paperbacks. Oxford University Press, 1986. https://books.google.de/books?id=Agvg1qD7lUkC.

Baskin, J. *Geoengineering, the Anthropocene and the End of Nature*. Springer International Publishing, 2019. https://books.google.de/books?id=_QeZDwAAQBAJ.

Baum, Seth. "A Survey of Artificial General Intelligence Projects for Ethics, Risk, and Policy." *Global Catastrophic Risk Institute Working Paper*, 17–1, 2017. https://papers.ssrn.com/sol3/papers.cfm?abstract_id=3070741.

BBC2. "Supervolcanoes." *BBC2*, 2000. www.billstclair.com/www.cheniere.org/misc/BBC%20-%20Horizon%20-%20Supervolcanoes%20-%20script.htm.

Beane, Silas R., Zohreh Davoudi, and Martin J. Savage. "Constraints on the Universe as a Numerical Simulation." *The European Physical Journal A* 50, no. 9 (2014): 1–9.

Beard, S.J. "Parfit Bio." *Personal Website*. 2019. https://sjbeard.weebly.com/parfit-bio.html.

Beard, Simon, and Patrick Kaczmarek. "On the Wrongness of Human Extinction." *Argumenta* 5, no. 1 (2020). https://doi.org/10.14275/2465-2334/20199.bea.

Beardsley, Monroe C. "Intrinsic Value." *Philosophy and Phenomenological Research* 26, no. 1 (1965): 1–17.

Beck, Ulrich. *Risk Society: Towards a New Modernity*. 1. Utg. Sage Publishing, 1986.

Beckstead, Nicholas. *On the Overwhelming Importance of Shaping the Far Future*. Rutgers the State University of New Jersey-New Brunswick, 2013a.

———. "A Proposed Adjustment to the Astronomical Waste Argument." *LessWrong Post*, 2013b.

———. "A Brief Argument for the Overwhelming Importance of Shaping the Far Future." In *Effective Altruism: Philosophical Issues*, edited by Hilary Greaves, and Theron Pummer, 80–98. Oxford University Press, 2019.

Beckstead, Nick, Peter Singer, and Matt Wage. "Preventing Human Extinction." *Effective Altruism*, 2013. https://forum.effectivealtruism.org/posts/tXoE6wrEQv7GoDivb/preventing-human-extinction.

Beckwith, Burnham Putnam. *The Next 500 Years: Scientific Predictions of Major Social Trends*. Exposition Press, 1967.

Beech, Martin. *Introducing the Stars: Formation, Structure and Evolution*. Springer International Publishing, 2019. https://books.google.de/books?id=uGqPDwAAQBAJ.

———. *The Physics of Invisibility: A Story of Light and Deception*. Springer Science & Business Media, 2011.

Beiser, F.C. *Weltschmerz: Pessimism in German Philosophy, 1860–1900*. Oxford University Press, 2016. https://books.google.de/books?id=drrmCwAAQBAJ.

Bell, Wendell. "Why Should We Care about Future Generations." In *Why Future Generations Now*, 40–62. Institute for the Integrated Study of Future Generations, 1993.

Benatar, D. *Better Never to Have Been: The Harm of Coming into Existence*. Oxford University Press, 2006. https://books.google.de/books?id=wwZREAAAQBAJ.

———. "No Life Is Good." *The Philosophers' Magazine*, no. 53 (2011): 62–66.

———. "Still Better Never to Have Been: A Reply to (More of) My Critics." *The Journal of Ethics* 17, no. 1 (2013): 121–51.

———. "Why It Is Better Never to Come into Existence." *American Philosophical Quarterly* 34, no. 3 (1997): 345–55.

Benatar, D., and D. Wasserman. *Debating Procreation: Is It Wrong to Reproduce?* Debating Ethics. Oxford University Press, 2015. https://books.google.de/books?id=LE3CBwAAQBAJ.

Benedick, Richard Elliot. "Montreal Protocol on Substances That Deplete the Ozone Layer." *International Negotiation* 1, no. 2 (1996/1989): 231–46.

Bennett, Jonathan. "On Maximizing Happiness." *Obligations to Future Generations* 61 (1978): 73.

Bentham, Jeremy. *An Introduction to the Principles of Morals and Legislation*. T. Payne and Son, 1789.

Bernal, J.D. *The World: The Flesh and the Devil; an Enquiry into the Future of the Three Enemies of the Rational Soul*. K. Paul, Trench, Trubner & Company, Limited, 1929. https://books.google.de/books?id=nk_wywEACAAJ.

Best, Shivali. "Tesla's Elon Musk Warns We Only Have 'a 5 to 10% Chance' of Preventing Killers Robots from Destroying Humanity." *The Daily Mail*, November 23, 2017. www.dailymail.co.uk/sciencetech/article-5110787/Elon-Musk-says-10-chance-making-AI-safe.html.

Bethe, Hans, Harrison Brown, Frederick Seitz, and Leo Szilard. "The Facts about the Hydrogen Bomb." *Bulletin of the Atomic Scientists* 6, no. 4 (1950): 106–9.

Bhandari, Rishabh, and Adrian Rodrigues. "Jonathan Schell, Thinker on Peace, Dies." *Yale News*, March 27, 2014. https://yaledailynews.com/blog/2014/03/27/jonathan-schell-thinker-on-peace-dies/.

Blake, Judith. "Population Policy for Americans: Is the Government Being Misled? Population Limitation by Means of Federally Aided Birth-Control Programs for the Poor Is Questioned." *Science* 164, no. 3879 (1969): 522–29.

Boecking, Otto, and E. Genersch. "Varroosis–The Ongoing Crisis in Bee Keeping." *Journal für Verbraucherschutz und Lebensmittelsicherheit* 3 (2008): 221–28.

Böhm, Monika, Ben Collen, Jonathan E.M. Baillie, Philip Bowles, Janice Chanson, Neil Cox, Geoffrey Hammerson, Michael Hoffmann, Suzanne R. Livingstone, and Mala Ram. "The Conservation Status of the World's Reptiles." *Biological Conservation* 157 (2013): 372–85.

Bolton, Henry. "A New Source of Heat: Radium." *Popular Science Monthly* 63 (May 1903). https://en.wikisource.org/wiki/Popular_Science_Monthly/Volume_63/May_1903/A_New_Source_of_Heat:_Radium.

Bostrom, Nick. *Anthropic Bias: Observation Selection Effects in Science and Philosophy*. Routledge, 2002a.

———. "Are We Living in a Computer Simulation?" *The Philosophical Quarterly* 53, no. 211 (2003a): 243–55.

———. "Astronomical Waste: The Opportunity Cost of Delayed Technological Development." *Utilitas* 15, no. 3 (2003b): 308–14.

———. "The Doomsday Argument Is Alive and Kicking." *Mind* 108, no. 431 (1999): 539–51.

———. "Existential Risk Prevention as Global Priority." *Global Policy* 4, no. 1 (2013): 15–31.

———. "Existential Risks: Analyzing Human Extinction Scenarios and Related Hazards." *Journal of Evolution and Technology* 9 (2002b).

———. "The Future of Human Evolution." In *Death and Anti-Death: Two Hundred Years After Kant, Fifty Years After Turing,* edited by Nick Bostrom, R. C. W. Ettinger, and Charles Tandy, 339–71. Ria University Press, 2004.

———. "The Future of Humanity." In *New Waves in Philosophy of Technology*, 186–215. Springer, 2009.

———. "A History of Transhumanist Thought." *Journal of Evolution and Technology* 14, no. 1 (2005a).

———. "Letter from Utopia." *Studies in Ethics, Law, and Technology* 2, no. 1 (2008/2020).

———. "Letter from Utopia." *Website,* 2005c. https://web.archive.org/web/20051124090502/www.nickbostrom.com/utopia.html.

———. "Pascal's Mugging." *Analysis* 69, no. 3 (2009): 443–45.

———. "Personal Website." 2000. https://web.archive.org/web/20020213221116/www.transhumanism.org/resources/faq.html#superintelligence.

———. "Personal Website." 2014. https://web.archive.org/web/20140221092418/https:/nickbostrom.com/.

———. "Personal Website." 2018. https://web.archive.org/web/20180708012512/https:/nickbostrom.com/.

———. "A Philosophical Quest for Our Biggest Problems." 2005b. www.Ted.Com/Talks/Nick_bostrom_on_our_biggest_problems.html (Accessed 25 February 2012).

———. "The Simulation Argument FAQ." 2011. https://www.simulation-argument.com/faq.

———. *Superintelligence: Paths, Dangers, Strategies*. Oxford University Press, 2014. https://books.google.de/books?id=7_H8AwAAQBAJ.

———. "The Superintelligent Will: Motivation and Instrumental Rationality in Advanced Artificial Agents." *Minds and Machines* 22, no. 2 (2012): 71–85.

———. "Transhumanist Values." In *Ethical Issues for the 21st Century*, edited by F. Adams. Philosophical Documentation Center Press, Charlottesville, 2003a.

———. "The Vulnerable World Hypothesis." *Global Policy* 10, no. 4 (2019): 455–76.

———. "What Is Transhumanism." *Nick Bostrom*, 2001/1998. https://nickbostrom.com/old/transhumanism.

———. "Why I Want to Be a Posthuman When I Grow Up." In *Medical Enhancement and Posthumanity*, 107–36. Springer, 2008.

Bostrom, N., and D.F.H.I.N. Bostrom. *Anthropic Bias: Observation Selection Effects in Science and Philosophy*. Studies in Philosophy: Outstanding Dissertations. Routledge, 2002a. https://books.google.de/books?id=TZ5FLwnCTMAC.

Bostrom, Nick, and Milan M. Ćirković. *Global Catastrophic Risks*. Oxford University Press, 2008.

Bostrom, Nick, and Toby Ord. "The Reversal Test: Eliminating Status Quo Bias in Applied Ethics." *Ethics* 116, no. 4 (2006): 656–79.

Bostrom, Nick, and Various other authors. "The Transhumanist FAQ." 1999. https://web.archive.org/web/20020213221116/www.transhumanism.org/resources/faq.html#superintelligence.

Bourget, David, and David Chalmers. "Philosophers on Philosophy: The 2020 Philpapers Survey." Unpublished Manuscript, 2021. https://philpapers.org/rec/BOUPOP-3.

Bowler, P.J. *Evolution: The History of an Idea*. University of California Press, 2003. https://books.google.de/books?id=gJXmS49Q7r0C.

———. *Progress Unchained: Ideas of Evolution, Human History and the Future*. Cambridge University Press, 2021.

Boyer, Paul. "American Intellectuals and Nuclear Weapons." *Revue Française d'études Américaines* (1986): 291–307.

———. *By the Bomb's Early Light: American Thought and Culture at the Dawn of the Atomic Age*. University of North Carolina Press, 1994. https://books.google.de/books?id=hsEBAwAAQBAJ.

Bradley, Ben. "Two Concepts of Intrinsic Value." *Ethical Theory and Moral Practice* 9, no. 2 (2006): 111–30.

Brake, Mark. *Revolution in Science: How Galileo and Darwin Changed Our World*. Springer, 2016.

Brand, Stewart. "The Clock and Library Projects." 2010. http://longnow.org/about.

———. "The Cite Site." n.d. https://thecitesite.com/authors/stewart-brand/#google_vignette.

Bremmer, J.N. *The Rise and Fall of the Afterlife*. Taylor & Francis, 2002. https://books.google.de/books?id=kfGEAgAAQBAJ.

Brin, Glen David. "The Great Silence—The Controversy Concerning Extraterrestrial Intelligent Life." *Quarterly Journal of the Royal Astronomical Society* 24 (1983): 283–309.

Bronson, Rachel. "Welcome to the Discussion, Professor Pinker." *Bulletin of the Atomic Scientists*, April 11, 2018. https://thebulletin.org/2018/04/welcome-to-the-discussion-professor-pinker.

Brooke, John Hedley. "Charles Darwin on Religion." *Perspectives on Science & Christian Faith* 61, no. 2 (2009).

Broome, J. *Climate Matters: Ethics in a Warming World (Norton Global Ethics Series)*. Norton Global Ethics Series. W. W. Norton, 2012. https://books.google.de/books?id=RjrYYEk8GYQC.

———. "Should We Value Population?*" *The Journal of Political Philosophy* 13, no. 4 (2005): 399–413.

Brower, David. "Introduction." In *The Population Bomb*. Buccaneer Books, 1968.

Browne, Malcolm. "The Debate over Dinosaur Extinctions Takes an Unusually Rancorous Turn." *New York Times*, January 19, 1988.

———. "Dinosaur Experts Resist Meteor Extinction Idea." *New York Times*, October 29, 1985.

Brundage, Miles, Shahar Avin, Jack Clark, Helen Toner, Peter Eckersley, Ben Garfinkel, Allan Dafoe, Paul Scharre, Thomas Zeitzoff, and Bobby Filar. "The Malicious Use of Artificial Intelligence: Forecasting, Prevention, and Mitigation." *ArXiv Preprint ArXiv:1802.07228*, 2018.

Brysse, Keynyn, Naomi Oreskes, Jessica O'Reilly, and Michael Oppenheimer. "Climate Change Prediction: Erring on the Side of Least Drama?" *Global Environmental Change* 23, no. 1 (2013): 327–37.

Buber, M. *Paths in Utopia*. Translated by R.F.C. Hull. Macmillan, 1949. https://books.google.de/books?id=4a13nAEACAAJ.

———. *Paths in Utopia*. Martin Buber Library. Syracuse University Press, 1996. https://books.google.de/books?id=MXGSnCcRaUwC.

Buck, Pearl. "The Bomb—The End of the World." *The American Weekly* 9, no. 8 (1959).

Buckley, M.J. *At the Origins of Modern Atheism*. Yale University Press, 1990. https://books.google.de/books?id=vg7PJQAACAAJ.

Buffon, G.L.L. de, G.L.C.A. Bexon, P.G. de Montbeillard, and L. Cépède. *Histoire Naturelle, Générale et Particulière: Histoire Naturelle Générale et Particulière. 1749–1767*. Histoire Naturelle, Générale et Particulière: Avec La Description Du Cabinet Du Roi. Impr. Royale, 1749. https://books.google.de/books?id=wM5CAQAAMAAJ.

Bulletin. "2007 Clock Statement." *Bulletin of the Atomic Scientists*, 2007. https://thebulletin.org/sites/default/files/2007%20Clock%20Statement.pdf.

———. "The Atomic Scientists of Chicago." *Bulletin of the Atomic Scientists* 1, no. 1 (December 10, 1945). https://books.google.de/books?id=-wsAAAAAMBAJ&printsec=frontcover&dq=bulletin+of+the+atomic+scientists+1945+volume+1&hl=en&sa=X&ved=2ahUKEwjLxZzZhZDzAhVZRvEDHaGUDnUQuwV6BAgIEAY#v=onepage&q=%22the%20public%20to%20a%20full%20understanding%20of%20the%20scientific%2C%20technological%2C%20and%20social%20problems%20arising%20from%20the%20release%20of%20nuclear%20energy%22&f=false.

———. "Bulletin FAQ." 2021. https://thebulletin.org/doomsday-clock/.

Bulletin FAQ. *Bulletin of the Atomic Scientists*, 2023. https://thebulletin.org/doomsday-clock/faq/.

Burchett, Wilfred. "The Atomic Plague." *Daily Express* 5 (1945): 34–36.

Burchfield, J.D. *Lord Kelvin and the Age of the Earth*. University of Chicago Press, 1990. https://books.google.de/books?id=s4AWPFdyrWIC.

———. "The Triumph of Limited Time." In *Lord Kelvin and the Age of the Earth*, 90–120. Springer, 1975.

Burd, Gene. "The Time Machine: An Invention by HG Wells (1895)." *Utopian Studies* 12, no. 2 (2001): 371–73.

Burge, Ryan. "How America's Youth Lost Its Religion in the 1990s." *National Catholic Reporter*, April 19, 2022. www.ncronline.org/news/opinion/how-americas-youth-lost-its-religion-1990s.

Burke, Edmund. *Reflections on the Revolution in France, and on the Proceedings in Certain Societies in London Relative to That Event*. James Dodsley, 1790.

Burns, William, and Andrew Strauss, eds. *Climate Change Geoengineering: Legal, Political and Philosophical Perspectives*. Cambridge University Press, 2013.

Butler, Samuel. "Darwin among the Machines." *June* 13, no. 1863 (1863): 205.

Callan, Curtis G., and Sidney Coleman. "Fate of the False Vacuum. II. First Quantum Corrections." *Physical Review D* 16, no. 6 (September 15, 1977): 1762–68. https://doi.org/10.1103/PhysRevD.16.1762.

Callendar, G.S. "The Artificial Production of Carbon Dioxide and Its Influence on Temperature." In *The Warming Papers: The Scientific Foundation for the Climate Change Forecast*, 261. Wiley, 2011.

Callicott, J. Baird. "Animal Liberation: A Triangular Affair." *Environmental Ethics* 2, no. 4 (1980): 311–38.

Campbell, Scott. "There's Nothing Anarchist About Eco-Fascism." *The Anarchist Library*, 2017. https://theanarchistlibrary.org/library/scott-campbell-there-s-nothing-anarchist-about-eco-fascism?v=1624030067.

Camus, A., and J. O'Brien. *The Myth of Sisyphus*. Penguin Modern Classics. Penguin Books Limited, 2013. https://books.google.de/books?id=zaPoAQAAQBAJ.

Cantor, Lee. "Thales—The 'First Philosopher'? A Troubled Chapter in the Historiography of Philosophy." *British Journal for the History of Philosophy* 30, no. 5 (2022): 727–50.

Capek, M. *The Concepts of Space and Time: Their Structure and Their Development*. Boston Studies in the Philosophy and History of Science. Springer Netherlands, 2014. https://books.google.de/books?id=OtnuCAAAQBAJ.

Čapek, Milič. *The Concepts of Space and Time: Their Structure and Their Development*. Springer, 1976.

Carnot, S. *Reflections on the Motive Power of Fire: And Other Papers on the Second Law of Thermodynamics*. Dover Books on Physics. Dover Publications, 2012. https://books.google.de/books?id=YdpQAQAAQBAJ.

Carrington, Damian. "Paul Ehrlich: 'Collapse of Civilisation Is a Near Certainty Within Decades'." *The Guardian*, March 22, 2018.

———. "What Is Biodiversity and Why Does It Matter to Us." *The Guardian*, March 12, 2018.

Carson, Rachel. "Silent Spring." *Houghton Mifflin*, September 27, 2009/1962.

Carter, Brandon. "Large Number Coincidences and the Anthropic Principle in Cosmology." In *Confrontation of Cosmological Theories with Observational Data*, 291–98. Springer, 1974.

Carus, T.L., A.E. Stallings, and R. Jenkyns. *The Nature of Things*. Penguin Classics. Penguin Publishing Group, 2007. https://books.google.de/books?id=84a0CvmAsKYC.

CEA. "Opening Keynote: Toby Ord & Will MacAskill, EA Global: San Francisco 2016." 2016. https://youtu.be/VH2LhSod1M4.

Ceci, Giovanni Mario. "A 'Historical Turn' in Terrorism Studies?" *Journal of Contemporary History* 51, no. 4 (2016): 888–96.

CERN. "Dark Matter." *CERN*, 2020. https://home.cern/science/physics/dark-matter#:~:text=Dark%20matter%20seems%20to%20outweigh,But%20what%20is%20dark%20matter%3F.

Chakrabarty, Dipesh. *Provincializing Europe*. Princeton University Press, 2000.

Chalmers, David J. *The Conscious Mind: In Search of a Fundamental Theory*. Oxford Paperbacks, 1996.

———. "The Singularity: A Philosophical Analysis (2010)." 2010. http://Consc.Net/Papers/Singularity.Pdf.

———. "The Singularity: A Philosophical Analysis." In *Science Fiction and Philosophy: From Time Travel to Superintelligence,* edited by Susan Schneider, 171–224. Wiley–Blackwell, 2016.

Chamberlin, J. Edward, and Sander L. Gilman. *Degeneration: The Dark Side of Progress.* Columbia University Press, 1985.

Chang, Jung, and Jon Halliday. *Mao: The Unknown Story.* Anchor, 2011.

Chidester, D. *Christianity: A Global History.* Penguin Books Limited, 2001. https://books.google.de/books?id=ttS-LUd7jRAC.

Choi, Anna L., Guifan Sun, Ying Zhang, and Philippe Grandjean. "Developmental Fluoride Neurotoxicity: A Systematic Review and Meta-Analysis." *Environmental Health Perspectives* 120, no. 10 (2012): 1362–68.

Chomanski, Bartlomiej Bartek. "Anti-Natalism and the Creation of Artificial Minds." *Journal of Applied Philosophy* 38, no. 5 (2021): 870–85.

Chomsky, Noam. "The Rationality of Collective Suicide." *Canadian Journal of Philosophy Supplementary Volume* 12 (1986): 23–39.

Chomsky, N., C. Derber, S. Moodliar, and P. Shannon. *Internationalism or Extinction.* Universalizing Resistance. Taylor & Francis, 2020. https://books.google.de/books?id=o3zADwAAQBAJ.

Christidis, Theodoros. "Ecpyrosis and Cosmos in Heraclitus." *Lyceum Journal, Philosophy Department of Saint Anselm College* 11, no. 1 (2009).

Christine, Korsgaard, and Christine Korsgaard. "Kant's Formula of Universal Law." *Pacific Philosophical Quarterly* 66 (1985): 24–47.

Churchill, Sir Winston. "Shall We Commit Suicide?" 1924. https://winstonchurchill.org/publications/finest-hour/finest-hour-094/shall-we-all-commit-suicide/.

Cioran, E.M. *The Trouble With Being Born.* Translated by R. Howard. Arcade Publishing, 1973.

Ćirković, Milan M. "Cosmological Forecast and Its Practical Significance." *Journal of Evolution and Technology* 12, no. 1 (2002b).

———. "Forecast for the Next Eon: Applied Cosmology and the Long-Term Fate of Intelligent Beings." *arXiv* (2002a). https://arxiv.org/abs/astro-ph/0211414.

———. "Forecast for the Next Eon: Applied Cosmology and the Long-Term Fate of Intelligent Beings." *Foundations of Physics* 34, no. 2 (2004): 239–61.

———. "Observation Selection Effects and Global Catastrophic Risks." In *Global Catastrophic Risks,* 120–45. Oxford University Press, 2008.

———. "Resource Letter: Pes-1: Physical Eschatology." *American Journal of Physics* 71, no. 2 (2003): 122–33.

Cirkovic, Milan, and Marina Radujkov. "On the maximal quantity of processed information in the physical eschatological context." arXiv preprint astro-ph/0112543 (2001).

Ćirković, Milan M., and Nick Bostrom. "Cosmological Constant and the Final Anthropic Hypothesis." *Astrophysics and Space Science* 274, no. 4 (2000): 675–87.

Clark, Peter U., Jeremy D. Shakun, Shaun A. Marcott, Alan C. Mix, Michael Eby, Scott Kulp, Anders Levermann, Glenn A. Milne, Patrik L. Pfister, and Benjamin D. Santer. "Consequences of Twenty-First-Century Policy for Multi-Millennial Climate and Sea-Level Change." *Nature Climate Change* 6, no. 4 (2016): 360–69.

Clark, S.R.L. *The Moral Status of Animals.* Oxford Palaeographical Handbooks. Oxford University Press, 1984. https://books.google.de/books?id=TQrXAAAAMAAJ.

Clarke, I.F. "The Pattern of Prediction: Forecasting: Facts and Fallibilities." *Futures* 3, no. 3 (1971): 302–5.

Clausius, Rudolf. "On Different Forms of the Fundamental Equations of the Mechanical Theory of Heat and Their Convenience for Application." *Annalen Der Physik Und Chemie* 124 (1865): 353–99.

———. "Über Den Zweiten Hauptsatz Der Mechanischen Wärmetheorie: Ein Vortrag, Gehalten in Einer Allgemeinen Sitzung Der 41." *Versammlung Deutscher Naturforscher Und Aerzte Zu Frankfurt a. M. Am 23. September 1867*. Vol. 3. Vieweg, 1867.

"Climate Change in the American Mind May 2017," *The Refutation of All Heresies (Complete)*. Library of Alexandria. Library of Alexandria, n.d. https://books.google.de/books?id=MX3If3ZRi14C.

Clynes, Manfred E., and Nathan S. Kline. *Cyborgs and Space L1. 2*. Astronautics, 1960. https://web.mit.edu/digitalapollo/Documents/Chapter1/cyborgs.pdf.

Coates, Ken. *Anti-Natalism: Rejectionist Philosophy from Buddhism to Benatar*. 1st ed. Design Pub., 2014.

Cohn, N. *The Pursuit of the Millennium*. Secker & Warburg, 1957. https://books.google.de/books?id=zinrDCwoBR0C.

Cohon, Rachel, "Hume's Moral Philosophy." In *The Stanford Encyclopedia of Philosophy*, edited by Edward N. Zalta, 2018. <https://plato.stanford.edu/archives/fall2018/entries/hume-moral/>.

Cole, Dandridge M., and Donald William Cox. *Islands in Space: The Challenge of the Planetoids*. Chilton Books, 1964.

Coleman, J.A. *The Dictionary of Mythology: An A-Z of Themes, Legends and Heroes*. Arcturus Publishing, 2007. https://books.google.de/books?id=FmVOyAEACAAJ.

Coleman, Sidney, and Frank De Luccia. "Gravitational Effects on and of Vacuum Decay." In *Euclidean Quantum Gravity*, 295–305. World Scientific, 1980.

Collison, Patrick, and Michael Nielsen. "Science Is Getting Less Bang for Its Buck." *The Atlantic*, 2018.

Condorcet, Antoine-Nicholas de. *Sketch for a Historical Picture of the Progress of the Human Mind. Translated from the French by June Barraclough (1955)*. Noonday Press, 1795.

Conway, John Horton. "Tomorrow Is the Day after Doomsday." *Eureka* 36 (1973): 28–31.

Cook, D. *Contemporary Muslim Apocalyptic Literature*. Religion and Politics. Syracuse University Press, 2008. https://books.google.de/books?id=5PjkU1gfTxIC.

Corvino, Fausto. "Samuel Scheffler, Why Worry About Future Generations?, (Oxford/New York: Oxford University Press), 2018 (Paperback Edition, 2020)." *Ethical Theory and Moral Practice* 24, no. 1 (March 1, 2021): 403–5. https://doi.org/10.1007/s10677-020-10152-6.

Cotton-Barratt, Owen, and Toby Ord. "Existential Risk and Existential Hope: Definitions." *Future of Humanity Institute: Technical Report* 1, no. 2015 (2015): 78.

Cousins, Norman. *Modern Man Is Obsolete*. The Viking Press, 1945.

Cowen, Tyler. "Caring about the Distant Future: Why It Matters and What It Means." *University of Chicago Law Review* 74 (2007): 5.

Coyne, L. *Hans Jonas: Life, Technology and the Horizons of Responsibility*. Bloomsbury Academic, 2020. https://books.google.de/books?id=CrfdzQEACAAJ.

Coyne, Lewis, and Michael Hauskeller, "Hans Jonas, Transhumanism, and What It Means to Live a Genuine Human Life." *Revue Philosophique de Louvain* 117, no. 2 (2019): 291–310.

CR. "About Us." *Club of Rome*, 2022. www.clubofrome.org/about-us/.

Cremer, Carla Zoe, and Luke Kemp. "Democratising Risk: In Search of a Methodology to Study Existential Risk." *ArXiv Preprint ArXiv:2201.11214* (2021).

Crisp, R. *Routledge Philosophy Guidebook to Mill on Utilitarianism*. Routledge Philosophy Guidebook to Mill on Utilitarianism. Routledge, 1997. https://books.google.de/books?id=DpV56X72594C.

———. "Would Extinction Be So Bad?" *The New Statesman*, August 10, 2021. www.newstatesman.com/ideas/agora/2021/08/would-extinction-be-so-bad.

Critchley, Simon. *Continental Philosophy: A Very Short Introduction*. Oxford University Press, 2001.

Crocker, L.G. *Diderot's Chaotic Order: Approach to Synthesis*. Princeton Legacy Library. Princeton University Press, 2015. https://books.google.de/books?id=ylx9BgAAQBAJ.

Cropper, William H. "Carnot's Function: Origins of the Thermodynamic Concept of Temperature." *American Journal of Physics* 55, no. 2 (February 1987): 120–29. https://doi.org/10.1119/1.15255.

Crouch, Will. "Deworming and Handwashing Can Offer Better Value than Immunisation." *FT.Com*, June 17, 2011. https://web.archive.org/web/20110721011207/www.ft.com/cms/s/0/b07be4c8-986f-11e0-94d7-00144feab49a.html#axzz1PWF9EuL5.

Crowe, M.J. *The Extraterrestrial Life Debate, 1750–1900*. Dover Publications, 2012. https://books.google.de/books?id=wUnCAgAAQBAJ.

Crowe, Michael J., and Matthew F. Dowd. "The Extraterrestrial Life Debate from Antiquity to 1900." In *Astrobiology, History, and Society*, 3–56. Springer, 2013.

Crowl, Adam, John Hunt, and Andreas M. Hein. "Embryo Space Colonization to Overcome the Interstellar Time/Distance Bottleneck." *Journal of the British Interplanetary Society* 65, no. 7 (2012): 283.

Crutzen, Paul J., and John W. Birks. "The Atmosphere after a Nuclear War: Twilight at Noon." In *Paul J. Crutzen: A Pioneer on Atmospheric Chemistry and Climate Change in the Anthropocene*, 125–52. Springer, 2016.

———. "Twilight at Noon: The Atmosphere after a Nuclear War." *Ambio* 11, no. 2–3 (1982): 114–25.

Crutzen, P.J., and E.F. Stoermer. "Global Change." *Newsletter* 41 (2000): 17–18.

Cudd, Ann, and Seena Eftekhari. "Contractarianism." *Stanford Encyclopedia of Philosophy*, 2021. https://plato.stanford.edu/cgi-bin/encyclopedia/archinfo.cgi?entry=contractarianism.

Curtis, G.T. *Creation Or Evolution?: A Philosophical Inquiry*. ATLA Monograph Preservation Program. D. Appleton, 1887. https://books.google.de/books?id=_hMH5R29JYUC.

Cusack, Sinéad. *Supervolcanoes*. BBC2. https://www.billstclair.com/www.cheniere.org/misc/BBC%20-%20Horizon%20-%20Supervolcanoes%20-%20script.htm.

Cuvier, G. "'Espèces Des Eléphans.' Translated in Rudwick M (1997) Georges Cuvier Fossil Bones and Geological Catastrophes, New Translations and Interpretations of the Primary Texts, 1796." *PNAS* 111, no. 52 (December 2014): 18405–6.

———. *Essay on the Theory of the Earth, 1813*. Taylor & Francis, 2018. https://books.google.de/books?id=l3l0DwAAQBAJ.

Cuvier, Georges, and Robert Jameson. *Essay on the Theory of the Earth; Translated From the French of M. Cuvier . . . by Robert Kerr . . .; With Mineralogical Notes and an Account of Cuvier's Geological Discoveries by Professor Jameson*. W. Blackwood, 1813.

Daley, Brian. "Eschatology in the Early Church Fathers." In The Oxford Handbook of Eschatology , edited by Jerry L. Walls. Oxford University Press, 2008.

Dalrymple, Theodore. "Contemplating Annihilation." *BMJ* 334, no. 7586 (2007): 211.

Darby, William J. "Silence, Miss Carson." *Chemical and Engineering News* 40, no. 1 (1962): 60–62.

Darwin, Charles. *Autobiographies*. Penguin, 2002.

———. *The Descent of Man, and Selection in Relation to Sex: In Two Volumes*. Murray, 1871. https://books.google.de/books?id=wyWKQPDR674C.

———. *The Origin of Species by Means of Natural Selection: Or the Preservation of Favored Races in the Struggle for Life*. Hurst, 1872. https://books.google.de/books?id=Crk4Tgz1ot4C.

———. *The Origin of Species by Means of Natural Selection, Or, The Preservation of Favoured Races in the Struggle for Life*. J. Murray, 1875. https://books.google.de/books?id=dZRVAAAAcAAJ.

Darwin, C., W. West, John Murray (Firm), William Clowes and Sons, and Bradbury & Evans. *On the Origin of Species by Means of Natural Selection, or the Preservation of Favoured Races in the Struggle for Life*. John Murray, Albemarle Street, 1859. https://books.google.de/books?id=jTZbAAAAQAAJ.

Davidson, Marc. "Why Worry about Future Generations?" *Environmental Values* 28, no. 2 (2019): 256–59.

Davis, Judith. *Effects of Population Growth on Natural Resources and the Environment: Hearings Before a Subcommittee of the Committee on Government Operations, House of Representatives, Ninety-First Congress, First Session. September 15 and 16, 1969*. Effects of Population Growth on Natural Resources and the Environment: Hearings Before a Subcommittee of the Committee on Government Operations, House of Representatives, Ninety-First Congress, First Session. September 15 and 16, 1969. U.S. Government Printing Office, 1969. https://books.google.de/books?id=plkkAAAAMAAJ.

Davis, Marc, Piet Hut, and Richard A. Muller. "Extinction of Species by Periodic Comet Showers." *Nature* 308, no. 5961 (1984a): 715–17.

———. *Extinction of Species by Periodic Comet Showers*. Pre-publication Draft, 1984b. https://escholarship.org/uc/item/9gm5c682.

Dawkins, R. *The Blind Watchmaker*. Norton, 1986. https://books.google.de/books?id=ZcWGSQAACAAJ.

———. *The God Delusion*. Bantam Press, 2006.

———. *The Greatest Show on Earth: The Evidence for Evolution*. Free Press, 2009.

———. "Progress." In *Keywords in Evolutionary Biology*, edited by Evelyn Fox Keller and Elisabeth Anne Lloyd. Harvard University Press, 1992.

Dawsey, Jason. "After Hiroshima: Günther Anders and the History of Anti-Nuclear Critique." In *Understanding the Imaginary War*, 140–64. Manchester University Press, 2016.

———. *The Limits of the Human in the Age of Technological Revolution: Günther Anders, Post-Marxism, and the Emergence of Technology Critique*. The University of Chicago, 2013.

DDH. "Terrifying Results of Hiroshima Blast Told." *Delphos Daily Herald*, August 8, 1945. www.newspapers.com/image/15213698/?fcfToken=eyJhbGciOiJIUzI1NiIsInR5cCI6IkpXVCJ9.eyJmcmVlLXZpZXctaWQiOjE1MjEzNjk4LCJpYXQiOjE2NjE2MzIzMzUsImV4cCI6MTY2MTcxODczNX0.CjfPr1fkpBi8jSVMnWMmjG7YYY2mp6eMy6JDi1aOeUQ.

Dean, Dennis R. "James Hutton on Religion and Geology: The Unpublished Preface to His Theory of the Earth (1788)." *Annals of Science* 32, no. 3 (1975): 187–93.

Decock, Paul B. "Origen: On Making Sense of the Resurrection as a Third Century Christian." *Neotestamentica* 45, no. 1 (2011): 76–91.

DeGroot, G. *The Bomb: A Life*. Random House, 2011. https://books.google.de/books?id=bc9hlNnxIq4C.

Delattre, Edwin. "Responsibilities and Future Persons." *Ethics* 82 (April 1972): 256.

Della Porta, Donatella. *Clandestine Political Violence*. Cambridge University Press, 2013.

Delord, Julien. "Can We Really Re-Create an Extinct Species by Cloning? A Metaphysical Analysis." *The Ethics of Animal Re-Creation and Modification: Reviving, Rewilding, Restoring* (2014): 22–39.

———. "The Nature of Extinction." *Studies in History and Philosophy of Science Part C: Studies in History and Philosophy of Biological and Biomedical Sciences* 38, no. 3 (2007): 656–67.

Delorme. *Atmospheric Carbon Dioxide (CO2) Concentrations from 1958 to 2021*, 2019. Wikipedia. https://en.wikipedia.org/wiki/Keeling_Curve#/media/File:Mauna_Loa_CO2_monthly_mean_concentration.svg.

d'Entreves, Maurizio Passerin. "Hannah Arendt." In *The Stanford Encyclopedia of Philosophy*, edited by Edward N. Zalta and Uri Nodelman, 2022. https://plato.stanford.edu/archives/fall2022/entries/arendt.

Deudney, D. *Dark Skies: Space Expansionism, Planetary Geopolitics, and the Ends of Humanity*. Oxford University Press, 2020. https://books.google.de/books?id=9LTRDwAAQBAJ.

———. "Going Critical: Toward a Modified Nuclear One Worldism." *Journal of International Political Theory* 15, no. 3 (2019): 367–85.

D'Hondt, Steven. "Theories of Terrestrial Mass Extinction by Extraterrestrial Objects." *Earth Sciences History* 17, no. 2 (1998): 157–73.

Dick, T. *The Sidereal Heavens and Other Subjects Connected with Astronomy*. Harper's Family Library. No. 99. Harper and Brothers, 1840. https://books.google.de/books?id=P3dTAAAAYAAJ.

Dick, S.J. *Plurality of Worlds: Origins of the Extraterrestrial Life Debate from Democritus to Kant*. Cambridge University Press, 1982. https://books.google.de/books?id=MbNyxgEACAAJ.

Dictionary, Oxford English. "omnicide, n." Oxford University Press, n.d. www.oed.com/view/Entry/246601.

Diderot, D., J. Stewart, and J. Kemp. *Diderot, Interpreter of Nature: Selected Writings*. Hyperion Press, 1979. https://books.google.de/books?id=ZQEQAQAAIAAJ.

Doherty, Thomas. *Cold War, Cool Medium: Television, McCarthyism, and American Culture*. Columbia University Press, 2005.

DOJ. "Individual Pleads Guilty to Participating in Internet-of-Things Cyberattack in 2016." *Department of Justice*, 2020. www.justice.gov/opa/pr/individual-pleads-guilty-participating-internet-things-cyberattack-2016.

Dorner, Isaak August. *History of the Development of the Doctrine of the Person of Christ*. T. & T. Clark, 1862.

Dörries, Matthias. "The 'Winter' Analogy Fallacy." *History of Meteorology* 4 (2008): 41–56.

Drexler, K. Eric. *Engines of Creation*. Anchor Books, 1986.

———. *Radical Abundance: How a Revolution in Nanotechnology Will Change Civilization*. PublicAffairs, 2013. https://books.google.de/books?id=eiE4DgAAQBAJ.

Driver, Julia. "The History of Utilitarianism." *Stanford Encyclopedia of Philosophy*, 2014. https://plato.stanford.edu/archives/win2014/entries/utilitarianism-history.

DS. *How To Do The Most Good: An Interview With Will MacAskill*. Daily Stoic, 2021.

Dufresne, Todd. "Simon Critchley, Continental Philosophy: A Very Short Introduction." *Philosophy in Review* 21 (2001).

Dye, LaVonne R. "The Marine Mammal Protection Act: Maintaining the Commitment to Marine Mammal Conservation." *Case Western Reserve Law Review* 43 (1992): 1411.

Dyson, Freeman J. "The Los Alamos Primer: The First Lectures on How to Build an Atomic Bomb." *Science* 256, no. 5055 (1992): 388–90.

———. "Time Without End: Physics and Biology in an Open Universe." *Reviews of Modern Physics* 51, no. 3 (1979): 447.

Earle, William. Review of *The Future of Mankind*, by Karl Jaspers and translated by E.B. Ashton. *Science* 133, no. 3460 (1961): 1236–38. https://doi.org/10.1126/science.133.3460.123.

Eaton, George. "Noam Chomsky:'We're Approaching the Most Dangerous Point in Human History.'" *New Statesman* (2022).

Ebeling, Gerhard. "The Message of God to the Age of Atheism." *Graduate School of Theology Bulletin* 9, no. 11 (1964).

Eddington, Arthur. *The Nature of the Physical World: The Gifford Lectures 1927*. Vol. 23. BoD—Books on Demand, 2019.

Edmond, Charlotte. *5 Amazing Animals We Thought Were Extinct*. World Economic Forum, 2017. https://www.weforum.org/agenda/2017/08/animals-we-thought-were-extinct/.

Edwards, Lin. "Humans Will Be Extinct in 100 Years Says Eminent Scientist." *Phys.Org*, June 23, 2010. https://phys.org/news/2010-06-humans-extinct-years-eminent-scientist.html.

Edwards, Matthew R. "Android Noahs and Embryo Arks: Ectogenesis in Global Catastrophe Survival and Space Colonization." *International Journal of Astrobiology* 20, no. 2 (2021): 150–8.

Edwards, P.N. *A Vast Machine: Computer Models, Climate Data, and the Politics of Global Warming*. Infrastructures. MIT Press, 2010. https://books.google.de/books?id=K9_LsJBCqWMC.

Eggleston, Ben. "Decision Theory." In *The Cambridge History of Moral Philosophy*, edited by Sacha Golub and Jens Timmermann. 706–17. Cambridge University Press, 2017.

Ehrlich, Paul R. *The Loss of Diversity*. National Academy Press, 1988.

Ehrlich, Paul R., and Anne H. Ehrlich. *The Population Bomb*. Ballantine Books, 1968. https://books.google.de/books?id=TGyDzgEACAAJ.

———. "The Population Bomb Revisited." *The Electronic Journal of Sustainable Development* 1, no. 3 (2009): 63–71.

Ehrman, Bart. "Was Resurrection a Zoroastrian Idea?" 2017. https://ehrmanblog.org/was-resurrection-a-zoroastrian-idea/.

Ehrman, D. *Heaven and Hell: A History of the Afterlife*. Simon & Schuster, 2021. https://books.google.de/books?id=uskfEAAAQBAJ.

———. *Misquoting Jesus: The Story Behind Who Changed the Bible and Why*. HarperCollins, 2009. https://books.google.de/books?id=xmJjSUiJtuQC.

Einstein, Albert. "Einstein's Letter to President Roosevelt—1939." 1939. www.atomicarchive.com (Accessed 30 April 2006) (et Delprojekt under National Science Digital Library).

———. *Ideas and Opinions*. Translated by Sonja Bargmann. New York: Bonanza Books, 1954.

———. *The Special and General Theory*. Prabhat Prakashan, 1948.

Einstein, Albert, and Leo Szilard. *Einstein-Szilard Letter*, 1939. https://en.wikipedia.org/wiki/Einstein%E2%80%93Szilard_letter#/media/File:Einstein-Roosevelt-letter.png.

Einstein, Albert, Harold C. Urey, Harrison Brown, T.R. Hogness, Joseph E. Mayer, Philip M. Morse, H.J. Muller, and Frederick Seitz. "A Policy for Survival: A Statement by the Emergency Committee of Atomic Scientists—April 12, 1948." *Bulletin of the Atomic Scientists* 4, no. 6 (1948): 176–88.

Eisenstein, Alex. "'The Time Machine' and the End of Man." *Science Fiction Studies* (1976): 161–65.

Ellis, John, and David N. Schramm. "Could a Nearby Supernova Explosion Have Caused a Mass Extinction?" *Proceedings of the National Academy of Sciences* 92, no. 1 (1995): 235–38.

Else, Jon H. "The Day After Trinity." 1980. www.youtube.com/watch?v=bTAjsB-yr-Y.

Encyclopaedia Britannica, Inc, and T. Pappas. *250th Anniversary Edition*. Encyclopaedia Britannica, 2018. https://books.google.de/books?id=zzJwDwAAQBAJ.

Engelhardt, Tom. "Suicide Watch on Planet Earth." *Le Monde diplomatique*, 2019. https://mondediplo.com/openpage/suicide-watch-on-planet-earth.

Ereshefsky, Marc, "Species," The Stanford Encyclopedia of Philosophy (Summer 2022 Edition), Edward N. Zalta (ed.), URL = <https://plato.stanford.edu/archives/sum2022/entries/species/>.

Estrada, Alejandro, Paul A. Garber, Anthony B. Rylands, Christian Roos, Eduardo Fernandez-Duque, Anthony Di Fiore, K. Anne-Isola Nekaris, Vincent Nijman, Eckhard W. Heymann, and Joanna E. Lambert. "Impending Extinction Crisis of the World's Primates: Why Primates Matter." *Science Advances* 3, no. 1 (2017): e1600946.

Eugene Merle Shoemaker. "Impact mechanics at Meteor crater, Arizona." *U.S. Atomic Energy Commission Open File Report*, 1959.

Event: The Future of World Religions, 2015. www.pewresearch.org/religion/2015/04/23/live-event-the-future-of-world-religions/.

Fahmy, Dalia. "Americans Are Far More Religious than Adults in Other Wealthy Nations." *Pew Research Center*. 2018. https://www.pewresearch.org/fact-tank/2018/07/31/americans-are-far-more-religious-than-adults-in-other-wealthy-nations/

Farrier, David. "Deep Time's Uncanny Future Is Full of Ghostly Human Traces." *Aeon*, October 31, 2016. https://aeon.co/ideas/deep-time-s-uncanny-future-is-full-of-ghostly-human-traces.

Fea, John. "Benjamin Franklin and His Religious Beliefs." *Pennsylvania Heritage*, 2011. http://paheritage.wpengine.com/article/benjamin-franklin-his-religious-beliefs/.

Fehige, Christoph. "A Pareto Principle for Possible People." In *Preferences,* edited by Christoph Fehige and Ulla Wessels, 508–43. Springer, 1998.

Feinberg, B., and R. Kasrils. *Bertrand Russell's America: His Transatlantic Travels and Writings. Volume Two 1945–1970*. Routledge Library Editions: Russell. Taylor & Francis, 2013. https://books.google.de/books?id=oWQe9nfmJmwC.

Feinberg, Joel. "The Rights of Animals and Unborn Generations." In *Environmental Rights,* 241–65. Routledge, 2017.

Feinberg, Joel. "Harm and Self-Interest," in *Law, Morality and Society: Essays in Honour of H. L. A. Hart,* ed. P. M. S. Hacker and J. Raz (Oxford: Clarendon Press, 1977), pp. 299–301.

———. "The Rights of Animals and Unborn Generations." In *Philosophy and Environmental Crisis,* edited by William Blackstone, 43–6, 49–50, 55, 57–63. University of Georgia Press, 1974.

———. *Rights, Justice, and the Bounds of Liberty: Essays in Social Philosophy*. Vol. 148. Princeton University Press, 2014.

Feinberg, Joel, and William T. Blackstone. "The Rights of Animals and Unborn Generations." *Ethica* (2013): 372.

Ferguson, James. *Astronomy Explained Upon Sir Isaac Newton's Principles and Made Easy to Those Who Have Not Studied*. 2nd ed. James Ferguson, 1757.

Ferris, Timothy. "Life Beyond Earth." *PBS*, 1999. www.pbs.org/lifebeyondearth/resources/intgottpop.html#:~:text=Our%20ancestor%2C%20Homo%20erectus%2C%20lasted,the%20Neanderthals%20lasted%20300%2C000%20years.

Feynman, Richard. *The Pleasure of Finding Things Out*. Perseus Books, 1999.

———. "There's Plenty of Room at the Bottom." 1959. www.ias.ac.in/public/Volumes/reso/016/09/0890-0905.pdf.

FHI. "About." *Future of Humanity Institute*, 2022. www.fhi.ox.ac.uk/about-fhi/.

———. "Home Page." *Future of Humanity Institute*, 2005. https://web.archive.org/web/20051013060521/www.fhi.ox.ac.uk/.

Fields, R.D. *Electric Brain: How the New Science of Brainwaves Reads Minds, Tells Us How We Learn, and Helps Us Change for the Better*. BenBella Books, 2020. https://books.google.de/books?id=Ju5KEAAAQBAJ.

Finkelman, Leonard. "De-Extinction and the Conception of Species." *Biology & Philosophy* 33, no. 5–6 (2018b): 32.

———. "Extinction, Resolved." *Extinct Blog*, 2018a. http://www.extinctblog.org/extinct/category/Leonard+Finkelman.

Finneron-Burns, Elizabeth. "Human Extinction and Moral Worthwhileness." *Utilitas* 34, no. 1 (2022): 105–12.

———. "What's Wrong with Human Extinction?" *Canadian Journal of Philosophy* 47, no. 2–3 (2017): 327–43.

Fisher, Adam. "Sam Bankman-Fried Has a Savior Complex: And Maybe You Should Too." *Sequoia*, September 22, 2022. https://archive.ph/xy4MR.

Fitzgerald, McKenna, Aaron Boddy, and Seth D. Baum. *2020 Survey of Artificial General Intelligence Projects for Ethics, Risk, and Policy*. Global Catastrophic Risk Institute, 2020.

Flammarion, C. *Omega: The Last Days of the World*. Cosmopolitan Publishing Company, 1894. https://books.google.de/books?id=I9QtAAAAMAAJ.

Flannery, F. *Understanding Apocalyptic Terrorism: Countering the Radical Mindset*. Cass Series on Political Violence. Routledge, 2016. https://books.google.de/books?id=1aInvgAACAAJ.

FLI. 2023. Pause Giant AI Experiments: An Open Letter. Future of Life Institute. https://futureoflife.org/open-letter/pause-giant-ai-experiments/.

Fontenelle, M. de, B.B. de Fontenelle, H.A. Hargreaves, and N.R. Gelbart. *Conversations on the Plurality of Worlds*. University of California Press, 1990. https://books.google.de/books?id=u6IwDwAAQBAJ.

Foot, Philippa. "The Problem of Abortion and the Doctrine of the Double Effect." *Oxford Review* 5 (1967).

Foreman, Dave. *Confessions of an Eco-Warrior*. Broadway Books, 1991.

———. *Ecodefense: A Field Guide to Monkeywrenching*. Earth First! Books, 1985. https://books.google.de/books?id=OsFKPgAACAAJ.

Foreman, Dave, and Bill Haywood. "Ecodefense: A Field Guide to Monkeywrenching, Chico." 1993. https://cernorudaprirucka.noblogs.org/files/2015/11/various-authors-ecodefense-a-field-guide-to-monkeywrenching.pdf.

Forge, John. "A Note on the Definition of 'Dual Use'." *Science and Engineering Ethics* 16, no. 1 (2010): 111–18.

Fotion, Nick, Nick Fotion, and J.C. Heller. *Contingent Future Persons: On the Ethics of Deciding Who Will Live, or Not, in the Future*. Springer Science & Business Media, 1997.

Fox, S.A. *Downwind: A People's History of the Nuclear West*. University of Nebraska Press, 2014. https://books.google.de/books?id=56RvBAAAQBAJ.

Frampton, P.H. "Vacuum Instability and Higgs Scalar Mass." *Physical Review Letters* 37, no. 21 (1976): 1378.

Francis, Matthew R. "When Carl Sagan Warned the World about Nuclear Winter." *Smithsonian Magazine* 15 (2017).

Franklin, B. *Poor Richard's Almanac for 1850–52*. Poor Richard's Almanac for 1850–52. J. Doggett jr., 1849. https://books.google.de/books?id=a9RJAAAAMAAJ.

Freud, Sigmund. *Civilization and Its Discontents*. Penguin Great Ideas. Penguin Books Limited, 2004. https://books.google.de/books?id=Wuhr78oOhpEC.

Freud, S., and SBP Editors. *Civilization and Its Discontents*. Samaira Book Publishers, 2018. https://books.google.de/books?id=JtKBDwAAQBAJ.

Frick, Johann. "Conditional Reasons and the Procreation Asymmetry." *Philosophical Perspectives* 34, no. 1 (2020): 53–87.

Frick, Johann. "'Making People Happy, Not Making Happy People': A Defense of the Asymmetry Intuition in Population Ethics." PhD Dissertation. Harvard University, 2014.

———. "On the Survival of Humanity." *Canadian Journal of Philosophy* 47, no. 2–3 (2017): 344–67.

Fried, Richard M. "One Nation Underground: The Fallout Shelter in American Culture." *The Journal of American History* 89, no. 2 (2002): 713.

Friesdorf, Rebecca, Paul Conway, and Bertram Gawronski. "Gender Differences in Responses to Moral Dilemmas: A Process Dissociation Analysis." *Personality and Social Psychology Bulletin* 41, no. 5 (2015): 696–713.

Frischknecht, Friedrich. "The History of Biological Warfare: Human Experimentation, Modern Nightmares and Lone Madmen in the Twentieth Century." *EMBO Reports* 4, no. S1 (2003): S47–52.

Frymer-Kensky, Tikva. "The Atrahasis Epic and Its Significance for Our Understanding of Genesis 1–9." *The Biblical Archaeologist* 40, no. 4 (1977): 147–55.

Gaia Liberation Front. "Statement of Purpose (A Modest Proposal)." *Church of Euthanasia*, 1994. www.churchofeuthanasia.org/resources/glf/glfsop.html.

Gallup, Jr., George. "Public Gives Organized Religion Its Lowest Rating." *Gallup*, 2003. https://news.gallup.com/poll/7534/public-gives-organized-religion-its-lowest-rating.aspx.

Gardiner, Stephen M. "Ethics and Global Climate Change." *Ethics* 114, no. 3 (April 2004): 555–600. https://doi.org/10.1086/382247.

———. *A Perfect Moral Storm: The Ethical Tragedy of Climate Change*. Environmental Ethics and Science Policy Series. Oxford University Press, 2011. https://books.google.de/books?id=A6yPX2y1RuAC.

Garreau, Joel. "From Internet Scientist, a Preview of Extinction." *The Washington Post*, March 12, 2000, p. 15.

Gebru, Timnit, Emily Bender, Angelina McMillan-Major, and Margaret Mitchell. 2023. Statement from the Listed Authors of Stochastic Parrots on the "AI Pause" Letter. DAIR. https://www.dair-institute.org/blog/letter-statement-March2023.

George, Andrew R. *The Babylonian Gilgamesh Epic: Introduction, Critical Edition and Cuneiform Texts.* Vol. 1. Oxford University Press, 2003.

———. *The Epic of Gilgamesh: The Babylonian Epic Poem and Other Texts in Akkadian and Sumerian; Translated and with an Introduction by Andrew George.* Penguin Classics. Allen Lane, 1999. https://books.google.de/books?id=ZdkXAQAAIAAJ.

George, Eaton. "Noam Chomsky: 'We're Approaching the Most Dangerous Point in Human History'." *New Statesman*, April 6, 2022.

Gimbel, Steven. "Albert Einstein: Scientist, Pacifist, Zionist." *Yale University Press*, March 17, 2015. https://yalebooks.yale.edu/2015/03/17/albert-einstein-scientist-pacifist-zionist/.

Glannon, Walter. "A World Without Us." *Journal of Medical Ethics Blog* (blog), December 6, 2021. https://blogs.bmj.com/medical-ethics/2021/12/06/a-world-without-us/.

Glanvill, Joseph. "Lux orientalis." In *Two Choice and Useful Treaties: The One Lux Orientalis, or An Enquiry into the Opinion of the Eastern Sages Concerning the Præexistence of Souls. Being a Key to Unlock the Grand Mysteries of Providence. In Relation to Man's Sin and Misery. The Other, a Discourse of Truth, by the Late Reverend Dr. Rust Lord Bishop of Dromore in Ireland. With Annotations on Them Both.* James Collins and S. Lowndes, 1682.

Glanvill, J., and B. Fabian. *Collected Works of Joseph Glanvill.* Collected Works of Joseph Glanvill. G. Olms, 1978. https://books.google.de/books?id=s-YnAAAAYAAJ.

Glasstone, S., and P. J. Dolan. "The Effects of Nuclear Weapons." *US Department of Defense and Department of Energy*, 1977. www.geengineeringsystems.com/ewExternalFiles/Fire-FollowingEarthquake.pdf.

Glen, W. *The Mass-Extinction Debates: How Science Works in a Crisis.* Stanford University Press, 1994. https://books.google.de/books?id=dePE-3YkQTEC.

Glover, J. *Causing Death and Saving Lives: The Moral Problems of Abortion, Infanticide, Suicide, Euthanasia, Capital Punishment, War and Other Life-or-Death Choices.* Pelican Books. Penguin Books Limited, 1990. https://books.google.de/books?id=MdOtOdFXR-MC.

Godwin, W. *Of Population: An Enquiry Concerning the Power of Increase in the Numbers of Mankind, Being an Answer to Mr. Malthus's Essay on That Subject.* Longman, Hurst, Rees, Orme, and Brown, 1820. https://books.google.de/books?id=7rc8AAAAcAAJ.

Godwin, W., and M. Philp. *An Enquiry Concerning Political Justice.* Oxford World's Classics. Oxford University Press, 2013. https://books.google.de/books?id=fYhuAAAAQBAJ.

Goertzel, Ben. "Superintelligence: Fears, Promises and Potentials: Reflections on Bostrom's Superintelligence, Yudkowsky's From AI to Zombies, and Weaver and Veitas's 'Open-Ended Intelligence'." *Journal of Ethics and Emerging Technologies* 25, no. 2 (2015): 55–87.

Goldberg, Leon. "How 'Longtermism' Is Shaping Foreign Policy." *UN Dispatch*, 2022. www.undispatch.com/how-longtermism-is-shaping-foreign-policy-will-macaskill/.

Good, Irving John. "Ethical Machines." In *Machine Intelligence.* Vol. 10, edited by J. E. Hayes, Donald Michie, and Y.-H. Pao, 555–60. Ellis Horwood, 1982.

———. "Speculations Concerning the First Ultraintelligent Machine." *Advances in Computers* 6 (1965). https://asset-pdf.scinapse.io/prod/1586718744/1586718744.pdf.

———. *Speculations on Perceptrons and Other Automata.* International Business Machines Corporation, 1959.

Gore, Al. *An Inconvenient Truth: The Planetary Emergency of Global Warming and What We Can Do about It.* Rodale, 2006.

Gott, J. Richard. "A Grim Reckoning—What Has a 16th-Century Astronomer Got to Do with the Defeat of Governments and the Possible Extinction of the Human Race? Answers in Fractions Please, Says J. Richard Gott III." *New Scientist*, November 15, 1997. www.newscientist.com/article/mg15621085-100/.

———. "Implications of the Copernican Principle for Our Future Prospects." *Nature* 363, no. 6427 (1993): 315–19.

Gould, Stephen J. "Is Uniformitarianism Necessary?" *American Journal of Science* 263, no. 3 (1965): 223–28.

———. *Time's Arrow, Time's Cycle: Myth and Metaphor in the Discovery of Geological Time*. Harvard University Press, 1996.

Graff, Garrett. "America's Decades-Old Obsession with Nuking Hurricanes (and More)." *Wired*, August 26, 2019. www.wired.com/story/nuking-hurricanes-polar-ice-caps-climate-change/.

Grainville, J.B.C. de, I.F. Clarke, and M. Clarke. *The Last Man*. Early Classics of Science Fiction. Wesleyan University Press, 2002. https://books.google.de/books?id=0F7sja1y77wC.

Grainville, J. B. C. de, S. Schiewe, and G. Poppenberg. *Der Letzte Mensch*. Matthes & Seitz Berlin Verlag, 2015. https://books.google.de/books?id=2HZ4DwAAQBAJ.

Granberry, Mike. "Octogenarian Coined 'Omnicide' During Lifelong Push for Peace." *Los Angeles Times*, November 30, 1986. www.latimes.com/archives/la-xpm-1986-11-30-vw-388-story.html.

Gray, Robert H. "The Fermi Paradox Is Neither Fermi's Nor a Paradox." *Astrobiology* 15, no. 3 (2015): 195–99.

Greaves, Hilary. "Population Axiology." *Philosophy Compass* 12, no. 11 (2017): e12442.

Greaves, Hilary, and William MacAskill. "The Case for Strong Longtermism." *Global Priorities Institute*, 2021. https://globalprioritiesinstitute.org/wp-content/uploads/The-Case-for-Strong-Longtermism-GPI-Working-Paper-June-2021-2-2.pdf.

Greaves, Hilary, William MacAskill, and Elliott Thornley. "The Moral Case for Long-Term Thinking." 2021. https://philarchive.org/archive/GRETMC-3.

Greaves, Hilary, and Toby Ord. "Moral Uncertainty about Population Axiology." *Journal of Ethics and Social Philosophy* 12 (2017): 135.

Greene, Preston. "The Termination Risks of Simulation Science." *Erkenntnis* 85, no. 2 (2020): 489–509.

Grehan, John. *Hitler's Wolfsschanze: The Wolf's Lair Headquarters on the Eastern Front – An Illustrated Guide*. Pen & Sword Books, 2021.

Griswold, Eliza. "How 'Silent Spring' Ignited the Environmental Movement." *The New York Times*, September 21, 2012.

Groenewold, H.J. "Modern Science and Social Responsibility." In *Induction, Physics and Ethics*, 359–78. Springer, 19681970.

Grooten, Monique, and Rosamunde E.A. Almond. *Living Planet Report-2018: Aiming Higher*. WWF International, 2018.

Grossman, Daniel. "High CO2 Levels Inside and Out: Double Whammy?" *Yale Climate Connections*, 2016. www.yaleclimateconnections.org/2016/07/Indoor-Co2-Dumb-and-Dumber.

Grove, Jairus Victor. "Savage Ecology." In *Savage Ecology*. Duke University Press, 2019.

———. *Savage Ecology: War and Geopolitics at the End of the World*. Duke University Press, 2019. https://books.google.de/books?id=F4tIvAEACAAJ.

GS. "Existential." *Oxford Languages*. Google, 2022. www.google.com/search?q=existential&oq=existential+&aqs=chrome.69i57j35i39j0i512j69i60j69i65j69i60l2j69i65.1842j1j4&sourceid=chrome&ie=UTF-8.

Gunn, Alistair. "The Restoration of Species and Natural Environments." *Environmental Ethics* 13, no. 4 (1991): 291–312.

GWWC. "About Us." *Giving What We Can*, 2007. https://web.archive.org/web/20070701205101/www.givingwhatwecan.org/.

———. "The Giving What We Can Team." *Giving What We Can*, 2017. https://web.archive.org/web/20170630175607/www.givingwhatwecan.org/about-us/team/.

———. "Recommended Charities." *Giving What We Can*, 2011a. https://web.archive.org/web/20110722095716/www.givingwhatwecan.org/resources/recommended-charities.php.

———. "Recommended Charities." *Giving What We Can*, 2011b. https://web.archive.org/web/20110813221852/www.givingwhatwecan.org/resources/recommended-charities.php.

H. "The New Monthly Magazine (NMM)." 1816. https://books.google.de/books?id=VDYaAQAAIAAJ.

Hacker, Peter, Michael Stephen, and Joseph Raz. *Law, Morality, and Society: Essays in Honour of HLA Hart*. Clarendon Press, 1977.

Häggström, O. *Here Be Dragons: Science, Technology and the Future of Humanity*. Oxford University Press, 2016. https://books.google.de/books?id=WWvQCgAAQBAJ.

Hahn, Otto, and Max Born. "Mainau Declaration." 1955. www.lindau-repository.org/permadocs/MainauDeclaration1955EN.pdf.

Haldane, John Burdon Sanderson. *Daedalus or Science and the Future*. EP Dutton, 1924.

Halsell, Grace. *Prophecy and Politics: Militant Evangelists on the Road to Nuclear War*. Hill, 1986.

Hamblin, James. "The Toxins That Threaten Our Brains." *The Atlantic*, March 18, 2014.

Hamrud, Eva. "Fact Check: Will the Oceans Be Empty of Fish by 2048, and Other Seaspiracy Concerns." *Science Alert*, April 30, 2021. www.sciencealert.com/no-the-oceans-will-not-be-empty-of-fish-by-2048.

Hand, Eric. "Acid Oceans Cited in Earth's Worst Die-Off." *Science* 348, no. 6231 (2015): 165–6.

Hansen, J. "Is There Still Time to Avoid Dangerous Anthropogenic Interference with Global Climate? The Importance of the Work of Charles David Keeling." *2005:U23D-01*, 2005. https://www.columbia.edu/~jeh1/2005/Keeling_20051206.pdf.

———. "Transcript of Dr. James Hansen's Testimony before the U.S. Senate Committee on Energy and Natural Resources on June 23, 1988." 1988. www.sealevel.info/1988_Hansen_Senate_Testimony.html.

Hanson, Robin. "Catastrophe, Social Collapse, and Human Extinction." *Global Catastrophic Risks* (2008): 363–78.

———. "The Great Filter-Are We Almost Past It." 1998. http://hanson.gmu.edu/Greatfilter.html.

———. "Personal Website." 2022. http://mason.gmu.edu/~rhanson/home.html.

Hansson, Sven Ove. "Risk." In *Stanford Encyclopedia of Philosophy*. Winter Edition, edited by Edward N. Zalta and Uri Nodelman. 2018. https://plato.stanford.edu/archives/win2022/entries/risk/.

Harrison, Timothy. "The Doomsday List." *Lighthouse Digest*, 2011. https://www.lighthousedigest.com/Digest/StoryPage.cfm?StoryKey=3468.

Haqq-Misra, Jacob, Sanjoy Som, Brendan Mullan, Rafael Loureiro, Edward Schwieterman, Lauren Seyler, Haritina Mogosanu, Adam Frank, Eric Wolf, and Duncan Forgan. "The Astrobiology of the Anthropocene." *ArXiv Preprint ArXiv:1801.00052*, 2017.

Harari, Y.N. *Homo Deus: A Brief History of Tomorrow*. Random House, 2016. https://books.google.de/books?id=dWYyCwAAQBAJ.

———. *Homo Deus: 'An Intoxicating Brew of Science, Philosophy and Futurism' Mail on Sunday*. Random House, 2016. https://books.google.de/books?id=dWYyCwAAQBAJ.

Haraway, D.J. *Staying with the Trouble: Making Kin in the Chthulucene*. Experimental Futures. Duke University Press, 2016. https://books.google.de/books?id=cND9jwEACAAJ.

Harris, John. *Genethics: Moral Issues in the Creation of People*. University of California Press, 1994.

Harris, S. *The End of Faith: Religion, Terror, and the Future of Reason*. W.W. Norton & Company, 2004. https://books.google.de/books?id=Lr8ytqlY9NgC.

Harrison, Peter, and Joseph Wolyniak. "The History of 'Transhumanism'." *Notes and Queries* 62, no. 3 (September 2015): 465–67. https://doi.org/10.1093/notesj/gjv080.

Harsanyi, John C. "Cardinal Utility in Welfare Economics and in the Theory of Risk-Taking." *Journal of Political Economy* 61, no. 5 (1953): 434–35.

———. "Morality and the Theory of Rational Behavior." *Social Research* 44, no. 4 (1977), 623–56.

———. "Rule Utilitarianism, Rights, Obligations and the Theory of Rational Behavior." In *Papers in Game Theory*, 235–53. Springer, 1982.

Hart, Michael H. "An Explanation for the Absence of Extraterrestrials." In *Extraterrestrials: Where Are They?*, edited by Ben Zuckerman and Michael Hart. Cambridge University Press, 1995/1975.

Hartmann, Eduard von. *Philosophy of the Unconscious: Speculative Results According to the Induction Method of the Physical Sciences*. Dunker, 1869.

Hawking, Stephen. "This Is the Most Dangerous Time for Our Planet." *The Guardian*, December 1, 2016, p. 14.

Häyry, Matti. "A Rational Cure for Prereproductive Stress Syndrome." *Journal of Medical Ethics* 30, no. 4 (2004): 377.

Hedgpeth, Joel W. "Pandora's Box." *Science* 103, no. 2669 (1946): 236–236. https://doi.org/10.1126/science.103.2669.236.

Herbers, John. "Religious Leaders Tell of Worry on Armageddon View Ascribed to Reagan." *The New York Times*, October 21, 1984.

Herman, Barbara. *The Practice of Moral Judgment*. Harvard University Press, 1993.

Hermansen, Marcia. "Eschatology." In *Classical Islamic Theology*, edited by Tim Winter. Cambridge University Press, 2008.

Hey, Jody. "The Mind of the Species Problem." *Trends in Ecology & Evolution* 16, no. 7 (2001): 326–29.

Heyd, David. "Genethics." In *Genethics*. University of California Press, 1992.

———. "The Intractability of the Nonidentity Problem." In *Intergenerational Justice*, 55–78. Routledge, 2017.

Hildebrand, Alan R., Glen T. Penfield, David A. Kring, Mark Pilkington, Antonio Camargo Z., Stein B. Jacobsen, and William V. Boynton. "Chicxulub Crater: A Possible Cretaceous/Tertiary Boundary Impact Crater on the Yucatan Peninsula, Mexico." *Geology* 19, no. 9 (1991): 867–71.

Hill, C.C. *In God's Time: The Bible and the Future*. Eerdmans Publishing Company, 2002. https://books.google.de/books?id=1mmpm5Gm9awC.

Hippolytus. *The Sacred Writings of Saint Hippolytus*. Jazzybee Verlag, 2012. https://books.google.de/books?id=xSDp6bOejVAC.

Hippolytus, Antipope. *The Refutation of All Heresies: Book I*. BoD—Books on Demand, 2022.

Hirose, Iwao, and Jonas Olson. *The Oxford Handbook of Value Theory*. Oxford University Press, 2015.

Hoffman, Bruce. "Terrorism Trends and Prospects." *Countering the New Terrorism* 7 (1999): 13.

Honderich, T. *The Presocratic Philosophers, Jonathan Barnes*. Routledge Taylor & Francis Group, 1982.

Hooke, R. *The Posthumous Works of Robert Hooke, . . . Containing His Cutlerian Lectures, and Other Discourses, Read at the Meetings of the Illustrious Royal Society. . . . Illustrated with Sculptures. To These Discourses Is Prefixt the Author's Life, . . . Publish'd by Richard Waller*. Sam. Smith and Benj. Walford, 1705. https://books.google.de/books?id=6xVTAAAAcAAJ.

Horgan, John. "AI Visionary Eliezer Yudkowsky on the Singularity, Bayesian Brains and Closet Goblins." 2016. https://blogs.scientificamerican.com/Cross-Check/Ai-Visionary-Eliezer-Yudkowsky-on-the-Singularity-Bayesian-Brains-and-Closet-Goblins (Дата Звернення: 10 July 2021).

Horn, E. (2014). The Last Man. The Birth of Modern Apocalypse in Jean Paul, John Martin and Lord Byron. In N. Lebovic, & A. Killen (Eds.), Catastrophes: A History and Theory of an Operative Concept (pp. 55–74). Walter de Gruyter. https://doi.org/10.1515/9783110312584.55

Hublin, Jean-Jacques, Abdelouahed Ben-Ncer, Shara E. Bailey, Sarah E. Freidline, Simon Neubauer, Matthew M. Skinner, Inga Bergmann, Adeline Le Cabec, Stefano Benazzi, and Katerina Harvati. "New Fossils from Jebel Irhoud, Morocco and the Pan-African Origin of *Homo sapiens*." *Nature* 546, no. 7657 (2017): 289–92.

Hulme, Mike. "Am I a Denier, a Human Extinction Denier?" *Personal Website*, May 27, 2019. https://mikehulme.org/am-i-a-denier-a-human-extinction-denier/.

Hume, D. *The Complete Works of David Hume. Illustrated: Treatise of Human Nature, An Enquiry Concerning Human Understanding, An Enquiry Concerning the Principles of Morals, Dialogues Concerning Natural Religion and Other*. Strelbytskyy Multimedia Publishing, 2021. https://books.google.de/books?id=0dczEAAAQBAJ.

———. *Writings on Economics*. Taylor & Francis, 2017. https://books.google.de/books?id=2yAuDwAAQBAJ.

Humphreys, Rachel. "Leaded Petrol, Acid Rain, CFCs: Why the Green Movement Can Overcome the Climate Crisis." *The Guardian*, 2020. https://www.theguardian.com/news/audio/2020/oct/19/leaded-petrol-acid-rain-cfcs-why-the-green-movement-can-overcome-the-climate-crisis.

Humphreys, Rachel, and Fiona Harvey. "Leaded Petrol, Acid Rain, CFCs: Why the Green Movement Can Overcome the Climate Crisis." October 19, 2020. www.theguardian.com/news/audio/2020/oct/19/leaded-petrol-acid-rain-cfcs-why-the-green-movement-can-overcome-the-climate-crisis.

Humphreys, W.J. "Volcanic Dust in Relation to Climate." *Eos, Transactions American Geophysical Union* 15, no. 1 (1934): 243–45.

Hunter, Robert. *Warriors of the Rainbow: A Chronicle of the Greenpeace Movement*. Holt, Rinehart and Winston, 1979.

Hurka, Thomas. "Moore's Moral Philosophy." *Stanford Encyclopedia of Philosophy*, 2021. https://plato.stanford.edu/entries/moore-moral/.

———. "Value and Population Size." *Ethics* 93, no. 3 (1983): 496–507.

Hut, Piet, and Martin J. Rees. "How Stable Is Our Vacuum?" *Nature* 302, no. 5908 (1983): 508–9.

Hutton, James. *Abstract of a Dissertation Read in the Royal Society of Edinburgh*, 1785.

———. *Theory of the Earth*. Vol. 1. Transactions of the Royal Society of Edinburgh. Royal Society of Edinburgh, 1788.

Huxley, J.S. *Religion Without Revelation: J.S. Huxley*. E. Benn, 1927. https://books.google.de/books?id=f023tAEACAAJ.

Huxley, Julian. "Knowledge, Morality, and Destiny: I †." *Psychiatry* 14, no. 2 (May 1951): 129–40. https://doi.org/10.1080/00332747.1951.11022818.

———. *New Bottles for New Wine: Essays*. Chatto & Windus, 1957. https://books.google.de/books?id=U4c0AAAAMAAJ.

———. *Religion Without Revelation*. Harper & Brothers Publishers, n.d. https://archive.org/details/in.ernet.dli.2015.90330/page/n5/mode/2up.

Huxley, Thomas Henry. *Evolution and Ethics, and Other Essays*. Macmillan, 1894.

Hyman, Gavin. *A Short History of Atheism*. Bloomsbury Publishing, 2010.

Impey, C. *How It Ends: From You to the Universe*. WW Norton, 2010. https://books.google.de/books?id=mc-w4E3jhCcC.

IPCC. "Climate Change 2001: Synthesis Report." *Intergovernmental Panel on Climate Change*, 2001. https://archive.ipcc.ch/ipccreports/tar/vol4/english/027.htm.

Isaacson, W. *Einstein: His Life and Universe*. Simon & Schuster, 2017. https://books.google.de/books?id=d2WZDgAAQBAJ.

IUCN 2005 Dhabi, Abu. "The Red List of Terrestrial Mammalian Species of the Abu Dhabi Emirate," 2005.

Jablow, Valerie. "A Tale of Two Rocks." *Smithsonian Magazine*, April 1998. www.smithsonianmag.com/science-nature/a-tale-of-two-rocks-151643588/.

Jackson, Frank. "Decision-Theoretic Consequentialism and the Nearest and Dearest Objection." *Ethics* 101, no. 3 (1991): 461–82.

Jacobs, M. "Declaration of the Rights of Animal and Plant Life." *Flora Malesiana Bulletin* 31, no. 1 (1977): 3048–3048.

Jacquet, Jennifer. "The Anthropocene." *The Edge*, 2017. www.edge.org/response-detail/27096.

Jamail, Dahr. "Will Humanity Become Extinct Within the Next Generation?" *History News Network*, December 17, 2013. https://historynewsnetwork.org/article/154243.

James, W. *Pragmatism – A New Name for Some Old Ways of Thinking*. Read Books Limited, 2015. https://books.google.de/books?id=AKF9CgAAQBAJ.

———. "The Pragmatic Method." *The Journal of Philosophy, Psychology and Scientific Methods* 1, no. 25 (December 8, 1904): 673. https://doi.org/10.2307/2012198.

———. *Der Pragmatismus*. Philosophisch-Soziologische Bücherei. Verlag nicht ermittelbar, 1928. https://books.google.de/books?id=uGUKAwAAQBAJ.

———. *Pragmatism, and Other Essays*. Meridian Books, 1955. https://books.google.de/books?id=VZQcjwEACAAJ.

Jaspers, Karl. *The Future of Mankind*. University of Chicago Press, 1961. https://books.google.de/books?id=cX1GDqK7NBsC.

Jeans, James. *The Universe Around US*. Cambridge University Press, 1929.

Jefferson, Thomas. "Instructions for Meriwether Lewis." 1803. https://founders.archives.gov/documents/Jefferson/01-40-02-0136-0005.

———. "Notes on the State of Virginia." 1785. https://xroads.virginia.edu/~Hyper/JEFFERSON/ch06.html.

Jerome, F. *The Einstein File: J. Edgar Hoover's Secret War Against the World's Most Famous Scientist*. St. Martin's Press, 2003. https://books.google.de/books?id=weECGK2rChcC.

JET. "A Short History of the Journal." *Journal of Evolution and Technology*, 2005. https://jetpress.org/history.html.

John, Tyler, and William MacAskill. "Longtermist Institutional Reform." In *The Long View: Essays on Policy, Philanthropy, and the Long-term Future*, edited by Natalie Cargill and Tyler John. Legal Priorities Project Working Paper Series 4-2021. First Strategic Insight Ltd., 2021.

Johnson, A.E., and K.K. Wilkinson. *All We Can Save: Truth, Courage, and Solutions for the Climate Crisis*. Random House Publishing Group, 2020. https://books.google.de/books?id=zbrWDwAAQBAJ.

Johnson, Robert and Adam Cureton, "Kant's Moral Philosophy", The Stanford Encyclopedia of Philosophy (Fall 2022 Edition), Edward N. Zalta & Uri Nodelman (eds.), URL = <https://plato.stanford.edu/archives/fall2022/entries/kant-moral/>.

Jonas, Hans. "Hannah Arendt: An Intimate Portrait." *New England Review* 27, no. 2 (2006): 133–42.

Jonas, H., and D. Herr. *The Imperative of Responsibility: In Search of an Ethics for the Technological Age*. Mersion: Emergent Village Resources for Communities of Faith Series. University of Chicago Press, 1979. https://books.google.de/books?id=sRP3uJkxydQC.

Jonas, H., and L. Vogel. *Mortality and Morality: A Search for Good after Auschwitz. Northwestern University Studies in Phenomenology & Existential Philosophy*. Northwestern University Press, 1996. https://books.google.de/books?id=A-3WAAAAMAAJ.

Jones, Christine Kenyon. "Religion." In *Byron in Context*, edited by Clara Tuite. Cambridge University Press, 2019. https://books.google.de/books?id=DZ_MDwAAQBAJ.

Jones, Larry. "Apocalyptic Eschatology in the Nuclear Arms Race." *Transformation* 5, no. 1 (1988): 25–27.

Joy, Bill. "Why the Future Doesn't Need Us, Wired Magazine," 2000. https://www.wired.com/2000/04/joy-2/.

Juergensmeyer, Mark. "Radical Religious Responses to Global Catastrophe." In *Exploring Emerging Global Thresholds: Toward 2030*. Orient Blackswan, 2017.

Kaczmarek, Patrick, and Simon Beard. "Human Extinction and Our Obligations to the Past." *Utilitas* 32, no. 2 (2020): 199–208.

Kaczynski, Theodore John. "Industrial Society and Its Future." *Washington Post*, September 19, 1995.

Kaempffert, Waldemar. "Rutherford Cools Atom Energy Hope." *New York Times*, September 12, 1933, p. 1.

———. "Rutherford Cools Atom Energy Hope; Sees 'Moonshine' in the Talk at Present of Releasing Power in Matter." *The New York Times*, 1933. www.nytimes.com/1933/09/12/archives/rutherford-cools-atom-energy-hope-sees-moonshine-in-the-talk-at.html?searchResultPosition=1.

Kagan, Shelly. "Rethinking Intrinsic Value." *The Journal of Ethics* 2, no. 4 (1998): 277–97.

Kahn, Herman. "Thinking about the Unthinkable New York." *Horizon* 164 (1962).

Kaku, M. *Parallel Worlds: The Science of Alternative Universes and Our Future in the Cosmos*. Penguin Science. Penguin Books, 2006. https://books.google.de/books?id=7-IUAAAACAAJ.

Kaku, Michio. Will Mankind Destroy Itself? *Big Think*, 2011. www.youtube.com/watch?v=7NPC47qMJVg.

Kaneda, Toshiko, and Carl Haub. "How Many People Have Ever Lived on Earth?" *Population Reference Bureau*, 2022. www.prb.org/articles/how-many-people-have-ever-lived-on-earth/.

Kant, Immanuel. *Critique of Judgment*. Hackett Publishing Company, Inc., 1987. https://monoskop.org/images/7/77/Kant_Immanuel_Critique_of_Judgment_1987.pdf.

———. *Grundlegung Zur Metaphysik Der Sitten*. Hartknoch, 1785. https://books.google.de/books?id=c9BgAAAAcAAJ.

———. *The Metaphysics of Morals*, 1797a.

———. *On a Supposed Right to Lie Because of Philanthropic Concerns*, 1797b. http://bgillette.com/wp-content/uploads/2011/08/KANTsupposedRightToLie.pdf.

———. *Universal Natural History and Theory of the Heavens*, 1755.

Kant, I., and J. H. Bernard. *The Critique of Judgment*. DigiCat, 2022. https://books.google.de/books?id=BflyEAAAQBAJ.

Kant, I., M.J. Gregor, and A.W. Wood. *Practical Philosophy. Kant, Immanuel, 1724–1804. Works. Engl. 1992*. Cambridge University Press, 1999. https://books.google.de/books?id=0hCsbUjFiBwC.

Karnofsky, Holden. "Some Comments on Recent FTX-Related Events." *EA Forum*, November 10, 2022. https://forum.effectivealtruism.org/posts/mCCutDxCavtnhxhBR/some-comments-on-recent-ftx-related-events.

Karnofsky, Holden. "Some Comments on Recent FTX-Related Events." *EA Forum* (blog), November 10, 2022. https://forum.effectivealtruism.org/posts/mCCutDxCavtnhxhBR/some-comments-on-recent-ftx-related-events.

Kateb, G. *The Inner Ocean: Individualism and Democratic Culture (Contestations)*. Cornell University Press, 2019. https://books.google.de/books?id=GXHxDwAAQBAJ.

Kavka, Gregory. "The Futurity Problem." In *Obligations to Future Generations*, edited by Richard I. Sikora and Brian M. Barry, 186–203. White Horse Press, 1978.

———. "The Paradox of Future Individuals." *Philosophy & Public Affairs* 11, no. 2 (1982): 93–112.

Kelvin, L. "On the Age of the Sun's Heat. Appendix E." In *Treatise on Natural Philosophy*. Cambridge University Press. First Published in Macmillan's Magazine, 1862.

Kemp, Luke. "'Stomp Reflex': When Governments Abuse Emergency Powers." *BBC Future*, 2021.

Kemp, Luke, Chi Xu, Joanna Depledge, Kristie L. Ebi, Goodwin Gibbins, Timothy A. Kohler, Johan Rockström, Marten Scheffer, Hans Joachim Schellnhuber, and Will Steffen. "Climate Endgame: Exploring Catastrophic Climate Change Scenarios." *Proceedings of the National Academy of Sciences* 119, no. 34 (2022): e2108146119.

Kennedy, John F. "Address Before the General Assembly of the United Nations." *JFK Library*, September 25, 1961. www.jfklibrary.org/archives/other-resources/john-f-kennedy-speeches/united-nations-19610925.

———. *Let Us Call a Truce to Terror*. Vol. 23. Office of Public Services, Bureau of Public Affairs, 1961.

Kennedy, R.F., and A.M. Schlesinger. *Thirteen Days: A Memoir of the Cuban Missile Crisis*. W. W. Norton, 2011. https://books.google.de/books?id=mWWAm0h5yP0C.

Keyes, Emilie. "Slightly Off the Record." *The Palm Beach Post-Times*, August 12, 1945, Vol. XII: No. 28 edition.

King, Alexander, and Bertrand Schneider. *The First Global Revolution*, 70, 115. The Club of Rome, 1991.

Kingsley, Scarlett, and Richard Parry. "Emedocles." *Stanford Encyclopedia of Philosophy*, 2020. https://plato.stanford.edu/archives/sum2020/entries/empedocles/.

Kirchin, S. *Reading Parfit: On What Matters*. Taylor & Francis, 2017. https://books.google.de/books?id=aEYlDwAAQBAJ.

Klein, Ezra. "Transcript: Ezra Klein Interviews William MacAskill." *New York Times*, August 9, 2022. www.nytimes.com/2022/08/09/podcasts/transcript-ezra-klein-interviews-will-macaskill.html.

Knight, Les. *About the Movement*. Voluntary Human Extinction Movement, 1997. https://www.vhemt.org/aboutvhemt.htm.

———. *These Exit Times*. Voluntary Human Extinction Movement, 1991. https://www.vhemt.org/TET1.pdf.

———. *Success*. Voluntary Human Extinction Movement, 1995. https://www.vhemt.org/success.htm.

Knipe, David. "Hindu Eschatology." In *The Oxford Handbook of Eschatology*, edited by Jerry Walls. Oxford University Press, 2008.

Knobe, Joshua. "Philosophical Intuitions Are Surprisingly Stable across Both Demographic Groups and Situations." *Filozofia Nauki* 29, no. 2 (114) (2021): 11–76.

Knutsson, Simon. "Permissible Moderate Paths to Human Extinction." *Working Draft*, 2022a. www.simonknutsson.com/permissible-moderate-paths-to-human-extinction/#_ednref5.

———. "Philosophical Pessimism: Varieties, Importance, and What to Do." *Blog of the APA* (blog), 2022b. https://blog.apaonline.org/2022/09/13/philosophical-pessimism-varieties-importance-and-what-to-do%ef%bf%bc/.

———. "Thoughts on Ord's 'Why I'm Not a Negative Utilitarian.'" 2022c. https://www.simonknutsson.com/thoughts-on-ords-why-im-not-a-negative-utilitarian.

———. "The World Destruction Argument." *Inquiry* 64, no. 10 (2021): 1004–23.

Koestler, Arthur. *The Ghost in the Machine*. Macmillan, 1967.

Kolbert, E. *The Sixth Extinction: An Unnatural History*. Henry Holt and Company, 2014. https://books.google.de/books?id=Ra9RAQAAQBAJ.

Konopinski, E.J., C. Marvin, and Edward Teller. "Ignition of the Atmosphere with Nuclear Bombs." *Report LA-602*. Los Alamos Laboratory, 1946.

Korda, Chris. *e-sermon #9*. Church of Euthanasia, 1994a. https://www.churchofeuthanasia.org/e-sermons/sermon9.html.

———. *e-sermon #7*. Church of Euthanasia, 1994b. https://www.churchofeuthanasia.org/e-sermons/sermon7.html.

———. *Prayer for a Good Death*. Church of Euthanasia, 1994c. https://www.churchofeuthanasia.org/snuffit3/prayer.html.

———. *Shuff It #5*. Internet Archive, 2019. https://archive.org/stream/SnuffIt5CoE/Snuff%20it%205%20combined%20pages-with%20photo%20spread_djvu.txt.

Kors, A.C. *D'Holbach's Coterie: An Enlightenment in Paris*. Princeton Legacy Library. Princeton University Press, 2015. https://books.google.de/books?id=m_l9BgAAQBAJ.

Korsgaard, Christine M. "Two Distinctions in Goodness." *The Philosophical Review* 92, no. 2 (1983): 169–95.

———. "Kant's Formula of Universal Law." *Pacific Philosophical Quarterly* 66, no. 1–2 (1985): 24–47.

Koscielniak, Maciej, Agnieszka Bojanowska, and Agata Gasiorowska. "Religiosity Decline in Europe: Age, Generation, and the Mediating Role of Shifting Human Values." *Journal of Religion and Health* (2022): 1–26.

Kovacs, M.G. *The Epic of Gilgamesh*. Penguin Classics. Stanford University Press, 1989. https://books.google.de/books?id=YYxEd9c0EUYC.

Kragh, H.S. *Entropic Creation: Religious Contexts of Thermodynamics and Cosmology*. Taylor & Francis, 2016. https://books.google.de/books?id=8ZUWDAAAQBAJ.

Kramers, H.A., and Helge Holst. *The Atom and the Bohr Theory of Its Structure*. Gyldendal, 1923.

Kuhlemann, Karin. "We Can't Tackle Overpopulation When the Time Comes—We Need to Talk about It Now." *Huffington Post*, January 24, 2018. www.huffingtonpost.co.uk/entry/lets-stop-thinking-we-can-tackle-it-when-the-time-comes-we-need-to-talk-about-overpopulation-now_uk_5a675db0e4b002283006fe0c.

Kuhn, Thomas S. "Carnot's Version of 'Carnot's Cycle'." *American Journal of Physics* 23, no. 2 (February 1955): 91–95. https://doi.org/10.1119/1.1933907.

Kumar, Rahul. "Contractualist Proposal." *Intergenerational Justice* (2009): 251.

———. "Samuel Scheffler, Why Worry about Future Generations?" *Journal of Moral Philosophy* 17, no. 5 (2020): 583–86.

Kunkle, Thomas, and Byron Ristvet. *Castle Bravo: Fifty Years of Legend and Lore. A Guide to Off-Site Radiation Exposures*. Defense Threat Reduction Information Analysis Center Kirtland AFB NM, 2013.

Kurzweil, R. *The Age of Spiritual Machines: When Computers Exceed Human Intelligence*. A Penguin Book. Viking, 1999. https://books.google.de/books?id=941QAAAAMAAJ.

———. *The Singularity Is Near: When Humans Transcend Biology*. Penguin Publishing Group, 2005. https://books.google.de/books?id=9FtnppNpsT4C.

Laërtius, Diogenes. *Lives of the Eminent Philosophers*. Vol. 1. Translated by Robert Drew Hicks. Harvard University Press, 1925.

Lafollette, Eva. *The International Encyclopedia of Ethics, 11 Volume Set*. John Wiley & Sons, 2021.

Lamb, Hubert Horace. "Volcanic Dust in the Atmosphere; with a Chronology and Assessment of Its Meteorological Significance." *Philosophical Transactions of the Royal Society of London. Series A, Mathematical and Physical Sciences* 266, no. 1178 (1970): 425–533.

Landau, Iddo. "Why Has the Question of the Meaning of Life Arisen in the Last Two and a Half Centuries?" *Philosophy Today* 41, no. 2 (1997): 263–69.

Lanier, Jaron. "The Social Dilemma." *Netflix*, 2020. https://www.netflix.com/de-en/title/81254224.

Lankester, Edwin Ray. *Degeneration: A Chapter in Darwinism*. Vol. 12. Macmillan and Company, 1880.

Lanouette, W., and B. Silard. *Genius in the Shadows: A Biography of Leo Szilard, the Man Behind the Bomb*. Skyhorse, 2013. https://books.google.de/books?id=idHawAEACAAJ.

Lanouette, W., B. Silard, and J. Salk. *Genius in the Shadows: A Biography of Leo Szilard, the Man Behind the Bomb*. Skyhorse Publishing, 2013. https://books.google.de/books?id=2y51EAAAQBAJ.

Lapin, Adam. *Coexistence or No Existence: Peace or H-Bomb Annihilation?* New Century Publishers, 1955.

Laqueur, Walter. "Fanaticism and the Arms of Mass Destruction." *The New Terrorism* 262 (1999).

Larson, Edward J., and Larry Witham. "Leading Scientists Still Reject God." *Nature* 394, no. 6691 (1998): 313–313.

Lasch, Christopher. *The Culture of Narcissism*, 39–40. Norton, 1978.

Lash, S., and B. Wynne. "Introduction." In *Risk Society—Towards a New Modernity*. Sage, 1992.

Lavenda, B.H. *A New Perspective on Thermodynamics*. Springer, 2009. https://books.google.de/books?id=UheDzjQmE8kC.

Lavenda, Bernard H. *A New Perspective on Thermodynamics*. Netherlands: Springer, 2010.

Lazari-Radek, K. de, and P. Singer. *Utilitarianism: A Very Short Introduction*. Very Short Introductions. Oxford University Press, 2017. https://books.google.de/books?id=HjsqDwAAQBAJ.

Leakey, R.E., and R. Lewin. *The Sixth Extinction: Patterns of Life and the Future of Humankind*. Knopf Doubleday Publishing Group, 1995. https://books.google.de/books?id=By_XQa87x1oC.

Lederberg, Joshua. *Hearings, Reports and Prints of the House Committee on International Relations*. U.S. Government Printing Office, 1975. https://books.google.de/books?id=79I1AAAAIAAJ.

———. *Statement of Dr. Joshua Lederberg, Professor of Genetics, Stanford University*. United States Congress, House Committee on Foreign Affairs, 1969. https://www.google.de/books/edition/Hearings/AAYaAQAAMAAJ?hl=en&gbpv=1&dq=%22However,+whatever+pride+I+might+wish+to+take+in+the+eventual+human+benefits+that+may+arise+from+my+own+research+is+turned+into+ashes+by+the+application+of+this+kind+of+scientific+insight+for+the+engineering+of+biological+warfare+agents%22&pg=RA5-PA87&printsec=frontcover.

Lehtipuu, O. *The Afterlife Imagery in Luke's Story of the Rich Man and Lazarus*. Novum Testamentum: Supplements. Brill, 2007. https://books.google.de/books?id=LyLrBidjIHEC.

Leiserowitz, Anthony, Edward Maibach, Connie Roser-Renouf, Seth Rosenthal, and Matthew Cutler. "Climate Change in the American Mind." *Yale Program on Climate Change Communication and George Mason University Center for Climate Change Communication*, 2017. https://climatecommunication.yale.edu/wp-content/uploads/2017/07/Climate-Change-American-Mind-May-2017.pdf.

Lenman, James. "On Becoming Extinct." *Pacific Philosophical Quarterly* 83, no. 3 (2002): 253–69.

Lennox, John. "Reflections on the Intelligent Design Debate." In *From 'Intelligent Design: Some Critical Reflections on the Current Debate' in Intelligent Design: William A. Dembski and Michael Ruse in Dialogue*, edited by Robert B. Stewart. Fortress Press, 2007.

Lenton, Timothy M., Johan Rockström, Owen Gaffney, Stefan Rahmstorf, Katherine Richardson, Will Steffen, and Hans Joachim Schellnhuber. "Climate Tipping Points—Too Risky to Bet Against." *Nature* 575 (2019): 592–95.

Lenton, Timothy M., and Hans Joachim Schellnhuber. "Tipping the Scales." *Nature Climate Change* 1, no. 712 (2007): 97–98.

Leopold, Aldo. *A Sand County Almanac*. Ballantine, 1949.

Leslie, John. "Anthropic Explanations in Cosmology." *Philosophy of Science Association* (1986): 87–95.

———. "Anthropic Principle, World Ensemble, Design." *American Philosophical Quarterly* 19, no. 2 (1982): 141–51.

———. *The End of the World: The Science and Ethics of Human Extinction*. Routledge, 1996. https://books.google.de/books?id=aWIU17K6JdEC.

———. "The Risk That Humans Will Soon Be Extinct." *Philosophy* 85, no. 4 (2010): 447–63.

———. "Observership in Cosmology: The Anthropic Principle." *Mind* 92, no. 368 (1983): 573–79.

———. "Why Not Let Life Become Extinct?" *Philosophy* 58, no. 225 (1983): 329–38.

Levin, S.B. *Posthuman Bliss?: The Failed Promise of Transhumanism*. Oxford University Press, Incorporated, 2020. https://books.google.de/books?id=HKkPEAAAQBAJ.

Lewis, Kevin N. "The Prompt and Delayed Effects of Nuclear War." *Scientific American* 241, no. 1 (1979): 35–47.

Lewis, Simon L., and Mark A. Maslin. "Defining the Anthropocene." *Nature* 519, no. 7542 (2015): 171–80.

Liessmann, Konrad Paul. "Reflexió Després d'Auschwitz i Hiroshima: Günther Anders i Hannah Arendt." *Enrahonar. An International Journal of Theoretical and Practical Reason* 46 (2011): 123–35.

Lifton, Robert Jay. "America in Vietnam—The Circle of Deception." *Trans-Action* 5, no. 4 (1968): 10–19.

———. "Beyond Psychic Numbing: A Call to Awareness." *American Journal of Orthopsychiatry* 52, no. 4 (1982): 619.

———. *Death in Life: Survivors of Hiroshima*. University of North Carolina Press, 2012.

Lifton, Robert Jay, and Richard Falk. *Indefensible Weapons*. Basic Books, 1982. https://books.google.de/books?id=L_f61e0sb8kC.

———. "Psychological Man in Revolution: The Struggle for Communal Resymbolization." In *Social Change and Human Behavior: Mental Health Challenges of the Seventies*, edited by George V. Coelho, Eli A. Rubinstein, and Elinor Stillman, 69–88. National Institute of Mental Health, 1972.

Ligotti, Thomas. *The Conspiracy Against the Human Race*. Hippocampus Press, 2010. https://i.4pcdn.org/tg/1518559287999.pdf.

Lindow, J. *Norse Mythology: A Guide to Gods, Heroes, Rituals, and Beliefs*. Oxford University Press, 2002. https://books.google.de/books?id=Y4gRDAAAQBAJ.

Linkola, Pentti. *Can Life Prevail? A Revolutionary Approach to the Environmental Crisis*. Arktos, 2011.

———. *Can Life Prevail? A Radical Approach to the Environmental Crisis*, edited by Sergio Knipe, translated by Eetou Rautio. Integral Tradition Publishing, 2009.

Lloyd, Marion, and Jeffrey Young. "Nanotechnologists Are Targets of Unabomber Copycat, Alarming Universities." *The Chronicle of Higher Education*, August 21, 2011. https://www.chronicle.com/article/nanotechnologists-are-targets-of-unabomber-copycat-alarming-universities/.

Locke, John. *Of Words or Language in General, Book III of Essays* [sic] *Concerning Human Under standing, with Notes*. London: William Tegg & Co., 1877. https://books.google.de/books?id=NZICAAAAQAAJ.

Lockwood, Jeffrey. "Six-Legged Soldiers." *The Scientist*, October 23, 2008. www.the-scientist.com/daily-news/six-legged-soldiers-44705.

Loeb, Zachary. "Life's a Glitch." *Real Life Magazine*, August 29, 2022. https://reallifemag.com/lifes-a-glitch/.

Lombroso, Patricia. "Chomsky: 'Republicans Are a Danger to the Human Species'." *Il Manifesto Global Edition*, February 25, 2016. https://chomsky.info/02252016/.

Long, Anthony A. "The Stoics on World-Conflagration and Everlasting Recurrence." *The Southern Journal of Philosophy* 23, no. Supplement (1984): 13–37.

Lorenz, Edward. *Predictability: Does the Flap of a Butterfly's Wing in Brazil Set off a Tornado in Texas?* American Association for the Advancement of Science, 1972. http://gym-portalen.dk/sites/lru.dk/files/lru/132_kap6_lorenz_artikel_the_butterfly_effect.pdf.

Lovejoy, Arthur O. *The Great Chain of Being: A Study of the History of an Idea*. Harvard University Press, 1936.

Lovejoy, A.O., and P.J. Stanlis. *The Great Chain of Being: A Study of the History of an Idea*. Transaction Publishers, 2011. https://books.google.de/books?id=ByHNG8GzUeAC.

Lowe, Adolph. "Prometheus Unbound? A New World in the Making." In *Organism, Medicine, and Metaphysics: Essays in Honor Hans Jonas on His 75th Birthday, May 10, 1978*, edited by Stuart F. Spicker, 1–10. D. Reidel Publishing Company, 1978.

Luper, Steven. "Death." In *The Stanford Encyclopedia of Philosophy*, edited by Edward N. Zalta. 2021. https://plato.stanford.edu/archives/win2021/entries/death/.

Lyell, C. *Principles of Geology, Being an Attempt to Explain the Former Changes of the Earth's Surface, by Reference to Causes Now in Operation: Vol. 3*. Murray, 1833. https://books.google.de/books?id=UAV7s0_PKl8C.

———. *Principles of Geology: Or, The Modern Changes of the Earth and Its Inhabitants Considered as Illustrative of Geology*. D. Appleton & Company, 1854. https://books.google.de/books?id=iCJDAAAAIAAJ.

Lyons, Leonard. "Loose-Leaf Notebook by Leonard Lyons." *Washington Post*, 1947.

Maas, Anthony. "General Resurrection." *The Catholic Encyclopedia*. Robert Appleton Company, 1911. www.newadvent.org/cathen/12792a.htm.

**MacAskill, William. "The History of the Term 'Effective Altruism'." *Effective Altruism Forum*, March 11, 2014a. https://forum.effectivealtruism.org/posts/9a7xMXoSiQs3EYPA2/the-history-of-the-term-effective-altruism.

———. "Replaceability, Career Choice, and Making a Difference." *Ethical Theory and Moral Practice* 17 (2014b): 269–83.

———. "Longtermism." *Effective Altruism Forum*, July 25, 2019. https://forum.effectivealtruism.org/posts/qZyshHCNkjs3TvSem/longtermism.

———. "Normative Uncertainty." PhD Thesis for the University of Oxford, 2014.

———. "Replaceability, Career Choice, and Making a Difference." *Ethical Theory and Moral Practice* 17, no. 2 (2014): 269–83.

———. *What We Owe the Future*. Basic Books, 2022. https://books.google.de/books?id=SaFTEAAAQBAJ.

———. "Why You Shouldn't Donate to Disaster Relief." *Observer*, July 28, 2015. https://observer.com/2015/07/why-you-shouldnt-donate-to-disaster-relief/.

MacAskill, William, D. Meissner, and R. Y. Chappell. "Elements and Types of Utilitarianism." *Utilitarianism.net*, 2022. www.utilitarianism.net/types-of-utilitarianism?rq=expectational#expectational-utilitarianism-versus-objective-utilitarianism.

Macfarlane, Robert. "Generation Anthropocene: How Humans Have Altered the Planet for Ever." *The Guardian*, April 1, 2016.

Mackie, John L. "Evil and Omnipotence." *Mind* 64, no. 254 (1955): 200–212.

Magnusson, Erik. "How to Reject Benatar's Asymmetry Argument." *Bioethics* 33, no. 6 (2019): 674–83.

Maharaj, Mark. "Les U. Knight of VHEMT." In *The Exploring Antinatalism Podcast*, #6. Internet Archive, 2020. https://www.youtube.com/watch?v=D_dV8ufb-FA.

Mainländer, Philipp. *The Philosophy of Redemption*, Vol. 1–2. Internet Archive, 1876/1886. https://archive.org/details/380501395thephilosophyofredemption.

Maistre, J.d. *Considerations on France*. Cambridge University Press, 1994.

Malthus, T., and R. Mayhew. *An Essay on the Principle of Population and Other Writings*. Penguin Books Limited, 2015. https://books.google.de/books?id=_Z0eBgAAQBAJ.

Manheim, David, and Anders Sandberg. "What Is the Upper Limit of Value?" 2021. https://www.researchgate.net/profile/David-Manheim/publication/348836201_What_is_the_Upper_Limit_of_Value/links/601297ee299bf1b33e2deef4/What-is-the-Upper-Limit-of-Value.pdf.

Mann, C.C. *The Wizard and the Prophet: Two Remarkable Scientists and Their Dueling Visions to Shape Tomorrow's World*. Knopf Doubleday Publishing Group, 2018. https://books.google.de/books?id=-cCtDgAAQBAJ.

Marilyn. *New Year, New Life for Old Lighthouses?* 2015. https://pathwayheart.com/new-year-new-life-for-old-lighthouses/.

Marin, Frédéric, and Camille Beluffi. "Computing the Minimal Crew: For a Multi-Generational Space Journey Towards Proxima Centauri B." *Journal of the British Interplanetary Society* 71, no. 2 (2018): 431–8.

Marvin, Ursula B. "Impact and Its Revolutionary Implications for Geology." In *Global Catastrophes in Earth History*, 147–54. Lunar and Planetary Institute, 1990. https://ntrs.nasa.gov/api/citations/19890011916/downloads/19890011916.pdf.

Marx, Karl. "Contribution to the Critique of Hegel's Philosophy of Right." *Deutsch-Französische Jahrbücher* 7, no. 10 (1844): 261–71.

———. *Critique of Hegel's "Philosophy of Right"*. Cambridge Studies in the History and Theory of Politics. Cambridge University Press, 1977. https://books.google.de/books?id=HZ_lzgEACAAJ.

Matashichi, O., and R.H. Minear. *The Day the Sun Rose in the West: Bikini, the Lucky Dragon, and I*. University of Hawaii Press, 2011. https://books.google.de/books?id=kFkEEAAAQBAJ.

Matheny, Jason G. "Reducing the Risk of Human Extinction." *Risk Analysis: An International Journal* 27, no. 5 (2007): 1335–44.

Matson, Wallace I. "Hegesias the Death-Persuader; Or, the Gloominess of Hedonism." *Philosophy* 73, no. 286 (1998): 553–7.

Maudsley, H. *Body and Will: Being an Essay Concerning Will in Its Metaphysical, Physiological, and Pathological Aspects*. D. Appleton, 1884. https://books.google.de/books?id=sSgFAQAAIAAJ.

MAW. "Dr. Fosdick at the Convocation." *Minnesota Alumni Weekly*, 1919. https://conservancy.umn.edu/bitstream/handle/11299/53834/1/umaaMag-018_3.pdf.

May, Todd. "Would Human Extinction Be a Tragedy?" *The New York Times*, 2018.

Mayhew, Peter. *Discovering Evolutionary Ecology: Bringing Together Ecology and Evolution*. Oxford University Press, 2006.

Mayor, Adrienne. *The First Fossil Hunters: Dinosaurs, Mammoths, and Myth in Greek and Roman Times*. Princeton University Press, 2011. https://books.google.de/books?id=9TwhfvU08UcC.

Mayr, E. *The Growth of Biological Thought: Diversity, Evolution, and Inheritance*. Belknap Press, 1982. https://books.google.de/books?id=pHThtE2R0UQC.

McFarland, Michael J., Matt E. Hauer, and Aaron Reuben. "Half of US Population Exposed to Adverse Lead Levels in Early Childhood." *Proceedings of the National Academy of Sciences* 119, no. 11 (2022): e2118631119.

McGregor, Rafe, and Ema Sullivan-Bissett. "Better No Longer to Be." *South African Journal of Philosophy= Suid-Afrikaanse Tydskrif Vir Wysbegeerte* 31, no. 1 (2012): 55–68.

McGuckin, J.A. *The Westminster Handbook to Origen*. The Westminster Handbooks to Christian Theology. Presbyterian Publishing Corporation, 2004. https://books.google.de/books?id=riEdrWEDFq0C.

McIntyre, J. Lewis. *Giordano Bruno*. Macmillan, 1903.

McLellan, Richard, Leena Iyengar, Barney Jeffries, and Natasja Oerlemans. *Living Planet Report 2014: Species and Spaces, People and Places*. WWF International, 2014.

McLeod, Hugh. "The Religious Crisis of the 1960s." *Journal of Modern European History* 3, no. 2 (2005): 205–30.

McMahan, Jeff. "Asymmetries in the Morality of Causing People to Exist." In *Harming Future Persons*, 49–68. Springer, 2009.

———. "Nuclear Deterrence and Future Generations." In *Nuclear Weapons and the Future of Humanity the Fundamental Questions*, edited by Avner Cohen and Steven Lee, 319–39. Rowman & Allanheld, 1986.

———. "Problems of Population Theory." In *Obligations to Future Generations*, edited by R. I. Sikora and Brian Barry. White Horse Press, 1981.

McMullen, Jay. *The Silent Spring of Rachel Carson*, 1963. www.imdb.com/title/tt0962224/.

McNeill, J. R., and P. Engelke. *The Great Acceleration: An Environmental History of the Anthropocene Since 1945*. Harvard University Press, 2014. https://books.google.de/books?id=9JG-CwAAQBAJ.

———. *The Great Acceleration: An Environmental History of the Anthropocene Since 1945*. Harvard University Press, 2016. https://books.google.de/books?id=9JG-CwAAQBAJ.

McPhee, John. *Basin and Range*. Farrar, Straus, Giroux, 1980.

McQueen, Alison. *Political Realism in Apocalyptic Times*. Cambridge University Press, 2017.

McTaggart, John McTaggart Ellis. *The Nature of Existence*. Vol. 2. Cambridge University Press, 1927.

McWhir, Anne. "Mary Shelley's Anti-Contagionism: 'The Last Man as' 'Fatal Narrative'." *Mosaic: A Journal for the Interdisciplinary Study of Literature* 35, no. 2 (2002): 23–38.

Meadows. *The Limits to Growth*. Club of Rome, n.d.

Mecklin, John. "It Is 100 Seconds to Midnight." *Bulletin of the Atomic Scientists*, 2020. https://thebulletin.org/wp-content/uploads/2020/01/2020-Doomsday-Clock-statement.pdf.

Medwin, T. *Conversations of Lord Byron: Noted During a Residence with His Lordship at Pisa, in the Years 1821 and 1822*. H. Colburn, 1824. https://books.google.de/books?id=CCQtAAAAYAAJ.

Meerloo, A.M. "Delusion and Mass Delusion." 1949. https://archive.org/stream/DelusionAndMassDelusion-ByAMMeerloo/DelusionAndMassDelusion-ByAMMeerloo_djvu.txt.

Meijers, Tim, and Angelieke L. Wolters. "Samuel Scheffler, Why Worry About Future Generations? (Oxford: Oxford University Press, 2018), Pp. Viii + 146." *Utilitas* 32, no. 4 (2020): 496–99. https://doi.org/10.1017/S0953820820000151.

Merkley, Eric, and Dominik Stecula. "Al Gore, Climate Change and An Inconvenient Truth about An Inconvenient Truth." *Newsweek*, August 17, 2017. www.newsweek.com/al-gore-climate-change-inconvenient-truth-651733.

Merriam-Webster. Extinct, 2021. https://www.merriam-webster.com/dictionary/extinct?utm_campaign=sd&utm_medium=serp&utm_source=jsonld.

Meyer, Lukas H. "More than They Have a Right to: Future People and Our Future-Oriented Projects." In *Contingent Future Persons: On the Ethics of Deciding Who Will Live, or Not, in the Future*, edited by N. Fotion and J.C. Heller, 137–56. Springer, 1997.

Michel, Jean-Baptiste, Yuan Kui Shen, Aviva Presser Aiden, Adrian Veres, Matthew K. Gray, Google Books Team, Joseph P. Pickett, Dale Hoiberg, Dan Clancy, and Peter Norvig. "Quantitative Analysis of Culture Using Millions of Digitized Books." *Science* 331, no. 6014 (2011): 176–82.

Migotti, Mark. "Schopenhauer's Pessimism in Context." In *The Oxford Handbook of Schopenhauer*, edited by Robert Wicks. Oxford University Press, 2020, 284.

Milbank, D. "In His Solitude, a Finnish Thinker Posits Cataclysms; What the World Needs Now, Pentti Linkola Believes, Is Famine and a Good War." *Wall Street Journal* (n.d.).

———. "A Strange Finnish Thinker Posits War, Famine as Ultimate 'Goods'." *The Wall Street Journal Asia*, 1, May 24, 1994.

Mill, John Stuart. *Utilitarianism*. Parker, Son and Bourn, 1863.

Miller, Boaz. "Is Technology Value-Neutral?" *Science, Technology, & Human Values* 46, no. 1 (2020): 53–80.

Miller, David. "Justice." *Stanford Encyclopedia of Philosophy*, 2021. https://plato.stanford.edu/entries/justice/#:~:text=The%20most%20plausible%20candidate%20for,render%20to%20each%20his%20due'.

Minsky, Marvin. "Afterword." In *True Names*. Bluejay Books, 1984.

MIRI. "Transparency and Financials." *Machine Intelligence Research Institute*, 2022. https://intelligence.org/transparency/.

Mitchell, Audra, and Aadita Chaudhury. "Worlding beyond 'the' 'End' of 'the World': White Apocalyptic Visions and BIPOC Futurisms." *International Relations* 34, no. 3 (2020): 309–32.

Moberly, R.W.L. "'Interpret the Bible Like Any Other Book'? Requiem for an Axiom." *Journal of Theological Interpretation* 4, no. 1 (2010): 91–110.

Mogensen, Andreas L. "Moral Demands and the Far Future." *Philosophy and Phenomenological Research* 103, no. 3 (2021): 567–85.

———. "Staking Our Future: Deontic Long-Termism and the Non-Identity Problem." *Working Paper*, 2019. https://globalprioritiesinstitute.org/andreas.

Molena, Francis. "Remarkable Weather of 1911." *Popular Mechanics*, March 1912.

Moltmann, J., and M. Kohl. *The Coming of God: Christian Eschatology*. Fortress Press, 2004. https://books.google.de/books?id=OtyDgOLxfvIC.

Montesquieu, Charles de Secondat baron de. *Persian Letters*. Oxford University Press, 1722/2008.

Moore, Gordon E. "Cramming More Components onto Integrated Circuits." 1965. http://www.computer-architecture.org/textual/Moore-Cramming-More-Components-1965.pdf.

———. *Principia Ethica*. Cambridge University Press, 1903.

Moore, J., and R. Moore. *Evolution 101*. Science 101. Greenwood Press, 2006. https://books.google.de/books?id=NlF5crD5RiIC.

Moorhouse, Fin. "Longtermism: An Introduction." *Effective Altruism*, January 27, 2021a. www.effectivealtruism.org/articles/longtermism.

———. "Longtermism Frequently Asked Questions." *longtermism.com*, 2021b. longtermism.com/faq.

Mora, Camilo, Randi L. Rollins, Katie Taladay, Michael B. Kantar, Mason K. Chock, Mio Shimada, and Erik C. Franklin. "Bitcoin Emissions Alone Could Push Global Warming above 2 C." *Nature Climate Change* 8, no. 11 (2018): 931–33.

Moravec, H. *Mind Children: The Future of Robot and Human Intelligence*. Harvard University Press, 1988. https://books.google.de/books?id=56mb7XuSx3QC.

More, Max. "Embrace, Don't Relinquish the Future". *Extropy*, May 7, 2000. https://www.kurzweilai.net/embrace-dont-relinquish-the-future.

———. "The Extropian Principles, v. 3.0: A Transhumanist Declaration." *The Extropian Principles* 3 (1998).

Morrisette, Peter M. "The Evolution of Policy Responses to Stratospheric Ozone Depletion." *Natural Resources Journal* (1989): 793–820.

Morris, Theresa. *Hans Jonas's Ethic of Responsibility: From Ontology to Ecology*. State University of New York Press, 2013. https://books.google.de/books?id=L1ma0LU8xZUC.

Morrison, David, Clark Chapman, and Paul Slovic. *The Impact Hazard*. Draft Chapter for University of Arizona Space Science Series Volume on the Hazards of Impacts by Comets and Asteroids, 1993. https://scholarsbank.uoregon.edu/xmlui/bitstream/handle/1794/22416/slovic_329.pdf?sequence=1.

Mounk, Yascha. "An Interview with T. M. Scanlon (Part VI)." *The Utopian*, 2011. www.the-utopian.org/T.M.-Scanlon-Interview-6.

Moyers, Bill. "Rachel Carson." *PBS*, 2007. www.pbs.org/moyers/journal/09212007/profile.html.

Moyers, B.D., and J. Campbell. *Joseph Campbell and the Power of Myth: With Bill Moyers*. Journal Graphics Incorporated, 1988. https://books.google.de/books?id=wvzGZwEACAAJ.

Moynihan, Thomas. "The End of Us." *Aeon*, 2019. https://aeon.co/essays/to-imagine-our-own-extinction-is-to-be-able-to-answer-for-it.

———. *X-Risk: How Humanity Discovered Its Own Extinction*. MIT Press, 2020. https://books.google.de/books?id=7oUBEAAAQBAJ.

Muehlhauser, Luke. "AI Risk & Opportunity: A Timeline of Early Ideas and Arguments." March 31, 2012. www.lesswrong.com/posts/Qdq2SKyMi8vf7Snxq/ai-risk-and-opportunity-a-timeline-of-early-ideas-and.

———. *Intelligence Explosion FAQ*. First Published. Machine Intelligence Research Institute, 2011. https://intelligence.org/ie-faq/.

Muehlhauser, Luke, and Louie Helm. "The Singularity and Machine Ethics." In *Singularity Hypotheses*, 101–26. Springer, 2012.

Muir, John. *A Thousand-Mile Walk to the Gulf*. Houghton Mifflin Harcourt, 1998.

Mukunda, Gautam, Kenneth A. Oye, and Scott C. Mohr. "What Rough Beast? Synthetic Biology, Uncertainty, and the Future of Biosecurity." *Politics and the Life Sciences* 28, no. 2 (2009): 2–26.

Mulgan, Tim. *Utilitarianism*. Elements in Ethics. Cambridge University Press, 2020. https://books.google.de/books?id=9wCaygEACAAJ.

Müller, Christopher John. "Hollywood, Exile, and New Types of Pictures: Günther Anders's 1941 California Diary 'Washing the Corpses of History'." *Modernism/modernity* 5, no. 4 (February 2021). https://doi.org/10.26597/mod.0185.

Müller, I. *A History of Thermodynamics: The Doctrine of Energy and Entropy*. Springer Berlin Heidelberg, 2007. https://books.google.de/books?id=u13KiGlz2zcC.

Müller, Ingo, and Wolf Weiss. "Thermodynamics of Irreversible Processes—Past and Present." *The European Physical Journal H* 37, no. 2 (2012): 139–236.

Müller, Vincent C., and Nick Bostrom. "Future Progress in Artificial Intelligence: A Poll among Experts." *AI Matters* 1, no. 1 (2014): 9–11.

Munster, R. van, and C. Sylvest. *The Politics of Globality since 1945: Assembling the Planet*. Taylor & Francis, 2016. https://books.google.de/books?id=ydAmDAAAQBAJ.

MUP. Most Expect ChatGPT Will Be Used for Cheating. Monmouth University Poll. February 15, 2023. https://www.monmouth.edu/polling-institute/reports/monmouthpoll_us_021523/.

Murti, Aditi. "All You Need to Know about Stormquakes, a Newly Discovered Natural Disaster." *The Swaddle*, December 11, 2019. https://theswaddle.com/what-are-stormquakes/.

Musk, Elon. "Tweet." *Twitter*, August 3, 2014. https://twitter.com/elonmusk/status/495759307346952192?lang=en.

Mutch, Thomas. *Volume 4 of 1981 NASA Authorization: Hearings Before the Subcommittee on Space Science and Applications of the Committee on Science and Technology, U.S. House of Representatives, Ninety-Sixth Congress, First Session, United States. Congress. House. Committee on Science and Technology. Subcommittee on Transportation, Aviation, and Communications.* 1981 NASA Authorization: Hearings Before the Subcommittee on Space Science and Applications of the Committee on Science and Technology, U.S. House of Representatives, Ninety-Sixth Congress, First Session. U.S. Government Printing Office, 1980. https://books.google.de/books?id=PIs9wP7oRagC.

Myers, Paul, and W. Parker Maudlin. *International Population Statistics Reports: Series P-90*. International Population Statistics Reports: Series P-90. U.S. Government Printing Office, 1952. https://books.google.de/books?id=1aMvAAAAYAAJ.

NA. *The Methods of Ethics*. Palgrave Macmillan, 2016. https://books.google.de/books?id=UUCxCwAAQBAJ.

Nagel, Thomas. "Death." *Noûs* 4, no. 1 (1970), 73–80.

———. "What Is It like to Be a Bat?" *The Philosophical Review* 83, no. 4 (1974): 435–50.

Naess, Arne. "The Shallow and the Deep: A Summary." *Inquiry* 16, no. 1 (1973).

Nakano-Okuno, Mariko. *Sidgwick and Contemporary Utilitarianism*. Springer, 2011.

Narveson, Jan. "Future People and Us." In *Obligations to Future Generations, edited by*. Richard I. Sikora and Brian M. Barry. White Horse Press, 1978.

———. "Moral Problems of Population." *The Monist* (1973): 62–86.

———. "Utilitarianism and New Generations." *Mind* 76, no. 301 (1967): 62–72.

Nath, Dipak. "The Darkness of Dark Matter and Dark Energy." *International Journal of Engineering and Applied Sciences* 5, no. 6 (2018): 257209.

Nathan, Otto, and Heinz Norden, eds. *Einstein on Peace*. Schocken Books, 1960.

Neall, Beatrice S. "Amillennialism Reconsidered." *Andrews University Seminary Studies (AUSS)* 43, no. 1 (2005): 17.

Newberry, Toby. "How Many Lives Does the Future Hold." In *Global Priorities Institute Technical Report*. 2021. https://globalprioritiesinstitute.org/wp-content/uploads/Toby-Newberry_How-many-lives-does-the-future-hold.pdf

Newell, Norman D. "Catastrophism and the Fossil Record." *Evolution* 10, no. 1 (1956): 97–101.

———. "Periodicity in Invertebrate Evolution." *Journal of Paleontology* (1952): 371–85.

Newhall, Christopher G., and Stephen Self. "The Volcanic Explosivity Index (VEI) an Estimate of Explosive Magnitude for Historical Volcanism." *Journal of Geophysical Research: Oceans* 87, no. C2 (1982): 1231–38.

Newport, Frank. "Five Key Findings on Religion in the US." *Gallup*, 2016. https://news.gallup.com/poll/200186/five-key-findings-religion.aspx.

Nietzsche, F., and T. Common. *The Gay Science*. Dover Philosophical Classics. Dover Publications, 2006. https://books.google.de/books?id=xj41AwAAQBAJ.

Ninkovich, Dragoslav, and William L. Donn. "Explosive Cenozoic Volcanism and Climatic Implications: Tectonic Plate Motion Modifies the Marine Record of Explosive Volcanism and Complicates Its Interpretation." *Science* 194, no. 4268 (1976): 899–906.

Ninkovich, Dragoslav, Nick J. Shackleton, Aboul A. Abdel-Monem, John D. Obradovich, and G. Izett. "K—Ar Age of the Late Pleistocene Eruption of Toba, North Sumatra." *Nature* 276, no. 5688 (1978a): 574–77.

Ninkovich, D., R.S.J. Sparks, and M.T. Ledbetter. "The Exceptional Magnitude and Intensity of the Toba Eruption, Sumatra: An Example of the Use of Deep-Sea Tephra Layers as a Geological Tool." *Bulletin Volcanologique* 41, no. 3 (1978b): 286–98.

Nobel. "Al Gore, Facts." *The Nobel Prize*, December 1, 2022. www.nobelprize.org/prizes/peace/2007/gore/facts/.

Nobel. *Al Gore, Facts*. The Nobel Prize, 2021. https://www.nobelprize.org/prizes/peace/2007/gore/facts/.

Nobelstiftelsen. *Chemistry: 1922–1941*. Chemistry. World Scientific, 1999. https://books.google.de/books?id=B8raAAAAMAAJ.

Nouri, Ali, and Christopher Chyba. "Biotechnology and Biosecurity." In *Global Catastrophic Risks*, edited by Nick Bostrom and Milan Ćirković. Oxford University Press, 2008, 450–80.

Nuttal, Chris. *Tech's Inconvenient Truth*. Financial Times, 2020. https://www.ft.com/content/8e0d7eef-00ad-4daa-8e05-28b53c783a97.

Nye, J.S. *Nuclear Ethics*. Free Press, 1986. https://books.google.de/books?id=vipGnwkTng4C.

NYT. "A Nameless Crime." *New York Times*, 1982. https://timesmachine.nytimes.com/timesmachine/1982/06/27/238548.html?pageNumber=167.

Oake, Roger B. "Montesquieu's Religious Ideas." *Journal of the History of Ideas* 14, no. 4 (1953): 548–60.

O'Brien, Patrick. *Atlas of World History*. Oxford University Press, 2010. https://books.google.de/books?id=sZ_ZRAAACAAJ.

———. *Philip's Atlas of World History*. George Philip, 2005.

OED. "Moral." *Online Etymology Dictionary*, 2022. www.etymonline.com/search?q=moral.

———. "Omnicide." *Oxford Engish Dictionary*, 2022. www.oed.com/view/Entry/246601#eid12254969.

Ogle, W.E. *An Account of the Return to Nuclear Weapons Testing by the United States after the Test Moratorium 1958–1961*. US DOE Publication NVO-291, 1985.

———. *Operation Castle. The Operation Plan Number 1–53. Task Group 7.1*. Kaman Tempo Santa Barbara CA, 1984.

O'Neill, John Joseph. *Almighty Atom: The Real Story of Atomic Energy*. Vol. I. Washburn, 1945.

Okasha, Samir. "Darwinian Metaphysics: Species and the Question of Essentialism." *Synthese* 131 (2002): 191–213.

Ord, Toby. "Opening Keynote." *Presented at the EA Global*, 2016. www.youtube.com/watch?v=VH2LhSod1M4&t=194s.

———. *The Precipice: Existential Risk and the Future of Humanity*. Hachette Books, 2020. https://books.google.de/books?id=tGCjDwAAQBAJ.

———. "Toby Ord on the Precipice and Humanity's Potential Futures." *80,000 Hours*, March 7, 2020. https://80000hours.org/podcast/episodes/toby-ord-the-precipice-existential-risk-future-humanity/.

———. "Why I'm Not a Negative Utilitarian. A Mirror Clear." 2013. http://www.amirrorclear.net/academic/ideas/negative-utilitarianism/.

Oremus, Will. 2013. "David Attenborough Calls Humans a 'Plague on the Earth.' Isn't He Sort of Right?" *Slate*, January 23, 2013. https://slate.com/technology/2013/01/david-attenborough-calls-humanity-a-plague-on-the-earth-isn-t-he-sort-of-right.html.

Oreskes, Naomi. "The Scientific Consensus on Climate Change." *Science* 306, no. 5702 (2004): 1686–1686.

Orlowski, D. "End Times: How the Antichrist Will Use Artificial Intelligence." 2014. www.christianrapturebooks.com/scripture-teachings/end-times-how-the-antichrist-will-use-artificial-intelligence/.

Orwell, George. "Lear, Tolstoy and the Fool." *Polemic* 7 (March 1947): 2–17.

Osborn, Fairfield. *Our Plundered Planet*. Little, Brown and Company, 1948. https://books.google.de/books?id=AKM1AAAAMAAJ.

OTA. *Technologies Underlying Weapons of Mass Destruction*. Office of Technology Assessment, 1993. https://ota.fas.org/reports/9344.pdf.

Pace, Eric. "E. M. Cioran, 84, Novelist and Philosopher of Despair." *The New York Times*, June 22, 1995. https://www.nytimes.com/1995/06/22/obituaries/e-m-cioran-84-novelist-and-philosopher-of-despair.html.

Paley, Morton D. *"The Last Man": Apocalypse Without Millennium*. Oxford University Press, 1999.

———. "Mary Shelley's The Last Man: Apocalypse Without Millennium." *The Keats-Shelley Review* 4, no. 1 (1989): 1–25.

Palmer, T. *Controversy Catastrophism and Evolution: The Ongoing Debate*. Springer, 2012. https://books.google.de/books?id=VQbTBwAAQBAJ.

Pamlin, Dennis, and Stuart Armstrong. "12 Risks That Threaten Human Civilisation: The Case for a New Risk Category." *Global Challenges Foundation*, 2015. www.academia.edu/12590781/Risks_that_threaten_human_civilisation.

Parfit, Derek. "Acts and Outcomes: A Reply to Boonin-Vail." *Philosophy & Public Affairs* 25, no. 4 (1996): 308–17.

———. "Lewis, Perry, and What Matters." In *The Identities of Persons*, 91–107. 1976.

———. *On What Matters: Volume One*. Oxford University Press, 2013.

———. *On What Matters: Volume Two*. Oxford University Press, 2011. https://books.google.de/books?id=ta0-AAAAQBAJ.

———. *On What Matters: Volume Three*. Oxford University Press, 2017.

———. *Reasons and Persons*. Oxford University Press, 1984.

Parfit, Derek, M.D. Bayles, J. Glover, John Robertson, and J. Feinberg. "On Doing the Best for Our Children." In *Population and Political Theory*, 68–80. Wiley-Blackwell, 2010.

Parsons, K.M., and R.A. Zaballa. *Bombing the Marshall Islands: A Cold War Tragedy*. Cambridge University Press, 2017. https://books.google.de/books?id=MLYrDwAAQBAJ.

Partridge, E. *Responsibilities to Future Generations: Environmental Ethics*. Prometheus Books, 1981. https://books.google.de/books?id=-pNkAAAAIAAJ.

Passmore, John. *Man's Responsibility for Nature,* 88–89. Scribners, 1974.

Pelton, Joseph. *Space Systems and Sustainability: From Asteroids and Solar Storms to Pandemics and Climate Change*. Springer, 2021. https://link.springer.com/content/pdf/10.1007/978-3-030-75735-9.pdf.

Penfield, G. T. "Definition of a major igneous zone in the central Yucatan platform with aeromagnetics and gravity." In *Soc. Explor. Geophys. Annu. Meeting, Tech. Progr. Abstracts*, 1981.

Pentti, Linkola. "The Doctrine of Survival and Doctor Ethics." n.d. www.penttilinkola.com/pentti_linkola/ecofascism_writings/translations/voisikoelamavoittaa_translation/VI%20-%20The%20World%20And%20We/.

Perry, R.B. *The Present Conflict of Ideals: A Study of the Philosophical Background of the World War*. Longmans, Green, 1922. https://books.google.de/books?id=-XdYAAAAMAAJ.

Pérez Cebada, Juan Diego. "An Editorial Flop Revisited: Rethinking the Impact of M. Bookchin's Our Synthetic Environment on Its Golden Anniversary." *Global Environment* 6, no. 12 (2013): 250–73.

Petersen, John L. *Out of the Blue: How to Anticipate Big Future Surprises*. Madison Books, 1999.

Peterson, M. *An Introduction to Decision Theory*. Cambridge Introductions to Philosophy. Cambridge University Press, 2009. https://books.google.de/books?id=qUBdAAAAQBAJ.

PEW. "America's Changing Religious Landscape." *Pew Research Center*, 2015. https://www.pewresearch.org/religion/2015/05/12/americas-changing-religious-landscape/.

———. "Americans Are Far More Religious than Adults in Other Wealthy Nations." 2018. https://www.pewresearch.org/fact-tank/2018/07/31/americans-are-far-more-religious-than-adults-in-other-wealthy-nations/.

Pew Research Center. "In US, Decline of Christianity Continues at Rapid Pace." *Pew Research Center's Religion & Public Life Project*, 2019. https://www.pewresearch.org/religion/2019/10/17/in-u-s-decline-of-christianity-continues-at-rapid-pace/.

Pimm, Stuart L., Gareth J. Russell, John L. Gittleman, and Thomas M. Brooks. "The Future of Biodiversity." *Science* 269, no. 5222 (1995): 347–50.

Pinch, Geraldine. *Handbook of Egyptian Mythology*. Abc-Clio, 2002.

Pinker, Steven. *The Better Angels of Our Nature: The Decline of Violence in History and Its Causes*. Penguin, 2011.

Pitt, William. 1835. No. VIII. *Blackwood's Edinburgh Magazine*. New York. https://www.google.de/books/edition/Blackwood_s_Edinburgh_Magazine/kghGAAAAcAAJ?hl=en&gbpv=1&dq=The+progress+of+inventive+cruelty+kept+pace+with+the+gory+necessities+of+the&pg=PA458&printsec=frontcover.

Poe, E.A. *The Conversation of Eiros and Charmion*. Feedbooks, 1839. https://books.google.de/books?id=9e4awQEACAAJ.

Poe, E.A., T.O. Mabbott, E.D. Kewer, and M.C. Mabbott. *Collected Works of Edgar Allan Poe: Tales and Sketches*, edited by Thomas Ollive Mabbott, Eleanor D. Kewer, and Maureen C. Mabbott. Belknap Press of Harvard University Press, 1978. https://books.google.de/books?id=ZJXsoAEACAAJ.

Pollack, Andrew. "Traces of Terror: The Science; Scientists Create a Live Polio Virus." *New York Times*, July 12, 2002.

Pope, Alexander. *Essay on Man*. Edited by Mark Pattison. Clarendon Press, 1879 [1733–1734].

Popper, K.R. *The Open Society and Its Enemies*. Routledge & Sons, Limited, 1945. https://books.google.de/books?id=S9wqzgEACAAJ.

Posner, R.A. *Catastrophe: Risk and Response*. Oxford University Press, 2004. https://books.google.de/books?id=bePiwAEACAAJ.

Powell, Corey S., and Diane Martindale. "20 Ways the World Could End." *Discover New York* 21, no. 10 (2000): 50–57.

Prochnau, Bill. "The Watt Controversy." *The Washington Post*, June 30, 1981. www.washingtonpost.com/archive/politics/1981/06/30/the-watt-controversy/d591699b-3bc2-46d2-9059-fb5d2513c3da/.

Protopapadakis, Evangelos D. "Environmental Ethics and Linkola's Ecofascism: An Ethics beyond Humanism." *Frontiers of Philosophy in China* 9, no. 4 (2014): 586–601.

PRRI. "The 2020 Census of American Religion." *Public Religion Research Institute*, 2020. www.prri.org/wp-content/uploads/2021/07/PRRI-Jul-2021-Religion.pdf.

Pugwash. "Joseph Rotblat." *Pugwash Conferences on Science and World Affairs*, 2022. https://pugwash.org/history/joseph-rotblat/.

QI. "Life Is a Sexually Transmitted Terminal Disease." *Quote Investigator*, 2017. https://quoteinvestigator.com/2017/01/29/life/.

Quaglia, Sofia. "'Fantastic Giant Tortoise' Species Thought Extinct for 100 Years Found Alive." *The Guardian*, 2022. https://www.theguardian.com/environment/2022/jun/09/galapagos-fantastic-giant-tortoise-species-thought-extinct-found-alive.

Rabenberg, Michael. "Harm." *Journal of Ethics and Social Philosophy* 8 (2014): viii.

Rabinowitch, Eugene. "Five Years After." *Bulletin of the Atomic Scientists* 7, no. 1 (1951): 3.

Rampino, Michael R., and Stephen Self. "Bottleneck in Human Evolution and the Toba Eruption." *Science* 262, no. 5142 (1993): 1955–1955.

Rampino, Michael R., Stephen Self, and Richard B. Stothers. "Volcanic Winters." *Annual Review of Earth and Planetary Sciences* 16 (1988): 73–99.

Randall, Lisa. "Dark Matter and the Dinosaurs: An Evening with Dr. Lisa Randall [Video]." *Encompass*, 2017. https://encompass.eku.edu/chautauquavideos/12/

Randle, Melanie, and Richard Eckersley. "Public Perceptions of Future Threats to Humanity and Different Societal Responses: A Cross-National Study." *Futures* 72 (2015): 4–16.

Ransom, Amy J. "The First Last Man: Cousin de Grainville's Le Dernier Homme." *Science Fiction Studies* 41, no. 2 (2014): 314–40.

Rasmussen, Norman, *et al.* "Reactor Safety Study." In *WASH-1400*. US Nuclear Regulatory Committee, 1975. https://www.nrc.gov/docs/ML1622/ML16225A002.pdf.

Raup, David M., and J. John Sepkoski Jr. "Mass Extinctions in the Marine Fossil Record." *Science* 215, no. 4539 (1982): 1501–3.

———. "Periodicity of Extinctions in the Geologic Past." *Proceedings of the National Academy of Sciences* 81, no. 3 (1984): 801–5.

Rawls, J. *Political Liberalism. Columbia Classics in Philosophy*. Columbia University Press, 2005. https://books.google.de/books?id=vlCsAgAAQBAJ.

———. *A Theory of Justice*. Harvard Paperback. Belknap Press of Harvard University Press, 1971. https://books.google.de/books?id=PMdsAAAAIAAJ.

Reardon, B.M.G. *Religious Thought in the Nineteenth Century: Illustrated from Writers of the Period.* Cambridge University Press, 1966. https://books.google.de/books?id=fRU0AAAAIAAJ.

Redfield, Robert. "Consequences of Atomic Energy." *The Phi Delta Kappan* 27, no. 8 (1946): 221–24.

Reed, Peter, and David Rothenberg. *Wisdom in the Open Air: The Norwegian Roots of Deep Ecology*. University of Minnesota Press, 1993.

Rees, Martin J. "The Collapse of the Universe: An Eschatological Study." *The Observatory* 89 (1969): 193–98.

———. *Our Final Hour: A Scientist's Warning: How Terror, Error, and Environmental Disaster Threaten Humankind's Future in This Century—On Earth and Beyond.* Basic Books, 2003. https://books.google.de/books?id=GqvgCDPFZ18C.

Reiss, Louise Zibold. "Strontium-90 Absorption by Deciduous Teeth: Analysis of Teeth Provides a Practicable Method of Monitoring Strontium-90 Uptake by Human Populations." *Science* 134, no. 3491 (1961): 1669–73.

Revelle. "Atmospheric Carbon Dioxide, in Restoring the Quality of Our Environment." *The Environmental Pollution Panel President's Science Advisory Committee*, November 1965. www-legacy.dge.carnegiescience.edu/labs/caldeiralab/Caldeira%20downloads/PSAC,%201965,%20Restoring%20the%20Quality%20of%20Our%20Environment.pdf.

Revkin, Andrew C. "Endless Summer: Living with the Greenhouse Effect." *Discover* 9, no. 10 (1988): 50–61.

———. "Special Report: Endless Summer—Living with the Greenhouse Effect." *Discover*, June 23, 2008. www.discovermagazine.com/environment/special-report-endless-summerliving-with-the-greenhouse-effect.

Rhodes, Richard. *The Making of the Atomic Bomb.* Simon & Schuster, 1986.

Risse, M. *Political Theory of the Digital Age.* Cambridge University Press, 2023. https://books.google.de/books?id=2k6hEAAAQBAJ.

Ringmar, Erik. "What Are Public Moods?" *European Journal of Social Theory* 21, no. 4 (2018): 453–69.

Rio. *Principle 15: The Precautionary Approach.* The Rio Declaration, 1992. https://www.gdrc.org/u-gov/precaution-7.html.

Roberts, M.A. "The Nonidentity Problem." In *The Stanford Encyclopedia of Philosophy*. Winter Edition, edited by Edward N. Zalta and Uri Nodelman, 2022. https://plato.stanford.edu/archives/win2022/entries/nonidentity-problem/.

Robertson, P. *The End of the Age.* Word Pub., 1995. https://books.google.de/books?id=56ewzgEACAAJ.

———. *The End of the Age.* Thomas Nelson Incorporated, 1998. https://books.google.de/books?id=cxrjy1xx_RkC.

Robock, Alan, Luke Oman, Georgiy L. Stenchikov, Owen B. Toon, Charles Bardeen, and Richard P. Turco. "Climatic Consequences of Regional Nuclear Conflicts." *Atmospheric Chemistry and Physics* 7, no. 8 (2007): 2003–12.

Robock, Alan, and Owen Brian Toon. "Self-Assured Destruction: The Climate Impacts of Nuclear War." *Bulletin of the Atomic Scientists* 68, no. 5 (2012): 66–74.

Rockström, Johan, Will Steffen, Kevin Noone, Åsa Persson, F. Stuart Chapin III, Eric Lambin, Timothy M. Lenton, Marten Scheffer, Carl Folke, and Hans Joachim Schellnhuber. "Planetary Boundaries: Exploring the Safe Operating Space for Humanity." *Ecology and Society* 14, no. 2 (2009b).

Rockström, Johan, Will Steffen, Kevin Noone, Åsa Persson, F. Stuart Chapin, Eric F. Lambin, Timothy M. Lenton, Marten Scheffer, Carl Folke, and Hans Joachim Schellnhuber. "A Safe Operating Space for Humanity." *Nature* 461, no. 7263 (2009a): 472–75.

Roffey, R., Anders Tegnell, and Fredrik Elgh. "Biological Warfare in a Historical Perspective." *Clinical Microbiology and Infection* 8, no. 8 (2002): 450–54.

Rome, Adam. "'Give Earth a Chance': The Environmental Movement and the Sixties." *The Journal of American History* 90, no. 2 (2003): 525–54.

Rønnow-Rasmussen, Toni. "Intrinsic and Extrinsic Value." In *The Oxford Handbook of Value Theory*, edited by Iwao Hirose and Jonas Olson. Oxford University Press, 2015, 29–43.

Rønnow-Rasmussen, Toni, and Michael J. Zimmerman. *Recent Work on Intrinsic Value*, Vol. 17. Springer Science & Business Media, 2006.

Root, T. "The 'Balance of Nature' Is an Enduring Concept. But It's Wrong." *National Geographic*, 2019.

Rose, K.D. *One Nation Underground: The Fallout Shelter in American Culture*. American History and Culture. New York University Press, 2004. https://books.google.de/books?id=DKsUCgAAQBAJ.

Rosenmeyer, T.G., and L.A. Seneca. *Senecan Drama and Stoic Cosmology*. University of California Press, 1989. https://books.google.de/books?id=PVuxQgAACAAJ.

Roser, Max. "Longtermism: The Future Is Vast—What Does This Mean for Our Own Life?" *Our World in Data*, March 15, 2022. https://ourworldindata.org/longtermism.

Rothberg, Peter. "When Jonathan Schell Introduced Me to Robert McNamara." *The Nation*, June 13, 2014. https://www.thenation.com/article/archive/when-jonathan-schell-introduced-me-robert-mcnamara/.

Rowe, Thomas, and Simon Beard. "Probabilities, Methodologies and the Evidence Base in Existential Risk Assessments." Working Paper. 2018. https://eprints.lse.ac.uk/89506/1/Beard_Existential-Risk-Assessments_Accepted.pdf.

Rubin, Charles. "Reading Rachel Carson." *The New Atlantis*, September 27, 2012. www.thenewatlantis.com/publications/reading-rachel-carson.

Rudwick, M.J.S. *Bursting the Limits of Time: The Reconstruction of Geohistory in the Age of Revolution*. University of Chicago Press, 2005. https://books.google.de/books?id=a5IlEAAAQBAJ.

Ruse, M. *Monad to Man: The Concept of Progress in Evolutionary Biology*. Harvard University Press, 1996. https://books.google.de/books?id=beXaAAAAMAAJ.

Russel, Paul, and Anders Kraal. "Hume on Religion." *Stanford Encyclopedia of Philosophy*. 2021. https://plato.stanford.edu/archives/win2021/entries/hume-religion.

Russell, Bertrand. "'Am I an Atheist or an Agnostic?'—Bertrand Russell (1947)." In *Voices of Unbelief*, edited by Dale Mcgowan. Greenwood, 2012, 143–46.

———. "The Atomic Bomb." *Online*, 1945. https://russell.humanities.mcmaster.ca/civbomb10.pdf.

———. "A Free Man's Worship." *Why I Am Not a*, 1903.

———. *Has Man a Future*? Penguin Books, 1961.

———. *Has Religion Made Useful Contributions to Civilization?* Rationalist Press Association, Limited, 1930.

———. *Icarus; or, the Future of Science*. Lulu.com, 2015. https://books.google.de/books?id=dhOfCgAAQBAJ.

———. "Man's Peril." *BBC*, 1954a. www.youtube.com/watch?v=oZzm6x_IMFE.

Russell, Bertrand, and Albert Einstein. "Russell-Einstein Manifesto." 1955. www.atomicheritage.org/key-documents/russell-einstein-manifesto.

Russell, B., and E. Bertrand Russell. *Human Society in Ethics and Politics*. Mentor Book. Allen & Unwin, 1954b. https://books.google.de/books?id=FR4tAAAAMAAJ.

Russell, B., and Bertrand Russell Supranational Society. *Bertrand Russell, the Social Scientist*. Bertrand Russell Supranational Society, 1973. https://books.google.de/books?id=RbAYAAAAIAAJ.

Russell, B., J.G. Slater, and P. Köllner. *A Fresh Look at Empiricism: 1927–42*. Russell, Bertrand: Selections, 1983. Routledge, 1996. https://books.google.de/books?id=oEoi0HnF7j0C.

Russell, F.A.R., and E.D. Archibald. "On the Unusual Optical Phenomena of the Atmosphere, 1883–1886, Including Twilight Effects, Coloured Suns, Moons, Etc." In *The Eruption of Krakatoa and Subsequent Phenomena*, edited by G.J. Symons, 151–463. Trübner & Company, 1888.

Russell, Josiah Cox. "Late Ancient and Medieval Population." *Transactions of the American Philosophical Society* 48, no. 3 (1958): 1–152.

Russell, Paul, and Anders Kraal. "Hume on Religion." In *The Stanford Encyclopedia of Philosophy*. Winter 2021 ed., edited by Edward N. Zalta, 2017. https://plato.stanford.edu/archives/win2021/entries/hume-religion/.

Russell, S. *Human Compatible: Artificial Intelligence and the Problem of Control*. Penguin Publishing Group, 2019. https://books.google.de/books?id=8vm0DwAAQBAJ.

Russill, Chris. "Climate Change Tipping Points: Origins, Precursors, and Debates." *Wiley Interdisciplinary Reviews: Climate Change* 6, no. 4 (2015): 427–34.

Sachs, J.R. *The Christian Vision of Humanity*. Zacchaeus Studies: New Testament. Liturgical Press, 2017. https://books.google.de/books?id=nF6tDwAAQBAJ.

Sade, D. *Philosophy in the Bedroom*. Phoemixx Classics Ebooks, 2021. https://books.google.de/books?id=FWJNEAAAQBAJ.

Sade, de. La Nouvelle Justine, Ou. "Les Malheurs de La Vertu: Suivie de L'histoire de Juliette, Sa Soeur." In *Archives of Sexuality and Gender: L'Enfer de La Bibliothèque Nationale de France*, 1797. https://books.google.de/books?id=xEL0xwEACAAJ.

Sade, Marquis de. *Juliette*. Translated by Austryn Wainhouse. Complete American Edition, 1968. Online Read Free Novel. https://onlinereadfreenovel.com/marquis-de-sade/69920-juliette_read.html.

Safari. "Eruvin 13b," 2017. www.sefaria.org/Eruvin.13b.15?ven=William_Davidson_Edition_-_English&vhe=William_Davidson_Edition_-_Vocalized_Aramaic&lang=bi&with=About&lang2=en.

Sagan, Carl. *Carl Sagan Discusses the Book "Contact"*, 1985. https://studsterkel.wfmt.com/programs/carl-sagan-discusses-book-contact?t=NaN%2CNaN&a=%2C.

———. *Cosmos*. Random House Publishing Group, 2011.

———. *Future Space Programs 1975: Hearings Before the Subcommittee on Space Science and Applications of the Committee on Science and Technology, U.S. House of Representatives, Ninety-Fourth Congress, First Session . . .* U.S. Government Printing Office, 1975. https://books.google.de/books?id=4xErAAAAMAAJ.

———. "Nuclear War and Climatic Catastrophe: Some Policy Implications." *Foreign Affairs* 62, no. 2 (1983a): 257–92.

———. "The Nuclear Winter: The World after Nuclear War." *Parade*, 1983b. www.e-reading-lib.com/bookreader.php/148584/The_Nuclear_Winter_:_The_World_After_Nuclear_War.pdf.

———. "The Quest for Extraterrestrial Intelligence." *Cosmic Search* 1, no. 2 (1979): 47.

Sagan, C., and R.P. Turco. *A Path Where No Man Thought: Nuclear Winter and the End of the Arms Race*. Random House, 1990. https://books.google.de/books?id=-LaAAAAMAAJ.

Sagan, C., and I.S. Shklovskii. *Intelligent Life in the Universe*. Random House Incorporated, 1980. https://books.google.de/books?id=uZ2EAAAACAAJ.

Saltus, Edgar Evertson. *The Philosophy of Disenchantment*. Belford Company, 1885.

Sambursky, S. "The Stoic Doctrine of Eternal Recurrence." In *The Concepts of Space and Time: Their Structure and Their Development*, edited by Milič Čapek. The Netherlands: Springer. 1976, 167–71.

Sample, Ian. "Pressure Points." *The Guardian*, Thursday, October 14, 2004.

Samuel, Sigal. "Effective Altruism's Most Controversial Idea." *Vox*, September 6, 2022. www.vox.com/future-perfect/23298870/effective-altruism-longtermism-will-macaskill-future.

Sandberg, Anders. "Ethics of Brain Emulations." *Journal of Experimental & Theoretical Artificial Intelligence* 26, no. 3 (2014): 439–57.

———. "The Five Biggest Threats to Human Existence." *The Conversation*, May 29, 2014. https://theconversation.com/the-five-biggest-threats-to-human-existence-27053.

———. *Grand Futures: Visions and Limits of What Can Be Achieved*, Forthcoming.

Sandberg, Anders, Stuart Armstrong, and Milan M. Cirkovic. "That Is Not Dead Which Can Eternal Lie: The Aestivation Hypothesis for Resolving Fermi's Paradox." *ArXiv Preprint ArXiv:1705.03394* (2017).

Sandberg, Anders, and Nick Bostrom. "Global Catastrophic Risks Survey." *Civil Wars* 98, no. 30 (2008): 4.

Sandberg, Anders, Eric Drexler, and Toby Ord. "Dissolving the Fermi Paradox." *ArXiv Preprint ArXiv:1806.02404* (2018).

Sandberg, Anders, Jason G. Matheny, and M.M. Ćirković. "How Can We Reduce the Risk of Human Extinction." *Bulletin of the Atomic Scientists* 9 (2008).

Scanlon, T.M. *What We Owe to Each Other*. Harvard University Press, 1998. https://books.google.de/books?id=FwuZcwMdtzwC.

Scheffler, Samuel. *Death and the Afterlife*. Oxford University Press, 2013.

———. "Immigration and the Significance of Culture." In *Nationalism and Multiculturalism in a World of Immigration*, 119–50. Springer, 2009.

———. "Immigration and the Significance of Culture." *Philosophy & Public Affairs* 35, no. 2 (2007): 93–125.

———. *Why Worry about Future Generations*? Uehiro Series in Practical Ethics. Oxford University Press, 2018. https://books.google.de/books?id=wqZTDwAAQBAJ.

Scheffler, S., and N. Kolodny. *Death and the Afterlife*. The Berkeley Tanner Lectures. Oxford University Press, 2013. https://books.google.de/books?id=5X-HAAAAQBAJ.

Schell, J. *The Fate of the Earth: And, the Abolition*. Stanford Nuclear Age Series. Stanford University Press, 1982/2000. https://books.google.de/books?id=tYKJsAEs1oQC.

Schell, Jonathan. "A Politics of Natality." *Social Research: An International Quarterly* 69, no. 2 (2002): 461–71.

Schelling, Thomas C. "Dynamic Models of Segregation." *Journal of Mathematical Sociology* 1, no. 2 (1971): 143–86.

Schilpp, Paul Arthur. "A Challenge to Philosophers in the Atomic Age." *Philosophy* 24, no. 88 (1949): 56–68.

———, ed. *The Library of Living Philosophers. Vol. 5, The Philosophy of Bertrand Russell*. Northwestern University, 1944.

Schopenhauer, Arthur. *The World as Will and Representation*. Vol. 1, 1818.

Schopenhauer, Arthur, 1851. "On the Sufferings of the World." In *The Complete Essays of Schopenhauer*, translated by T. Bailey Saunders (trans.), New York: Willey Book Co., 1942.

Schopenhauer, Arthur. "On the Vanity of Existence." In *Exploring the Meaning of Life: An Anthology and Guide*, edited by J.W. Seachris, T. Metz, J.G. Cottingham, G. Thomson, E.J. Wielenberg, and J.M. Fischer. Wiley, 2012, 227–29. https://books.google.de/books?id=WaQmA44YD-kC.

Schorr, Daniel. "Reagan Recants: His Path from Armageddon to Detente." *Los Angeles Times*, January 3, 1988. www.latimes.com/archives/la-xpm-1988-01-03-op-32475-story.html.

Schröder, K.-P., and Robert Connon Smith. "Distant Future of the Sun and Earth Revisited." *Monthly Notices of the Royal Astronomical Society* 386, no. 1 (2008): 155–63.

Schubert, Stefan, Lucius Caviola, and Nadira S. Faber. "The Psychology of Existential Risk: Moral Judgments About Human Extinction." *Scientific Reports* 9, no. 1 (2019): 15100.

Schultz, Bart. *Essays on Henry Sidgwick*. Cambridge University Press, 2002. https://books.google.de/books?id=VgJwwE9imlwC.

Schwartz, John. "Robert Jastrow, Who Made Space Understandable, Dies at 82." *New York Times*, February 12, 2008.

Schwartz, Thomas. "Obligations to Posterity." In *Obligations to Future Generations*, edited by R. I. Sikora and Brian Barry. White Horse Press, 1978.

Schwarz, Joel. "Humans Have Feared Comets, Other Celestial Phenomena Through the Ages." *NASA*, 1997. www2.jpl.nasa.gov/comet/news59.html.

Schweitzer, A., W. Montgomery, and F.C. Burkitt. *The Quest of the Historical Jesus*. Dover Publications, 2005. https://books.google.de/books?id=TMYqAwAAQBAJ.

Schwöbel, Christoph. "Last Things First: The Century of Eschatology in Retrospect." In *The Future as God's Gift*, edited by David Fergusson. Dover Publications, 2000, 217–41.

Scranton, R. *Learning to Die in the Anthropocene: Reflections on the End of a Civilization*. City Lights Open Media Series. City Lights Books, 2015. https://books.google.de/books?id=QLXBwAEACAAJ.

Seachris, Joshua. "Death, Futility, and the Proleptic Power of Narrative Ending." *Religious Studies* 47, no. 2 (2011): 141–63.

Secondat baron de Montesquieu, C. de. *Persian Letters*. ReadHowYouWant.com, Limited, 2008. https://books.google.de/books?id=caInBgAAQBAJ.

Segal, A. *Life After Death: A History of the Afterlife in Western Religion*. Crown Publishing Group, 2004. https://books.google.de/books?id=owd9zig7i1oC.

Segrè, Emilio. "Enrico Fermi: Physicist." *Bulletin of the Atomic Scientists*, 1970. https://books.google.de/books?id=EwcAAAAAMBAJ&pg=PA38&lpg=PA38&dq=%22I+believe+that+for+a+moment+I+thought+the+explosion+might+set+fire+to+the+atmosphere+and+thus+finish+the+earth,+even+though+I+knew+that+this+was+not+possible.%22&source=bl&ots=Is5Or_xGFs&sig=ACfU3U2ybf4PDk_85ojuguizE8aJP9JczA&hl=en&sa=X&ved=2ahUKEwiV0MCcs5XzAhVo_7sIHYX8BUIQ6AF6BAgJEAM#v=onepage&q=%22I%20believe%20that%20for%20a%20moment%20I%20thought%20the%20explosion%20might%20set%20fire%20to%20the%20atmos

phere%20and%20thus%20finish%20the%20earth%2C%20even%20though%20I%20 knew%20that%20this%20was%20not%20possible.%22&f=false.

———. *Enrico Fermi, Physicist: Emilio Segrè*. University of Chicago Press, 1970. https://books.google.de/books?id=wS_2swEACAAJ.

Sekerci, Yadigar, and Sergei Petrovskii. "Mathematical Modelling of Plankton—Oxygen Dynamics under the Climate Change." *Bulletin of Mathematical Biology* 77, no. 12 (2015): 2325–53.

Sepkoski, D. *Catastrophic Thinking: Extinction and the Value of Diversity from Darwin to the Anthropocene*. Science. Culture (CHUP) Series. University of Chicago Press, 2020. https://books.google.de/books?id=4er5DwAAQBAJ.

Serber, Robert. *The Los Alamos Primer: The First Lectures on How to Build an Atomic Bomb*. University of California Press, 1992.

Servigne, P., R. Stevens, and A. Brown. *How Everything Can Collapse: A Manual for Our Times*. Wiley, 2020. https://books.google.de/books?id=u7d1ygEACAAJ.

Shaw, Bill. "A Virtue Ethics Approach to Aldo Leopold's Land Ethic." *Environmental Ethics* 19, no. 1 (1997): 53–67.

Shelley, M. *The Collected Works of Mary Shelley*. Illustrated Edition: Novels, Short Stories, Plays & Travel Books, Including Biography of the Author. e-artnow, 2018. https://books.google.de/books?id=ptGSDwAAQBAJ.

Shelley, M.W., and S. Jansson. *Frankenstein, Or, The Modern Prometheus*. Classics Library. Wordsworth Classics, 1993. https://books.google.de/books?id=Rc6OG65y-yAC.

Shelley, M.W., and M.D. Paley. *The Last Man*. Oxford World's Classics. Oxford University Press, 2008/1826. https://books.google.de/books?id=OWEVDAAAQBAJ.

Sheridan. "Superintelligence in the Flesh. AI and the Antichrist." 2015. https://aiantichrist.blogspot.com/2015/09/superintelligence-in-flesh.html (Accessed 13 February 2018).

Shields, C.W. *The Final Philosophy: Or, System of Perfectible Knowledge Issuing from the Harmony of Science and Religion*. Scribner, Armstrong & Company, 1889. https://archive.org/details/philosophiaultim02shie/page/n7/mode/2up?q=%22describes+the+awful+catastrophe+which+must+ensue+when+the+last+man+shall+gaze+upon+the+frozen+earth%22.

———. "The History of the Sciences and the Logic of the Sciences." In *Philosophia Ultima: Or, Science of the Sciences*. C. Scribner's, 1889. https://books.google.de/books?id=94YuAAAAYAAJ.

Shiller, Derek. "In Defense of Artificial Replacement." *Bioethics* 31, no. 5 (2017): 393–99.

Shils, E. *The Torment of Secrecy: The Background and Consequences of American Security Policies*. Ivan R. Dee, 1996. https://books.google.de/books?id=MAwFGNy505MC.

Sidgwick, H. *The Methods of Ethics*. Donald F. Koch American Philosophy Collection. Macmillan, 1874. https://books.google.de/books?id=KVAtAAAAYAAJ.

Siipi, Helena, and Leonard Finkelman. "The Extinction and De-Extinction of Species." *Philosophy & Technology* 30 (2017): 427–41.

Sikora, Richard I., and Brian M. Barry. *Obligations to Future Generations*. White Horse Press, 1978.

Singer, Peter. *Animal Liberation: A New Ethics for Our Treatment of Animals*. Random House, 1975.

———. "Famine, Affluence, and Morality." *Philosophy and Public Affairs* 1, no. 3 (1972): 229–43.

———. "The Hinge of History. Project Syndicate." *Project Syndicate*, 2021. https://www.project-syndicate.org/commentary/ethical-implications-of-focusing-on-extinction-risk-by-peter-singer-2021-10.

———. "Killing Humans and Killing Animals." *Inquiry* 22, no. 1–4 (1979): 145–56. https://doi.org/10.1080/00201747908601869.

———. *One World: The Ethics of Globalization*. The Terry Lectures Series. Yale University Press, 2002. https://books.google.de/books?id=9DxPaVGw3koC.

———. "Right to Life?" *The New York Review*, August 14, 1980.

Slaughter, Richard A. "Why we should care for future generations now." *Futures* 26, no. 10 (1994): 1077–85.

Slovic, Paul. "If I Look at the Mass I Will Never Act: Psychic Numbing and Genocide." In *Emotions and Risky Technologies*, 37–59. Springer, 2010.

———. "Psychic Numbing and Genocide." *Psychological Science Agenda* 29, no. 8 (August 2015). https://www.apa.org/science/about/psa/2007/11/slovic.html.

SLPD. "Atomic Destruction in Hiroshima." *St. Louis Post-Dispatch*, September 12, 1945. www.newspapers.com/clip/47464197/picture-of-atomic-destruction-after-the/.

———. "A Decision for Mankind." *St. Louis Post-Dispatch*, August 7, 1945.

Smart, J.J.C. *Ethics, Persuasion and Truth*. Routledge Library Editions: Ethics. Routledge, 2020. https://books.google.de/books?id=vLjwDwAAQBAJ.

———. *Outlines of a Utilitarian System of Ethics*. Melbourne, 1961.

Smart, Roderick Ninian. "Negative Utilitarianism." *Mind* 67, no. 268 (1958): 542–43.

Smith, G.S. *Faith and the Presidency from George Washington to George W. Bush*. Oxford University Press, 2006. https://books.google.de/books?id=IH48DwAAQBAJ.

Smyth, Nicholas. "What Is the Question to Which Anti-Natalism Is the Answer?" *Ethical Theory and Moral Practice* 23, no. 1 (2020): 71–87.

SN. "Failing Phytoplankton, Failing Oxygen: Global Warming Disaster Could Suffocate Life on Planet Earth." *Science News*, December 1, 2015. www.sciencedaily.com/releases/2015/12/151201094120.htm.

Snyder, Ryan. "A Proliferation Assessment of Third Generation Laser Uranium Enrichment Technology." *Science & Global Security* 24, no. 2 (2016): 68–91.

Soddy, F. *Wealth, Virtual Wealth and Debt: The Solution of the Economic Paradox*. CreateSpace Independent Publishing Platform, 1926. https://books.google.de/books?id=-OeizgEACAAJ.

Solomon, H.M. *The Rape of the Text: Reading and Misreading Pope's Essay on Man*. University of Alabama Press, 1993. https://books.google.de/books?id=klqFEFXnk0wC.

Somerville, John. "The Catholic Bishops' Peace Revolution." *Peace Research* 15, no. 1 (1983): 34–36.

———. "The Last Inquest: A Preventable Nightmare in One Act." *Peace Research* 13, no. 2 (1981): 73–88.

———. "Philosophy of Peace Today: Preventive Eschatology." *Peace Research* 12, no. 2 (1980): 61–66.

———. "Scientific—Technological Progress and the New Problem of Preventing the Annihilation of the Human World." *Peace Research* 11, no. 1 (1979): 11–18.

———. "The UNESCO Approach to Interrelations of Cultures: Principles and Practices." *Peace Research* 16, no. 1 (1984): 25–29.

———. "War, Omnicide and Sanity: The Lesson of the Cuban Missile Crisis." *Dialectics and Humanism* 16, no. 2 (1989): 37–46.

Sotala, Kaj, and Roman V. Yampolskiy. "Responses to Catastrophic AGI Risk: A Survey." *Physica Scripta* 90, no. 1 (2014): 018001.

Souder, William. "Silent Spring Didn't Condemn Millions to Death." *New Scientist*, September 6, 2012. www.newscientist.com/article/dn22245-silent-spring-didnt-condemn-millions-to-death/.

Sparrow, Robert. "A Not-So-New Eugenics: Harris and Savulescu on Human Enhancement." *The Hastings Center Report* 41, no. 1 (2011): 32–42.

Spengler, Joseph J. "Malthus on Godwin's of Population." *Demography* 8, no. 1 (1971): 1–12.

Spicker, Stuart F. *Organism, Medicine, and Metaphysics: Essays in Honor of Hans Jonas on His 75th Birthday, May 10, 1978*. Vol. 7. Springer Science & Business Media, 2012.

Spiegel, Der. "Second Hand Apocalypses?" October 10, 1982. https://www.spiegel.de/kultur/apokalypsen-aus-zweiter-hand-a-ebe86a2a-0002-0001-0000-000014350592.

Srinivasan, Amia. "Remembering Derek Parfit." *London Review of Books*, 2017.

Srinivasan, R. *Whose Global Village?: Rethinking How Technology Shapes Our World*. New York University Press, 2018. https://books.google.de/books?id=3JhVDwAAQBAJ.

Stanley, Steven M. "Estimates of the Magnitudes of Major Marine Mass Extinctions in Earth History." *Proceedings of the National Academy of Sciences* 113, no. 42 (2016): E6325–34.

Stark, R. *The Rise of Christianity: A Sociologist Reconsiders History*. Princeton University Press, 1996. https://books.google.de/books?id=HcFSaGvgKKkC.

Steel, D. *Rogue Asteroids and Doomsday Comets: The Search for the Million Megaton Menace That Threatens Life on Earth*. Wiley, 1997. https://books.google.de/books?id=AoA1DhZSJXoC.

Steffen, W., et al. "1950 Marked the Beginning of a Massive Acceleration in Human Activity and Large-Scale Changes in the Earth System." In *Global Change and the Earth System*. Springer, 2004. www.igbp.net/download/18.56b5e28e137d8d8c09380001694/1376383141875/SpringerIGBPSynthesisSteffenetal2004_web.pdf.

Steffen, Will, Jacques Grinevald, Paul Crutzen, and John McNeill. "The Anthropocene: Conceptual and Historical Perspectives." *Philosophical Transactions of the Royal Society A: Mathematical, Physical and Engineering Sciences* 369, no. 1938 (2011): 842–67.

Steffen, Will, Katherine Richardson, Johan Rockström, Sarah E. Cornell, Ingo Fetzer, Elena M. Bennett, Reinette Biggs, Stephen R. Carpenter, Wim De Vries, and Cynthia A. De Wit. "Planetary Boundaries: Guiding Human Development on a Changing Planet." *Science* 347, no. 6223 (2015): 1259855.

Steffen, Will, Johan Rockström, Katherine Richardson, Timothy M. Lenton, Carl Folke, Diana Liverman, Colin P. Summerhayes, Anthony D. Barnosky, Sarah E. Cornell, and Michel Crucifix. "Trajectories of the Earth System in the Anthropocene." *Proceedings of the National Academy of Sciences* 115, no. 33 (2018): 8252–59.

Stern, Nicholas. *Stern Review: The Economics of Climate Change*. Government of the United Kingdom, 2006. http://mudancasclimaticas.cptec.inpe.br/~rmclima/pdfs/destaques/sternreview_report_complete.pdf.

Sternglass, Ernest. "The Death of All Children." *Esquire*, September 1, 1969.

Stevens, William K. "Balance of Nature? What Balance Is That?" *New York Times C* 4 (1991).

Stich, Stephen P., and Edouard Machery. "Demographic Differences in Philosophical Intuition: A Reply to Joshua Knobe." *Review of Philosophy and Psychology* (2022), 1–34.

Stirrat, Michael, and R. Elisabeth Cornwell. "Eminent Scientists Reject the Supernatural: A Survey of the Fellows of the Royal Society." *Evolution: Education and Outreach* 6 (2013): 1–5.

Stitzinger, James F. "The Rapture in Twenty Centuries of Biblical Interpretation." *The Master's Seminary Journal* 13 (2002): 149–72.

Stoll, Mark. "The US Federal Government Responds." *Environment & Society Portal*, 2020. www.environmentandsociety.org/exhibitions/rachel-carsons-silent-spring/us-federal-government-responds.

Stolz, Jörg. "Secularization Theories in the Twenty-First Century: Ideas, Evidence, and Problems. Presidential Address." *Social Compass* 67, no. 2 (2020): 282–308.

Stone, M. "Lifetime and Decay of 'Excited Vacuum' States of a Field Theory Associated with Nonabsolute Minima of Its Effective Potential." *Physical Review D* 14, no. 12 (December 15, 1976): 3568–73. https://doi.org/10.1103/PhysRevD.14.3568.

Stucky, H.J. *August 6, 1965: The Impact of Atomic Energy*. American Press, 1964. https://books.google.de/books?id=JA4JAQAAMAAJ.

Sturm, Tristan. "Hal Lindsey's Geopolitical Future: Towards a Cartographic Theory of Anticipatory Arrows." *Journal of Maps* 17, no. 1 (2021): 39–45.

Swatos, William H., and Kevin J. Christiano. "Secularization Theory: The Course of a Concept." *Sociology of Religion* 60, no. 3 (1999): 209. https://doi.org/10.2307/3711934.

Swinburne, Algernon Charles, and Toni Savage. *The Garden of Proserpine*. Pandora Press, 1961.

Szilard, L., B.T. Feld, G.W. Szilard, S.R. Weart, H.S. Hawkins, and G.A. Greb. *The Collected Works of Leo Szilard: Leo Szilard: His Version of the Facts: Selected Recollections and Correspondence*, 1978. https://books.google.de/books?id=9N8PjwEACAAJ.

Tamny, Martin. "Newton, Creation, and Perception." *Isis* 70, no. 1 (1979): 48–58.

Tanner, L., and S. Calvari. *Volcanoes: Windows on the Earth*. New Mexico Museum of Natural History and Science, 2012. https://books.google.de/books?id=tGBLCgAAQBAJ.

Tännsjö, Torbjörn. Identifiering av värdet av ett möjligt utfall av beslut i kärnavfallsfrågan. In *Osäkerhet och beslut*, 2019. https://inis.iaea.org/collection/NCLCollectionStore/_Public/22/067/22067094.pdf.

———. "Who Cares? The COVID-19 Pandemic, Global Heating and the Future of Humanity." *Journal of Controversial Ideas* 1, no. 1 (2021).

———. "Why We Ought to Accept the Repugnant Conclusion." In *The Repugnant Conclusion*, 219–37. Springer, 2004.

Tappolet, Christine. "Evaluative vs. Deontic Concepts." *International Encyclopedia of Ethics*, 2013.

———. "The Normativity of Evaluative Concepts." In *Mind, Values, and Metaphysics: Philosophical Essays in Honor of Kevin Mulligan*, edited by Anne Reboul. Vol. 2, 39–54. Springer, 2014.

Tarbuck, Edward J., Frederick K. Lutgens, Dennis Tasa, and Dennis Tasa. *Earth: An Introduction to Physical Geology*. Pearson/Prentice Hall Upper Saddle River, 2005.

Tarsney, Christian. *The Epistemic Challenge to Longtermism*. Global Priorities Institute, 2019. https://philpapers.org/archive/TARTEC-2.pdf; https://globalprioritiesinstitute.org/wp-content/uploads/2020/Christian_Tarsney_epistemic_challenge.pdf.

Taube, Karl. *Aztec and Maya Myths (The Legendary Past)*. British Museum Press, 1993.

Taylor, R.P. *Death and the Afterlife: A Cultural Encyclopedia*. ABC-CLIO E-Books. ABC-CLIO, 2000. https://books.google.de/books?id=zhnXAAAAMAAJ.

Tegmark, M. *Life 3.0: Being Human in the Age of Artificial Intelligence*. Knopf Doubleday Publishing Group, 2017. https://books.google.de/books?id=2hIcDgAAQBAJ.

———. "Top Myths about Advanced AI." *Future of Life Institute*, 2016. https://web.archive.org/web/20160812071218/https:/futureoflife.org/background/aimyths/.

Thaler, Richard. "The Premortem." *The Edge*, 2017. www.edge.org/response-detail/27174.

Thomas, Edward. "Atomic Bomb Smashes Nagasaki in Inferno of Smoke and Flame." *Freeport Journal-Standard*, August 10, 1945. www.newspapers.com/clip/47464614/atomic-bomb-smashes-nagasaki-in/.

Thomas, Gillian. "Lifton's Law and the Teaching of Literature." *The Dalhousie Review* 66, nos. 1 and 2 (1986).

Thomas, P.J., C.F. Chyba, and C.P. McKay. *Comets and the Origin and Evolution of Life*. Springer, 2013. https://books.google.de/books?id=h0L2BwAAQBAJ.

Thompson, Thomas H. "Are We Obligated to Future Others." In *Responsibilities to Future Generations*, edited by Ernest Partridge. Prometheus Books, 1981, 195–202.

Thomson, William. "2. On a Universal Tendency in Nature to the Dissipation of Mechanical Energy." *Proceedings of the Royal Society of Edinburgh* 3 (1857): 139–42.

———. "On the Age of the Sun's Heat." *Macmillan's Magazine* 5, no. March (1862a): 288–93.

———. "Physical Considerations Regarding the Possible Age of the Sun's Heat." *The London, Edinburgh, and Dublin Philosophical Magazine and Journal of Science* 23, no. 152 (1862b): 158–60.

———. "XV.—On the Dynamical Theory of Heat, with Numerical Results Deduced from Mr Joule's Equivalent of a Thermal Unit, and M. Regnault's Observations on Steam." *Earth and Environmental Science Transactions of the Royal Society of Edinburgh* 20, no. 2 (1853): 261–88.

Thorsett, S.E. "Terrestrial Implications of Cosmological Gamma-Ray Burst Models." *ArXiv Preprint Astro-Ph/9501019* (1995).

Thorstad, David. "The Scope of Longtermism." *GPI Working Paper*. Global Priorities Institute, 2021. https://globalprioritiesinstitute.org/wp-content/uploads/David-Thorstad-the-scope-of-longtermism.pdf.

Thunberg, Greta. "'Our House Is on Fire': Greta Thunberg, 16, Urges Leaders to Act on Climate." *The Guardian*, January 25, 2019. www.theguardian.com/environment/2019/jan/25/our-house-is-on-fire-greta-thunberg16-urges-leaders-to-act-on-climate.

Tigay, J.H. *The Evolution of the Gilgamesh Epic*. Bolchazy-Carducci, 2002. https://books.google.de/books?id=cxjuHTH6I2sC.

Timmermann, Jens. "V—What's Wrong with 'Deontology'?" *Wiley Online Library* 115 (2015): 75–92.

Tipler, Frank J. "Extraterrestrial Intelligent Beings Do Not Exist." *Quarterly Journal of the Royal Astronomical Society* 21 (1980): 267–81.

Tiseo, Ian. "Historic Average Carbon Dioxide (CO2) Levels in the Atmosphere Worldwide from 1959 to 2021 (in Parts per Million)*." *Statistica*, June 21, 2022. www.statista.com/statistics/1091926/atmospheric-concentration-of-co2-historic/.

TM. "Omnicide: Trademark Information." n.d. https://trademark.trademarkia.com/omnicide-71379979.html.

———. "Omnicide Trademark Information." *Trademarkia*, 2022. www.oed.com/view/Entry/246601#eid12254969.

Tolkien, J. R. R. "Letter 89." November 7, 1944. https://tolkiengateway.net/wiki/Letter_89.

Tollefson, Jeff. "Humans Are Driving One Million Species to Extinction." *Nature* 569, no. 7755 (2019): 171–72.

Tomasik, Brian. "Strategic Considerations for Moral Antinatalists." *Reducing Suffering*, 2018. https://reducing-suffering.org/strategic-considerations-moral-antinatalists/.

———. *What Are Suffering Subroutines?* Self-Published. https://reducing-suffering.org/what-are-suffering-subroutines/.

Tonn, Bruce E. "500-Year Planning: A Speculative Provocation." *Journal of the American Planning Association* 52, no. 2 (1986): 185–93.

———. *Anticipation, Sustainability, Futures and Human Extinction: Ensuring Humanity's Journey into the Distant Future*. Routledge, 2021.

———. "Beliefs about Human Extinction." *Futures* 41, no. 10 (2009): 766–73.

———. "Integrated 1000-Year Planning." *Futures* 36, no. 1 (2004): 91–108.

———. "Obligations to Future Generations and Acceptable Risks of Human Extinction." *Futures* 41, no. 7 (2009): 427–35.

Tonn, Bruce E., and Jenna Tonn. "A Literary Human Extinction Scenario." *Futures* 41, no. 10 (2009): 760–65.

Toon, Owen Brian, Alan Robock, and Richard P. Turco. "Environmental Consequences of Nuclear War." *Physics Today* 61, no. 12 (2008): 37.

Topol, Sarah. "How to Save Mankind from the New Breed of Killer Robots." *Buzzfeed*, August 26, 2016. www.buzzfeed.com/sarahatopol/how-to-save-mankind-from-the-new-breed-of-killer-robots.

Tornau, Christian. "Saint Augustine." *Stanford Encyclopedia of Philosophy*, 2019. https://plato.stanford.edu/cgi-bin/encyclopedia/archinfo.cgi?entry=augustine.

Torres, Émile. "Beyond 'New Atheism': Where Do People Alienated by the Movement's Obnoxious Tendencies Go from Here?" 2017. www.salon.com/2017/08/07/beyond-new-atheism-where-do-people-alienated-by-the-movements-obnoxious-tendencies-go-from-here/.

———. "Can We Clean up the Mess We've Created? We Have to Do It Now, or Face Extinction." *Salon*, September 5, 2021. www.salon.com/2021/09/05/can-we-clean-up-the-mess-weve-created-we-have-to-do-it-now-or-face-extinction/.

———. "Godless Grifters: How the New Atheists Merged with the Far Right." 2021. www.salon.com/2021/06/05/how-the-new-atheists-merged-with-the-far-right-a-story-of-intellectual-grift-and-abject-surrender/.

———. *Nick Bostrom, Longtermism, and the Eternal Return of Eugenics*. Truthdig, 2023. https://www.truthdig.com/dig/nick-bostrom-longtermism-and-the-eternal-return-of-eugenics/.

———. "Selling 'Longtermism': How PR and Marketing Drive a Controversial New Movement." *Salon*, 2022. www.salon.com/2022/09/10/selling-longtermism-how-pr-and-marketing-drive-a-controversial-new-movement/.

———. "Some Problems with 'X-Risk: How Humanity Discovered Its Own Extinction'." *Medium*, 2021. https://philosophytorres.medium.com/some-problems-with-x-risk-how-humanity-discovered-its-own-extinction-58de1265e72d.

———. "Steven Pinker's Fake Enlightenment: His Book is Full of Misleading Claims and False Assertions." *Salon*, January 26, 2019. www.salon.com/2019/01/26/steven-pinkers-fake-enlightenment-his-book-is-full-of-misleading-claims-and-false-assertions.

———. "Were the Great Tragedies of History 'Mere Ripples'? The Case Against Longtermism." Unpublished manuscript, n.d.

———. "Why an Existential Risk Expert Finds Hope in Humanity's Certain Doom." *OneZero*, 2019. https://onezero.medium.com/rebelling-against-extinction-d7e112979bed.

Torres, Phil. "Agential Risks: A Comprehensive Introduction." *Journal of Ethics and Emerging Technologies* 26, no. 2 (2016): 31–47.

———. "Agential Risks and Information Hazards: An Unavoidable but Dangerous Topic?" *Futures* 95 (2018a): 86–97.

———. "Can Anti-Natalists Oppose Human Extinction? The Harm-Benefit Asymmetry, Person-Uploading, and Human Enhancement." *South African Journal of Philosophy* 39, no. 3 (2020): 229–45.

———. "Existential Risks: A Philosophical Analysis." *Inquiry* (2019): 1–26.

———. "Facing Disaster: The Great Challenges Framework." *Foresight* 21, no. 1 (2018b): 4–34.

———. "From the Enlightenment to the Dark Ages: How 'New Atheism' Slid into the Alt-Right." *Salon*, July 29, 2017b.

———. "How Elon Musk Sees the Future: His Bizarre Sci-Fi Vision Should Concern Us All." *Salon*, July 17, 2022. www.salon.com/2022/07/17/how-elon-musk-sees-the-future-his-bizarre-sci-fi-vision-should-concern-us-all/.

———. "'New Atheist' Sam Harris—Still Deeply Wrong on Islamic Extremism and Terrorism." *Salon*, 2017c. www.salon.com/2017/07/09/new-atheist-sam-harris-still-deeply-wrong-on-islamic-extremism-and-terrorism/.

———. "Scared Straight: How Prophets of Doom Might Save the World." *Bulletin of the Atomic Scientists*, May 27, 2021. https://thebulletin.org/2021/05/scared-straight-how-prophets-of-doom-might-save-the-world.

———. "Superintelligence and the Future of Governance: On Prioritizing the Control Problem at the End of History." In *Artificial Intelligence Safety and Security*, 357–74. Chapman and Hall/CRC, 2018c.

———. "Who Would Destroy the World? Omnicidal Agents and Related Phenomena." *Aggression and Violent Behavior* 39 (2018d): 129–38.

Torres, P., and Russell Blackford. *The End: What Science and Religion Tell Us about the Apocalypse*. Pitchstone Publishing, 2016. https://books.google.de/books?id=tGpbrgEACAAJ.

Torres, P., and Martin Rees. *Morality, Foresight, and Human Flourishing: An Introduction to Existential Risks*. Pitchstone Publishing, 2017a. https://books.google.de/books?id=DDPZAQAACAAJ.

Trisel, Brooke Alan. "How Best to Prevent Future Persons from Suffering: A Reply to Benatar." *South African Journal of Philosophy* 31, no. 1 (2012): 79–93.

———. "Human Extinction, Narrative Ending, and Meaning of life." *Journal of Philosophy of Life* 6, no. 1 (2016): 1–22.

Troeltsch, E. *Glaubenslehre: Nach Heidelberger Vorlesungen Aus Den Jahren 1911 Und 1912*. Edition Classic. VDM, Müller, 2006. https://books.google.de/books?id=yCnxMQAACAAJ.

Truman, Harry. "Truman Statement on Hiroshima." August 6, 1945. www.atomicheritage.org/key-documents/truman-statement-hiroshima.

Tuite, C. *Byron in Context*. Literature in Context. Cambridge University Press, 2019. https://books.google.de/books?id=DZ_MDwAAQBAJ.

Turco, Richard P., Owen B. Toon, Thomas P. Ackerman, James B. Pollack, and Carl Sagan. "Nuclear Winter: Global Consequences of Multiple Nuclear Explosions." *Science* 222, no. 4630 (1983): 1283–92.

Turco, R.P., O.B. Toon, J.B. Pollack, and C. Sagan. "Global Consequences of Nuclear Warfare." *EOS* 63, no. 1018 (1982).

Turing, Alan. "Intelligent Machinery, a Heretical Theory." *The '51 Society*, 1951. https://terrorgum.com/tfox/books/turingtest_verbalbehaviorasthehallmarkofintelligence.pdf#page=120.

Turing, Alan M., and J. Haugeland. "Computing Machinery and Intelligence." In *The Turing Test: Verbal Behavior as the Hallmark of Intelligence*, edited by S. Shieber, 29–56. MIT Press, 1950.

UNFCCC. "First Anniversary of Pope Francis' Encyclical 'Laudato Si'." 2016. https://newsroom.unfccc.int/news/first-anniversary-of-pope-francis-encyclical-laudato-si.

———. "Islamic Declaration on Climate Change." *United Nations Climate Change*, 2015. https://unfccc.int/news/islamic-declaration-on-climate-change.

UNICEF. "The State of Food Security and Nutrition in the World 2021." *ReliefWeb*, 2021. https://reliefweb.int/report/world/state-food-security-and-nutrition-world-2021-transforming-food-systems-food-security?gclid=Cj0KCQiA0oagBhDHARIsAI-Bbgcc-eYn6bSLl0oI_Xys2AoYNxoBEsZPFDDMqBmkyK7v4a8a3kGaeJcaAnN-GEALw_wcB.

UN News. *UN Aid Chief Urges Global Action as Starvation, Famine Loom for 20 Million Across Four Countries*. United Nations, 2017. https://news.un.org/en/story/2017/03/553152.

United Nations. *Report of the United Nations Conference on Environment and Development*. Vol. 1, *Resolutions Adopted by the Conference*. Rio de Janeiro, June, 3–14, 1992.

United States. "Congress. House. Committee on Science and Technology. Subcommittee on Space Science and Applications, and Aviation United States. Congress. House. Committee on Science and Technology. Subcommittee on Transportation and Communications." *1981 NASA Authorization: Hearings Before the Subcommittee on Space Science and Applications of the Committee on Science and Technology, U.S. House of Representatives, Ninety-Sixth Congress, First Session* . . . 1981 NASA Authorization: Hearings Before the Subcommittee on Space Science and Applications of the Committee on Science and Technology, U.S. House of Representatives, Ninety-Sixth Congress, First Session. U.S. Government Printing Office, 1980. https://books.google.de/books?id=RVYrAAAAMAAJ.

———. "President's Science Advisory Committee. Environmental Pollution Panel." *Restoring the Quality of Our Environment: Report*. Restoring the Quality of Our Environment: Report of the Environmental Pollution Panel, President's Science Advisory Committee. White House, 1965. https://books.google.de/books?id=LAwEAQAAIAAJ.

UN News. "UN Aid Chief Urges Global Action as Starvation, Famine Loom for 20 Million Across Four Countries." *United Nations*, n.d. https://news.un.org/en/story/2017/03/553152.

Urrutia-Fucugauchi, Jaime, Antonio Camargo-Zanoguera, and Ligia Pérez-Cruz. "Discovery and Focused Study of the Chicxulub Impact Crater." *Eos, Transactions American Geophysical Union* 92, no. 25 (2011): 209–10.

Urrutia-Fucugauchi, Jaime, Antonio Camargo-Zanoguera, Ligia Pérez-Cruz, and Guillermo Pérez-Cruz. "The Chicxulub Multi-Ring Impact Crater, Yucatan Carbonate Platform, Gulf of Mexico." *Geofísica Internacional* 50, no. 1 (2011): 99–127.

USCD. *Introduction to Radioactive Fallout*. U.S. Government Printing Office, 1955. https://books.google.de/books?id=NmXk7c0dXAIC.

USGS. "Today in Earthquake History." *USGS*, 2022. https://earthquake.usgs.gov/learn/today/index.php?month=11&day=1&submit=View+Date.

Van Wylen, Gordon J., and Richard E. Sonntag. *Fundamentals of Classical Thermodynamics*, 1985.

Velikovsky, I. *Worlds in Collision*. Doubleday & Company, 1950. https://books.google.de/books?id=FJst27kSVBgC.

Verdoux, Philippe. "Transhumanism, Progress and the Future." *Journal of Evolution and Technology* 20, no. 2 (2009): 49–69.

Verne, J. *Five Weeks in a Balloon: Journeys and Discoveries in Africa by Three Englishmen: Easyread Large Bold Edition*. CreateSpace, 2008. https://books.google.de/books?id=Z_L-i5K-6YoC.

Vetter, Hermann. "Discussion." In *Induction, Physics, and Ethics*, edited by Paul Weingartner and Gerhard Zecha. D. Reidel Publishing Company, 1968.

———. "IV. The Production of Children as a Problem of Utilitarian Ethics." *Inquiry: An Interdisciplinary Journal of Philosophy* 12, no. 1–4 (1969): 445–447.

———. "Utilitarianism and New Generations." *Mind* 80, no. 318 (1971): 301–2.

Vidal, Céline M., Christine S. Lane, Asfawossen Asrat, Dan N. Barfod, Darren F. Mark, Emma L. Tomlinson, Amdemichael Zafu Tadesse, Gezahegn Yirgu, Alan Deino, and William Hutchison. "Age of the Oldest Known *Homo sapiens* from Eastern Africa." *Nature* 601, no. 7894 (2022): 579–83.

Villiers, M. de. *The End: Natural Disasters, Manmade Catastrophes, and the Future of Human Survival*. St. Martin's Publishing Group, 2010. https://books.google.de/books?id=MlDhGmsV880C.

Vinding, Magnus. "Antinatalism and Reducing Suffering: A Case of Suspicious Convergence." *Personal Website*, n.d. https://magnusvinding.com/2021/02/20/antinatalism-and-reducing-suffering/.

———. *Suffering-Focused Ethics: Defense and Implications*. Amazon Digital Services LLC – KDP Print US, 2020.

Vinge, Vernor. "The Coming Technological Singularity: How to Survive in the Post-Human Era." 1993. https://edoras.sdsu.edu/~vinge/misc/singularity.html.

Vogel, Lawrence. "Does Environmental Ethics Need a Metaphysical Grounding?" *The Hastings Center Report* 25, no. 7 (1995): 30–39.

———. "Hans Jonas's Exodus: From German Existentialism to Post-Holocaust Theology." In *Introduction to Mortality and Morality: A Search for Good After Auschwitz*, edited by Hans Jonas, 1–40. Northwestern University Press, 1996.

Vogt, William. "On Man the Destroyer." *Natural History*, 1963. https://archive.org/details/naturalhistory72newy/page/n13/mode/2up?q=%22two+books%22.

———. *Road to Survival*. Vol. 67. LWW, 1949.

Vogt, W., B.M. Baruch, and S.I. Freeman. *Road to Survival*. W. Sloane Associates, 1948. https://books.google.de/books?id=MJUiAAAAMAAJ.

Von Neumann, J. "The General and Logical Theory of Automata, Papers of John von Neumann on Computing and Computer Theory." In *Collected works of John von Neumann*, Pregamon Press, 1948.

———. *Health and Safety Problems and Weather Effects Associated with Atomic Explosions: Hearings Before the United States Joint Committee on Atomic Energy, Eighty-Fourth Congress, First Session, on Apr. 15, 1955*. U.S. Government Printing Office, 1955. https://books.google.de/books?id=QOREAQAAMAAJ.

———. "Statement of Dr. John Von Neumann, Commissioner, Atomic Energy Commission." *United States Congress*, 1955. www.google.de/books/edition/Hearings/Wha6qUxvg2YC?hl=en&gbpv=1&dq=%22to+bring+back+the+conditions+of+the+last+ice+age%22&pg=RA7-PA36&printsec=frontcover.

Wade, Lisa. "Are College Professors Less Religious than the General Population?" *The Society Pages*, April 12, 2010. https://thesocietypages.org/socimages/2010/04/12/are-college-professors-less-religious-than-the-general-population/.

Wade, Nicholas. "CO2 in Climate: Gloomsday Predictions Have No Fault." *Science* 206, no. 4421 (1979): 912–13.

Walker, Mark. "H+: Ship of Fools: Why Transhumanism Is the Best Bet to Prevent the Extinction of Civilization." *Metanexus*, 2009. https://metanexus.net/h-ship-fools-why-transhumanism-best-bet-prevent-extinction-civilization/.

———. "Ship of Fools: Why Transhumanism Is the Best Bet to Prevent the Extinction of Civilization." In *Transhumanism and Its Critics*, edited by Gregory R. Hansell and William Grassie, 94–111. Metanexus Institute, 2011.

Wallace, Robert. *A Dissertation on the Numbers of Mankind, in Ancient and Modern Times . . .* 2nd edition, revised and corrected. Archibald Constable & Company, 1809.

Waller, James. *Confronting Evil: Engaging Our Responsibility to Prevent Genocide*. Oxford University Press, 2016.

Walls, Jerry. *The Oxford Handbook of Eschatology*. Oxford University Press, 2008.

Walsh, Bryan. "The Case for Genetically Engineering Ethical Humans." *OneZero*, July 26, 2018. https://onezero.medium.com/the-case-for-genetically-engineering-ethical-humans-b44c17b9e3d6.

Walters, Gregory J. "Karl Jaspers on the Role of 'Conversion' in the Nuclear Age." *Journal of the American Academy of Religion* 56, no. 2 (1988): 229–56.

Ward, P.D., and D. Brownlee. *Rare Earth: Why Complex Life Is Uncommon in the Universe*. Copernicus Series. Springer, 2000. https://books.google.de/books?id=SZVV26vCSi8C.

Ware, James. "Paul's Understanding of the Resurrection in 1 Corinthians 15: 36–54." *Journal of Biblical Literature* 133, no. 4 (2014): 809–35.

Warren, Wagar. *Title: Terminal Visions: The Literature of Last Things*. Indiana University Press, 1982.

Waters, Colin N., Jan A. Zalasiewicz, Mark Williams, Michael A. Ellis, and Andrea M. Snelling. "A Stratigraphical Basis for the Anthropocene?" *Geological Society, London, Special Publications* 395, no. 1 (2014): 1–21.

Watson, Justin. "How Pat Finally Gets Even: Apocalyptic Asteroids and American Politics in Pat Robertson's The End of the Age." *Journal of Millennial Studies* 2, no. 2. (Winter 2000). http://www.mille.org/publications/jrnlpast.html.

Weart, Spencer R. *The Discovery of Global Warming*. Harvard University Press, 2003.

———. *The Discovery of Global Warming*. Harvard University Press, 2008. https://books.google.de/books?id=5zrjngEACAAJ.

———. "The Heyday of Myth and Cliché." *Bulletin of the Atomic Scientists* 41, no. 7 (1985): 38–43.

———. *Nuclear Fear: A History of Images*. Harvard University Press, 1988. https://books.google.de/books?id=yL-kOJzt0hwC.

———. "The Public and Climate Change." In *The Discovery of Global Warming*, 2022. https://history.aip.org/climate/public.htm.

———. *The Rise of Nuclear Fear*. Harvard University Press, 2012.

Weart, Spencer, and Gertrud Weiss Szilard, eds. *Leo Szilard: His Version of the Facts*. The MIT Press, 1978.

———. "Leo Szilard: His Version of the Facts." *Bulletin of the Atomic Scientists*, 1979. https://books.google.de/books?id=7goAAAAAMBAJ&pg=PA57&lpg=PA57&dq=%E2%80%9CIn+certain+circumstances+it+might+become+possible+to+set+up+a+nuclear+chain+reaction,+liberate+energy+on+an+industrial+scale,+and+construct+atomic+bombs.%E2%80%9D&source=bl&ots=2wP90FhmZL&sig=ACfU3U1ZMl8fUQhZTfY93glEJIM8PXyidQ&hl=en&sa=X&ved=2ahUKEwiNy9ftrIjzAhVsgv0HHddFDo8Q6AF6BAgEEAM#v=onepage&q=%E2%80%9CIn%20certain%20circumstances%20it%20might%20become%20possible%20to%20set%20up%20a%20nuclear%20chain%20reaction%2C%20liberate%20energy%20on%20an%20industrial%20scale%2C%20and%20construct%20atomic%20bombs.%E2%80%9D&f=false.

Wei-Haas, Maya. "New Seismic Phenomenon Discovered, Named Stormquakes." *National Geographic*, October 16, 2019. www.nationalgeographic.com/science/article/new-seismic-phenomenon-discovered-named-stormquakes.

Weinberg, Alvin M. "Impact of Large-Scale Science on the United States: Big Science Is Here to Stay, but We Have Yet to Make the Hard Financial and Educational Choices It Imposes." *Science* 134, no. 3473 (1961): 161–64.

Weissmann, Jordan. "An Interview With Robin Hanson, the Sex Redistribution Professor." *Slate*, 2018. https://slate.com/business/2018/05/robin-hanson-the-sex-redistribution-professor-interviewed.html.

Wells, H.G. *Certain Personal Matters*. Outlook Verlag, 2018. https://books.google.de/books?id=E9RRDwAAQBAJ.

———. *The Discovery of the Future: A Discourse Delivered to the Royal Institution on January 24, 1902*. T. Fisher Unwin, 1902.

———. "The Extinction of Man." *Certain Personal Matters*, 1983. www.online-literature.com/wellshg/certain-personal-matters/24/.

———. *God, the Invisible King*. Macmillan, 1917.

———. *The Shape of Things to Come*. Hutchinson, 1933.

———. *The Time Machine: An Invention*. Henry Holt and Company, 1895.

———. "Wanted—Professors of Foresight." *Futures Research Quarterly* 3, no. 1 (1932): 89–91.

———. *The World Set Free*. Electric Umbrella Publishing, 2021. https://books.google.de/books?id=fqJIEAAAQBAJ.

Wells, W. *Apocalypse When?: Calculating How Long the Human Race Will Survive*. Springer Praxis Books. Springer, 2009. https://books.google.de/books?id=h8SuSY4v9sYC.

Wensveen, Louke van. Dirty Virtues: The Emergence of Ecological Virtue Ethics. Prometheus Books, 2000.

Wheelwright, P. *Heraclitus*. Atheneum, 1968. https://books.google.de/books?id=mNl1DwAAQBAJ.

Whisenant, Edgar. *88 Reasons Why the Rapture Will Be in 1988*." World Bible Society, 1988.

Whittlestone, Jess. "The Long-Term Future." *Effective Altruism*, November 16, 2017. https://web.archive.org/web/20181020232825/www.effectivealtruism.org/articles/cause-profile-long-run-future/.

Wiblin, Robert. "Toby Ord on Why the Long-Term Future of Humanity Matters More Than Anything Else, and What We Should Do about It." September 6, 2017. https://80000hours.org/podcast/episodes/why-the-long-run-future-matters-more-than-anything-else-and-what-we-should-do-about-it/.

Wicks, Robert. "Arthur Schopenhauer." *The Stanford Encyclopedia of Philosophy*, 2021. https://plato.stanford.edu/archives/fall2021/entries/schopenhauer.

William H. Desvousges, F. Reed Johnson, Richard W. Dunford, Kevin J. Boyle, Sara P. Hudson, and K. Nicole Wilson. *Measuring Nonuse Damages Using Contingent Valuation: An Experimental Evaluation of Accuracy*, Technical Report. RTI International, 2010.

William, James. *Pragmatism, a New Name for Some Old Ways of Thinking: Popular Lectures on Philosophy*. Library of American Civilization. Longmans, Green, and Company, 1907. https://books.google.de/books?id=1lWouTG6oYwC.

Williams, C. *Terminus Brain: The Environmental Threats to Human Intelligence*. Global Issues. Cassell, 1997. https://books.google.de/books?id=Gv7aAAAAMAAJ.

Wilson, Edward O. "Beware the Age of Loneliness." *The Economist*, November 18, 2013.

———. "Editor's Foreword." In *Biodiversity*. National Academy Press, 1988. https://nap.nationalacademies.org/read/989/chapter/1.

Winchell, A. *Sketches of Creation: A Popular View of Some of the Grand Conclusions of the Sciences in Reference to the History of Matter and of Life. Together with a Statement of the Intimations of Science Respecting the Primordial Condition and the Ultimate Destiny of the Earth and the Solar System*. Harper & Brothers, 1870. https://books.google.de/books?id=r-1jaU6Is4QC.

Winchell, Walter. "No Title." *The Times*, September 23, 1946. www.newspapers.com/image/210569782/?terms=%22I%20dunno%22%20he%20said%20%22but%20in%20the%20war%20after%20the%20next%20war%20sure%20as%20Hell%20they%27ll%20be%20using%20spears%21%22&match=1.

Winner, L. *Autonomous Technology: Technics-out-of-Control as a Theme in Political Thought*. Autonomous Technology. MIT Press, 1977. https://books.google.de/books?id=uNIG0gi4b40C.

Winter, T. *The Cambridge Companion to Classical Islamic Theology*. Cambridge Collections Online. Cambridge University Press, 2008. https://books.google.de/books?id=rSPVnQEACAAJ.

Wittes, B., and G. Blum. *The Future of Violence: Robots and Germs, Hackers and Drones-Confronting A New Age of Threat*. Basic Books, 2015. https://books.google.de/books?id=iFc4DgAAQBAJ.

Witzel, M. *The Origins of the World's Mythologies*. Oxford University Press, 2012. https://books.google.de/books?id=UALji7FE-1UC.

Wolf, Clark. "Person-Affecting Utilitarianism and Population Policy; or, Sissy Jupe's Theory of Social Choice." In *Contingent Future Persons*, 99–122. Springer, 1997.

Wood, Lewis. "Steel Tower 'Vaporized' in Trial of Mighty Bomb." *New York Times*, August 7, 1945.

Woodhouse, K.M. *The Ecocentrists: A History of Radical Environmentalism*. Columbia University Press, 2018. https://books.google.de/books?id=J7NGtAEACAAJ.

Worm, Boris, Edward B. Barbier, Nicola Beaumont, J. Emmett Duffy, Carl Folke, Benjamin S. Halpern, Jeremy B.C. Jackson, Heike K. Lotze, Fiorenza Micheli, and Stephen R. Palumbi. "Impacts of Biodiversity Loss on Ocean Ecosystem Services." *Science* 314, no. 5800 (2006): 787–90.

Wormald, Benjamin. "The Future of World Religions: Population Growth Projections, 2010–2050." *Pew Research Center's Religion & Public Life Project*, 2015. https://www.pewresearch.org/religion/2015/04/02/religious-projections-2010-2050/.

Wright, M.R. *Empedocles, the Extant Fragments*. Yale University Press, 1981. https://books.google.de/books?id=4qtSF3BUbjAC.

Wright, Nicholas. "AI & Global Governance: Three Distinct AI Challenges for the UN." *United Nations University*, July 12, 2018. https://cpr.unu.edu/publications/articles/ai-global-governance-three-distinct-ai-challenges-for-the-un.html.

Wright, T.I.D. *An Original Theory or New Hypothesis of the Universo, Founded Upon the Laws of Nature, and Solving by Mathematical Principles the General Phaenomena of the Visible Creation and Particularly the Via Lactea.* Chapelle, 1750. https://books.google.de/books?id=80VZAAAAcAAJ.

WS. *Chemistry: 1922–1941.* World Scientific, 1999. https://books.google.de/books?id=B8raAAAAMAAJ.

WWF. "Living Planet Report 2020." *World Wildlife Fund*, 2020. www.zsl.org/sites/default/files/LPR%202020%20Full%20report.pdf.

———. "Living Planet Report 2022." *World Wildlife Fund*, 2022. https://livingplanet.panda.org/.

Yampolskiy, Roman V. "Minimum Viable Human Population with Intelligent Interventions." 2018. https://www.researchgate.net/profile/Roman-Yampolskiy/publication/329012008_Minimum_Viable_Human_Population_with_Intelligent_Interventions/links/5befa8204585150b2bbc7401/Minimum-Viable-Human-Population-with-Intelligent-Interventions.pdf.

Yeo, Richard R. "The Principle of Plenitude and Natural Theology in Nineteenth-Century Britain." *The British Journal for the History of Science* 19, no. 3 (1986): 263–82.

Young, David B. "Libertarian Demography: Montesquieu's Essay on Depopulation in the Lettres Persanes." *Journal of the History of Ideas* 36, no. 4 (1975): 669–82.

Yudkowsky, Eliezer. "Artificial Intelligence as a Positive and Negative Factor in Global Risk." *Global Catastrophic Risks* 1, no. 303 (2008): 184.

———. "Pascal's Mugging: Tiny Probabilities of Vast Utilities. Less Wrong." 2007. https://www.lesswrong.com/posts/a5JAiTdytou3Jg749/pascal-s-mugging-tiny-probabilities-of-vast-utilities.

———. "Shut Up and Multiply." *LessWrong*, 2021. www.lesswrong.com/tag/shut-up-and-multiply?version=1.25.0.

———. *The Singularitarian Principles, Version 1.0.* Self Published, 2000. https://museum.netstalking.ru/cyberlib/lib/critica/sing/singprinc.html.

———. Pausing AI Development Isn't Enough. We Need to Shut it All Down. TIME, 2023. https://time.com/6266923/ai-eliezer-yudkowsky-open-letter-not-enough/.

Yunkaporta, T. *Sand Talk: How Indigenous Thinking Can Save the World.* HarperCollins, 2020. https://books.google.de/books?id=-7moDwAAQBAJ.

Zaitchik, Alexander. "The Heavy Price of Longtermism." *New Republic*, October 24, 2022. https://newrepublic.com/article/168047/longtermism-future-humanity-william-macaskill.

Zapffe, Peter Wessel. "The Last Messiah." 1933/1993. https://openairphilosophy.org/wp-content/uploads/2019/06/OAP_Zapffe_Last_Messiah.pdf.

Zapffe, Peter Wessel, Sigmund Hoftun, and Bernt Vestre. *Essays og epistler.* Gyldendal, 1967.

Zhou, David. "BOOKENDS: 'Forgetful Prof Parks Girl, Takes Self Home'." *The Harvard Crimson*, May 4, 2005. www.thecrimson.com/article/2005/5/4/bookends-forgetful-prof-parks-girl-takes/.

Zimmer, Dan. "The Existential Anthropocene: Taking Total Risk as a Chronic Condition." *Harvard University*, October 20, 2021. https://drive.google.com/file/d/1q2cvtH4beg9iyztCXbpY8f4t4pFyQ6f-/view.

———. "The Immanent Apocalypse: Humanity and the Ends of the World." Doctoral Dissertation, Cornell University, 2022. https://ecommons.cornell.edu/handle/1813/112107.

———. "Kainos Anthropos: Existential Precarity and Human Universality in the Earth System Anthropocene." Draft Manuscript (under review), n.d.

Zoellner, T. *Uranium: War, Energy, and the Rock That Shaped the World*. Penguin Publishing Group, 2009. https://books.google.de/books?id=XM67WMGwugYC.

Zuber, Stéphane, Nikhil Venkatesh, Torbjörn Tännsjö, Christian Tarsney, H. Orri Stefánsson, Katie Steele, Dean Spears, et al. "What Should We Agree on about the Repugnant Conclusion?" *Utilitas* 33, no. 4 (2021): 379–83. https://doi.org/10.1017/S0953820821000011X.

Index